Advanced Antenna Array Engineering for 6G and Beyond Wireless Communications

IEEE Press
445 Hoes Lane
Piscataway, NJ 08854

Advanced Antenna Array Engineering for 6G and Beyond Wireless Communications

Y. Jay Guo
University of Technology Sydney (UTS), Ultimo, Australia

Richard W. Ziolkowski
University of Technology Sydney (UTS), Ultimo, Australia

Published by John Wiley & Sons, Inc., Hoboken, New Jersey.
Published simultaneously in Canada.

For general information on our other products and services or for technical support, please contact our Customer Care Department within the United States at (800) 762-2974, outside the United States at (317) 572-3993 or fax (317) 572-4002.

Wiley also publishes its books in a variety of electronic formats. Some content that appears in print may not be available in electronic formats. For more information about Wiley products, visit our web site at www.wiley.com.

Library of Congress Cataloging-in-Publication Data applied for:

ISBN: 9781119712909

Cover design by Wiley
Cover Image: © Photo presented by Jean-Philippe Chessel; © africanpix/iStock/Getty Images; © FiledIMAGE/iStock Editorial/Getty Images

Contents

Author Biographies

Y. Jay Guo received a Bachelor Degree and a Master Degree from Xidian University in 1982 and 1984, respectively, and a PhD Degree from Xian Jiaotong University in 1987, all in China. His research interest includes antennas, mm-wave, and THz communications and sensing systems, and beyond 5G mobile communication networks. He has published four books, over 550 research papers including over 280 journal papers, most of which are in IEEE Transactions, and he holds 26 patents.

Prof. Guo is a Fellow of the Australian Academy of Engineering and Technology, a Fellow of IEEE, and a Fellow of IET. He was a member of the College of Experts of Australian Research Council (ARC, 2016–2018). He has won a number of most prestigious Australian Engineering Excellence Awards (2007, 2012) and CSIRO Chairman's Medal (2007, 2012). He was named one of the most influential engineers in Australia in 2014 and 2015, respectively, and one of the top researchers in Australia in 2020.

Prof. Guo has over 30 years of international academic, industrial, and government research experience. Currently, he is a Distinguished Professor and the Director of Global Big Data Technologies Centre (GBDTC) at the University of Technology Sydney (UTS), Australia. Prior to this appointment in 2014, he served as a Director in the Commonwealth Scientific Industrial Research Organization (CSIRO) for over nine years, leading the research on advanced information and wireless communication technologies. Before joining CSIRO in 2005, he held various senior technology leadership positions in Fujitsu, Siemens, and NEC in the UK.

Richard W. Ziolkowski received the B.Sc. (magna cum laude) degree (Hons.) in physics from Brown University, Providence, RI, USA in 1974; the MS and PhD degrees in physics from the University of Illinois at Urbana-Champaign, Urbana, IL, USA in 1975 and 1980, respectively; and an Honorary Doctorate degree from the Technical University of Denmark, Kongens Lyngby, Denmark in 2012.

Prof. Ziolkowski was the recipient of the 2019 IEEE Electromagnetics Award (IEEE Technical Field Award). He is a Life Fellow of the Institute of Electrical and Electronics Engineers (IEEE Fellow, 1994) and a Fellow of the Optical Society of America (OSA, 2006) and the American Physical Society (APS, 2016). He served as the

President of the IEEE Antennas and Propagation Society in 2005. He is also actively involved with the URSI, OSA, and SPIE professional societies. He was the Australian DSTO Fulbright Distinguished Chair in Advanced Science and Technology from 2014 to 2015. He was a 2014 Thomson-Reuters Highly Cited Researcher.

He is currently a Distinguished Professor in the Global Big Data Technologies Centre in the Faculty of Engineering and Information Technologies (FEIT) at the University of Technology Sydney, Ultimo NSW, Australia. He became a Professor Emeritus at the University of Arizona in 2018, where he was a Litton Industries John M. Leonis Distinguished Professor in the Department of Electrical and Computer Engineering in the College of Engineering and was also a Professor in the College of Optical Sciences. He was the Computational Electronics and Electromagnetics Thrust Area Leader with the Engineering Research Division of the Lawrence Livermore National Laboratory before joining The University of Arizona, Tucson, AZ, USA in 1990. His current research interests include the application of new mathematical and numerical methods to linear and nonlinear problems dealing with the interaction of electromagnetic and acoustic waves with complex linear and nonlinear media, as well as metamaterials, metamaterial-inspired structures, nanostructures, and other classical and quantum application-specific configurations.

Acknowledgments

Antennas are a significant, fundamental, and practical research area of electromagnetics. Unfortunately, they have been considered by many academic administrators and government funding agencies simply as being well established, i.e. "old stuff." However, the wireless communications and sensors community is well aware that antennas are the key enabling technology of all things wireless.

Wireless technologies have become ubiquitous and truly critical in many ways to our everyday lives. These facts have become exceptionally clear now during this 2020 COVID pandemic. Whether you are a homemaker ordering foodstuffs to sustain your family via your cell phone and its network or you are a child learning and doing schoolwork online in your room while your academic parent is lecturing via Zoom from a home office, both being enabled by their computer's WiFi connection to the family's MIMO-based router, or you are a new grandparent seeing the newest member of your family remotely for the first time with FaceTime on your mobile platform or you are an antenna engineer interacting with company colleagues through Microsoft Teams on your handheld device to practice proper social-distancing protocols or even if you are two authors writing a book on antenna array technologies and are separated by a 19-hour time difference and a mere 13,000-km, wireless has meant that we can continue to perform tasks that need to be accomplished and can communicate and interact with family, friends, and colleagues on a regular basis.

Consequently, there have been very real and intense industry pushes and market pulls for various modern antenna systems to empower current fifth-generation (5G) and future sixth-generation (6G), and beyond wireless devices, applications, and their associated ecosystems. Scientific and engineering progress in array technologies has particularly benefited from user and stakeholder cravings for higher data rates and lower latencies. Antenna arrays will continue to play a major role in all future wireless generations. Pioneering wireless array research typically stresses advanced features such as steerable beams, multi-beams, multiband antenna coexistence, antenna reconfiguration, low-cost feed networks, and conformity to platforms. The various conundrums associated with the evolving land, air, and space networks associated with them will challenge all of us to develop fundamental and applied electromagnetics breakthroughs to solve them.

Under this backdrop, we have had the great privilege of working with a number of very talented PhD students, postdoctoral fellows, visiting scholars, and international collaborators. Our mutual interest and joint research efforts in antennas and antenna arrays for current 5G (fifth-generation) and future 6G, and beyond wireless ecosystems have deepened our understanding of their fundamentals, as well as their practical considerations necessary to successfully deliver useful systems for commercial applications that actually satisfy most of their generally overambitious, initial performance goals.

Our presentation of antenna and antenna arrays for current 5G and evolving 6G, and beyond systems in this book is organized into eight logical chapters that reflect our thoughts and the findings generated in those endeavors. Consequently, we are deeply indebted to our colleagues for their dedication and great contributions to the state of the art which are highlighted in these chapters. In particular, we would like to acknowledge specific inputs to them as follows:

Chapter 2: Ji-Wei Lian, Visiting Student, University of Technology Sydney (UTS), Australia

Chapter 3: Prof. Ming-Chun Tang, Chongqing University, China

Chapter 4: Dr. Can Ding, Lecturer, UTS, Australia and Dr. Hai-Han Sun, postdoctoral researcher, Nanyang Technology University (NTU), Singapore

Chapter 5: Dr. He Zhu, postdoctoral researcher, UTS, Australia

Chapter 6: Dr. Pei-Yuan Qin, Senior Lecturer, and Ph.D student Li-Zhao Song, UTS, Australia

Chapter 7: Dr. Stanley (Shulin) Chen, postdoctoral researcher, UTS, Australia; Dr. Debabrata K. Karmokar, Lecturer, University of South Australia, Australia; Prof. José Luis Gómez Tornero, Technical University of Cartagena, Spain; and Ji-Wei Lian, visiting student at UTS, Australia; Prof. Zheng Li, Beijing Jiaotong University, China.

Chapter 8: Prof. Yanhui Liu, Research Principal, UTS, Australia, and Ming Li, PhD student, UTS, Australia.

We thank them all for their invaluable time and efforts and wish them even greater successes in their future endeavors and careers.

We would also like to express our gratitude to University of Technology Sydney (UTS) for their whole-hearted support to our antennas research team.

Finally, we happily acknowledge our wives, Clare Guo and Lea Ziolkowski, and thank our lucky stars for their endless understanding, support and patience, particularly when we disappear for uncountable hours on cosmic efforts such as this :-)

1

A Perspective of Antennas for 5G and 6G

The roll-out of the fifth generation (5G) of wireless and mobile communications systems has commenced, and the technology race on the sixth-generation (6G) mobile and wireless communications systems has started in earnest [1, 2]. 5G promises significantly increased capacity, massive connections, low latency, and compelling new applications. For example, device-to-device (D2D) and vehicle-to-vehicle (V2V) communication systems will help facilitate the realization of autonomous transport. The rapid access to and exchange of "Big Data" will increasingly impact real-time economic and political decisions. Similarly, highly integrated, accessible "infotainment" systems will continue to alter our social relationships and communities. Wireless power transfer will replace cumbersome, weighty, short-life batteries enabling widespread health, agriculture, and building monitoring sensor networks with much less waste impact on the environment. 6G networks aim to achieve a number of new features such as full global coverage, much greater data rates and mobility, and higher energy and cost efficiency. These will usher in new services based on virtual reality/augmented reality and artificial intelligence [3].

At the core of wireless devices, systems, networks, and ecosystems are their antennas and antenna arrays. Antennas enable the transmission and reception of electromagnetic energy. Antenna arrays enhance our abilities to direct and localize the desired energy and information transfer. To achieve the many stunning and amazing 5G and 6G promises, significant advances in antenna and antenna array technologies must be accomplished.

1.1 5G Requirements of Antenna Arrays

One of the most important features of 5G is the employment of massive antenna arrays, with the size of the array currently varying from 64 to 128 and 256 elements. Such a large number of antenna elements in an array provide an unprecedented variety of possibilities. These include a means to increase the network capacity; the distance and data rates of individual links between the base station and mobile users; and the reduction of interference between different users and cells.

1.1.1 Array Characteristics

Generally speaking, there are three ways to exploit the benefits of antenna arrays in 5G wireless communication systems [4, 5], namely diversity, spatial multiplexing, and beamforming. These concepts are explained as follows.

Advanced Antenna Array Engineering for 6G and Beyond Wireless Communications, First Edition.
Y. Jay Guo and Richard W. Ziolkowski.

a) Diversity and Diversity Combining

It is a fact that mobile wireless communication channels typically suffer from both temporal fading and frequency fading. As a consequence, the quality of the channel varies with time and across different frequencies. Thus, the specific characteristics of the two propagation channels observed between any two pairs of transmitting and receiving antennas are usually different due to the variation in the scattering along the corresponding propagation paths. The peaks and troughs of the strength of the received signal at one antenna would be different from those at another antenna in a rich scattering environment. If the correlation between those two signals is low, one can combine them through so-called *diversity combining* to obtain a greater signal-to-interference-and-noise ratio (SINR). The latter is also known as diversity gain. A simple viewpoint is that diversity combining techniques aim to improve the quality of the individual links between the base stations and the user terminals by increasing the SINR.

From an antenna point of view, *diversity* can be obtained by exploiting either the distance between adjacent antennas, i.e., their positions, or different polarizations at the receiver and the transmitter. However, a fundamental requirement is that the mutual coupling between these diversity antennas must be low. Most modern base station antennas employ polarization diversity, i.e., each antenna element is dual-polarized typically with two pairs of slanted dipole "arms" in the $\pm 45°$ directions. In 5G millimeter-wave (mm-wave) systems, for example, a popular antenna configuration is to have beamforming antenna arrays with $\pm 45°$ polarizations, respectively.

b) Spatial Multiplexing

Multiplexing is the process of combining multiple digital or analog signals into a data stream for their transmission over a common medium, thus sharing a scarce resource. *Spatial multiplexing* aims to establish separate data streams in parallel using the same time/frequency resources. Thus, the space dimension is reused, i.e., multiplexed.

The simplest spatial multiplexing scheme is to employ sectorized antennas, a conventional technique for frequency reuse. More advanced spatial multiplexing schemes employ spatial–temporal (or frequency) coding by virtue of multiple input and multiple output (MIMO) antennas. A MIMO system requires the use of multiple antennas at least at the base stations. MIMO is implemented with two basic schemes as described below.

The first spatial multiplexing scheme is known as single user MIMO (SU-MIMO). By virtue of multiple antennas at both the base station and the user terminals, SU-MIMO first splits the data stream transmitted toward a specific user into multiple data streams. It then recombines them together at the user terminal to improve the information throughput and system capacity. One major challenge to SU-MIMO is the need for the tightly packed multiple antennas in the terminals to be decoupled.

The second spatial multiplexing scheme is known as multiuser MIMO (MU-MIMO). MU-MIMO aims to maximize the overall data throughput between all of the users and their associated base station. While it employs an antenna array at the base station, only one or a few antenna elements are present at each user terminal. Since user terminals are typically well dispersed within a radio cell and their individual channels are likely to be uncorrelated, the benefits of MU-MIMO are easier to achieve.

Both SU-MIMO and MU-MIMO protocols are intended for implementation in most 5G systems.

c) Beamforming

Spatial filtering can be regarded as a simple version of MU-MIMO. *Beamforming* achieves this spatial filtering by coherently combining the fields radiated by the array elements to direct their radiated energy into particular directions. These multiple beams are created at the base station to communicate with different users simultaneously.

Beamforming offers two benefits to a communication system. The first is capacity. If there is no overlap of the beams, simultaneous communications can take place in the same frequency band and at the same time without causing much interference. The second is the gain of the antenna array. Higher gain translates into information exchange over greater distances or higher data rates due to increased SINR values. Unlike 3G and 4G antenna arrays that provide coverage with fixed beam patterns and directivity, 5G arrays must support on-demand beam coverage according to real-time application scenarios and user distributions. Moreover, they must be able to support beam management in order to deliver precise coverage in target areas while significantly suppressing interference in other areas.

For beamforming to be effective, large antenna arrays are necessary to generate narrow beams and produce scattering from mobile users with small angular spreads. The latter is to ensure that the majority of the signals transmitted and received from a mobile platform is covered by a narrow base station antenna beam. These requirements, in conjunction with wide bandwidths, support the use of millimeter wave (mm-wave) communications for 5G. In particular, mm-waves propagate in a pseudo-light fashion so the scattering of the signals to and from a mobile platform is highly localized. Furthermore, since their wavelengths are small, an electrically large mm-wave array can be fit easily into a physically small space.

1.1.2 Frequency Bands

Another major challenge associated with 5G antenna arrays is the simultaneous support of all allotted frequency bands [6]. As the number of bands being considered to meet current and future 5G needs increases, significant antenna array innovations are required to support all of them. Moreover, existing 4G bands must be supported as well [7].

Owing to the stringent requirements placed on the radiation patterns produced by cellular systems and on the levels of their impedance matching to sources to maximize their realized gains, the mobile communication industry has so far adopted an approach of using different antennas to support different frequency bands. However, because of the limited space at base station antenna sites and in mobile platforms, the coexistence of these different antennas has posed serious challenges already. It is extremely difficult to maintain low coupling levels between antennas operating over the same band and even harder to suppress the scattering interactions between antennas that operate over different bands. The latter can cause significant distortions to the radiation patterns. It is with this background that the decoupling and de-scattering issues will be addressed in Chapters 2 and 3, respectively.

1.1.3 Component Integration and Antennas-in-Package (AiP)

Clearly, the number of antenna ports and radios for 5G systems will grow dramatically with the increasing numbers of massive antenna arrays and operating bands. This growth implies that the number of cables that connect the radios to the antennas would increase accordingly. This increase necessarily leads to increased fabrication complexities, losses in the cables and connectors, and difficulties in the control of passive intermodulation (PIM) and testing. To mitigate these problems, one needs to change antenna system design methodologies to introduce much higher levels of integration. To this end, there has been a high expectation that 5G antennas, the mm-wave band antennas in particular, will become highly integrated systems.

Integrated antenna and radio systems eliminate the need for multiple cables between the radios and antennas, thus increasing their reliability by reducing part counts and handling, and simplifying their testing and installation. As a result, there has been an increasing need for effective antenna-in-package (AiP) solutions. In addition to managing the radiation performance of the

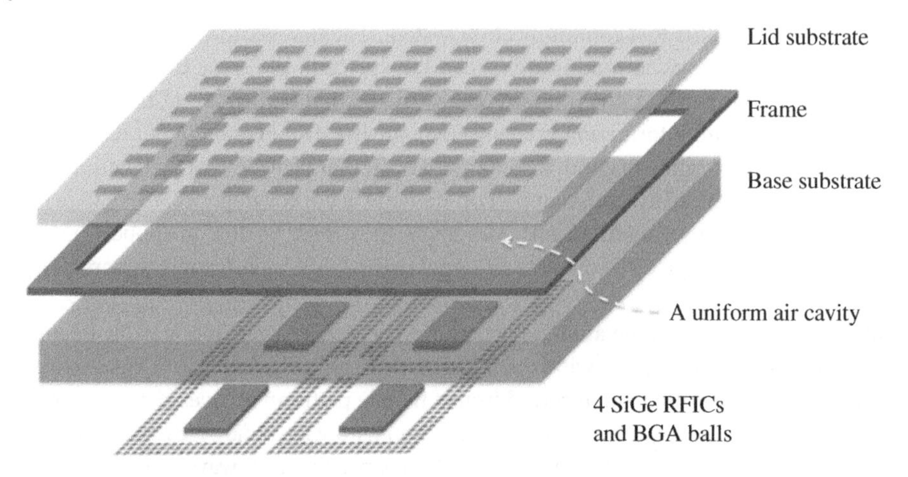

Figure 1.1 An illustration of a 64-element antenna-in-package (AiP) assembly breakout. *Source*: From [8] / with permission of IEEE.

antenna elements and arrays, one must consider several issues for AiP designs. These include, for instance, the materials; process selection and control; power and heat management; and new testing techniques. As an example, Figure 1.1 shows a 64-element AiP system at 28 GHz. It has four flip-chip-mounted transceiver ICs that support its dual-polarized operation [8]. For clarity, the heat sink below the ball-grid-array (BGA) interface is not shown.

One particular new challenge associated with highly integrated 5G antenna arrays is obtaining accurate antenna beam patterns. Depending on their actual implementation, methods for testing active antennas vary. Current examples include the following [4]:

a) Sample Testing

This approach involves the fabrication of a number of fixed analog beamforming circuits that provide the requisite amplitude and phase excitations to the antenna array to produce the desired beams including narrow beams for user traffic and broad beams for user management. Each circuit produces one specific beam. This allows one to sample each of the desired beam types and steering directions. For practical reasons, it is difficult to perform a comprehensive test of all of the possible beams generated by a large array. Therefore, only those beams of greatest interest are likely to be tested.

b) Element-by-Element Testing

The far-field vectorial pattern of each element, i.e., the amplitude and phase distribution in the far-field of the array, can be measured with respect to a common reference. Any beamforming pattern can then be synthesized numerically by adding all the element patterns with the corresponding appropriate complex weights. This approach is the most flexible method since all possible patterns can be tested. Nevertheless, one can argue realistically that the synthesized beam patterns may differ from the real ones to a certain extent because all of the actual interactions are not explicitly included.

c) Employ Beam Testers

Beam testers are effectively flexible beamforming networks. By connecting a beam tester to an antenna array, one can test a variety of the beams defined by the beam tester using a traditional method for antenna pattern testing. The 3rd Generation Partnership Project (3GPP), which unites

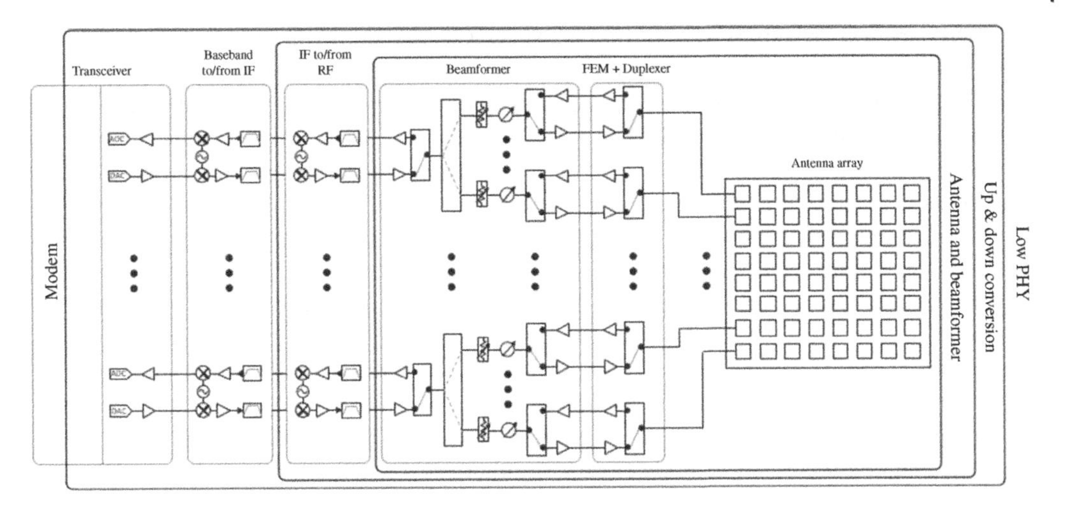

Figure 1.2 Three levels of AiP implementation by TMYTECH. *Source*: From [9] / with permission of TMY Technology Inc.

seven telecommunications standard development organizations (ARIB, ATIS, CCSA, ETSI, TSDSI, TTA, and TTC), has defined three Over-the-Air (OTA) test methods for MIMO antennas: the direct far-field (DFF) method using a far-field chamber, the indirect far-field (IFF) method using a compact range, and the near-field to far-field transform (NFTF) method using a near-field chamber. All three OTA approaches are conventional methods familiar to antenna engineers.

It must be recognized that when active electronics are added to a radiating aperture to form a MIMO antenna, the antenna ports are now embedded in the system. As a result, it becomes much more difficult to measure the true gain and antenna efficiency. Because a massive MIMO antenna has a large number of antenna elements and its radiating aperture can be excited in many ways to create different beams, both narrow and broad, it is truly difficult to fully test and validate beam performance in terms of conventional figures of merit, e.g., pattern characteristics, beam shapes, beam steering, side lobe levels, and null locations. Testing is further complicated because measurements for both the transmit case and the receive case must be performed to understand the operating characteristics of both RF chains.

To facilitate the manufacturing and adoption of large antenna arrays in 5G and beyond systems, the wireless industry is pushing to increase the level of integration of the system frontend modules (FEM). Figure 1.2 shows the AiP roadmap of the TMY Technology (TMYTEK) company for their 5G mm-wave products [9]. Each enclosure block represents one particular level of component integration. The industry trend is to integrate the antenna arrays with all of the radio frequency (RF) and intermediate frequency (IF) modules into one package. Characterization of all of the beams produced by such modules is undoubtedly a new challenge for antenna designers.

1.2 6G and Its Antenna Requirements

5G mobile and wireless systems are ground-based. Consequently, they have coverage requirements similar to earlier generations of terrestrial networks. In contrast, space-communication networks provide vast coverage for people and vehicles at sea and in the air, as well as in remote and rural areas. They are complementary to terrestrial networks. Clearly, future information networks must seamlessly integrate space networks with terrestrial networks to achieve significant advances beyond 5G. This integrated wireless ecosystem may become one of the most ambitious targets of 6G systems [10]. It is currently envisaged that 6G wireless systems will support truly global wireless communications, anywhere and anytime. An integrated space and terrestrial network (ISTN) is expected to be at the core of beyond 5G communication systems. As a consequence, the development of the technologies to achieve a high-capacity, yet low-cost, ISTN is of significant importance to all of the emerging 6G wireless communication systems.

Currently, there are a number of commercial and government spaceborne and airborne platforms that support various applications in communications and sensing. These include geostationary Earth orbit (GEO), medium Earth orbit (MEO), and low Earth orbit (LEO) satellites. As their names indicate, they operate at different altitudes relative to the Earth's center. Various airborne platforms also operate at different altitudes such as high-altitude platforms (HAPs), airplanes, and unmanned aerial vehicles (UAVs, otherwise known as drones). It is anticipated that any eventual 6G and beyond mobile wireless communication networks will thus consist of three network layers, namely the space network layer, the airborne network layer, and the terrestrial network layer. An illustration of a potential ISTN architecture is shown in Figure 1.3. Figure 1.3 clearly suggests that there will be a huge number of dynamic nodes constituting the mobile airborne networks, in addition to the dynamic nodes of the ground and space (satellite) networks [10].

Airborne networks have a number of unique characteristics. First, most of their nodes would have multiple links to achieve network reliability, high capacity, and low latency. Second, most of them will be mobile. Therefore, both their network links and topologies will vary with time, some faster than others. Third, the distances between any two adjacent nodes will vary significantly, from hundreds of meters to tens of kilometers. Fourth, the power supplied to any node would be limited. Consequently, as in the case for terrestrial networks, the energy efficiency of each node not only impacts the operation costs, but also the commercial viability of the entire network. Fifth, it is highly desirable for antennas on most airborne platforms to be conformal in order to meet their aerodynamic requirements and to maintain their mechanical integrity.

All of the noted, desirable ISTN features pose a number of significant and interesting challenges for future 6G antennas and antenna arrays. The antennas, for example, must be compact, conformal, and high-gain. They must be reliable, lightweight, and low-cost. The corresponding arrays must provide individually steerable multiple beams; dynamic reconfiguration of their patterns, polarizations, and frequencies to cope with the movement of the platforms; and overall high energy efficiency. The biggest challenge among all of them is arguably the reduction of the overall energy consumption. One promising solution is to employ analog steerable multi-beam antennas. Hybrid beamforming is another. Since beamforming and beam scanning can be done by antenna reconfiguration through electronic switching or tuning, the energy required is negligible in comparison to employing a full digital beamforming approach.

1.3 From Digital to Hybrid Multiple Beamforming

There are several ways to form multiple beams from an array. Major schemes can be categorized into digital, analog, and crossover strategies. We begin by describing digital beamforming and a major crossover of much recent excitement, hybrid beamforming.

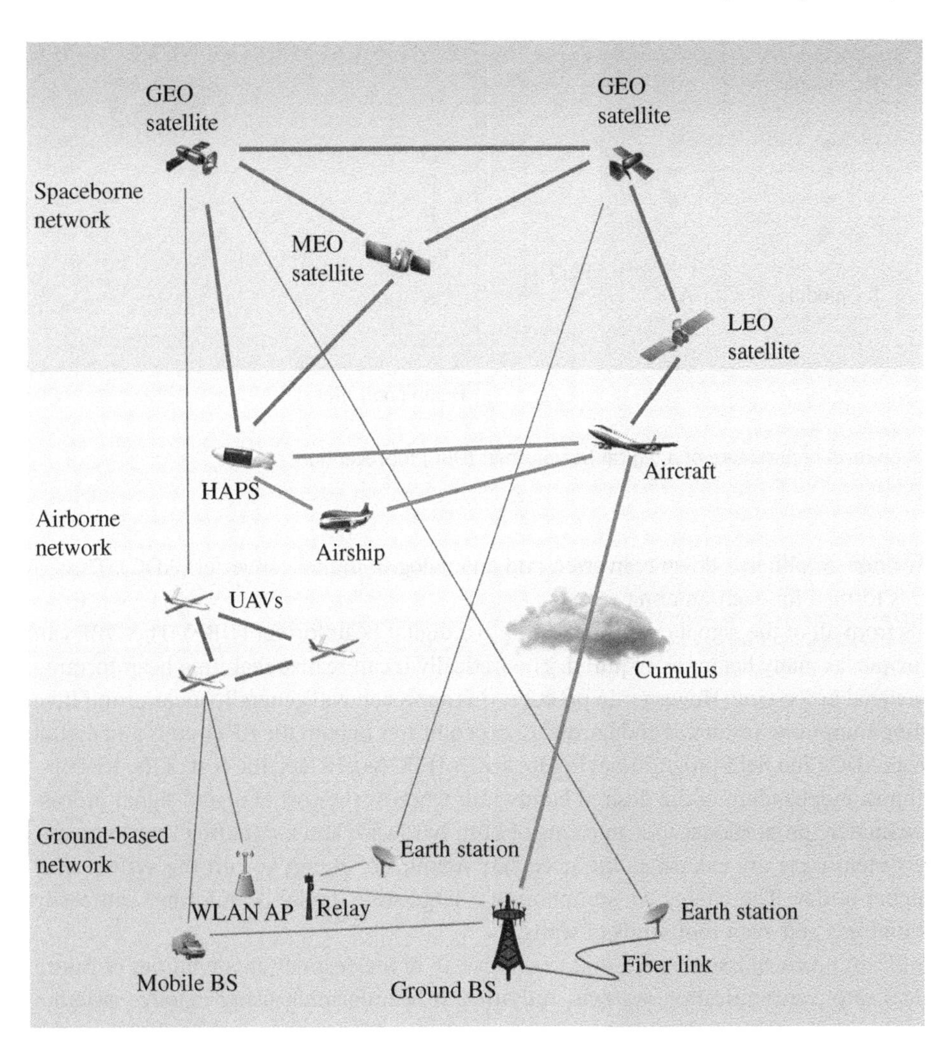

Figure 1.3 An illustration of a potential ISTN architecture for 6G and beyond. *Source*: From [10] / with permission of IEEE.

1.3.1 Digital Beamforming

Given an antenna array, digital beamforming is the ultimate way to achieve optimal performance. It is the most flexible approach to generating individually steerable and high-quality multiple beams. With a single antenna array of large enough size and the same set of RF circuits, one can effectively create as many beams as desired by applying different complex weights (amplitude and phase) to each element of the array in the digital domain. More advanced digital beamforming schemes employ algorithms such as eigen-beamforming to obtain the maximum SINR values [11]. Fully digital beamforming with massive antenna arrays serves as a powerful technology to meet some of the most challenging desired features of future wireless communication networks including capacity, latency, data rates, and security.

A high-level digital beamformer for reception is shown in Figure 1.4. It consists of an array of antennas, each antenna element being connected with an RF receiver. The RF receiver includes

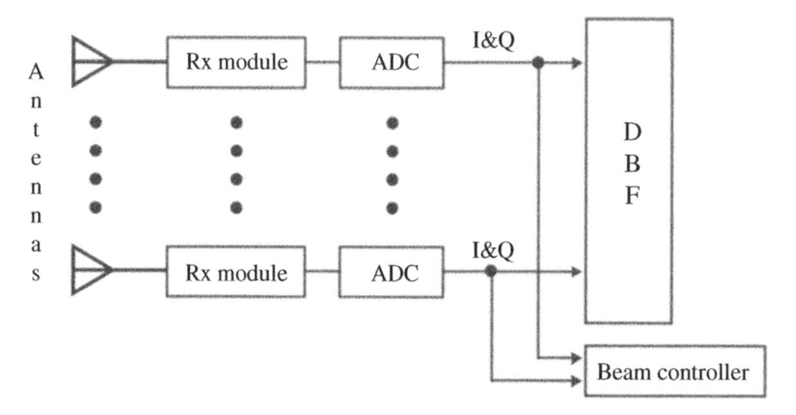

Figure 1.4 High-level architecture of a digital beamformer (DBP) for reception.

a filter, a low noise amplifier, a down converter, and an analog-to-digital converter (ADC). Thus, a signal chain is formed for each antenna.

The signals from all of the signal chains are fed into a digital beamformer (DBF). The DBF can form, in principle, as many beams as required. Theoretically it can realize real-time beamforming via real-time signal processing. However, in practice, this approach will generally incur prohibitive costs, including computing resources and hardware expenditures in both the RF circuits and digital devices such as ADCs and field-programmable gate arrays (FPGAs). In fact, the cost of the RF components is almost independent of the desired bandwidth whereas the cost of digital signal processing is approximately proportional to it in terms of both hardware and computing requirements. While those system costs are extensive, the necessary amount of energy to run the system may be even a higher outlay. The energy consumption of a large scale digital beamformer can easily amount to hundreds and even thousands of watts.

These significant practical issues mean that to achieve all of the desired functionalities of future ultra-high data rate communication systems, fully digital beamforming using massive antenna arrays is simply unaffordable for most application scenarios. Moreover, it is actually not even acceptable for many base station antennas for 5G with the current state of the art of device technologies [9]. These factors lead to the conclusion that some kind of hybrid system based on both digital and analog beamforming might serve as a good solution to large scale antenna arrays with multiple steerable beams in the foreseeable future.

1.3.2 Hybrid Beamforming

Hybrid beamforming is a strategy that combines the advantages of both analog and digital beamforming techniques. The motivation for employing hybrid beamformers is now clear. One wants to reduce hardware costs and processing complexities while retaining nearly the optimal performance that is achievable with optimized digital designs.

The hybrid beamforming approach does not treat every antenna element as a completely independent one. The key concept is to partition a large antenna array into smaller subarrays. This type of array is also known in the 5G literature as an array of subarrays (AOSA) [4]. Each subarray consists of a conventional analog antenna array that forms its beam in the analog domain [12, 13]. The number of sub-arrays into which the whole array is partitioned determines its degrees of freedom.

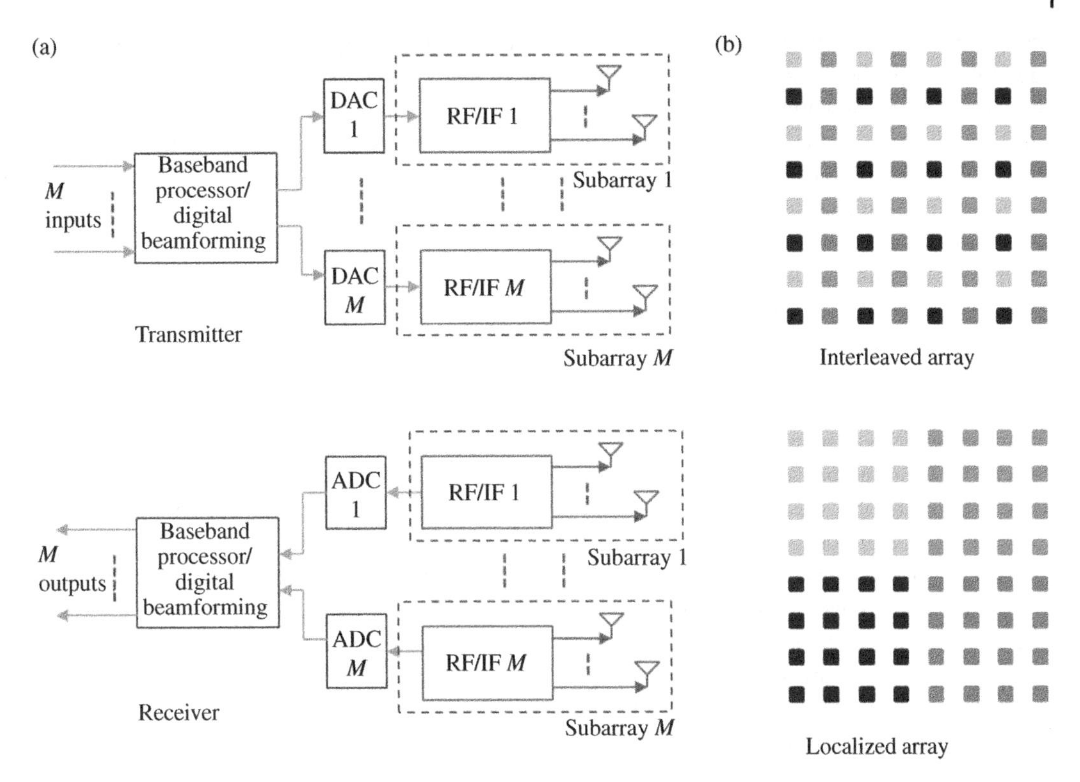

Figure 1.5 Hybrid antenna arrays. (a) The basic architectures of transmitter and receiver systems. (b) Two types of array configurations for uniform square hybrid arrays: interleaved (upper) and localized (bottom). Each square represents an antenna element and squares with the same color represent antenna elements in the same analog subarray. *Source*: From [12] / with permission of IEEE.

When analog beamforming is performed using analog phase shifters and other equivalent devices, significant cost reductions can be achieved immediately due to the decrease in the number of complete RF chains required to form the beams. However, the number of simultaneously supported data streams or beams in a hybrid array is lower in comparison to a full-blown digital array. In practice, the actual antenna array design depends on the beamforming capabilities required along with the system's total complexity and budget considerations, both issues being influenced directly by factors such as the number of steerable beams and costs. Although reducing the number of RF chains also limits the number of data streams, per-user performance can be designed to come close to that attained with a fully digital beamformer. Owing to the nature of line of sight radio propagation and smaller numbers of users per cell, the hybrid beamforming strategy is definitely the more practical beamforming approach for mm-wave systems in the near future [4, 11].

Figure 1.5a shows the basic architectures of both transmitting and receiving hybrid arrays. Their schematics illustrate the whole array being divided into many analog subarrays [12]. Each subarray includes N antennas and an RF/IF (intermediate frequency) unit. These components can be shared by different antenna elements in different ways, depending on their actual implementations. For convenience, we have simply denoted an array with M subarrays with N antenna elements in each subarray as an $N \times M$ hybrid array. Typically, given the dimension of the whole array, the decision on the size of the subarray, or the selection of N and M, is a trade-off between the system cost and performance. If N is large, a high antenna gain can be achieved at a lower cost. If N is too large, however, the number of users the array can support would be limited. The distance between corresponding elements in adjacent subarrays is called the *subarray spacing*. It is determined by the desired multiple beam performance and the allowed physical area of the array. Each subarray is connected to a baseband processor via a digital-to-analog convertor (DAC) in the transmitter and an analog-to-digital convertor (ADC) in the receiver. The signals from all of the subarrays are interconnected and processed centrally in the baseband processor.

Signals in the analog subarray and in the digital processor can be processed in different domains and in different ways. A signal in each subarray can be simply weighted in the analog domain mainly for the purpose of achieving array gain and beam steering. The signal for each antenna element in a subarray can be varied in both its magnitude and phase, typically with limited resolution. In the simplest case, only a phase shifter is applied and the signal is weighted by a discrete phase shift value from a quantized set of values. The size of the set is typically represented by the number of quantized bits. For example, a 3-bit quantization means eight discrete values are uniformly distributed over the angular interval $[-\pi, \pi]$. In the digital processor, signals from/to all of the subarrays are jointly processed. Advanced techniques which are similar to those utilized in conventional MIMO systems, such as spatial precoding/decoding, can be implemented.

Antenna elements in a hybrid array can be configured in various ways to form different topologies. Each of them has respective advantages and disadvantages. A configuration is typically fixed at the fabrication stage. The typical two types of regular configurations are interleaved and localized arrays. They are illustrated in Figure 1.5b for a 16×4 uniform square hybrid array. The antenna elements in each subarray in an interleaved array are distributed uniformly over the whole array. On the other hand, they are adjacent to each other in a localized array.

The analog subarrays in Figure 1.5 can be implemented in four different configurations depending on where the phase shifters are placed for beamforming. They are illustrated in Figure 1.6. Figure 1.6a shows the conventional phased array architecture for a receive analog array. Only the phase shifters and antennas are independent; all of the rest of its components are shared by all elements in each analog subarray. This passive power combining architecture incurs losses in the phase shifters and power combiners which increase with the number of antenna elements and operating frequency. These power losses could make large passive arrays impractical. A modification of this architecture is shown in Figure 1.6b. An individual LNA is applied to each antenna element before the phase shifter. This modification reduces the noise significantly and provides increased receiver sensitivity. This architecture can be implemented using either a shared frequency converter (with individual RF chains combined at the input to the mixer) or individual frequency conversion and combining in the IF unit. Figures 1.6c and 1.6d depict more advanced configurations in which the phase shift is implemented in the IF unit and local oscillator (LO) circuits, respectively.

It must be noted that commercial 6-bit digital phase shifter mm-wave integrated circuits (MMICs) are available for a range of LO and IF frequencies suitable for mm-wave arrays. These devices provide 360° of phase change with a least significant bit (LSB) of 5.625°. This resolution allows analog beamforming with a scan angle accuracy to a fraction of a degree. The system configuration in Figure 1.5d is particularly attractive since the devices in the LO path are typically operated in saturation. Consequently, variable losses that usually change with any phase shift are avoided in this scheme.

It is should be pointed out that a more elegant and highly desirable solution to forming multiple beams in a hybrid fashion is to employ analog multi-beam antennas rather than using subarrays of antenna elements. In principle, the entire antenna aperture can be shared by all the users. However, the generation of multiple individually steerable analog beams is in itself a huge challenge. Unfortunately, there exist only a very limited number of solutions that can be incorporated into the hybrid beamforming configurations discussed above. A number of the remaining chapters in this book will explore various ideas to fill such current technology gaps.

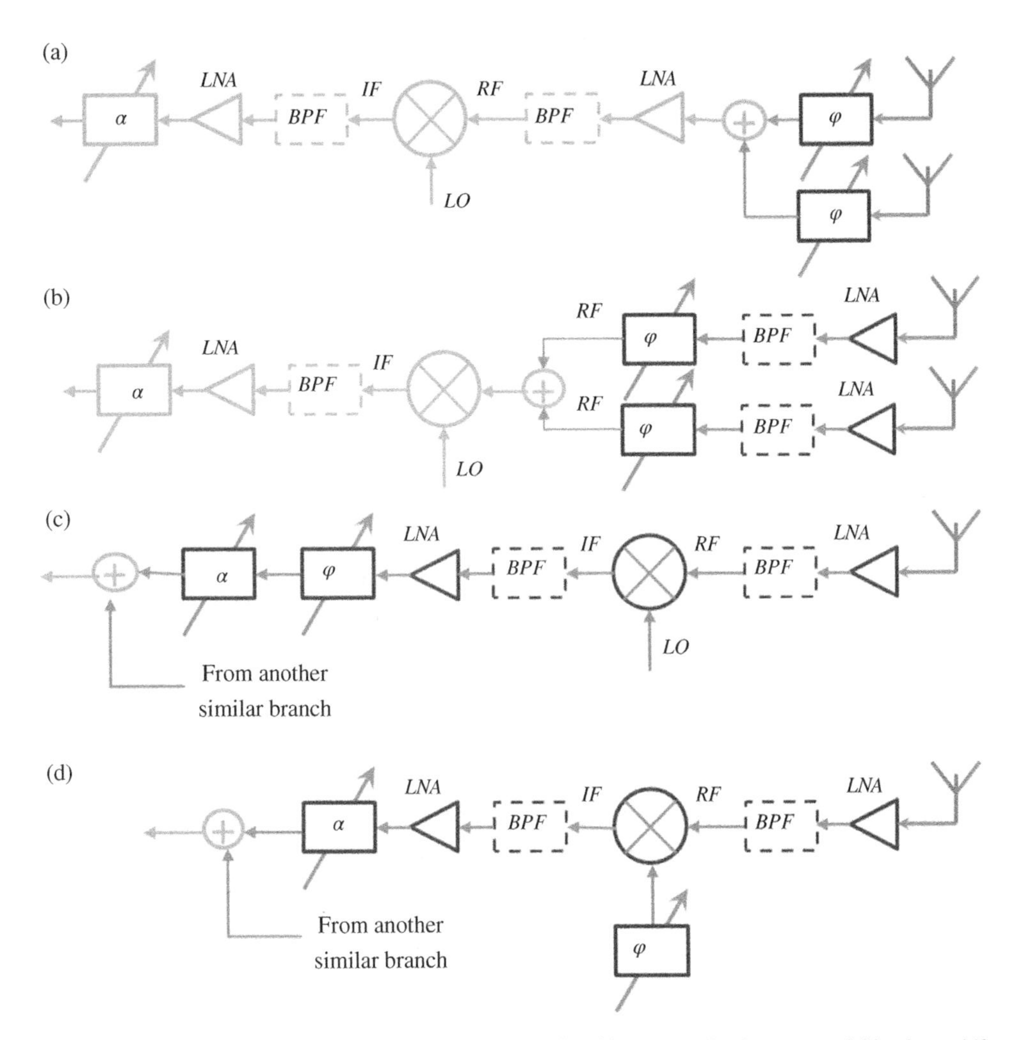

Figure 1.6 Options for implementing analog subarrays. The blocks φ and α denote a variable phase shifter and magnitude attenuator, respectively. Blocks in green represent those able to be shared by the antenna elements in a subarray. The circle with a cross in it denotes where signals from individual antenna elements are combined to be delivered to the shared components. The subfigures show the four main options. (a) Phase shifter at the RF element before an LNA. (b) Phase shifter at the RF element after an LNA. (c) Phase shifter at the IF unit. (d) Phase shifter at the LO unit.

Notice that it is expected that both fully digital beamformers and hybrid beamformers will be employed for 5G deployments. Some of the anticipated use cases envisaged by industry are listed in Table 1.1 [4].

1.4 Analog Multiple Beamforming

There are a number of ways to create steerable antenna beams in an analog manner. These include the use of circuit-type beamformers, reflectors, lenses, and phased arrays. These and other more advanced methods will be presented in later chapters. We review some of the basic concepts here.

Table 1.1 Use cases for digital and hybrid beamforming.

Beamforming type	Use cases
Digital beamforming	• Sub-6 GHz massive MIMO: MU-MIMO • Sub-6 GHz macro cell • 2D beamforming • Fixed wireless access
Hybrid beamforming	• mm-wave based systems • Sub-6 GHz small cells/hot spot coverage • Fixed wireless access • Massive MIMO macro cells

Source: From [4] / with permission of 5G Americas.

The most common analog beamforming antennas are phased arrays. The original technology dates back to the mid-twentieth century. It remained primarily as a military technology until the 5G era. In a phased array, the same signal is fed to each antenna element. The amplitudes of the elements are weighted according to the desired shape of the beam, i.e., the shape of the radiated pattern, and then phase-shifters are used to steer the beam emitted by the array into the desired direction. In order to save cost, the current commercial 5G mm-wave systems employ phased arrays to conduct analog beamforming. Both the base station and the user equipment (UE) use a number of fixed weight settings, or sets of phase-shifting values, to produce different beams pointed in specific directions. There are only two beams pointed in the same direction at any given point of time, each for the horizontal and vertical polarizations, respectively. Consequently, current base stations steer their beams sequentially in different directions to provide the desired coverage. The system capacity could be significantly improved by introducing multiple beams. However, it remains a major technological challenge to provide sufficient flexibility to achieve multiple beam directions with an analog beamforming system.

Phased arrays are inherently suited for producing single beams. Because one signal is fed to all of its elements, a phased array constitutes only one antenna port per beam. The beam is steered to follow the intended user by controlling the values of its phase shifters. Some sacrifices have to be made to produce individually steerable multiple beams with a phased array. They include partitioning the array aperture for different beams; and, hence, this limits the overall performance of each generated beam. In the following subsections, we present two multiple analog beamforming techniques that are currently popular for cellular systems: Butler matrices, and Luneburg lenses.

1.4.1 Butler Matrix

One traditional method of producing multiple beams is to utilize Butler matrices [14]. These multiple beams can be steered together in principle, but not independently. Therefore, Butler matrices are almost exclusively used for fixed beams. A Butler matrix is an RF circuit consisting of couplers, delay lines, crossovers, and transition parts. An n-way Butler matrix has n inputs and n outputs. A signal applied to a given input will lead to outputs of equal amplitude but with a uniform phase gradient, thus leading to a single steered beam. The phase increment between adjacent outputs is a multiple of $\frac{360^\circ}{n}$ depending on which input is fed. The phase increment across the outputs, that occurs if input i is fed, is $\frac{360^\circ}{n}i$, where i can take on integer values from 0 to $n-1$. If the n outputs of the Butler matrix are connected to a linear array of n equally spaced radiating elements, a set of n

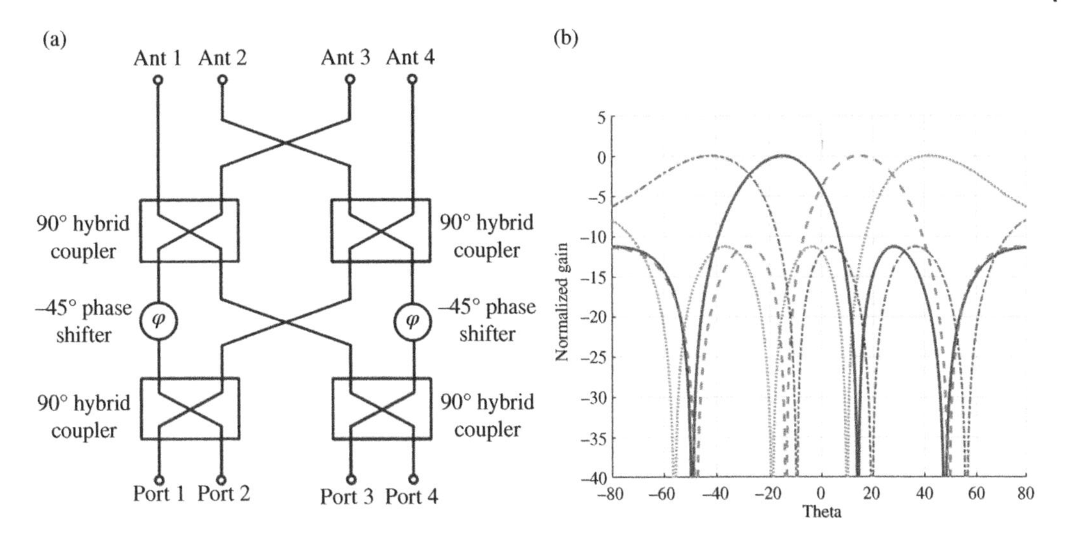

Figure 1.7 Typical implementation of a 4 × 4 Butler matrix (BM) connected to 4 radiating elements and the 4 beams it produces.

beams equally spaced in angle will be generated if all of the inputs are fed. Figure 1.7 shows the configuration of a 4×4 Butler matrix and the 4 beams it produces with 4 radiating elements.

Unfortunately, multiple beamforming employing a Butler matrix has a number of disadvantages. First, the beams are fixed. Consequently, it is only a switched beam solution for tracking mobile users. Second, owing to the losses in the Butler matrix's circuits, a major challenge for large antenna arrays is keeping the overall losses small, especially at millimeter-wave frequencies. Third, a 2D Butler matrix would be required for two-dimensional (2D) beamforming. However, the conventional structure is generally too bulky and too lossy owing to the complicated requisite crossovers. Fourth, a complete system engineering approach is required to achieve wideband operation with a Bulter matrix. These issues are only some of the challenges facing the antenna research community. They and some recently developed solutions will be addressed in several later chapters.

1.4.2 Luneburg Lenses

A simple, yet powerful, analog method to create steerable and multiple beams is to employ a spherical Luneburg lens. A Luneburg lens in its simplest form consists of a radially inhomogeneous sphere with a well-defined graded dielectric constant that varies from 2.0 at the center of the sphere to 1.0 at its outer surface. The gradation is given by the equation: $\varepsilon_r = 2 - (r/a)^2$, where ε_r is the relative dielectric constant at radius r and a is the outer radius of the sphere. The resulting structure serves to transform rays incident on one side to parallel rays on the opposite side. An antenna feed located on the surface of the lens produces a steered beam if the element moves around the surface as illustrated in Figure 1.8a. The low dielectric constant near the lens surface ensures that no energy is reflected back to the feed. In order to accommodate feeds whose phase centers cannot be placed at the surface of the Luneburg lens, such as a horn antenna, one can modify the distribution of the dielectric constant within the lens [15]. Figure 1.8b shows the corresponding cylindrical version, which is known as a *cylindrical Luneburg lens*.

The beamwidth of a Luneburg lens is approximately the same as that of a linear array whose length equals the diameter of the lens. Nevertheless, the nulls are considerably deeper. If one places

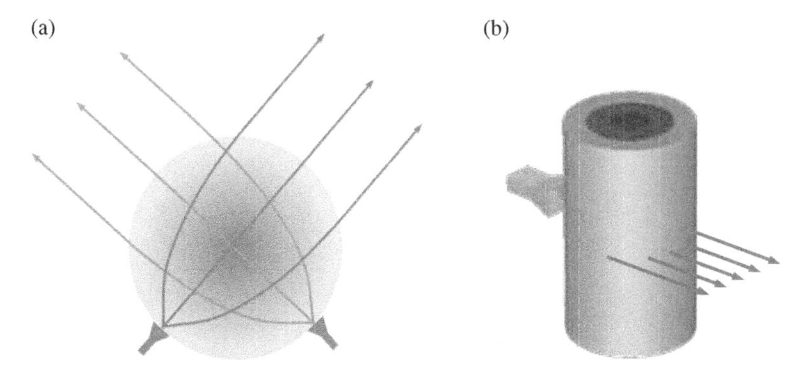

Figure 1.8 Illustration of Luneburg lenses. (a) Spherical. (b) Cylindrical.

a number of feeds along the surface of a Luneburg lens, one can produce a multiple beam antenna, one beam per feed. These multi-beam antennas can be employed for data distribution or broadcasting in 5G networks.

It is very difficult and very costly to produce an ideal Luneburg lens. As a practical alternative, one can employ several separate shells to replace the theoretical continuous gradation of the dielectric constant with a discrete approximation to it. Many such versions have been deployed in a variety of current systems.

The main advantages of Luneburg lenses over antenna arrays based on beamforming networks can be summarized as follows [14]:

- A great simplification in component count and inherent low passive intermodulation (PIM).
- Reduction of network losses.
- Beam crossover levels can be selected arbitrarily by choosing the spacing of the source elements.
- Isolation between elements is generally superior to that obtained with beamforming networks.

The relative disadvantage of a Luneburg lens antenna is its three-dimensional bulk compared with planar forms of the array antennas. Nevertheless, some mobile operators are currently showing strong interest in Luneburg lenses due to their low cost in hardware and low energy consumption.

1.5 Millimeter-Wave Antennas

To date, every new generation of mobile wireless communication has been allocated its own dedicated spectrum. This is again true for 5G networks. Given the fact that the radio spectrum is a worldwide limited resource, the mobile wireless communication industry has been "forced" to start using the mm-wave spectrum to accommodate some portion of its 5G networks, known as 5G mm-wave. Application examples include small cells for data-hungry hot spots and fixed wireless access services where line of sight (LoS) propagation is easier to be guaranteed. Moving forward to 6G, it is expected that some airborne and satellite systems will also embrace the mm-wave spectrum. Compared with the microwave frequency bands, the propagation of mm-waves is negatively impacted by higher attenuation rates and severe weather.

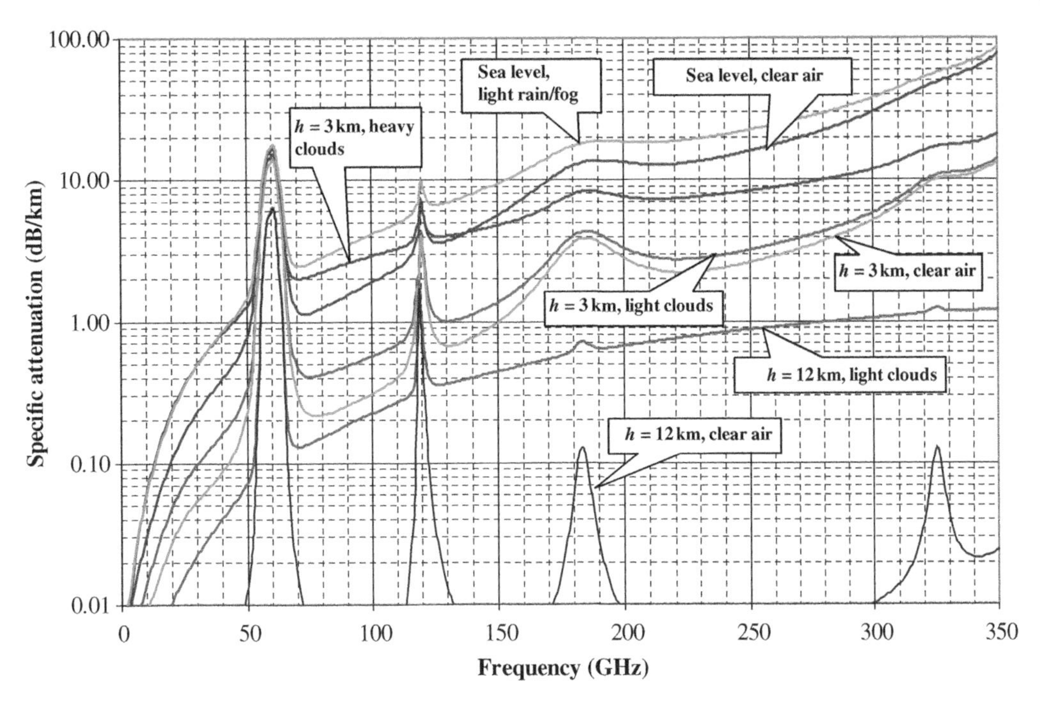

Figure 1.9 Specific atmospheric attenuation (dB/km) at the indicated altitude *h* and for several exemplary weather and air conditions. *Source*: Based on [16] / IEEE.

To emphasize this issue, Figure 1.9 shows the attenuation of electromagnetic waves from DC into the low terahertz (THz) range as functions of the propagation distance, altitude, and weather conditions. Notice that there are some windows in these spectra where the atmospheric attenuation is high, such as around 60 GHz, and, conversely, much lower. The former are clearly not suitable for long-distance communication. The latter are targeted for many applications. Also notice that the propagation losses are reduced at higher altitudes where the air is thinner. Examining Figure 1.9 more closely, it is little wonder that the current "first choice" for commercial 5G rollouts of mm-wave systems is at the lower end of the mm-wave range, i.e., around 28 GHz.

Certain important advantages for 5G operations are offered by mm-wave systems. One is that high-gain mm-wave antenna arrays can be realized over physically small areas because the associated wavelengths are small (recall that the gain of an aperture antenna – Gain = 4π Area/λ^2). In fact, given the inherent high propagation losses of their radiated fields, high-gain antennas are needed for virtually all mm-wave communication systems. As a result, it has become imperative to develop mm-wave beamforming networks to support multi-beam mm-wave antennas. In the current 3GPP standards for 5G mm-wave, for example, user equipment (UE) or terminals are required to have an array antenna with between 8 and 64 elements [17].

1.6 THz Antennas

With 6G data rates promised to be even higher than those of 5G [1–3], a much wider spectrum is needed to accommodate 6G expectations. Unfortunately, a large currently unoccupied spectrum does not exist below 100 GHz. Consequently, it is widely expected that 6G will occupy a significant part of the THz spectrum [2]. Along with terrestrial-based communication systems, it is anticipated that THz systems will also play a major role in space-based communications [18, 19].

Currently, the most common definition of the THz band is that it consists of frequencies from 0.3 to 3.0 THz. Recall that the wavelength at 0.3 THz (300 GHz) is just 1.0 mm. Owing to the fact that THz wavelengths are even smaller than the mm-wave ones, very narrow multiple beams with low probability of intercept (LPI) can be generated from very physically small areas. Beam steering and target tracking again will be indispensable features for THz antennas.

Referring to Figure 1.9, signal attenuation in the lower portion of the THz range is even more severe than in the mm-wave band. Thus, high-gain antenna arrays are even more necessary for anticipated 6G operations. Other important related THz technologies that must also be developed to address 6G expectations are high power sources and highly sensitive receivers [20]. Feeding a large array of THz antenna elements of 0.5λ in size using a corporate network is a daunting engineering task. Therefore, it has not been favoured to date. Instead, a more promising approach is to employ an electrically large lens fed by a simple radiating element such as a dipole or a slot or even a small array. To ease the problem of the precise alignment of the antenna and lens, one could integrate the antenna feed with the lens. Antennas with this characteristic are known as integrated lens antennas [20–22].

1.7 Lens Antennas

A number of different types of lens antennas operating in the mm-wave and THz bands have been reported [21–25]. These include the elliptical lens, extended hemispherical lens, and Fresnel zone lens. Each has its own unique physical and performance characteristics.

A homogeneous elliptical lens has two focal points. It can transform the radiation pattern of a feed placed at one focal point into a plane wave exterior to it propagating in the direction of the second focal point. Assuming a represents the major semiaxis, b represents the minor semiaxis, L represents the distance between the focal point of the feed to the centre of the ellipsoid, and n is the index of refraction of the dielectric from which the lens is fabricated, one has the following relationships:

$$a = b/\sqrt{\left(1 - \frac{1}{n^2}\right)} \tag{1.1}$$

$$L = a/n \tag{1.2}$$

An integrated elliptical lens antenna is obtained by cutting off the part of the dielectric below the bottom focal point and placing the feeding antenna beneath it. As depicted in Figure 1.10, only rays that hit the surface of the elliptical lens above the plane of its maximum diameter, denoted herein as its waist, are collimated. The portion of the radiated fields intersecting the lens below its waist is not collimated, but rather propagates along undesired directions or excites surface wave modes, thus giving rise to side lobes or other perturbations in the lens' radiation pattern [20]. One solution to solve this problem is to control the beamwidth of the feed in order that the majority of its radiated energy falls within the angular range above the waist of the lens.

Another issue arising from the internal reflections at the surface of an elliptical lens is the matching of the feed. One inherent characteristic of elliptical lenses is that all of the reflected rays that pass through the second focal point are reflected back to the first focal point. This reflected power causes a substantial mismatch to the feed impedance. A classical method to address this issue is to

Figure 1.10 Illustration of (a) an integrated elliptical lens antenna and (b) an extended hemispherical lens antenna.

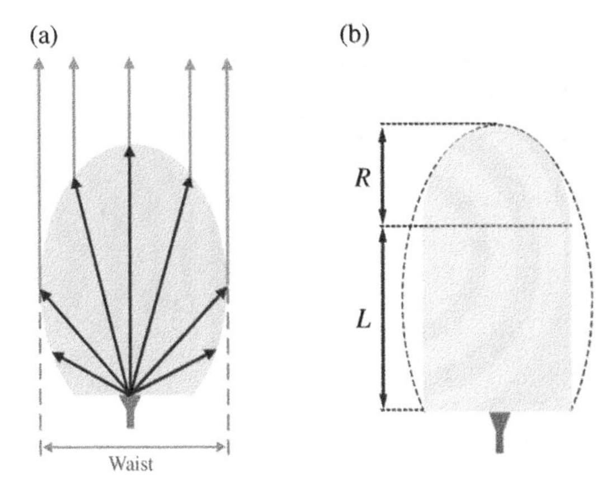

enclose the elliptical length with a matching shell that is a quarter-wave thick. The shell dimensions are specified according to the following equations:

$$n_{\text{match}} = \sqrt{n_1 n_2} \tag{1.3}$$

$$h = \frac{\lambda}{4n_{\text{match}}} \tag{1.4}$$

where n_1, n_2, and n_{match} represent the refraction indexes of the lens, the air, and the matching shell. The main drawback of this approach is that the improved matching performance can only be maintained within a relatively narrow bandwidth. To improve the bandwidth, one can incorporate multiple consecutive matching layers to perform a gradual transition between the two dielectric constants across each interface.

Since the collimation from an elliptical lens only occurs for the portion of the wave front that impinges on its front surface, the part below its waist can be replaced with a cylinder. Furthermore, the top elliptical part, the hemi-ellipse, can be approximated by a hemisphere. This modification significantly reduces the fabrication complexity. The difference in the height of the hemi-ellipse and the hemisphere can be compensated by the height of the cylindrical extension. This new lens is known as an extended hemispherical lens. It turns out to be a rather good approximation to a true elliptical lens, although it tends to present a slightly lower directivity compared to one having the same diameter. The relationship between the radius of the hemisphere, R; the height of the cylinder under it, L; and the refraction index of the lens material, n, is given by

$$L = R/(n-1) \tag{1.5}$$

A lens similar to the extended hemispherical lens is known as a hyper hemispherical lens. In contrast to Eq. (1.5), the cylindrical extension length is now given by [23]:

$$L = R/n \tag{1.6}$$

The rays at the output of the hyper hemispherical lens are not collimated. Therefore, the beam that it generates is much broader than that of the extended hemispherical lens. Nevertheless, it does sharpen the beam radiated by the feed antenna and increases its gain by a factor of n^2. However, unlike the collimating lenses, the directivity of this lens does not increase with the lens size, i.e., its aperture size. The hyper hemispherical length satisfies the Abbe sine condition so the lens itself is free from coma aberration when the feed is transversely displaced from the lens axis [25]. Therefore, it is well suited for beam steering.

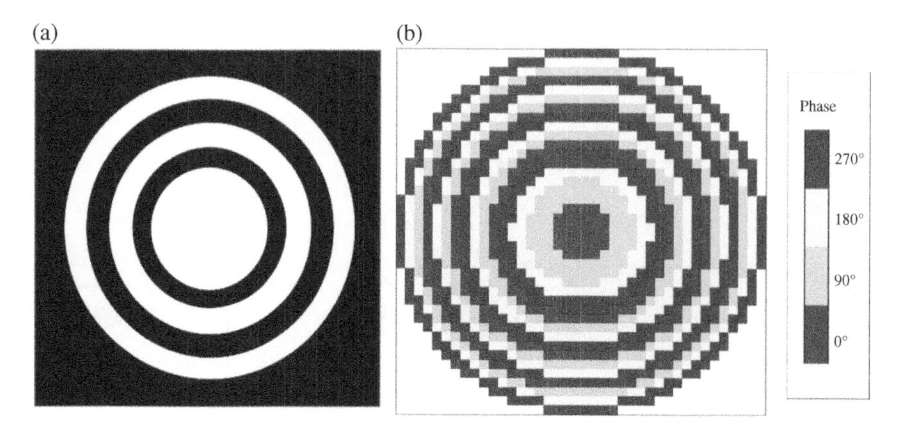

Figure 1.11 Illustration of Fresnel lenses. (a) Original Fresnel lens. (b) Circular phase correcting version [29]. *Source*: Modified from [25] / IEEE.

It must be noted that Eqs. (1.5) and (1.6) are based simply on geometrical optics and point source assumptions. For actual antenna designs at mm-wave and THz frequencies, optimization of the cylindrical extension would be required to achieve the optimal radiation performance [26, 27].

Another type of lenses for mm-wave and THz operations is the Fresnel lens [28]. A Fresnel lens consists of a number of alternately transparent and opaque half-wave zones. The source of the antenna is placed at its focal point. The opaque zones are attained by covering the corresponding portions of the lens with conducting or absorbing materials. Figure 1.11a shows a circular Fresnel lens. It consists of a series of zonal openings in a finite conducting sheet. To increase the antenna efficiency and reduce the sidelobe level, one can introduce phase correcting elements into the zones as depicted in Figure 1.11b [29]. The resulting radiator is effectively a transmit array.

One salient advantage of the Fresnel lens is its low profile. On the other hand, its main disadvantage is its relatively narrow bandwidth. Nevertheless, substantial progress has been made to increase the bandwidth of transmit arrays in recent years [30]. It should be pointed out that although a Fresnel lens can be made flat, it still needs a feed typically placed many wavelengths away from the lens. If the required beamwidth is not too narrow, one can achieve a completely flat version, i.e., a metasurface-based antenna that is created by placing a metasurface above an antenna backed by a ground plane [31].

1.8 SIMO and MIMO Multi-Beam Antennas

Before we end this chapter, we would like to clarify the concept of multi-beam antennas. A conventional antenna has only one input port for one single beam with a specified polarization. We call such an antenna a single input and single output (SISO) antenna. There are two options to create multiple beams. The first option is that a single signal is fed into one port and then is split or distributed to sets of radiating elements and, hence, into a number of beams. We call this type of antenna a single input and multiple output (SIMO) multi-beam antenna; it is illustrated in Figure 1.12a. The second option is that multiple signals are fed into multiple ports, separately, and then each input signal is delivered to a specific set of radiating elements to produce one dedicated beam. We call this type of antenna a multiple input and multiple output (MIMO) multi-beam

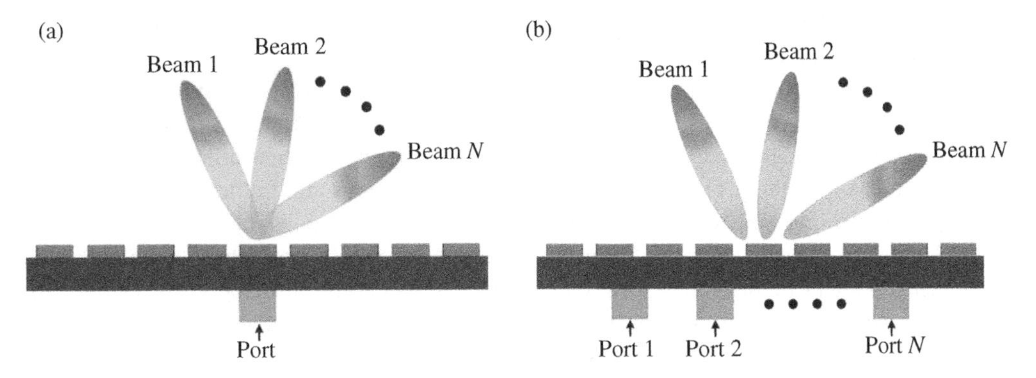

Figure 1.12 Illustration of (a) SIMO and (b) MIMO multi-beam antennas.

antenna; it is illustrated in Figure 1.12b. SIMO multi-beam antennas are useful for data distribution and targeted broadcasting services. On the other hand, MIMO multi-beam antennas are useful for multiuser communications in which each user is at a different location and communicates different information. The multiple beams created by SIMO and MIMO antennas can be fixed or steerable. The latter is much harder to achieve and would serve as a major research direction for the future. This topic is addressed further in Chapters 5 and 6.

Notice that the SIMO and MIMO multi-beam antenna concepts presented above are substantially different from the concepts of SIMO and MIMO in wireless communication systems. All of the inputs and outputs in the former reside in one transmitter or receiver system, typically in the base stations. The multi-beams produced by the antennas are distinct beam patterns. In contrast, all of the inputs and outputs of the latter reside separately in the transmitter, typically at the base station, and the receivers, typically in the user terminals. The transmitted RF signal may not have distinct conventional beam patterns; certain types of multiuser detection or spatial–temporal decoding algorithms are employed at the receivers with no regard to specific beam patterns.

1.9 In-Band Full Duplex Antennas

As frequency resources become more and more scarce, the issue of spectral efficiency has become a top priority for future generations of wireless communication systems. Consequently, in-band full-duplex (IBFD) radios are widely regarded as a key technology for the evolution of 5G and 6G systems. IBFD radios allow signal transmission and reception in the same frequency band and at the same time [32]. IBFD radios can double the data rate without using more frequency bands or more time, thus resulting in unprecedented spectrum efficiency enhancement. However, one major issue existing in full-duplex radios is the suppression of in-band self-interference between the transmitters and receivers caused by mismatching at their ports, the mutual coupling between their antenna elements, and the scattering from objects in the environment in which they actually must work.

To realize a practical IBFD radio, the self-interference from the colocated transmitter must be canceled first as it is typically much stronger than the intended received signal. For IBFD communication systems to operate, it usually requires more than 110–130 dB isolation between the transmitter and the receiver [33]. However, to cancel the self-interference in IBFD systems satisfactorily, one needs a three-stage solution in the antenna domain (or propagation domain), the analog domain, and the digital domain.

Clearly, no digital circuits can operate without adequate isolation and appropriate cancelations in the antenna and analog domains to bring the signal-to-noise-and-interference ratio down to an acceptable level. To this end, major efforts have been made in analog cancelation methods using adaptive circuits and antennas [34–37]. Reported antenna solutions aim to increase the isolation between the transmitter and receiver ports by virtue of spatial and polarization separation, use of metamaterials, and beam squinting. In principle, an ideal solution would be a combination of antenna-decoupling techniques to be discussed in Chapter 3 and self-interference cancelation circuits. Major challenges facing antenna researchers and engineers are wide bandwidth, limited antenna space, and low-loss circuit designs.

1.10 Conclusions

Up until the emergence of the third- and fourth-generation mobile wireless communication networks, the focus of most antenna researchers was largely on antennas for radar and satellite communications. On the other hand, antenna designers working in the mobile communication industry were faced with "engineering" challenges largely ignored by the majority of academic antenna researchers. The collocation and coexistence of antennas for 3G and 4G, as well as the demand for antenna miniaturization and stringent specifications, posed serious research challenges to the antenna community. However, judging by the number of publications, one may argue that base station antennas and terminal antennas did not receive the attention they deserved from academic researchers. This lack of attention might have been partly attributed to the unique global industrial landscape formed in that period; the industry was consolidated to only a few players in the end. Moving forward to 5G and 6G, the technology competition among national governments and industries from all around the globe is rapidly gathering pace, thus attracting the widespread interest of the international antennas community. As a result, research on 5G and 6G antennas has started taking center stage globally. In this chapter, we have provided our own perspectives for 5G and 6G antennas. We have outlined some of the major challenges facing antenna researchers and designers, and have enunciated possible technology pathways. In the following chapters, we shall present detailed overviews, and our own studies to address some of the main technical challenges associated with 5G and beyond antenna arrays. We have mainly focused on antennas for base stations and large platforms. Given the potentially vast scope of 5G and 6G systems, we make no claim that all the antenna topics for 5G and beyond have been covered.

References

1. Dang, S., Amin, O., Shihada, B., and Alouini, M.-S. (2020). What should 6G be? *Nat. Electron.* **3**: 20–29.
2. Saad, W., Bennis, M., and Chen, M. (2020). A vision of 6G wireless systems: applications, trends, technologies, and open research problems. *IEEE Network* **34** (3): 134–142.
3. You, X., Wang, C.-X., Huang, J. et al. (2021). Towards 6G wireless communication networks: vision, enabling technologies, and new paradigm shifts. *Sci. China Inform. Sci.* **64**: 110301. doi. https://doi.org/10.1007/s11432-020-2955-6.

4. https://www.5gamericas.org/wp-content/uploads/2019/08/5G-Americas_Advanced-Antenna-Systems-for-5G-White-Paper.pdf (accessed 16 January 2020).
5. https://www.ericsson.com/en/reports-and-papers/white-papers/advanced-antenna-systems-for-5g-networks (accessed 16 January 2020).
6. https://carrier.huawei.com/~/media/CNBGV2/download/products/antenna/New-5G-New-Antenna-5G-Antenna-White-Paper-v2.pdf (accessed 16 January 2020).
7. Han, H., Ding, C., Jones, B., and Guo, Y.J. (2019). Suppression of cross-band scattering in multiband antenna arrays. *IEEE Trans. Antennas Propag.* **67** (4): 2379–2389.
8. Gu, X., Liu, D., Baks, C., et al. (2017). A multilayer organic package with 64 dual-polarized antennas for 28 GHz 5G communication. *Proceedings of the IEEE International Microwave Symposium (IMS)*, Honololu, HI, 4–9 June 2017, pp. 1899–1901.
9. 5G NR Antenna-in-Package (AiP) Technology. White paper. www.tmytek.com (accessed 16 January 2020).
10. Huang, X., Hanzo, L., Zhang, A. et al. (2019). Airplane-aided integrated networking for 6G wireless: will it work? *IEEE Veh. Technol. Mag.* **14** (3): 84–91.
11. Heath, R.W. Jr., González-Prelcic, N., Rangan, S. et al. (2016). An overview of signal processing techniques for millimeter wave MIMO systems. *IEEE J. Sel. Topics Signal Process.* **10** (3): 436–453.
12. Zhang, J.A., Huang, X., Dyadyuk, V., and Jay Guo, Y. (2015). Massive hybrid antenna array for millimetre wave cellular communications. *IEEE Wireless Commun.* **22** (1): 79–87.
13. Huang, X., Guo, Y.J., and Bunton, J. (2010). A hybrid adaptive antenna array. *IEEE Trans. Wireless Commun.* **9** (5): 1770–1779.
14. Guo, Y.J. and Jones, B. (2018). Base station antennas. In: *Antenna Engineering Handbook*, fifthe (ed. J.L. Volakis). New York: McGraw Hill, Chapter 40.
15. Ansari, M., Jones, B., Zhu, H. et al. (2020). A dual polarized 3D Luneburg lens multi-beam antenna system. *IEEE Antennas Propag.* https://doi.org/10.1109/TAP.2020.3044638.
16. Guo, Y.J., Huang, X., and Dyadyuk, V. (2012). A hybrid adaptive antenna array for long range mm-wave communications. *IEEE Antennas Propag. Mag.* **54** (2): 271–282.
17. R4–1807849, 3GPP TSG-RAN WG4 Meeting #87.
18. O'Hara, J.F., Ekin, S., Choi, W., and Song, I. (2019). A perspective on terahertz next-generation wireless communications. *MDPI Technol.* **7** (2): 43.
19. Mehdi, I., Siles, J., Chen, C.P., and Jornet, J.M. (2018). THz technology for space communications. *Proceedings of the 2018 Asia-Pacific Microwave Conference (APMC)*, Kyoto, Japan, 6–9 November 2018, pp. 76–78.
20. Pasqualini, D. and Maci, S. (2004). High-frequency analysis of integrated dielectric lens antennas. *IEEE Trans. Antennas Propag.* **52** (3): 840–847.
21. Gao, X., Zhang, T., Du, J., and Guo, Y.J. (2020). 340-GHz double-sideband mixer based on antenna-coupled high-temperature superconducting Josephson junction. *IEEE Trans. THz Sci. Technol.* **10** (1): 21–31.
22. Costa, J.R., Fernandes, C.A., Godi, G. et al. (2008). Compact Ka-band lens antennas for LEO satellites. *IEEE Trans. Antennas Propag.* **56** (5): 1251–1258.
23. Costa, J.R., Lima, E.B., and Fernandes, C.A. (2009). Compact beam-steerable lens antenna for 60-GHz wireless communications. *IEEE Trans. Antennas Propag.* **57** (10): 2926–2933.
24. Filipovic, D.F., Gearhart, S., and Rebeiz, G.M. (1993). Double-slot antennas on extended hemispherical and elliptical silicon dielectric lenses. *IEEE Trans. Microw. Theory Techn.* **41** (10): 1738–1749.
25. Rebeiz, G.M. (1992). Millimeter-wave and terahertz integrated circuit antennas. *Proc. IEEE* **80** (11): 1748–1770.

26. Born, M. and Wolf, E. (1959). *Principles of Optics*. New York: Pergamon Press.
27. Konstantinidis, K., Feresidis, A.P., Constantinou, C.C. et al. (2017). Low-THz dielectric lens antenna with integrated waveguide feed. *IEEE Trans. THz Sci. Technol.* **7** (5): 572–581.
28. Wu, G.-B., Zeng, Y.-S., Chan, K.F. et al. (2019). 3-D printed circularly polarized modified Fresnel lens operating at terahertz frequencies. *IEEE Trans. Antennas Propag.* **67** (7): 4429–4437.
29. Guo, Y.J. and Barton, S.K. (2013). *Fresnel Zone Antennas*. Berlin, Germany: Kluwer Academic Publishers.
30. Song, L.Z., Qin, P.-Y., Maci, S., and Guo, Y.J. (2020). Ultrawideband transmitarray employing connected slot-bowtie dipole elements. *Proceedings of the European Conference on Antennas and Propagation (EuCAP2020)*, Copenhagen, Denmark, 15–20 March 2020, paper A19.P1.060.
31. González-Ovejero, D., Minatti, G., Chattopadhyay, G., and Maci, S. (2017). Multibeam by metasurface antennas. *IEEE Trans. Antennas Propag.* **65** (6): 2923–2930.
32. Zhang, Z., Long, K., Vasilakos, A.V., and Hanzo, L. (2016). Full-duplex wireless communications: challenges, solutions, and future research directions. *Proc. IEEE.* **104** (7): 1369–1409.
33. Zhou, J., Reiskarimian, N., Diakonikolas, J. et al. (2017). Integrated full duplex radios. *IEEE Commun. Mag.* **55** (4): 142–151.
34. Le, A.T., Tran, L.C., Huang, X. et al. (2021). Analog least mean square adaptive filtering for self-interference cancellation in full-duplex radios. *IEEE Wireless Commun. Mag.* **28**: 12–18.
35. Huang, X. and Guo, Y.J. (2017). Radio frequency self-interference cancellation with analog least mean square loop. *IEEE Trans. Microw. Theory Techn.* **65** (9): 3336–3350.
36. Alrabadi, O.N., Tatomirescu, A.D., Knudsen, M.B. et al. (2013). Breaking the transmitter–receiver isolation barrier in mobile handsets with spatial duplexing. *IEEE Trans. Antennas Propag.* **61** (4): 2241–2251.
37. Acimovic, I., McNamara, D.A., and Petosa, A. (2008). Dual-polarized microstrip patch planar array antennas with improved port-to-port isolation. *IEEE Trans. Antennas Propag.* **56** (11): 3433–3439.

2

Millimeter-Wave Beamforming Networks

A beamforming network (BFN) is a physical layer element of an array system that combines signals with the requisite amplitudes and phases required to produce a desired angular distribution of emitted radiation, i.e., one or more beams pointed in prescribed directions. Energy delivered to a particular input port is thus associated with a beam radiated by the antenna elements connected to the output ports. The radio frequency (RF) BFN is a critical component of analog MIMO multi-beam antenna arrays as discussed in Chapter 1. There are mainly two types of RF BFNs, i.e., the circuit-type and the quasi-optical type.

Circuit-type BFNs are composed of some basic circuit components, namely, couplers, crossovers, and phase shifters. The most popular circuit-type BFN is the Butler matrix (BM) that is briefly discussed in Chapter 1. Circuit-type BFNs are suited to microwave and lower millimeter-wave (mm-wave) bands. In contrast, quasi-optical-type BFNs are based on optical principles; and, as a consequence, they are more suited for mm-wave and THz systems. Three popular quasi-optical-type BFNs are Luneburg lens, Rotman lens, and reflectors.

An overview of both circuit-type BFNs and quasi-optical-type BFNs that have been developed for mm-wave frequency systems is provided in this chapter. The underlining structure is the substrate integrated waveguide (SIW). These SIW-based BFNs provide solutions to fixed analog multi-beam antennas. As demonstrated in Chapter 7, they can also be employed in hybrid systems to realize steerable multi-beam antennas.

2.1 Circuit-Type BFNs: SIW-Based Butler and Nolen Matrixes

The commonly used microstrip-line-based BFNs that are popular for microwave systems suffer from high radiation and transmission losses at mm-wave frequencies. Because of its low profile, ease of fabrication, low insertion losses, and compatibility with other planar circuits, the SIW has attracted widespread attention as the basic guided wave structure for mm-wave BFNs and mm-wave circuits in general. A review of the most recent research progress in mm-wave SIW-based BFNs is presented next.

2.1.1 Butler Matrix for One-Dimensional Multi-Beam Arrays

The classic BM is an $N \times N$ network that generates uniform amplitude and linear phase distributions at N output ports given signals at N input ports. Those outputs are then generally applied to excite the N radiating elements of an antenna array. The BM provides a specific phase distribution

Advanced Antenna Array Engineering for 6G and Beyond Wireless Communications, First Edition.
Y. Jay Guo and Richard W. Ziolkowski.

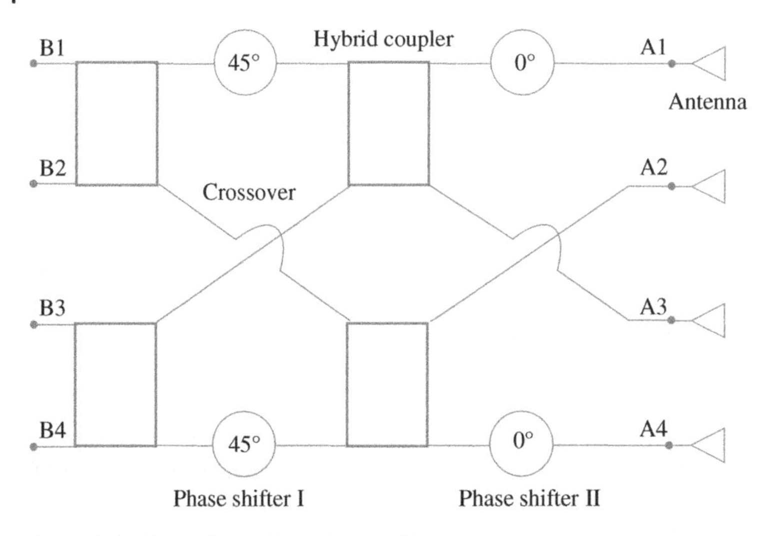

Figure 2.1 Topology of a classic 4 × 4 BM.

on those elements for signals applied to a given input port. The array will then radiate a beam into a particular direction. Consequently, by choosing a different input port and, hence, a different output phase distribution applied to the array elements, a different beam will be radiated into some other direction. A multi-beam array is thus realized if all of the input ports are driven simultaneously.

BMs have been generally used to excite a linear, i.e., one-dimensional (1-D), antenna array and to control the directions of the radiated output beams in terms of *either* its azimuth or zenith directions. Many researchers in recent years have combined stacked BMs to build two-dimensional (2-D) multi-beam arrays whose beam directions can be controlled in *both* the azimuth and zenith directions. To avoid any confusion in terminology in this chapter, the terms "1-D multi-beam" and "'2-D multi-beam" arrays will mean arrays that can radiate multiple beams and that can scan those beams along only one direction or in any direction.

The topology of a classic 4 × 4 BM driving a 4-element 1-D array is shown in Figure 2.1.

A systematic approach to the design of a 60-GHz SIW BM based on this topology was developed in [1]. Its four input ports are labeled as beam ports B1–B4, and its four output ports are labeled array ports A1–A4. The BM consists of four 90° hybrid couplers, two crossovers, two 45° phase shifters, and two 0° phase shifters. These basic components of the BM play different roles to achieve its beamforming features. A 90° hybrid coupler equally divides the power with a phase difference of ±90°. Crossovers are introduced to address circuit path overlaps. Phase shifters are required to change the phase distribution at the array ports. With appropriate combinations of these components, one can obtain the desired output amplitude and phase distributions. Theoretically, a 4 × 4 BM provides equal amplitude and with phase differences of ±45°, and ±135° at the array ports. Fed by such a 4 × 4 BM, the angles of the multiple beams radiated by a 1-D array with respect to the boresight direction of the array are as follows:

$$\theta_m = \arcsin\left[-\frac{(\Delta\varphi)_m}{kd}\right] \tag{2.1}$$

where θ_m is the beam angle of *m-th* beam, $(\Delta\varphi)_m$ is the phase difference provided by the *m-th* beam port, k is the free-space wave number, and d is the distance between adjacent antenna elements of the array.

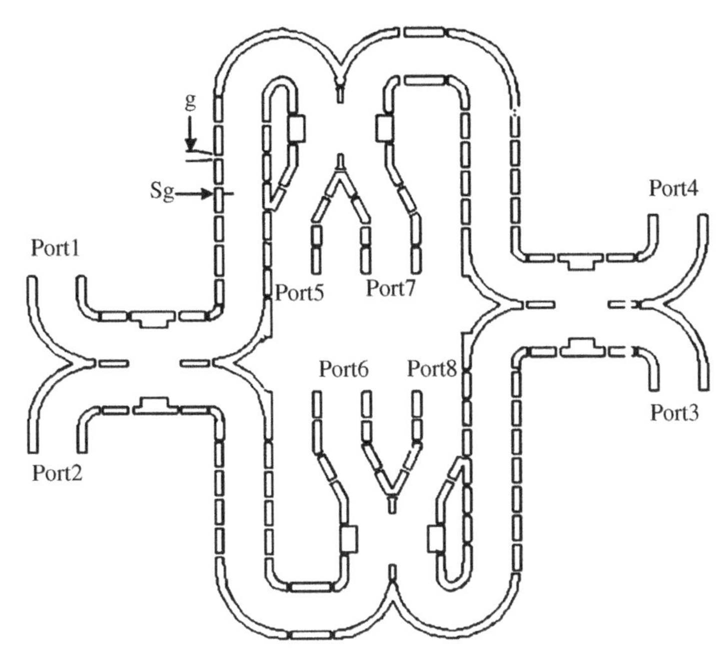

Figure 2.2 Model of the modified 4 × 4 BM that has no crossovers. *Source:* From [2] / with permission of IEEE.

The BM in Figure 2.1 actually contains a large number of components. Operating at mm-wave and higher frequencies, the losses associated with these basic components, which include both dielectric and conductor, cannot be ignored. One alternative to decrease the loss in a BM is to reduce the number of required components. Hybrid couplers and phase shifters, however, are indispensable to the functions of a BM because of their ability to manipulate phases and amplitudes. On the other hand, the crossovers, which simply handle circuit path overlaps, are a potential target for removal.

A modified topology of the classic 4 × 4 BM used in [1] was developed in [2] to avoid the use of any crossovers. This configuration is shown in Figure 2.2. This modified topology is different from the traditional left-to-right arrangement of the basic components. It places the input ports on the outer side of the layout and the output ports on its inner side. Thus, there are no overlaps of the four signal paths. This topology was extended to a 4 × 8 BM in [3] by introducing four power splitters in order to excite an eight-element array with the intent to produce beams with higher gain.

Another drawback of having an excessive number of components in a BM is the large footprint associated with its layout. This issue can be resolved by using a multilayer configuration that is facilitated by the development of SIW technologies. Since an SIW is a closed structure, several of them can be directly stacked on top of each other without influencing the transmission performance. To demonstrate the reduction of the dimensions of a 4 × 4 BM, the dual-layer configuration shown in Figure 2.3 was developed in [4, 5]. There are two main advantages of using this dual-layer configuration. In addition to eliminating the crossovers and, hence, reducing the losses, it also reduces the footprint of the BM by half.

The SIW-based 4 × 4 BMs considered above excited array elements that radiated linearly polarized (LP) fields. SIW-based BMs have also been used to excite end-fire circularly polarized (CP) arrays [6, 7]. An example of such a BM is shown in Figure 2.4. Furthermore, a 4 × 4 BM was developed in [8] to realize antenna arrays that generate beams with ±45° dual LP and dual CP fields.

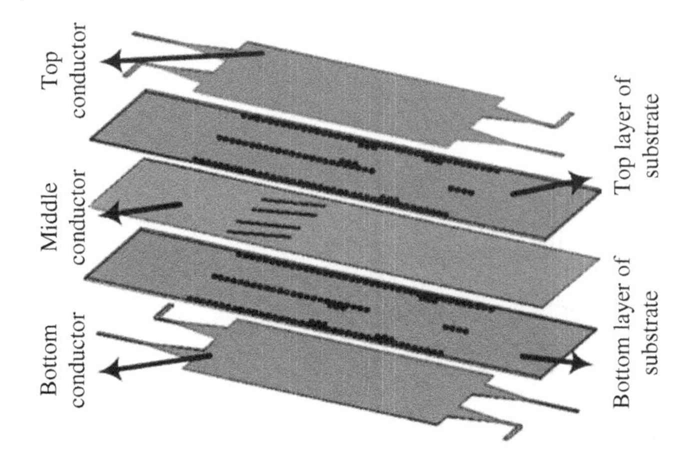

Figure 2.3 Simulated model of the dual-layer 4 × 4 BM. *Source:* From [5] / with permission of IEEE.

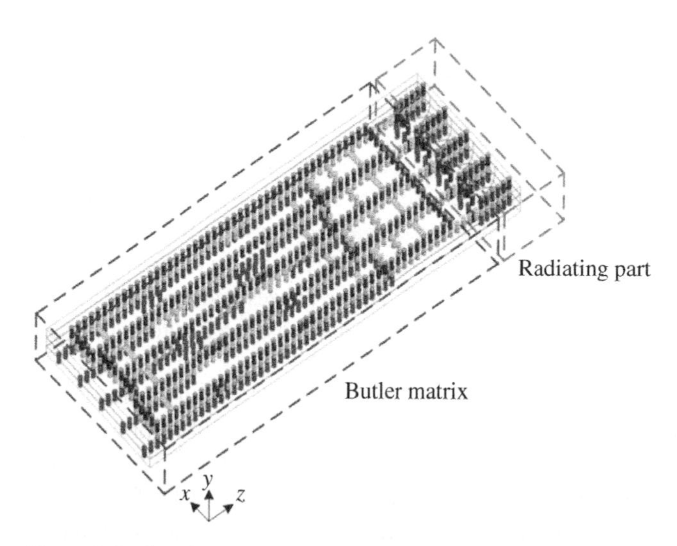

Figure 2.4 Simulated model of the circularly polarized multi-beam array fed by a 4 × 4 BM. *Source:* From [6] / with permission of IEEE.

While a 4 × 4 BM can be used to feed an antenna array to generate four beams, more beams or higher gain may be required for some applications. A higher-order BM with eight inputs and eight outputs is helpful in those cases. An 8 × 8 BM can equally divide the power from any input into eight outputs with phase differences of ±22.5°, ±67.5°, ±112.5°, and ±157.5°. An example of an 8 × 8 BM based on the traditional left-to-right topology was developed using single-layer SIW technology in [9]. Its layout is shown in Figure 2.5. Notice that the 8 × 8 BM is much more complicated than a 4 × 4 BM since many more components are required to provide the desired phase differences. As illustrated above, a dual-layer configuration can help remove some of the crossovers to decrease the losses and to improve the compactness of its layout. An example of dual-layer SIW 8 × 8 BM is shown in Figure 2.6; it was developed in [10].

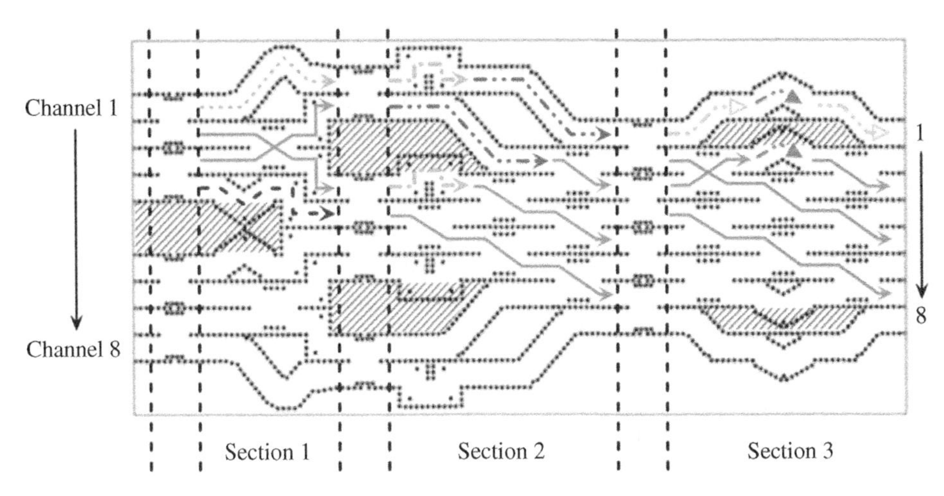

Figure 2.5 Simulated model of a SIW-based 8 × 8 BM. *Source:* From [9] / with permission of IEEE.

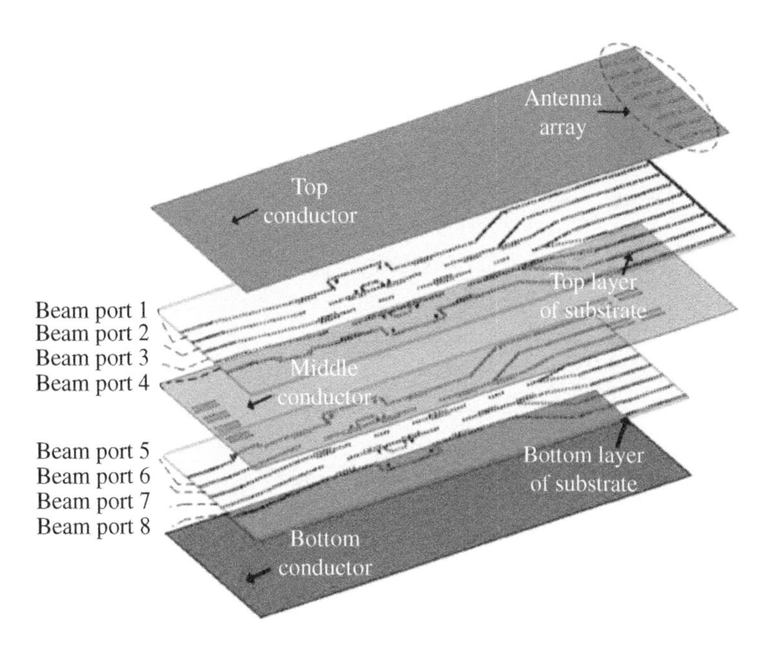

Figure 2.6 Simulated model of the dual-layer 8 × 8 BM. *Source:* From [10] / with permission of IEEE.

2.1.2 Butler Matrix for a 1-D Multi-Beam Array with Low Sidelobes

As discussed above, classic BMs are $N \times N$ networks that produce uniform amplitude distributions. A linear array excited with uniform amplitudes theoretically yields the maximum directivity for its size and has a sidelobe level (SLL) of approximately −13 dB. Unfortunately, because of mutual coupling and feeding errors, i.e., phase and amplitude errors, the realized SLLs in practical designs are higher and can easily deteriorate to −10 dB or worse. Such high SLL values can cause serious interference issues. Arrays with such performance characteristics are not suitable for most 5G and beyond multi-beam array applications since, as discussed in Chapter 1, much lower SLLs are desired.

One well-known approach to decrease the SLLs of the radiation patterns of any array is to taper the excitation amplitudes along it. The cost of lower SLLs is a reduced gain and wider main lobe. For example, a Chebyshev distribution of amplitudes allows one to achieve SLLs below a specified maximum. To realize these nonuniform amplitudes with a BM BFN, one can introduce N unequal power dividers into the classic $N \times N$ BM and extend it to an $N \times 2N$ network. As an example, the topology of a 4 × 8 BM is shown in Figure 2.7. Two different unequal power dividers, D1 and D2, are connected to a 4 × 4 BM. By changing the power dividing ratio, one can obtain any desired tapered amplitude distribution. Nevertheless, some crossovers and phase shifters are required in this design to maintain the original phase distributions.

Two recently developed SIW-based 4 × 8 BMs demonstrate the progress made in the realization of BFNs that can distribute nonuniform amplitudes to an eight-element array. Similar to the 4 × 4 and 8 × 8 BMs introduced above, an effective approach to reduce losses and minimize the footprint of the BM layout is again to reduce the number of required crossovers. Such a simplified 4 × 8 BM configuration was developed in [11]. It is shown in Figure 2.8. One of the five crossover sections

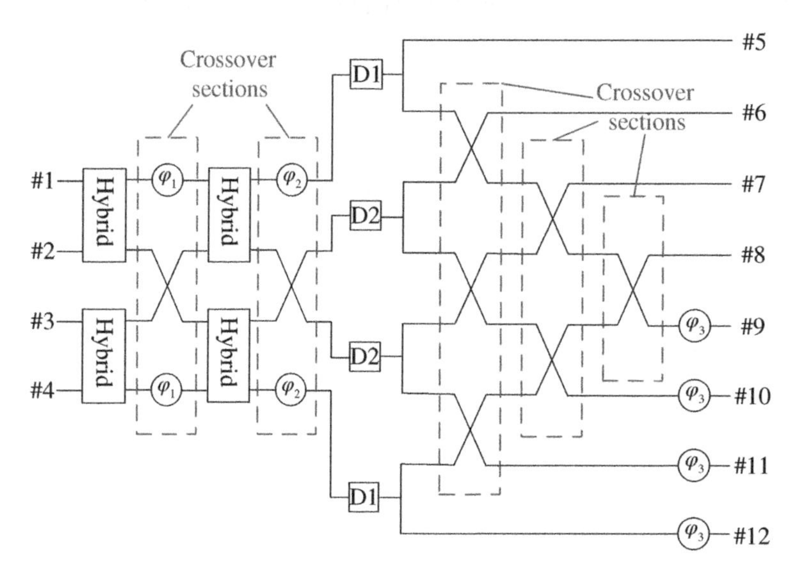

Figure 2.7 Topology of a 4 × 8 BM that delivers nonuniform amplitudes at its output ports. *Source:* From [12] / with permission of IEEE.

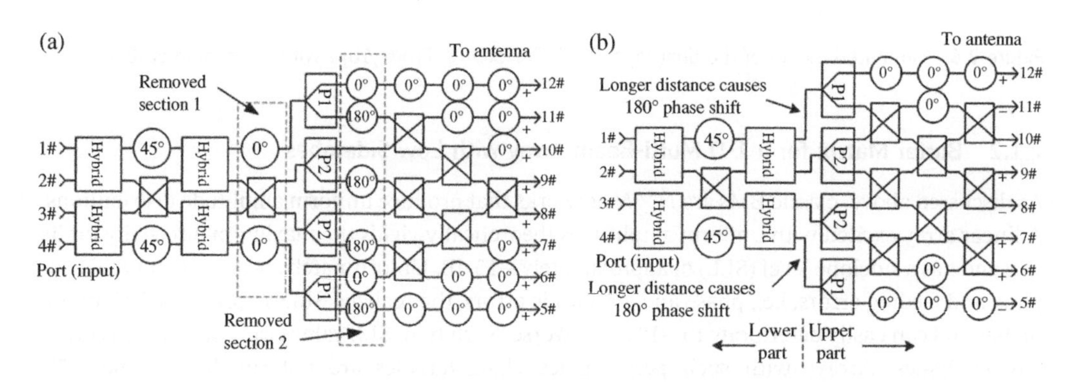

Figure 2.8 Classic and innovative 4 × 8 BM topologies. (a) Classic configuration. (b) Simplified configuration. *Source:* From [11] / with permission of IEEE.

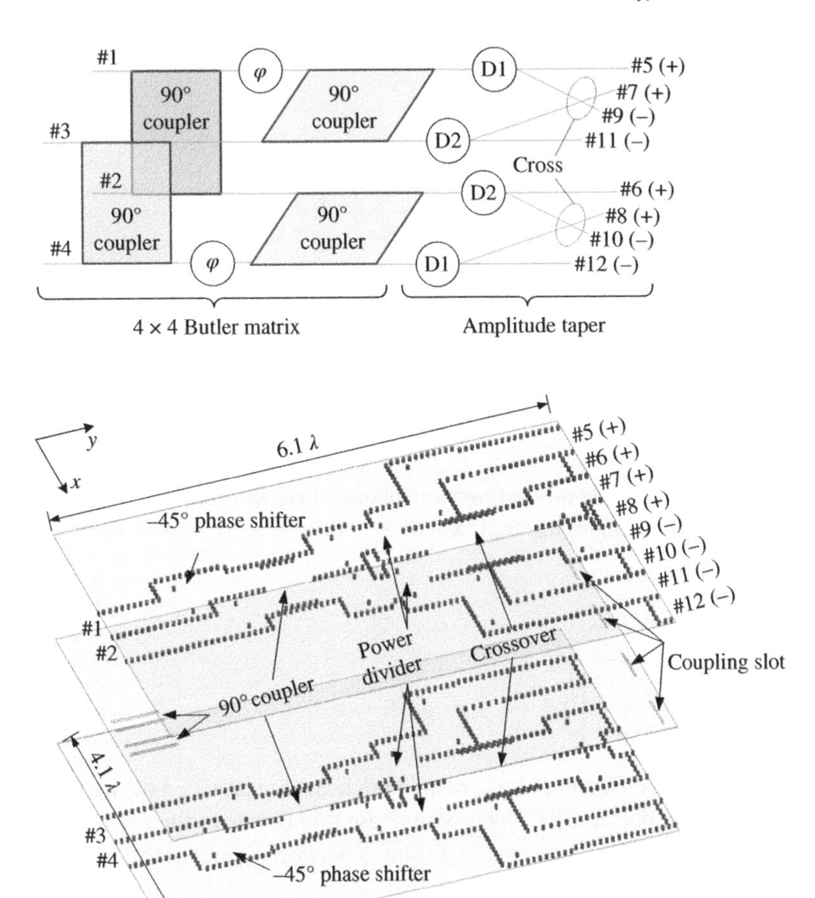

Figure 2.9 Topology and simulation model of a dual-layer 4 × 8 BM that distributes nonuniform amplitudes to the radiators of an eight-element linear array. *Source:* From [12] / with permission of IEEE.

and one phase shifter section were removed from the classic design. Because crossovers are usually introduced to obtain a desired phase distribution, this modified BM configuration had to maintain the original phase differences. As demonstrated in [11], half of the array ports in the modified design, denoted by the "–" signs, provide a reversed current direction to the antennas connected to them. With these additional 180° phase shifts, the simplified configuration thus provides the same phase distribution as the classic one.

The dual-layer version of this 4 × 8 BM topology has also been realized in [12]. It is depicted in Figure 2.9. This dual-layer configuration reduces the required crossover sections from five to one. This reduction in the number of crossovers significantly decreases the losses and improves the compactness of this BFN.

2.1.3 Butler Matrix for 2-D Multi-Beam Arrays

As indicated above, researchers have been attempting to combine 1-D BMs to create BFNs for 2-D multi-beam arrays. The simplest example of a BFN for a 2-D multi-beam array would be a 4 × 4 2-D BFN that supports a 2 × 2 array that generates four beams. The classic topology for such a BFN is shown in Figure 2.10. Two sets of sub-BFNs are orthogonally connected to each other; they realize

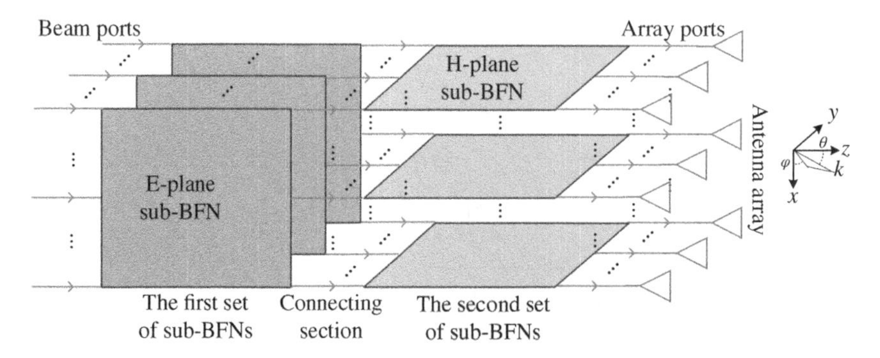

Figure 2.10 Topology of a BFN for a 2-D multi-beam array.

beamforming properties in both the horizontal and vertical planes. Fed with such a BFN, the radiated beam direction is given by the angle (θ_m, φ_m), where:

$$\varphi_m = \tan^{-1}\left[\frac{(\Delta\varphi_y)_m d_x}{(\Delta\varphi_x)_m d_y}\right] \tag{2.2a}$$

$$\theta_m = \sin^{-1}\sqrt{\left[\frac{(\Delta\varphi_x)_m}{kd_x}\right]^2 + \left[\frac{(\Delta\varphi_y)_m}{kd_y}\right]^2} \tag{2.2b}$$

The beam angle (φ_m, θ_m) and the phase difference $(\Delta\varphi)_m$ of the *m-th* beam are provided by the *m-th* beam port. The terms $\Delta\varphi_x$ and $\Delta\varphi_y$ represent the progressive phase differences in the *x*-direction and *y*-direction, respectively. The terms d_x and d_y are the distances between the adjacent antenna elements in the *x*-direction and *y*-direction, respectively.

This classic topology is composed of two sets of 1-D BMs placed as illustrated in Figure 2.10. It provides phase gradients of $\pm 90°$ along both the *x*- and *y*-directions. This 4×4 2-D BFN topology has the same number of components as a 4×4 BM does when its phase shifters are set to produce 0° phase shifts. As demonstrated in [13], this feature allows one to modify a traditional 4×4 BM as shown in Figure 2.11. Similar to the design in [2], this structure avoids the use of crossovers to improve the layout size. A major difference is that the output ports feed a 2×2 planar array while the one in [2] was designed for a 1×4 linear array. Similar 4×4 2-D BFNs can be found in [14] and [15], where those designs excite a magnetoelectric dipole array and a cavity-backed patch array, respectively.

Recalling the discussions of the classic 1-D BFNs, classic 2-D BFNs can only provide equal amplitudes to the array elements, and, as a consequence, the beams they support also suffer from relatively high SLLs when they point away from boresight. Nevertheless, 2-D BFNs can also be designed to provide the tapered amplitude distributions to the array elements required to achieve SLLs that meet the requirements for practical applications. The 2-D BFN developed in [16] is a good example; it is shown in Figure 2.12. The hybrid coupler in each sub-BFN is again replaced with a 2×4 BM by resorting to unequal power dividers. Furthermore, replacing the hybrid couplers in a classic 4×4 2-D BFN with 2×4 BMs, one can extend it to a 4×16 2-D BFN that produces a tapered amplitude distribution.

The multilayer version of the planar 4×16 2-D BFN in [16] placed the second set of sub-BFNs, i.e., four E-plane 2×4 BMs, underneath the radiation portion to excite the radiating elements. The first set of sub-BFNs, i.e., two H-plane 2×4 BMs, were placed on the lateral sides of the first set of

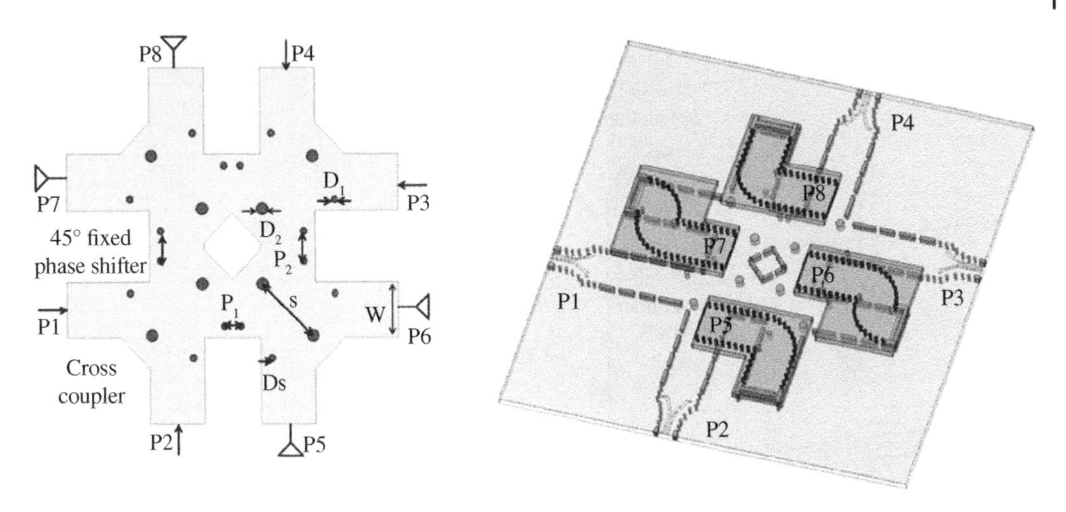

Figure 2.11 BFN layout (left) that feeds the 2 × 2 planar array (right), which generates four beams. *Source:* From [13] / with permission of IEEE.

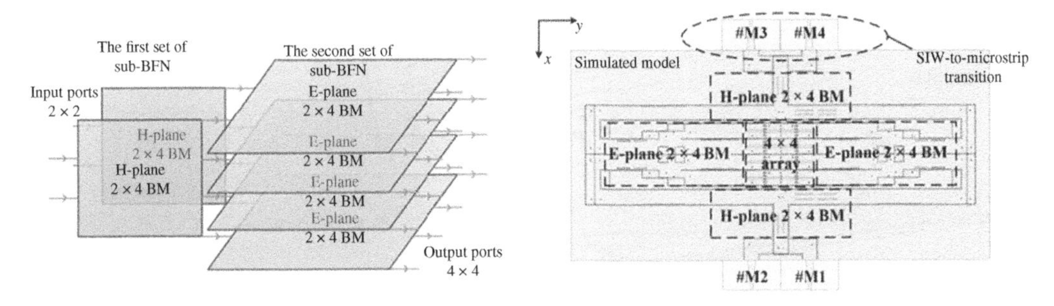

Figure 2.12 Multilayer BFN that delivers nonuniform amplitudes to the elements of a 2 × 2 multi-beam array for SLL suppression. *Source:* From [16] / with permission of IEEE.

two E-plane sub-BFNs. Some coupling slots and interconnections were introduced between these two sets of sub-BFNs. This design successfully distributed tapered amplitude distributions that suppressed the SLL of the associated 2-D multi-beam array.

If more beams are required, the number of components in a 2-D BFN would have to increase. An 8 × 8 2-D BFN producing eight (2 × 4) beams can be realized with four hybrid couplers and two 4 × 4 BMs. As with the smaller multi-beam arrays, there are two main design tendencies in the open literature. One is to modify the topology of the 2-D BFN and integrate it into a planar design using, for example, SIW technologies. The other one is to use a multilayer design to reduce the footprint and make the system more compact. Two examples of an 8 × 8 2-D BFN deployed in a multilayer SIW configuration were realized in [17, 18]. Both designs used folded 4 × 4 BMs to reduce their overall size. Their overall configurations were composed of a six-layer SIW, as shown in Figure 2.13.

If a 16 × 16 2-D BFN is used to excite a 4 × 4 multi-beam array, the main difficulty in its design would be maintaining its planarity. This complication arises because eight 4 × 4 BMs would have to be connected spatially to each other in a manner like the design shown in Figure 2.10. In a traditional 16 × 16 2-D BFN, one would usually design eight identical 4 × 4 BMs, four of which are placed vertically to function as the E-plane sub-BFNs, while the rest would be placed horizontally as

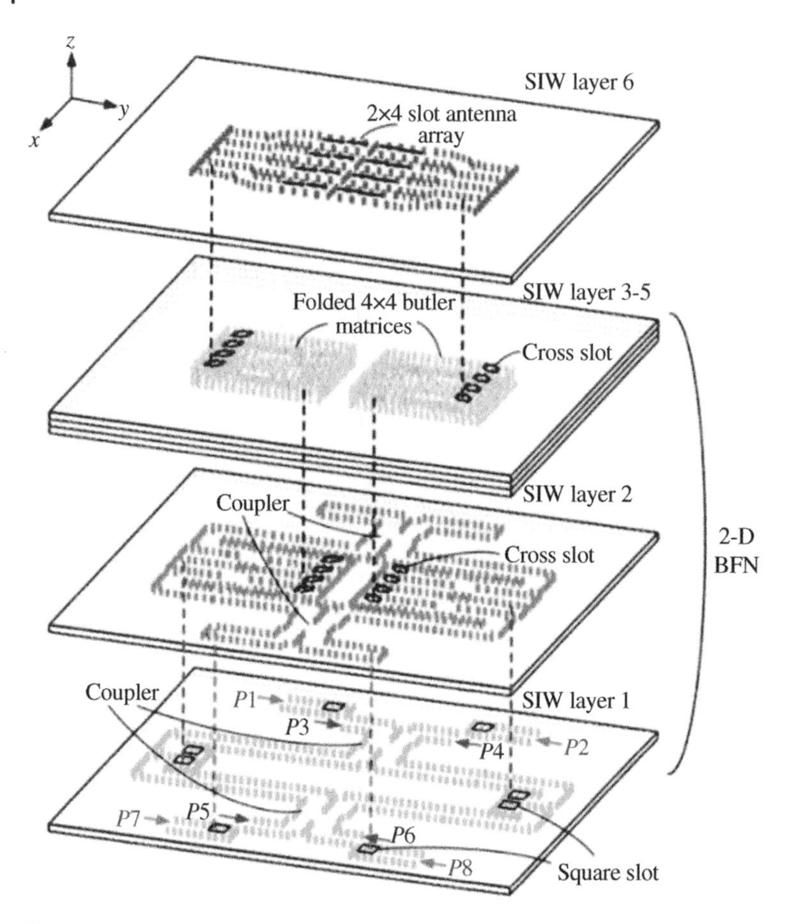

Figure 2.13 Multilayer 2-D BFN that delivers nonuniform amplitudes to the elements of a 2 × 4 multi-beam array for SLL suppression. *Source:* From [18] / with permission of IEEE.

H-plane sub-BFNs. Cables or connectors are then required to connect these two sets of sub-BFNs. Thus, traditional 16 × 16 2-D BFNs are bulky in size. Some researchers have recently addressed the design of planar 16 × 16 2-D BFNs. Three examples, which followed different design approaches, are discussed below.

The first one is a single-layer planar design realized in [19] and is shown in Figures 2.14 and 2.15. The main idea was the development of planar 2-D components. As shown in Figure 2.14, all the couplers were placed within one plane and connected to each other. Phase shifters with different phase shift values $\alpha 1$–$\alpha 4$ were introduced between the couplers to achieve the desired phase distribution at the output ports. However, a classic single-layer configuration without any modification would introduce an excessive number of path overlaps; and thus, many crossovers would be required. To avoid such an excessive number of crossovers, a new topology was developed from Figures 2.14a and 2.14b by introducing two eight-port crossovers. These crossover components allow four paths to cross over at the same intersection. With them, the total number of path intersections in the 16 × 16 2-D BFN was reduced from 16 to only 4.

Based on the topology illustrated in Figure 2.14b, the single-layer SIW-based 16 × 16 BFN shown in Figure 2.15a was achieved. A photo of the fabricated prototype is shown in Figure 2.15b. It was

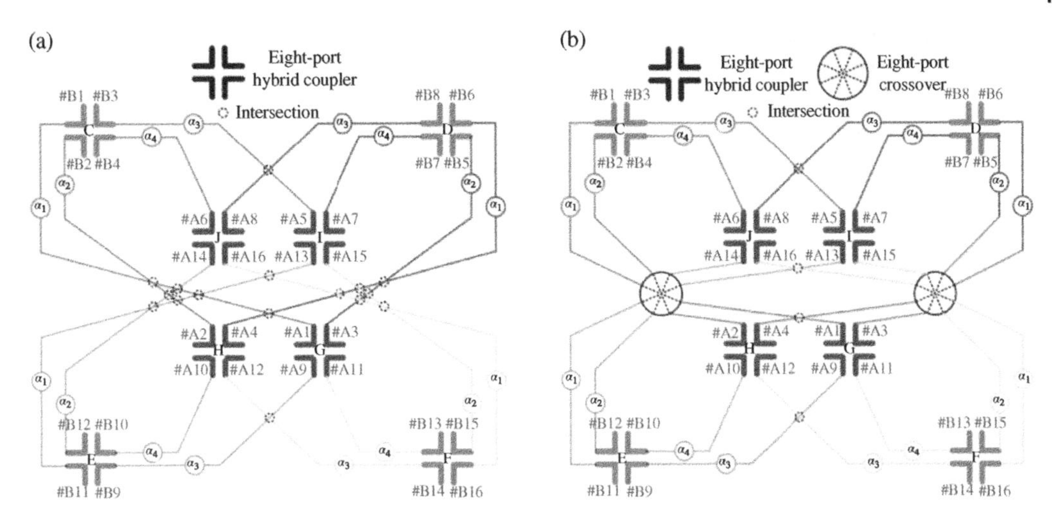

Figure 2.14 Developed topologies of a planar 16 ×16 2-D BFN. (a) Without the eight-port crossover. (b) With the eight-port crossover. *Source:* From [19] / with permission of IEEE.

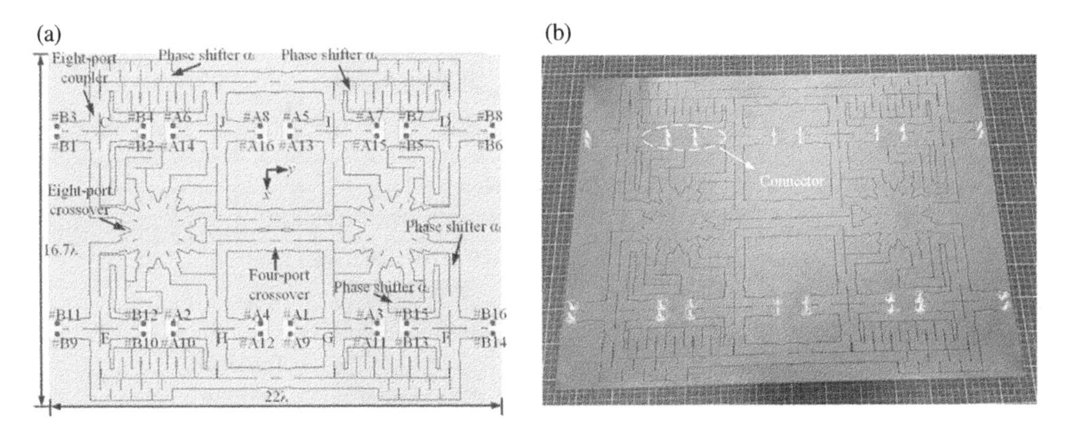

Figure 2.15 16 × 16 2-D BFN developed in [19]. (a) Simulated model. (b) Fabricated prototype of the optimized design. *Source:* From [19] / with permission of IEEE.

designed to operate with a center frequency of 10 GHz with a F4B substrate having a dielectric constant of 2.55 and a loss tangent of 0.001 at that frequency. Despite its successful operation, this BFN suffers from a relatively large footprint since all of its components are integrated into a single layer. In those cases where a reduced size is important, multilayer designs, such as the two examples described next, can be used, but with a higher cost and a more complex fabrication process.

The first multilayer 2-D BFN example [20] is shown in Figure 2.16. This 16 × 16 design was focused on the E-plane sub-BFN, which was a 4 × 4 BM in a four-layer SIW configuration. The H-plane sub-BFN was realized with a traditional planar H-plane 4 × 4 BM. The two sub-BFNs were directly connected without resorting to any connectors or connecting networks.

The second example [21] was focused on the interconnections between the two sub-BFN sections. It is illustrated in Figure 2.17. Eight H-plane 4 × 4 BMs were designed to be placed according to the modified configuration of the 2-D BFN shown in Figure 2.15. The first set of sub-BFNs was placed underneath the radiating elements, while the second set of sub-BFNs was arranged laterally. Some

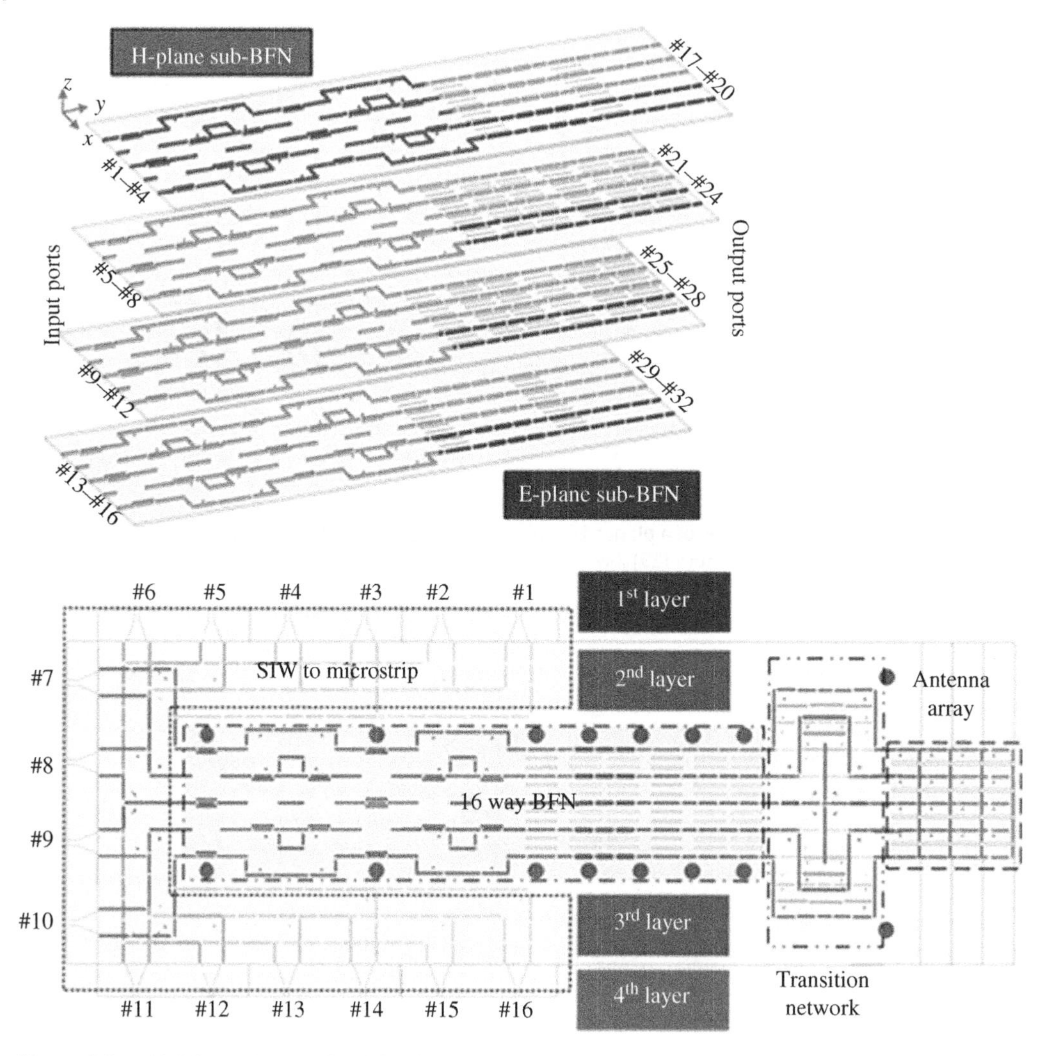

Figure 2.16 Multilayer 16 ×16 2-D BFN that delivers nonuniform amplitudes to the elements of a 4 × 4 multi-beam array for SLL suppression. *Source:* From [20] / with permission of IEEE.

vertical interconnections and horizontal connecting sections were carefully developed to seamlessly connect the two sets of sub-BFNs to realize the desired 2-D beamforming system.

2.1.4 Nolen Matrix

A BFN based on a BM feeds an array in a parallel fashion. In contrast, one based on a Nolen matrix feeds it in a serial fashion. In comparison to an $N \times N$ BM, an $N \times N$ Nolen matrix needs more components and may lead to a narrower band system. On the other hand, it does not need any crossovers and offers flexibility in both the dimension of the array, which can be M (inputs) $\times$ N (outputs) with $M \leq N$, and the distribution of the antenna weights. In other words, a BFN based on a Nolen matrix can be designed to feed a $1 \times N$ antenna array to produce up to M beams with controllable sidelobes. Like the BM, the Nolen matrix is lossless if $M \leq N$ [22].

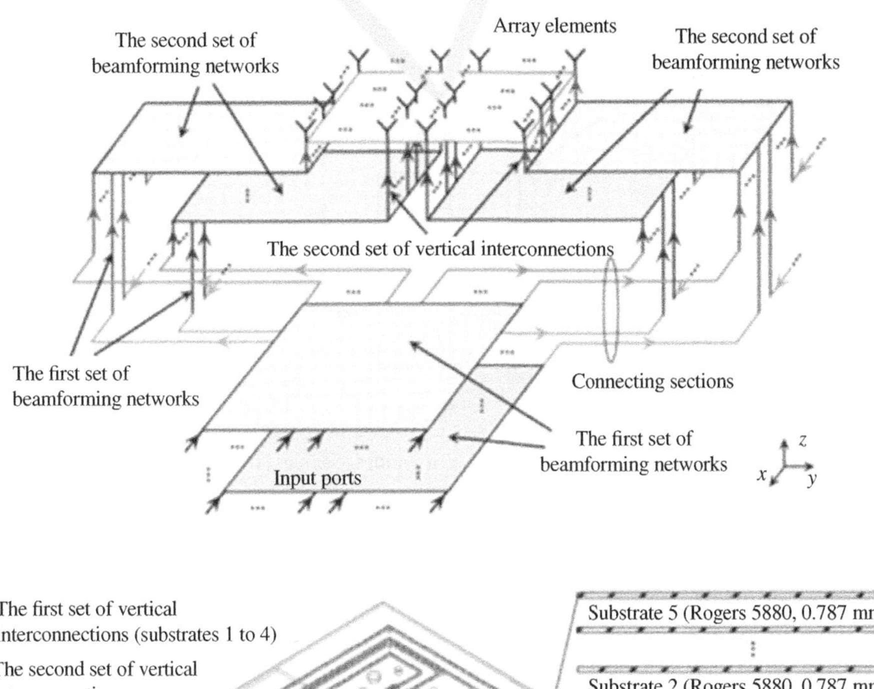

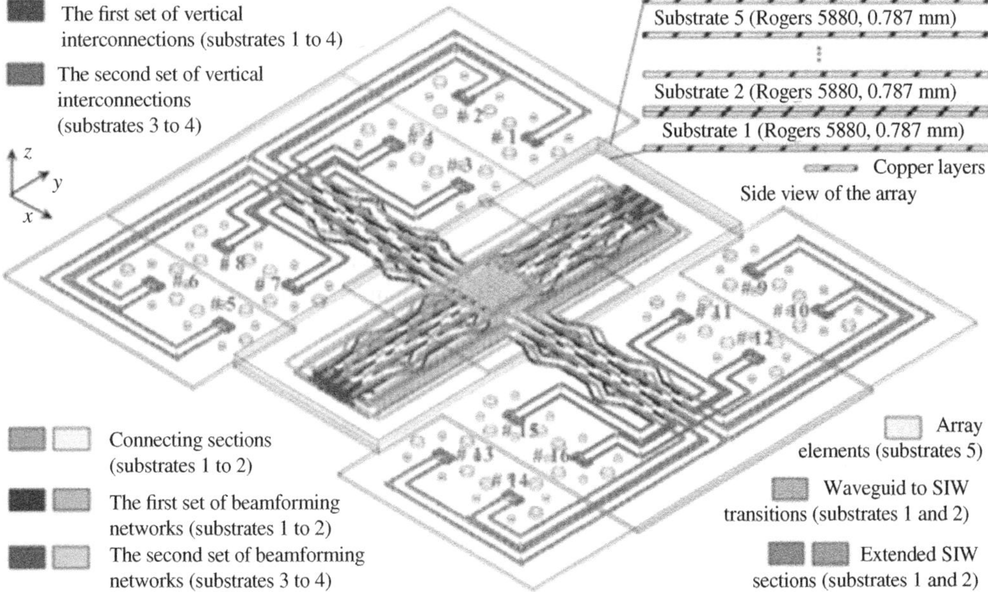

Figure 2.17 16 ×16 2-D BFN that delivers nonuniform amplitudes to the elements of a 4 × 4 multi-beam array for SLL suppression. *Source:* From [21] / with permission of IEEE.

Figure 2.18 shows a basic Nolen matrix with a_i, $i = 1,\ldots, M$ as its input signals and b_j, $j = 1, \ldots, N$ as its output signals. A node of the Nolen matrix is shown next to it. It consists of a directional coupler with coupling coefficient θ_{ij} and a phase shifter with phase-shifting value φ_{ij}. Given the desired antenna output weight, the phase-shifter values and the coupling coefficients can be calculated

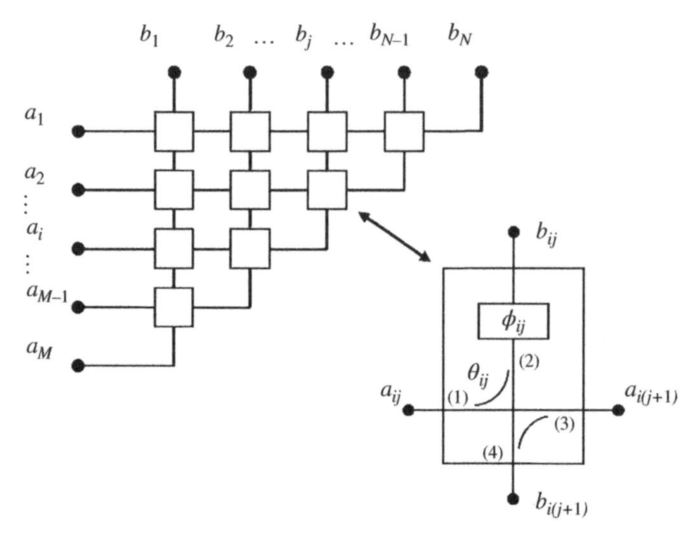

Figure 2.18 Illustration of a Nolen matrix and one of its nodes. *Source:* From [23] / with permission of IEEE.

recursively [22]. An SIW-based Nolen matrix operating at mm-wave frequencies was introduced in [23]. Most recently, the Nolen matrix has been employed to feed an antenna array for MIMO applications [24].

2.2 Quasi Optical BFNs: Rotman Lens and Reflectors

2.2.1 Rotman Lens

Although originally based on optical principles, the Rotman lens is normally realized using a set of transmission lines. The configuration of the traditional Rotman lens is shown in Figure 2.19. It consists of a focal arc on which multiple feeds are placed. A pickup array along a separate surface is known as the "inner lens contour." The straight line, called the "outer lens contour," is where the antenna elements are placed. Electrical and geometrical constraints are imposed on the lens system to uniquely define its configuration: a straight front face, two symmetrical off-axis focal points F_1 and F_2, and an on-axis focal point G_0. These three focal points determine the fields that are radiated into the angles: $-\alpha$, α, and 0°, respectively.

The shape of the focal arc and lens contour can be determined by the following constraint equations. The parameters are indicated in Figure 2.19.

1) Electrical constraints:

$$\overline{F_1P}\sqrt{\varepsilon_r} + W + N\sin\beta = \sqrt{\varepsilon_r}F + W_0 \tag{2.3a}$$

$$\overline{F_2P}\sqrt{\varepsilon_r} + W - N\sin\beta = \sqrt{\varepsilon_r}F + W_0 \tag{2.3b}$$

$$\overline{GP}\sqrt{\varepsilon_r} + W = \sqrt{\varepsilon_r}G + W_0 \tag{2.3c}$$

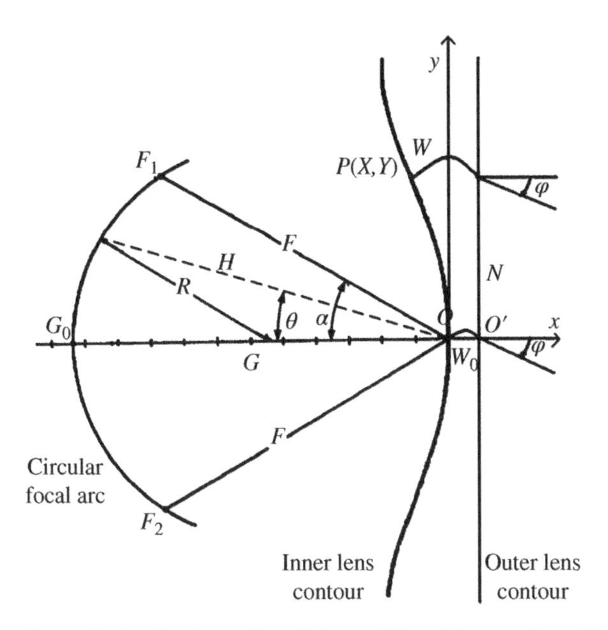

Figure 2.19 Configuration of the traditional Rotman lens. *Source:* From [25] / with permission of IEEE.

2) Geometrical constraints:

$$\left(\overline{F_1P}\right)^2 = F^2 + X^2 + Y^2 + 2FX\cos\alpha - 2FY\sin\alpha \tag{2.4a}$$

$$\left(\overline{F_2P}\right)^2 = F^2 + X^2 + Y^2 + 2FX\cos\alpha + 2FY\sin\alpha \tag{2.4b}$$

$$\left(\overline{GP}\right)^2 = (G + X)^2 + Y^2 \tag{2.4c}$$

The Rotman lens shown in Figure 2.20 was realized with SIW technologies in [25]. The original Rotman lens topology employed fixed cables with different electrical lengths. To achieve a more compact configuration, an SIW phase-shifting network, which is equivalent to the fixed-cable design over a moderate bandwidth, was introduced. A dual-layer SIW-based Rotman lens was developed in [26]. The lens was implemented in two layers using a new transition based on several star-shaped coupling slots and an SIW-integrated reflector. Compared to the standard rectangular coupling slot transitions, it is broadband and maximizes the power transfer between the two layers of the lens regardless of the position of the beam port along the focal arc. This design was improved by the same research group by introducing ridged delay lines to reduce its footprint [27], as shown in Figure 2.21. The coupling elements along the array port contours were implemented with several cylindrical vias connected to ridged waveguide delay lines. The use of cylindrical vias with azimuthal symmetry for the coupling transition and delay lines in ridged waveguides allowed an improvement in bandwidth over a larger field of view.

Similar to BMs, the beams radiated by a multi-beam array fed with a Rotman lens also suffer from relatively high SLLs, mainly due to unbalanced amplitude and phase distributions. While the amplitude distributions realized with a BM can be manipulated by changing the dividing ratio of its power dividers, this approach is not feasible for a Rotman lens. However, there are several

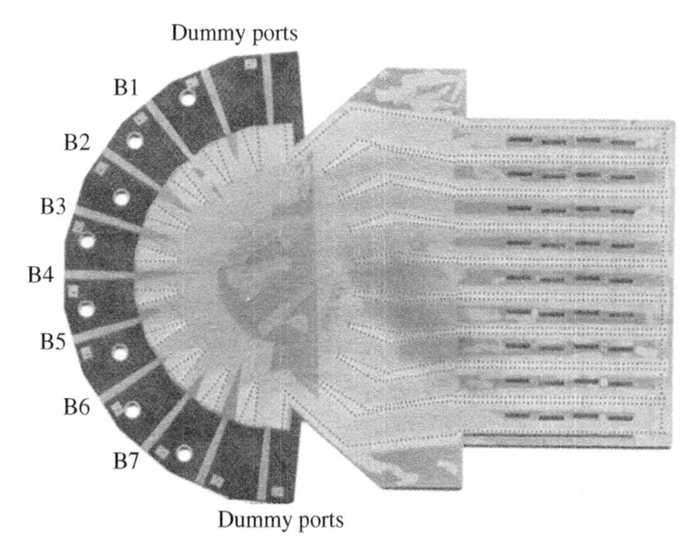

Figure 2.20 SIW-based Rotman lens. *Source:* From [25] / with permission of IEEE.

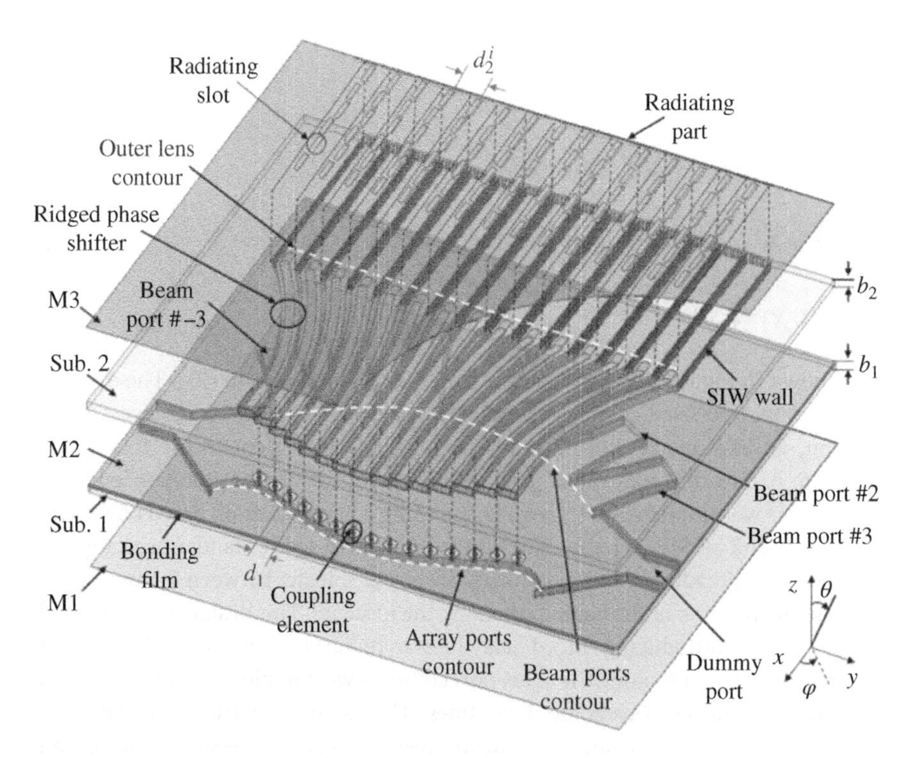

Figure 2.21 SIW-based dual-layer Rotman lens with ridged delay lines. *Source:* From [26] / with permission of IEEE.

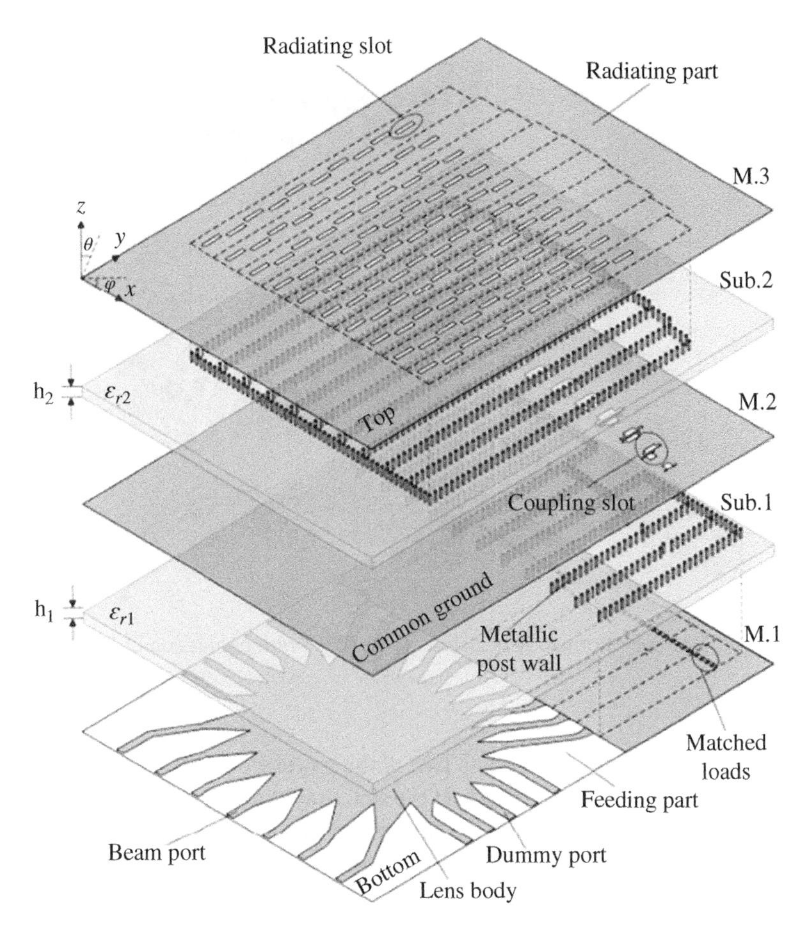

Figure 2.22 SIW-based dual-layer Rotman lens with reduced SLL that delivers nonuniform amplitudes to the elements of a multi-beam slot array for SLL suppression. *Source:* From [29] / with permission of IEEE.

common ways that SLL reduction is attained for multi-beam arrays fed by a Rotman lens [28]. They include the following:

1) Increasing the aperture size of each beam port to decrease the edge taper of the array ports.
2) Introducing a small lens as the prime feeding network to excite a large lens.
3) Employing a dual port feeding method.
4) Being terminated with lossy networks.
5) Using active circuits like power amplifiers or phase shifters to change the distribution of the amplitudes or phases.

A lossy network at the output of the Rotman lens to taper the amplitude distribution was introduced in [28], as shown in Figure 2.22, to reduce the SLLs. The amplitude distribution was controlled with a series of double-stepped SIW-based slot couplers that connected the Rotman lens to the radiating elements. These couplers regulated the coupling power and the associated phase values to achieve low sidelobes by changing the geometrical parameters. This solution achieved some antenna size reduction but at the expense of a relatively complex dielectric stack-up.

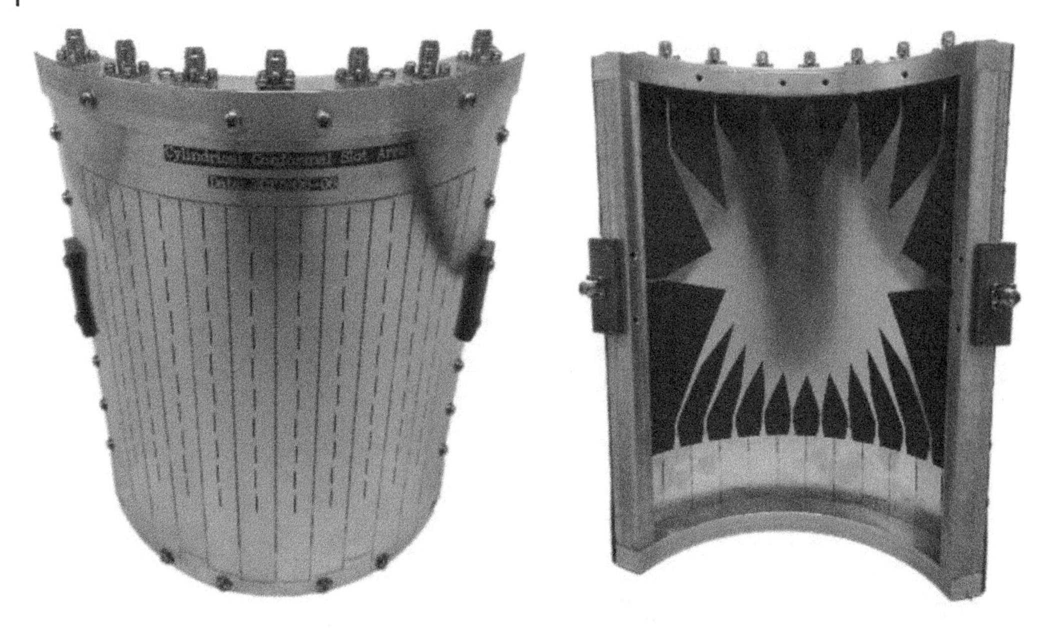

Figure 2.23 Conformal SIW-based dual-layer Rotman lens. *Source:* From [29] / with permission of IEEE.

The double-stepped slot coupler was designed so that a part of the power from the array port in the lens layer would be coupled to the antenna layer to feed the slot array through double-stepped slots, while the remaining power would be absorbed. An SIW-based matching load was used to absorb the excess power in each SIW-based feed line in the lens layer. Each one consisted of a transverse slot, four chip resistors, and a short-circuited wall.

A similar configuration was also developed to feed a cylindrically conformal slot array antenna in [29]. Photos of the prototype are shown in Figure 2.23. The conformal slot array consists of 10 × 10 radiation slots. A conformal Rotman lens that was mounted onto a cylindrical surface was developed to be integrated with the conformal array. The Rotman lens feeds the slot array through the coupling slots in the broad sides of the SIW waveguides. Double-layer SIW phase shifters are introduced between the Rotman lens and the slot array to manipulate the phase distribution on the slots. While the phase shifter design in [29] is similar to that in [28], the chip resistors were not used. Consequently, the SLLs were no longer suppressed, but the overall efficiency of the multi-beam system was increased.

2.2.2 Reflectors

The types of reflectors employed in the realization of multiple beams can be divided into two simple categories: single reflectors and dual reflectors. When using a single reflector, for example, a cylindrical parabolic, a feed positioned at the focus *F* of the parabolic produces a cylindrical wave which, upon reflection, is converted into a plane wave traveling in the direction perpendicular to the parabola's aperture. Similarly, a cylindrical wave is converted into a plane wave after the reflections from two reflectors in a dual-reflector system. Thus, the desired in-phase output is achieved after the reflections encountered in either reflector system. The plane waves generated from the reflectors can then be used to excite an antenna array. While this is the case when a single source is placed

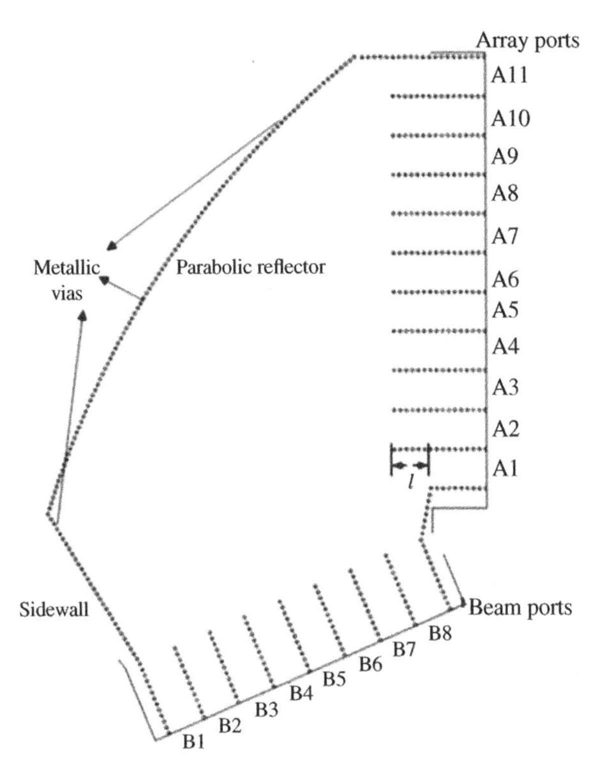

Figure 2.24 SIW-based offset-fed parabolic reflector lens. *Source:* From [30] / with permission of IEEE.

at the focal point, a multi-beam array requires multiple excitations placed at other positions away from the focal point. When the excitation is not at the focal point, the wave front and phase relationships change and a tilted beam is realized. Details of both types of reflector systems are described below.

2.2.2.1 Single Reflectors

Two types of single reflectors, i.e., offset-fed and pillbox configurations, are typically employed for BFNs. The scattering characteristics of a single parabolic-shaped reflector begin with its geometrical definition, i.e., its shape is defined relative to an x–y coordinate system as:

$$y^2 = 4fx \tag{2.5}$$

where f is its focal length. Both aperture blockage caused by physical support structures and the feed can be eliminated with an offset-fed parabolic reflector design. Following this concept, a parabolic reflector lens was realized with SIW technologies in [30]. It is illustrated in Figure 2.24. The feed network consists of a cylindrical parabolic reflector realized with a series of metallic vias and multiple open SIWs, which generate the cylindrical wave. An SIW slot array was designed and connected to this feed network.

The above reflector is generally referred to as a pillbox reflector, which is a cylindrical reflector sandwiched between two metal walls. A pillbox reflector can be easily integrated with SIW-based circuits and radiators. Another interesting example is a slot antenna array fed by the mechanically

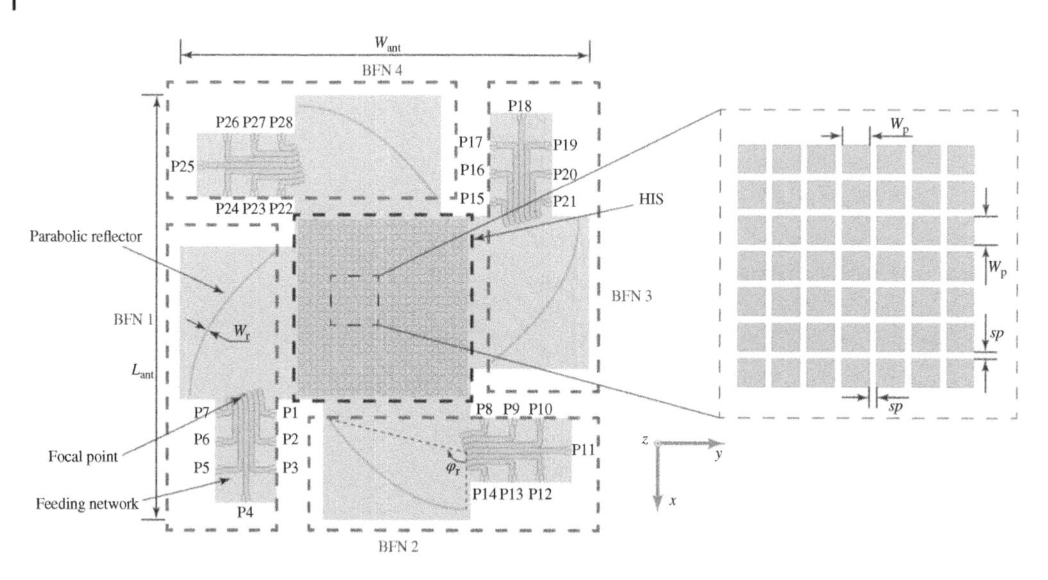

Figure 2.25 SIW-based offset-fed parabolic reflector lens that realizes full azimuth coverage. *Source:* From [31] / with permission of IEEE.

scanning pillbox reported in [31] as illustrated in Figure 2.25. Its feed network part is similar to those in [30], but four of them are employed to illuminate four reflectors. Each reflector covers one quarter of the intended angular range by launching a surface wave to a high-impedance surface from a different direction. The high-impedance surface made of microstrip patches serves as the main radiator. This unique low-profile configuration facilitated the realization of a multi-beam antenna that produced full-azimuth coverage [31].

The offset-fed configuration, however, causes unbalanced phase and amplitude distributions that lead to undesirable features in the patterns of the output beams. To overcome this asymmetry problem, the pillbox configuration shown in Figure 2.26 was developed in [32]. It integrates a 2D parabolic reflector realized with metallic pins and side metal walls into a dual-layer structure to yield symmetric patterns radiated by an SIW-based slot array. The feeding and radiating parts of this dual-layer configuration are hosted by two dielectric substrates. The bottom layer is the pillbox reflector with a mechanically rotating feed. The top layer consists of 23 waveguides, each having eight radiating slots. The coupling between the pillbox reflector and the radiating layer is realized using coupling slots etched in the conductor between them. The energy in the cylindrical wave produced in the bottom layer of this configuration is totally transmitted through the coupling slots into the upper layer; the coupling layer ideally causes no reflections. This type of a symmetric structure can be used to obtain more balanced phase and amplitude distributions on the radiation portion of the system. Several multi-beam arrays have been realized with extensions of this pillbox configuration [33–37].

The SLL and beam crossover levels are interdependent in a multi-beam antenna system using a single radiating aperture. Low SLLs lead to low-beam crossover levels, and vice versa [38]. For many applications such as cellular networks, however, this feature is undesirable for cell coverage. It was proposed in [38] to overcome this limitation using the so-called split aperture decoupling method, which employs two radiating apertures. Each aperture is associated with a pillbox quasi-optical system with several integrated feed horns in its focal plane. Interleaving beams generated by two separated BFNs result in the flexibility in determining the SLL and the beam crossover level independently. The simulated model is shown in Figure 2.27.

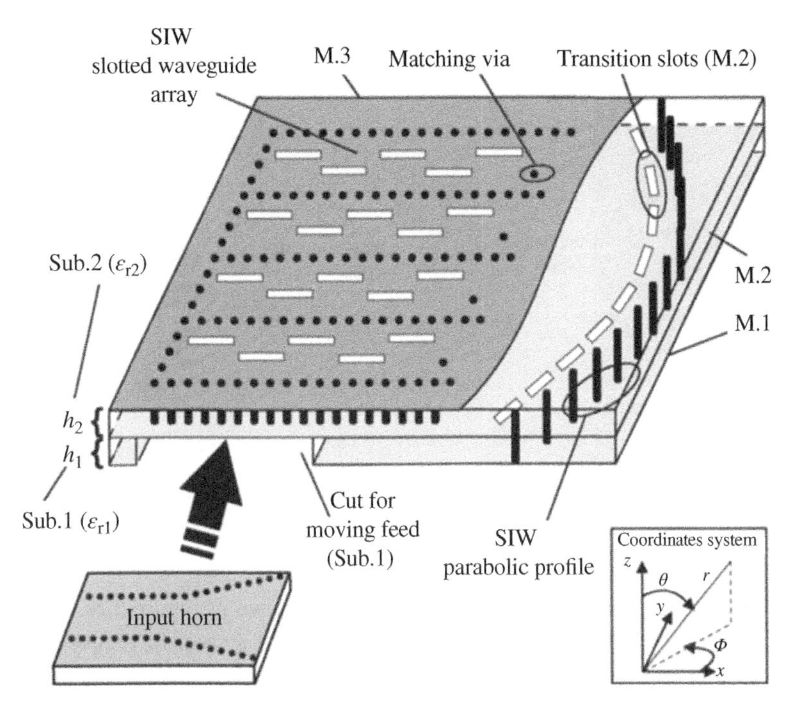

Figure 2.26 SIW-based pillbox-configured multi-beam slot array. There are three metal layers denoted as M_i. M_1 is the bottom wall of the pillbox reflector. M_2 is the top wall of the pillbox, as well as the host of the coupling slots. M_3 is the radiating layer hosting the slots. *Source:* From [32] / with permission of IEEE.

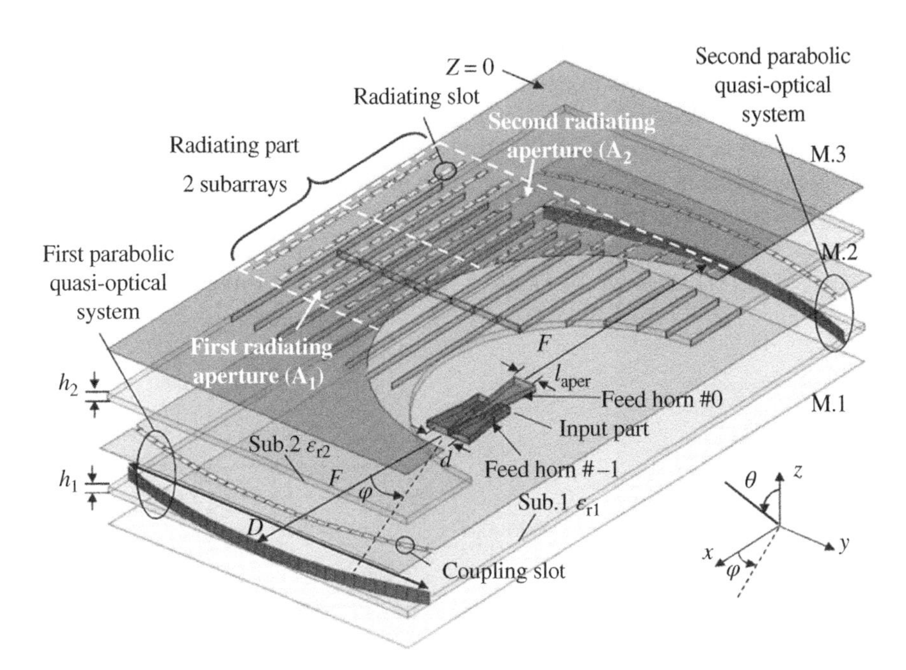

Figure 2.27 SIW-based pillbox-configured multi-beam slot array realized with the split aperture decoupling method. *Source:* From [38] / with permission of IEEE.

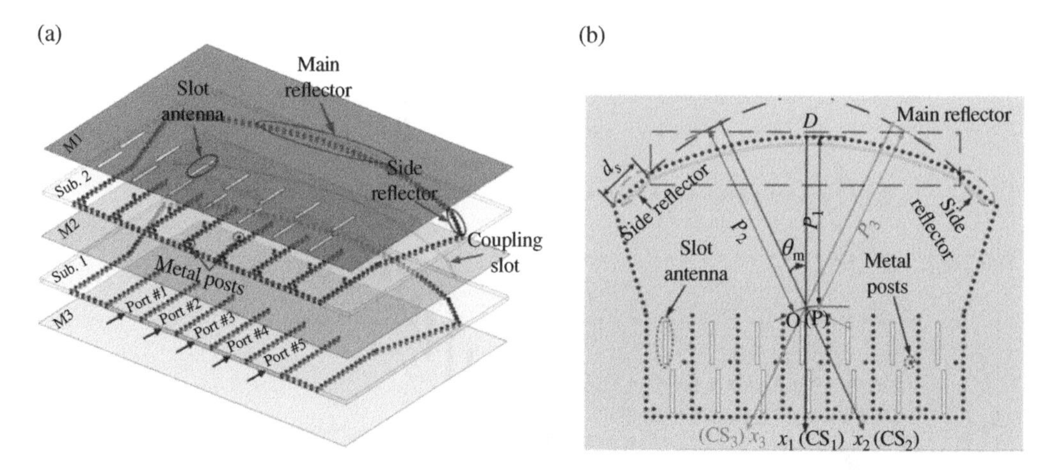

Figure 2.28 SIW-based modified pillbox reflector-fed multi-beam slot array with reduced SLLs. (a) Perspective view. (b) Top view. *Source:* From [39] / with permission of IEEE.

The modified pillbox reflector multi-beam array developed in [39] is shown in Figure 2.28. Two extra parabolic reflectors were introduced into the parabolic quasi-optical system to reduce the SLLs. They were placed on both sides of the central parabolic reflector. Moreover, metal posts were added into the antenna array section to achieve a further reduction in the SLLs. The additional metal posts helped cancel the reflections arising from the slot array.

2.2.2.2 Dual Reflectors

The single reflector examples presented above included offset-fed and pillbox configurations. Both can be extended to dual-reflector systems. Two examples of dual-reflector systems employing different configurations are shown in Figures 2.29 and 2.30. The first one is a dual offset Gregorian reflector system that is used to feed a leaky-wave antenna [40]. The perspective and top views of this design are shown in Figure 2.29. The design is a simple 2-D version of the classic 3-D Gregorian configuration; more details can be found in [42] and [43]. Note that only one point feed was applied and only one beam was generated in [40]. The beam-steering properties of this multi-beam array were realized by using leaky-wave antennas.

Another example was derived from the single-reflector pillbox configuration. The folded Cassegrain lens developed in [41] is shown in Figure 2.30. It consisted of three layers. Similar

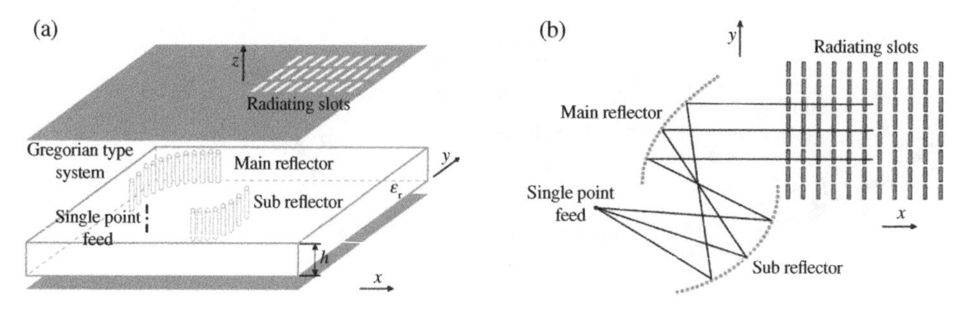

Figure 2.29 Multi-beam slot array fed by a dual offset Gregorian reflector system. (a) Perspective view. (b) Top view. *Source:* From [40] / with permission of IEEE.

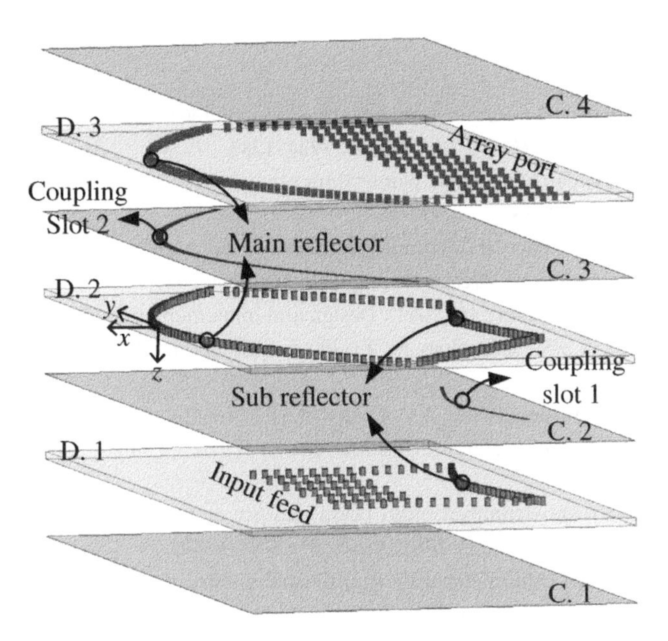

Figure 2.30 SIW-based Cassegrain lens BFN for a multi-beam array. The symbols "D." and "C." represent dielectric and conductor layers, respectively. *Source:* From [41] / with permission of IEEE.

to the pillbox configuration, the electromagnetic wave transmission between adjacent layers is realized by introducing coupling slots in the common conductor layer (C.2 and C.3). As shown in Figure 2.30, the coupling slot 1 is etched into C.2, which is in charge of the transmission between D.1 and D.2. Similarly, the coupling slot 2 in C.3 is responsible for the transmission between D.2 and D.3. By using this folded configuration, the Cassegrain lens is symmetric and achieves more balanced amplitude and phase distributions. More details of the Cassegrain configuration can be found in [43].

2.3 Conclusions

With the development of 5G communications, mm-wave SIW BFNs have attracted more and more attention. They have the advantage of compact size, low insertion loss, and ease of fabrication. This chapter briefly reviewed the recent development of mm-wave SIW-based BFNs, including those associated with circuit-type and quasi-optics-type BFNs. As the most popular type of circuit-type BFN, BMs have been extensively applied to multi-beam array designs, mainly focusing on 1-D and 2-D multi-beam systems. Its serial counter-part, the Nolen matrix, has attracted some interest in recent years [44]. A popular quasi-optical BFN is the Rotman lens. The biggest change in SIW-based Rotman lens design is the dual-layer configuration. The SIW-based single and dual reflector array feeds are also becoming popular for mm-wave arrays. The pillbox configuration is the most attractive structure, and it has been widely applied to various types of multi-beam antenna designs. We look forward to seeing more engineering applications using these antennas in future 5G and beyond array systems.

References

1. Chen, C.-J. and Chu, T.-H. (2010). Design of a 60-GHz substrate integrated waveguide Butler matrix – a systematic approach. *IEEE Trans. Microwave Theory Tech.* **58** (7): 1724–1733.
2. Djerafi, T. and Wu, K. (2012). A low-cost wideband 77-GHz planar Butler matrix in SIW technology. *IEEE Trans. Antennas Propag.* **60** (10): 4949–4954.
3. Cao, Y., Chin, K., Che, W. et al. (2017). A compact 38 GHz multibeam array array with multifolded Butler matrix for 5G applications. *IEEE Antennas Wireless Propag. Lett.* **16**: 2996–2999.
4. Ali, A.A.M., Fonseca, N.J.G., Coccetti, F., and Aubert, H. (2011). Design and implementation of two-layer compact wideband butler matrices in SIW technology for Ku-band applications. *IEEE Trans. Antennas Propag.* **59** (2): 503–512.
5. Karamzadeh, S., Rafii, V., Kartal, M., and Virdee, B.S. (2015). Compact and broadband 4 × 4 SIW Butler matrix with phase and magnitude error reduction. *IEEE Microwave. Wireless Compon. Lett.* **25** (12): 772–774.
6. Cheng, X., Yao, Y., Tomura, T. et al. (2018). A compact multi-beam end-fire circularly polarized septum antenna array for millimeter-wave applications. *IEEE Access* **6**: 62784–62792.
7. Wu, Q., Hirokawa, J., Yin, J. et al. (2018). Millimeter-wave multibeam endfire dual-circularly polarized antenna array for 5G wireless applications. *IEEE Trans. Antennas Propag.* **66** (9): 4930–4935.
8. Cheng, Y.J., Bao, X.Y., and Guo, Y.X. (2013). 60-GHz LTCC miniaturized substrate integrated multibeam array antenna with multiple polarizations. *IEEE Trans. Antennas Propag.* **61** (12): 5958–5967.
9. Li, Y. and Luk, K.-M. (2016). A multibeam end-fire magnetoelectric dipole antenna array for millimeter-wave applications. *IEEE Trans. Antennas Propag.* **64** (7): 2894–2904.
10. Zhong, L.-H., Ban, Y.-L., Lian, J.-W. et al. (2017). Miniaturized SIW multibeam array array fed by dual-layer 8 × 8 Butler matrix. *IEEE Antennas Wireless Propag. Lett.* **16**: 3018–3021.
11. Chen, P., Hong, W., Kuai, Z. et al. (2009). A multibeam antenna based on substrate integrated waveguide technology for MIMO wireless communications. *IEEE Trans. Antennas Propag.* **57** (6): 1813–1821.
12. Lian, J.-W., Ban, Y.-L., Xiao, C., and Yu, Z.-F. (2018). Compact substrate integrated 4 × 8 Butler matrix with sidelobe suppression for millimeter wave multibeam application. *IEEE Antennas Wireless Propag. Lett.* **17** (5): 928–932.
13. Guntupalli, A.B., Djerafi, T., and Wu, K. (2014). Two-dimensional scanning antenna array driven by integrated waveguide phase shifter. *IEEE Trans. Antennas Propag.* **62** (3): 1117–1124.
14. Li, Y. and Luk, K.-M. (2016). 60-GHz dual-polarized two-dimensional switch-beam wideband antenna array of aperture-coupled magneto-electric dipoles. *IEEE Trans. Antennas Propag.* **64** (2): 554–563.
15. Mohamed, I.M. and Sebak, A.-R. (2019). 60 GHz 2-D scanning Multibeam cavity-backed patch array fed by compact SIW beam forming network for 5G application. *IEEE Trans. Antennas Propag.* **67** (4): 2320–2331.
16. Lian, J.-W., Ban, Y.-L., Zhu, J.-Q. et al. (2019). Planar 2-D scanning SIW multibeam array with low Sidelobe level for millimeter wave applications. *IEEE Trans. Antennas Propag.* **67** (7): 4570–4578.
17. Wang, J., Li, Y., Ge, L. et al. (2017). A 60 GHz horizontally polarized magneto-electric dipole antenna array with 2-D multibeam end-fire radiation. *IEEE Trans. Antennas Propag.* **65** (11): 5837–5845.
18. Yang, W., Yang, Y., Che, W. et al. (2017). 94-GHz compact 2-D multibeam LTCC antenna based on multifolded SIW beam-forming network. *IEEE Trans. Antennas Propag.* **65** (8): 4328–4333.

19. Lian, J.-W., Ban, Y.-L., Zhu, H., and Guo, Y.J. (2020). Uniplanar beam-forming network employing eight-port hybrid couplers and crossovers for 2-D multibeam array antennas. *IEEE Trans. Microwave Theory Tech.* https://doi.org/10.1109/TMTT.2020.2992026.
20. Lian, J.-W., Ban, Y.-L., Yang, Q.-L. et al. (2018). Planar millimeter-wave 2-D beam-scanning multibeam array antenna fed by compact SIW beam-forming network. *IEEE Trans. Antennas Propag.* **66** (3): 1299–1310.
21. Li, Y., Wang, J., and Luk, K.-M. (2017). Millimeter-wave multibeam aperture-coupled magnetoelectric dipole array with planar substrate integrated beamforming network for 5G applications. *IEEE Trans. Antennas Propag.* **65** (12): 6422–6431.
22. Fakoukakis, F.E. and Kyriacou, G.A. (2013). Novel nolen matrix based beamfroming networks for series-fed low SLL multibeam antennas. *Prog. Electromagn. Res. B* **51** (33): 64.
23. Djerafi, T., Fonseca, N.J.G., and Wu, K. (2011). Broadband substrate integrated waveguide 4 × 4 nolen matrix based on coupler delay compensation. *IEEE Trans. Microwave Theory Tech.* **59** (7).
24. Ren, H., Zhang, H., Li, P. et al. (2019). A novel planar nolen matrix phased array for MIMO applications. *2019 IEEE International Symposium on Phased Array System & Technology (PAST).*
25. Cheng, Y.J., Hong, W., Wu, K. et al. (2008). Substrate integrated waveguide (SIW) Rotman lens and its Ka-band multibeam array antenna applications. *IEEE Trans. Antennas Propag.* **56** (8): 2504–2513.
26. Tekkouk, K., Ettorre, M., Le Coq, L., and Sauleau, R. (2016). Multibeam SIW slotted waveguide antenna system fed by a compact dual-layer Rotman lens. *IEEE Trans. Antennas Propag.* **64** (2): 504–514.
27. Tekkouk, K., Ettorre, M., and Sauleau, R. (2018). SIW Rotman lens antenna with ridged delay lines and reduced footprint. *IEEE Trans. Microwave Theory Tech.* **66** (6): 3136–3144.
28. Liu, Y., Yang, H., Jin, Z. et al. (2017). Compact Rotman lens-fed slot array antenna with low sidelobes. *IET Microwave Antennas Propag.* **12** (5): 656–661.
29. Liu, Y., Yang, H., Jin, Z. et al. (2018). A multibeam cylindrically conformal slot array antenna based on a modified Rotman lens. *IEEE Trans. Antennas Propag.* **66** (7): 3441–3452.
30. Cheng, Y.J., Hong, W., and Wu, K. (2008). Millimeter-wave substrate integrated waveguide multibeam antenna based on the parabolic reflector principle. *IEEE Trans. Antennas Propag.* **56** (9): 3055–3058.
31. Ma, Z.L. and Chan, C.H. (2017). A novel surface-wave-based high-impedance surface multibeam antenna with full azimuth coverage. *IEEE Trans. Antennas Propag.* **65** (4): 1579–1588.
32. Gandini, E., Ettorre, M., Casaletti, M. et al. (2013). SIW slotted waveguide array with pillbox transition for mechanical beam scanning. *IEEE Antennas Wireless Propag. Lett.* **11**: 1572–1575.
33. Ettorre, M., Manzillo, F.F., Casaletti, M. et al. (2015). Continuous transverse stub array for Ka-band applications. *IEEE Trans. Antennas Propag.* **63** (11): 4792–4800.
34. Ettorre, M., Sauleau, R., and Coq, L.L. (2011). Multi-beam multi-layer leaky-wave SIW pillbox antenna for millimeter-wave applications. *IEEE Trans. Antennas Propag.* **59** (4): 1093–1100.
35. F. F. Manzillo, Śmierzchalski M., Coq L.L. *et al.*, "A wide-angle scanning switched-beam antenna system in LTCC technology with high beam crossing levels for V-band communications," *IEEE Trans. Antennas Propag.*, vol. **67**, no. 1, pp. 541–553, 2019.
36. Tekkouk, K., Ettorre, M., Le Coq, L., and Sauleau, R. (2015). SIW pillbox antenna for monopulse radar applications. *IEEE Trans. Antennas Propag.* **63** (9): 3918–3927.
37. Tekkouk, K., Ettorre, M., and Sauleau, R. (2018). Multi-beam Pillbox antenna integrating amplitude-comparison monopulse technique in the 24-GHz band for tracking applications. *IEEE Trans. Antennas Propag.* **66** (5): 2616–2621.
38. Tekkouk, K., Ettorre, M., Gandini, E., and Sauleau, R. (2015). Multibeam pillbox antenna with low sidelobe level and high-beam crossover in SIW technology using the split aperture decoupling method. *IEEE Trans. Antennas Propag.* **63** (11): 5209–5215.

39. Yan, S.-P., Zhao, M.-H., Ban, Y.-L. et al. (2019). Dual-layer SIW multibeam pillbox antenna with reduced sidelobe level. *IEEE Antennas Wireless Propag. Lett.* **18**: 541–545.

40. Ettorre, M., Neto, A., Gerini, G., and Maci, S. (2008). Leaky-wave slot array antenna fed by a dual reflector system. *IEEE Trans. Antennas Propag.* **56** (10): 3143–3149.

41. Lian, J.-W., Ban, Y.-L., Chen, Z. et al. (2018). SIW folded Cassegrain lens for millimeter-wave multibeam application. *IEEE Antennas Wireless Propag. Lett.* **17** (4): 583–586.

42. Rahmat-Samii, Y. (1988). Reflector antennas. In: *Antenna Handbook* (eds. Y.T. Lo and S.W. Lee), 15-1–15-124. New York: Van Nostrand Reinhold.

43. Rush, W.V.T., Prata, A., Rahmat-Samii, Y., and Shore, R.A. (1990). Derivation and application of the equivalent paraboloid for classical offset Cassegrain and Gregorian antennas. *IEEE Trans. Antennas Propag.* **38**: 1141–1149.

44. Guo, Y.J., Ansari, M., and Fonseca, N.J.G. (2021). Circuit type multiple beamforming networks for antenna arrays in 5G and 6G terrestrial and non-terrestrial networks. *IEEE J Microwave* **1** (2). https://doi.org/10.1109/JMW.2021.3072873.

3

Decoupling Methods for Antenna Arrays

With the rapid development of fifth-generation (5G) communication systems, massive multiple-input-multiple-output (massive MIMO) arrays have become an essential technology to meet the service requirements. Larger data capacity, higher spectrum efficiency, and more system adaptability, to name a few major characteristics, are advantages associated with them. As a result, it has become a significant challenge to maintain high isolation between adjacent antenna elements in these arrays to achieve the noted performance enhancements with a large number of antenna elements. This is particularly true when the antenna elements are closely packed into a mobile platform because of critical space restrictions. A variety of approaches are being employed to realize desirable high isolation levels between the antenna elements in densely packed arrays. This chapter will illustrate the most widely used mutual coupling reduction strategies, i.e., electromagnetic bandgap (EBG) structures, defected ground structures (DGSs), neutralization lines, polarization rotators, decoupling surfaces, metamaterial structures, and parasitic resonators. Typical examples related to each of them are presented.

3.1 Electromagnetic Bandgap Structures

Densely packed arrays with element separations of less than a half wavelength suffer from strong mutual coupling effects caused by inter-element surface waves as well as space waves. Surface waves can exist on the boundary between any two dissimilar media. Notably, for arrays, they can occur at the interface between metal or dielectric regions and free space. Their names arise from the fact that they are bound to the surface defined by the interface, i.e., their fields decay exponentially into the surrounding materials. While these fields can extend many wavelengths into the surrounding regions at radio frequencies, they are often described in terms of surface currents. Consequently, they can be modeled as an effective surface impedance [1].

In order to reduce the strong mutual E-plane coupling that occurs when two microstrip antennas lie on a thick substrate with high permittivity, mushroom-like EBG structures, such as the one depicted in Figure 3.1a [1], have been inserted between the two adjacent antenna elements [2]. The mushroom-like EBG structure consists of four parts: a ground plane, a dielectric substrate that lies on the ground plane, metallic patches on top of the substrate, and vias connecting the patches to the ground plane. For waves normally or slightly obliquely incident on this structure from the air region, it acts as an artificial magnetic conductor (AMC) [1]. On the other hand, when its parameters are properly designed, it becomes an EBG structure that forms a stopband for the waves

Advanced Antenna Array Engineering for 6G and Beyond Wireless Communications, First Edition.
Y. Jay Guo and Richard W. Ziolkowski.

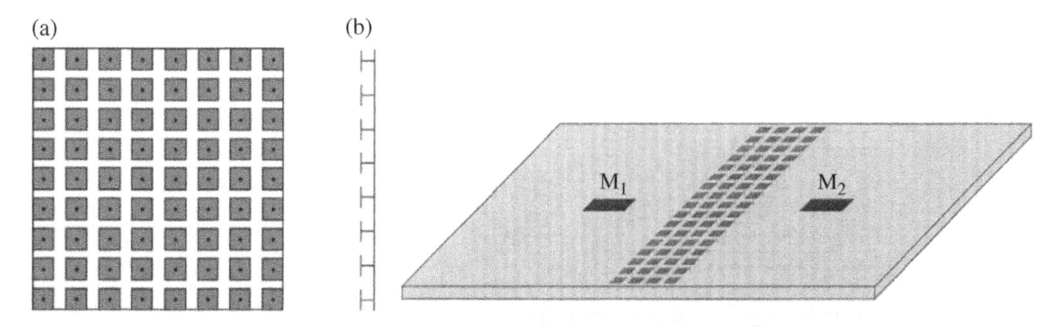

Figure 3.1 Mushroom EBG structure [1]. (a) Top and side views. (b) When it is properly designed and placed between two patch antennas, it can reduce the mutual coupling between them that arises from surface waves. *Source*: From [2] / with permission of IEEE.

propagating along the dielectric-air interface. Thus, the surface waves can be strongly suppressed within a certain frequency range and low mutual coupling is achieved [2–8].

The operating mechanisms of such an EBG structure can be explained as an *LC* filter network, i.e., the inductance *L* arises from the induced currents flowing through the vias and the capacitance *C* arises from the effects of the gap between the adjacent metal patches and between the patches and the ground plane. A sheet impedance can then be assigned to the surface which is equal to the impedance of a parallel resonant circuit that consists of the sheet capacitance *C* and sheet inductance *L* as:

$$Z = \frac{j\omega L}{1 - \omega^2 LC} \tag{3.1}$$

Consequently, the surface impedance is inductive at low frequencies and capacitive at high frequencies. Furthermore, the impedance becomes very high near the resonance frequency:

$$\omega_0 = \frac{1}{\sqrt{LC}} \tag{3.2}$$

The frequency interval in which the high impedance occurs is associated with the forbidden bandgap. The sizes and number of the gaps in the two-layer geometry shown in Figure 3.1 determine the sheet value of *C* and the radii of the vias and the thickness of the dielectric determines the sheet value of *L*.

In order to verify the decoupling effectiveness of the EBG structure between the two antenna elements, two pairs of microstrip antenna arrays with and without the EBG structure were fabricated and measured. The size of the antenna element is 6.8 mm × 5.0 mm; the distance between the edges of the antennas was 38.8 mm (0.75 times the free space wavelength at 5.8 GHz). The size of the ground plane was 100 mm × 50 mm. The size of each patch in the mushroom surface was 3.0 mm and the gap between each of the patches was 0.5 mm from edge to edge. The measured results are shown in Figure 3.2. Both antennas resonated at 5.86 GHz and the return loss was better than 10 dB. The mutual coupling for the antennas without the EBG structure was −16.8 dB at 5.86 GHz. In comparison, the mutual coupling between the antennas with the EBG structure was reduced to −24.6 dB. Thus, the EBG approach realized approximately an 8 dB reduction of the mutual coupling at the resonance frequency, which demonstrates that mushroom-like EBG structures are an effective option for reducing the mutual coupling between antenna elements in a patch-based array.

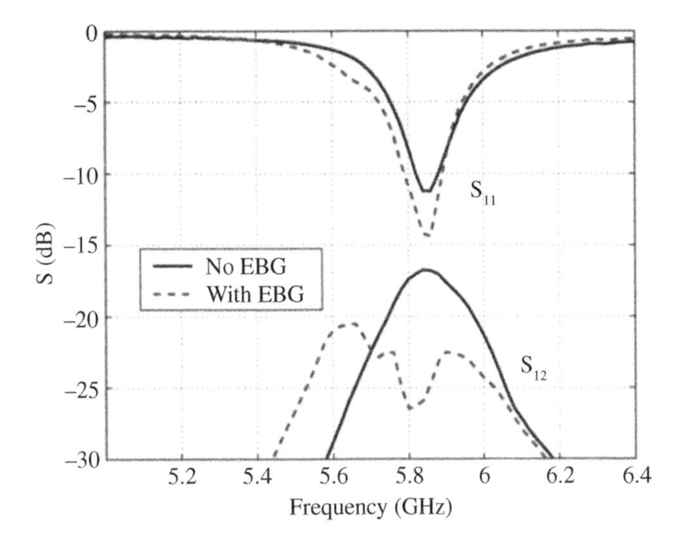

Figure 3.2 Measured results of the two-element microstrip patch antenna array with and without the mushroom EBG structure. *Source*: From [2] / with permission of IEEE.

3.2 Defected Ground Structures

DGSs are also commonly utilized to reduce the mutual coupling between antenna elements in an array. They generally are realized as periodic or aperiodic grid structures etched into the metallic ground plane of the array [9–15]. The operating mechanisms are similar, i.e., they introduce combinations of inductances and capacitances to achieve bandstop characteristics. Once a simple DGS is integrated between the antenna elements in an array, the isolation between those elements is enhanced effectively, but typically at the cost of a radiation pattern with an inherent lower front-to-back ratio.

The mutual coupling suppression in microstrip arrays was studied in [11] using DGSs. The single U-shaped, dumbbell-shaped, and back-to-back U-shaped DGS shown in Figure 3.3 were analyzed and compared. As shown in Figure 3.3a, the DGS was etched at the center of the ground plane between two weakly coupled microstrip lines with 50-Ω characteristic impedance to test its decoupling performance. The microstrip substrate had a relative permittivity of 10.2 and a thickness of 2.0 mm, which is greater than $0.3\lambda_0/(2\pi/\sqrt{\varepsilon_r}) = 0.152\ \lambda_0$, and consequently, a pronounced surface wave was excited.

The single U-shaped DGS shown in Figure 3.3b was designed first and optimized to support a wide bandgap around 6.0 GHz. In comparison, the dumbbell-shaped DGS in Figure 3.3c was designed to have the same bandgap. Considering that multiple circuit units are usually cascaded to improve the bandstop level in bandstop filter designs, then the two back-to-back U-shaped DGS units shown in Figure 3.3d were cascaded to increase the rejection bandwidth. The distance L between two open ports of the weakly coupled microstrip lines was 20.0 mm. This distance was enough to avoid a strong coupling between the DGS and the open microstrip lines. The DGS performance was characterized by defining the rejection bandwidth as the frequency range over which $|S_{21}|$ was suppressed by more than 3 dB compared with that of the structure without the DGS. The simulated results of the structure with and without DGS are shown in Figure 3.4. It is recognized that the single U-shaped and dumbbell-shaped DGSs have similar bandgap characteristics, i.e. about 3.5% fractional rejection bandwidth, while the back-to-back U-shaped DGS doubles the fractional rejection bandwidth to about 7% around 6.0 GHz.

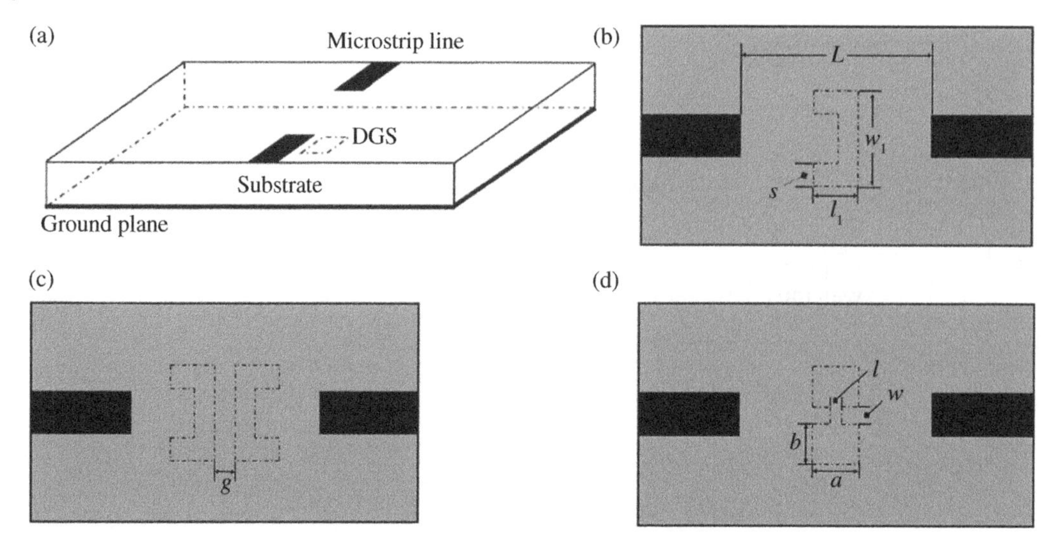

Figure 3.3 Configurations with different types of DGSs. (a) 3-D view of two weakly coupled microstrip lines over a DGS. (b) Single U-shaped DGS. (c) Dumbbell-shaped DGS. (d) Back-to-back U-shaped DGS. *Source*: From [11] / with permission of The Institution of Engineering and Technology.

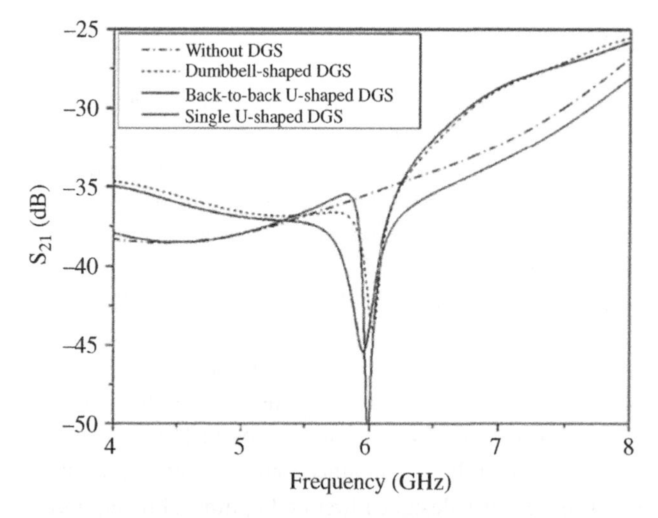

Figure 3.4 Simulated $|S_{21}|$ values with and without the DGS between two weakly coupled microstrip line ports as functions of the source frequency. *Source*: From [11] / with permission of The Institution of Engineering and Technology.

Figure 3.5 shows a coaxial-line-fed E-plane-coupled two-element microstrip array over the back-to-back U-shaped DGS. The DGS is etched at the center of the ground plate halfway between the antenna elements. This choice achieved a significantly low level of mutual coupling. The measured results for the two-element antenna array with and without the DGS are given in Figure 3.6. The two microstrip antennas are spaced with a center-to-center distance of 0.5 λ_0, λ_0 being the operational wavelength in free space at the center frequency. The simulation and experimental results demonstrate that the utilization of a DGS can achieve a large mutual coupling reduction, but at the cost of a perforated ground plate that leads to larger backward radiation.

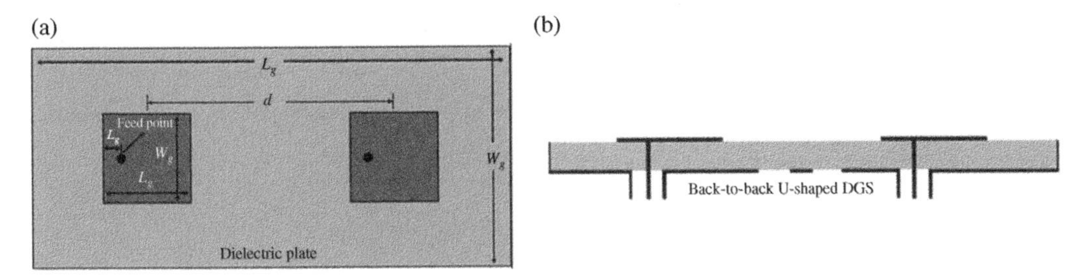

Figure 3.5 Configuration of the microstrip decoupling array. (a) Top view. (b) Side view. *Source*: From [11] / with permission of The Institution of Engineering and Technology.

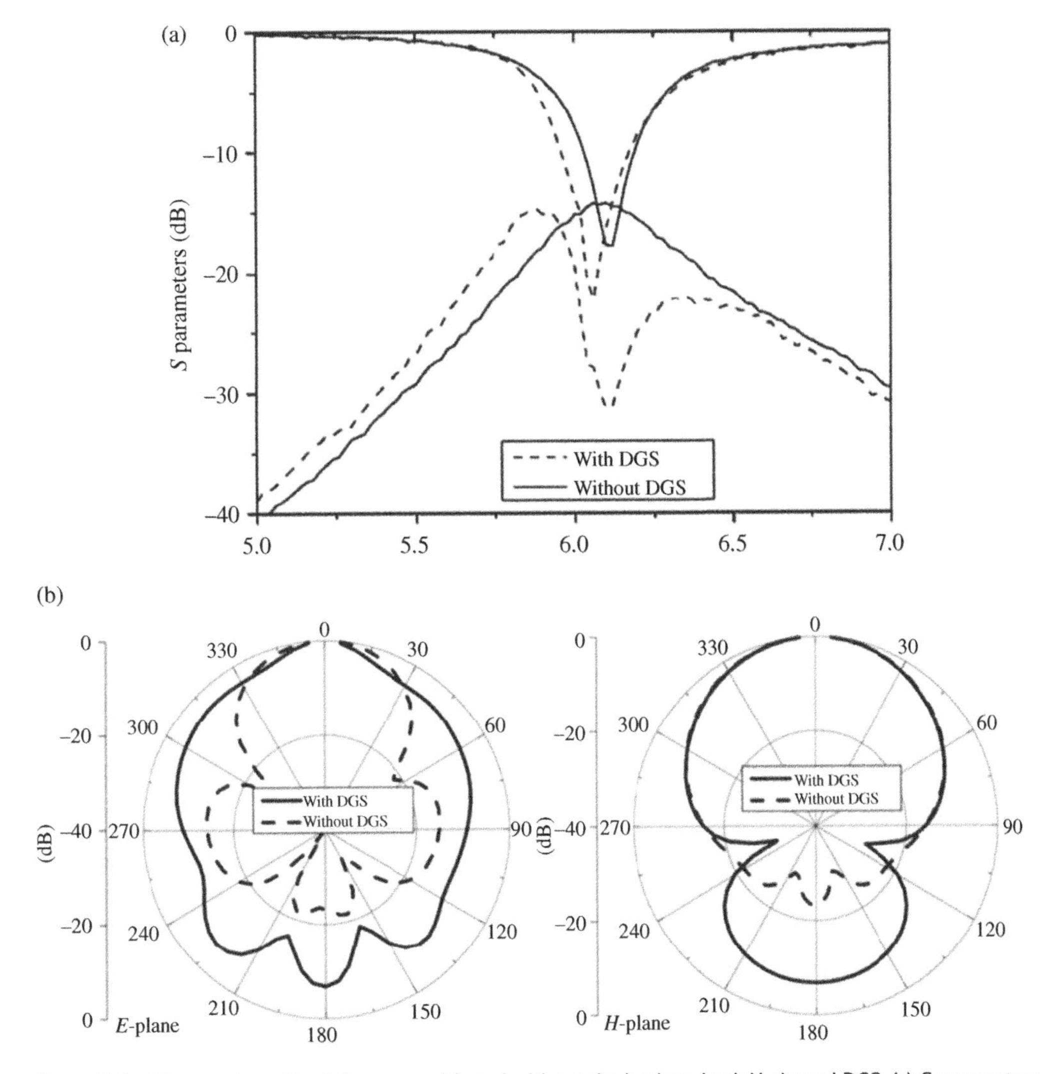

Figure 3.6 Measured results of the array with and without the back-to-back U-shaped DGS. (a) *S*-parameters as functions of the source frequency. (b) Radiation patterns at the resonance frequency. *Source:* From [11] / with permission of The Institution of Engineering and Technology.

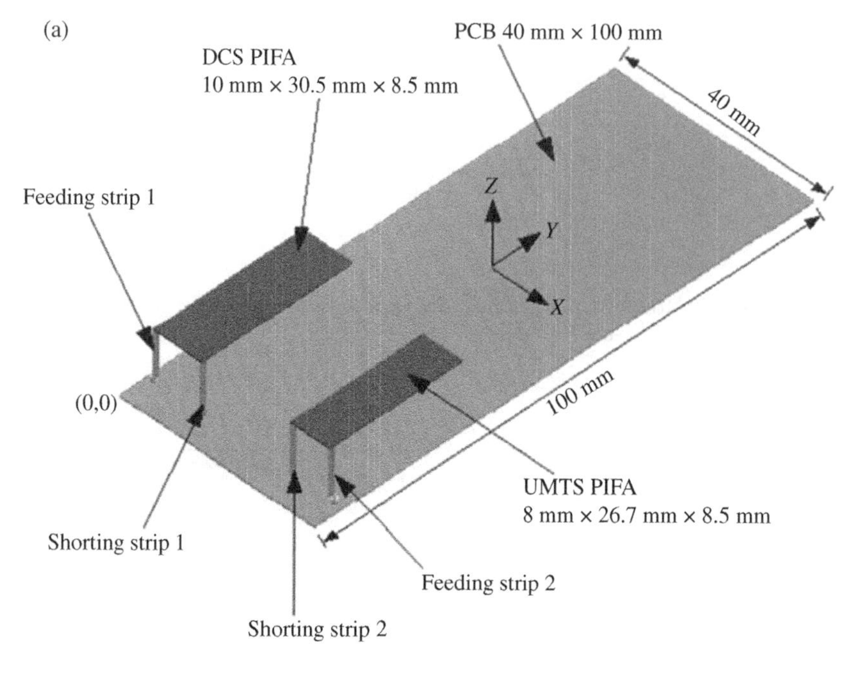

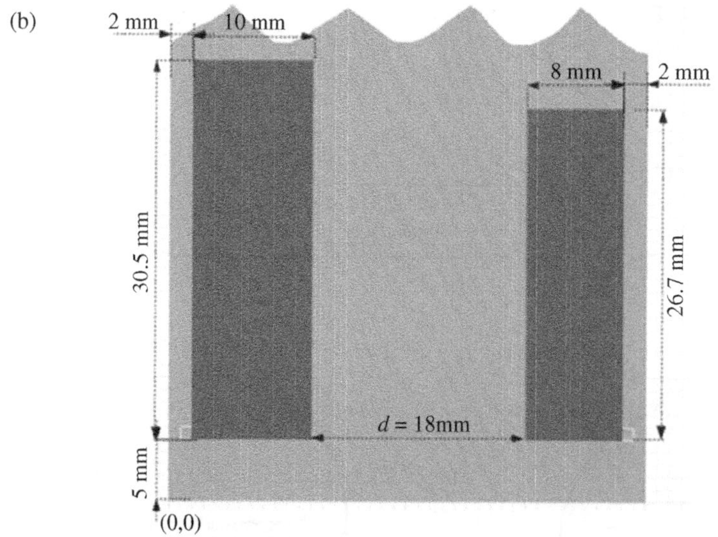

Figure 3.7 The DCS and UMTS PIFAs are arranged on the same side of the PCB. (a) Three-dimensional view. (b) Top view. *Source*: From [16] / with permission of IEEE.

3.3 Neutralization Lines

Neutralization lines can be added between the array elements to compensate for the existing complex electromagnetic coupling pathways. They provide alternative pathways for the induced currents to mitigate the coupling [16–23]. The design principles for this method are tied to impedance-loaded traces that are introduced to connect pairs of antennas in the array.

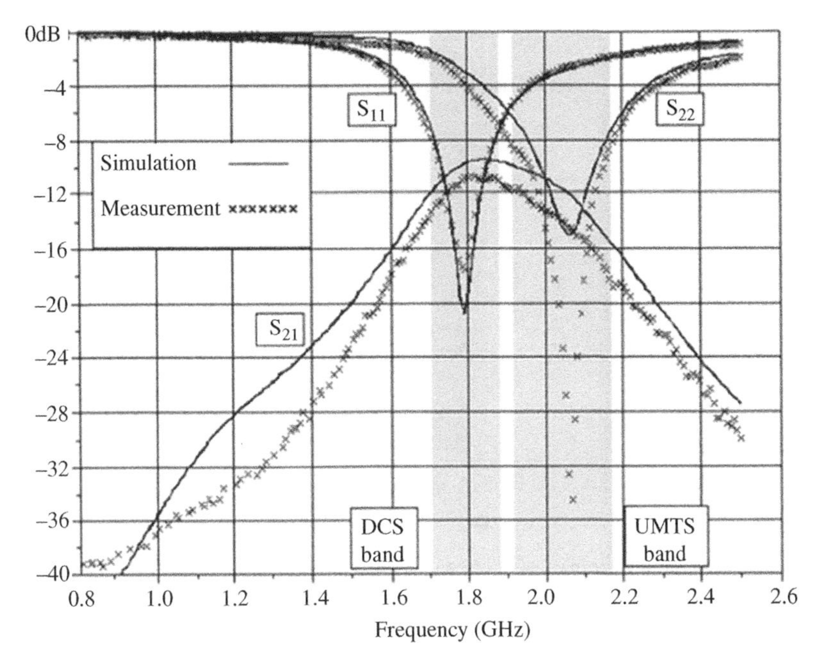

Figure 3.8 Simulated and measured values of $|S_{11}|$, $|S_{22}|$, and $|S_{21}|$ as functions of the source frequency driving the DCS/UMTS PIFAs (d = 18 mm). *Source*: From [16] / with permission of IEEE.

Consider the two planar inverted-F antenna (PIFA) elements shown in Figure 3.7. One is designed to operate in the DCS1800 band and the other in the Universal Mobile Telecommunications System (UMTS) band. They both reside on the same side of the indicated printed circuit board (PCB) and each has its own feed. Port 1 feeds the distributed control system (DCS) element; port 2 feeds the UMTS element [16]. The antennas are placed edge-to-edge with an 18.0-mm separation distance. The simulated and measured reflection coefficients for each antenna are shown in Figure 3.8. The level of mutual coupling is also given. Very good agreement between all of the simulated and measured results was obtained. The maximum magnitude of the measured $|S_{21}|$ values, −10.6 dB, was reached at 1.81 GHz. This minimum isolation level occurred approximately where the $|S_{11}|$ and $|S_{22}|$ curves crossed.

In order to reduce the mutual coupling between the two PIFAs, a suspended microstrip line was introduced to connect them as shown in Figure 3.9a. Its size was 18×0.5 mm^2; its height above the PCB was the same as the horizontal strips of the PIFAs. Its orientation was orthogonal to their feeding strips. This neutralization line caused a blue shift of their resonance frequencies of less than 4% without any degradation of their bandwidths. The simulation results for this case are presented in Figure 3.10a. It is clear that the introduction of the neutralization line significantly reduced the magnitude of the $|S_{21}|$ values, especially near the frequencies at which the deep nulls were observed. Slightly more complicated versions were also considered. The length of the linking line was increased from 18 to 47 mm while keeping its width at 0.5 mm. The 47 mm case is shown in Figure 3.9b. The corresponding simulation results are presented in Figure 3.10a. The simulation results presented in Figure 3.10b were obtained by varying the width of the link from 0.1 to 2.0 mm while keeping its length at 18 mm. These results clearly show that the width and the length of the line have a great influence on the amount that the coupling is decreased. When the length of the line increases or its width decreases, the effective inductance increases and thus the observed

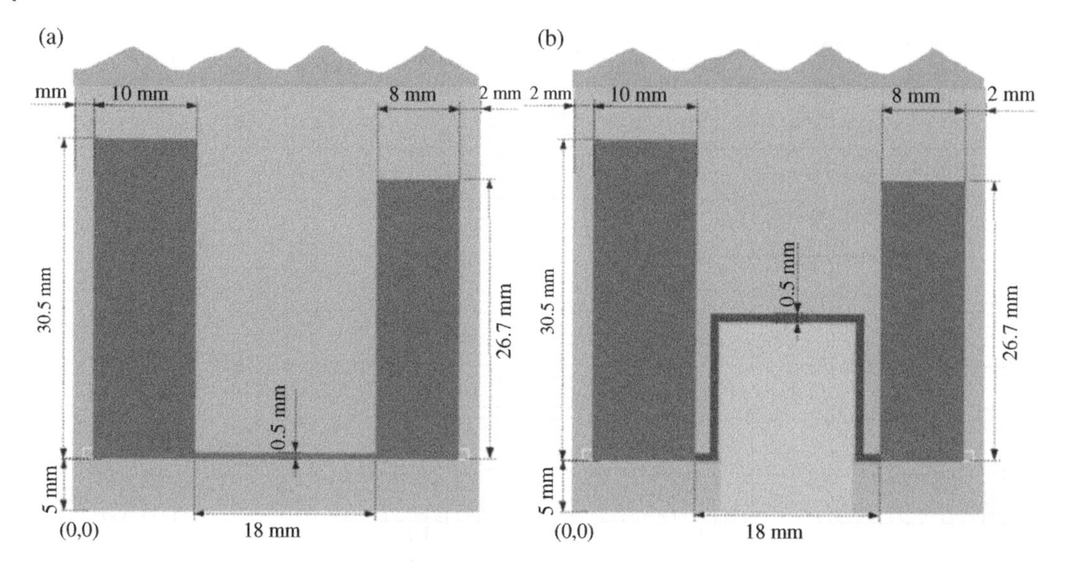

Figure 3.9 Top view of the arrangement of the PIFAs on the PCB when the feeding strips are oriented face-to-face and linked by a suspended line whose width is 0.5 mm. (a) Line length = 18 mm. (b) Line length = 47 mm. *Source*: From [16] / with permission of IEEE.

nulls clearly move toward lower frequencies. Furthermore, this feature nicely corresponds to an increase of the characteristic impedance, $Z_0 = \sqrt{L/C}$, of the neutralization line.

Several series of simulations were conducted to determine the optimal location of this line. It was determined that it should be connected to a low impedance region of the PIFAs. This location is far away from the open end of the line where the voltage and charge densities are maximum. Furthermore, it is close to the feed region and shorting strips where the currents have their highest density. Moreover, these locations of the linkage line connections do not affect the resonance frequencies or the bandwidths of those antennas.

The impact of introducing the linkage line between both shorting strips of the PIFAs was also investigated. This configuration is illustrated in Figure 3.11. Two specific cases are provided. The first introduces a shorting strip between the DCS antenna and the feeding strip of the UMTS PIFA. The second introduces the linkage between both shorting strips of the PIFAs. The S-parameter results for both cases in which the microstrip link was 0.5 mm wide and 18 mm long are given in Figure 3.12.

Both antennas in the first case achieved good matching around 1.95 GHz with a very high isolation value. The simulated maximum improvement was nearly 22 dB. The measured maximum improvement was actually 34 dB. Nevertheless, the isolation was not improved over the entire operational bandwidths of the antennas. The $|S_{21}|$ parameter values in the second case have a flat shape over the entire operational bandwidth; their magnitude level is always below −20 dB instead of having deep nulls at its center frequency. In comparison to the results of the initial configuration in Figure 3.10, a minimum improvement of 10 dB was observed over the entire bandwidth.

In summary, while the microstrip linkages that connect either the feeding strips or the shorting strips act as neutralization elements, they yield quite different isolation behaviors. The microstrip link in the first case was arranged to connect the feeding strips of both PIFAs, which have a 50-Ω impedance. Unfortunately, the resulting impedance varied when the frequency changed, i.e., it was far from being constant. Consequently, any signal radiated will have amplitudes and phases that

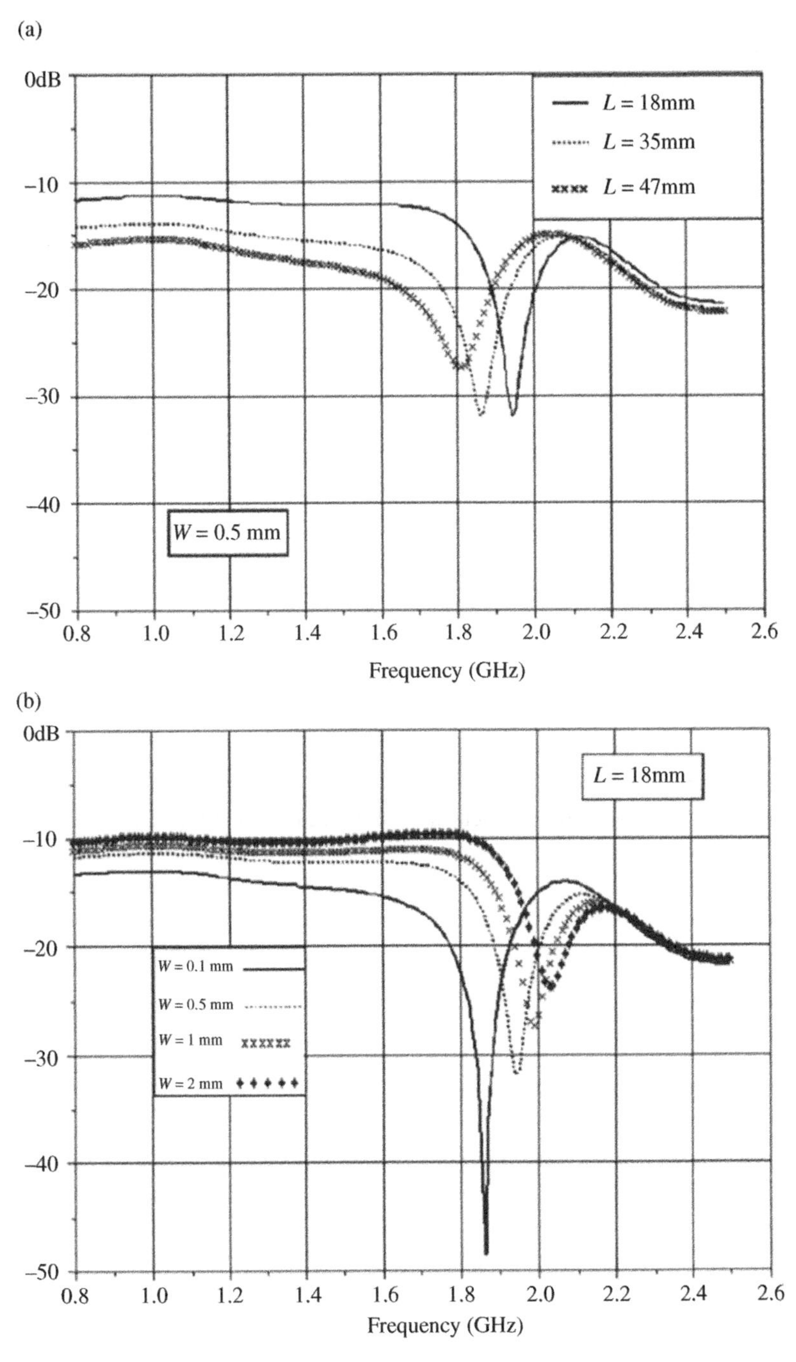

Figure 3.10 Simulated $|S_{21}|$ values as functions of the source frequency when the feeds of the PIFAs are connected with a suspended microstrip link whose parameters are varied. (a) Width $W = 0.5$ mm, length L is varied. (b) $L = 18$ mm, width W is varied. *Source*: From [16] / with permission of IEEE.

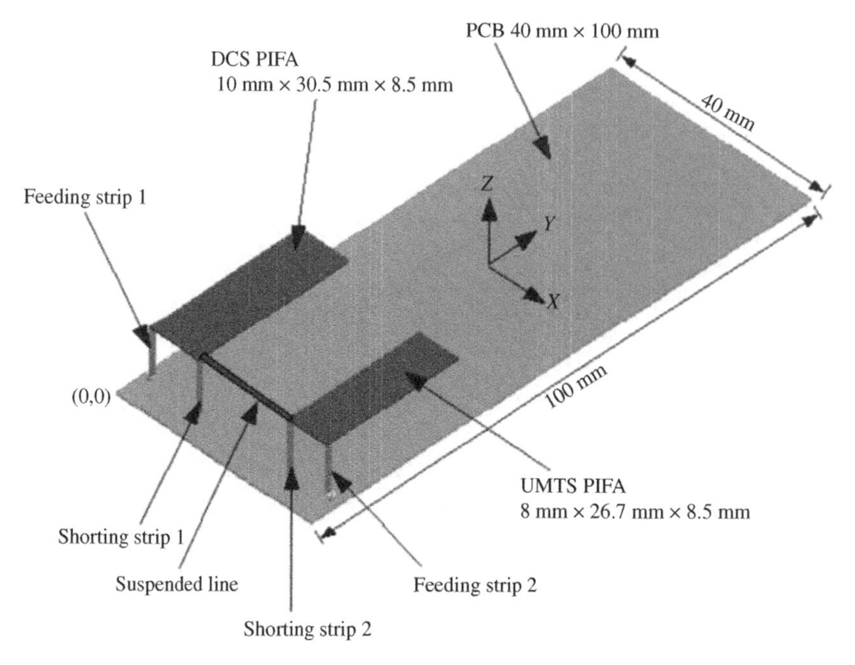

Figure 3.11 Arrangement of the PIFAs on the same side of the PCB when the shorting strips are parallel and linked by a suspended microstrip line. *Source*: From [16] / with permission of IEEE.

change along with those impedance variations. This behavior is the main reason the efficiency of the neutralization line approach is very high only near a specific frequency point. On the other hand, the microstrip link in case 2 connects the shorting strips of both PIFAs in their very low impedance regions where the resulting inductance and voltage values are low. Thus, the impedance varies little with frequency. Hence, it follows the amplitude and phase variations of any radiated signal. As a result, this choice of neutralization linkage is effective over a wide frequency bandwidth.

3.4 Array-Antenna Decoupling Surfaces

An array-antenna decoupling surface (AADS) is a thin surface that is placed less than a half wavelength above the ground plane of the array [24, 25]. It is typically composed of a set of electrically small metallic patches that reflect any incident waves. A novel AADS was designed to be a partially reflecting surface. It thus creates reflected waves that cancel the coupling space waves from adjacent antenna elements. When the distance between the AADS and the array elements and the sizes of the reflecting elements are adjusted appropriately, the out-of-phase and equal amplitude conditions necessary to create the desired destructive interference fields are achieved. A high degree of cancelation of the unwanted mutual coupling waves then occurs. Figure 3.13 depicts an array antenna augmented with a generic AADS [24].

As depicted in Figure 3.13, it is observed that the energy radiated from the array consists of four main components when the AADS is present. These are: (i) the waves being radiated outward from the array elements, (ii) the waves reflected by the elements in the AADS, (iii) the waves coupling back into the array elements, and (iv) the waves being radiated outward from the entire system into

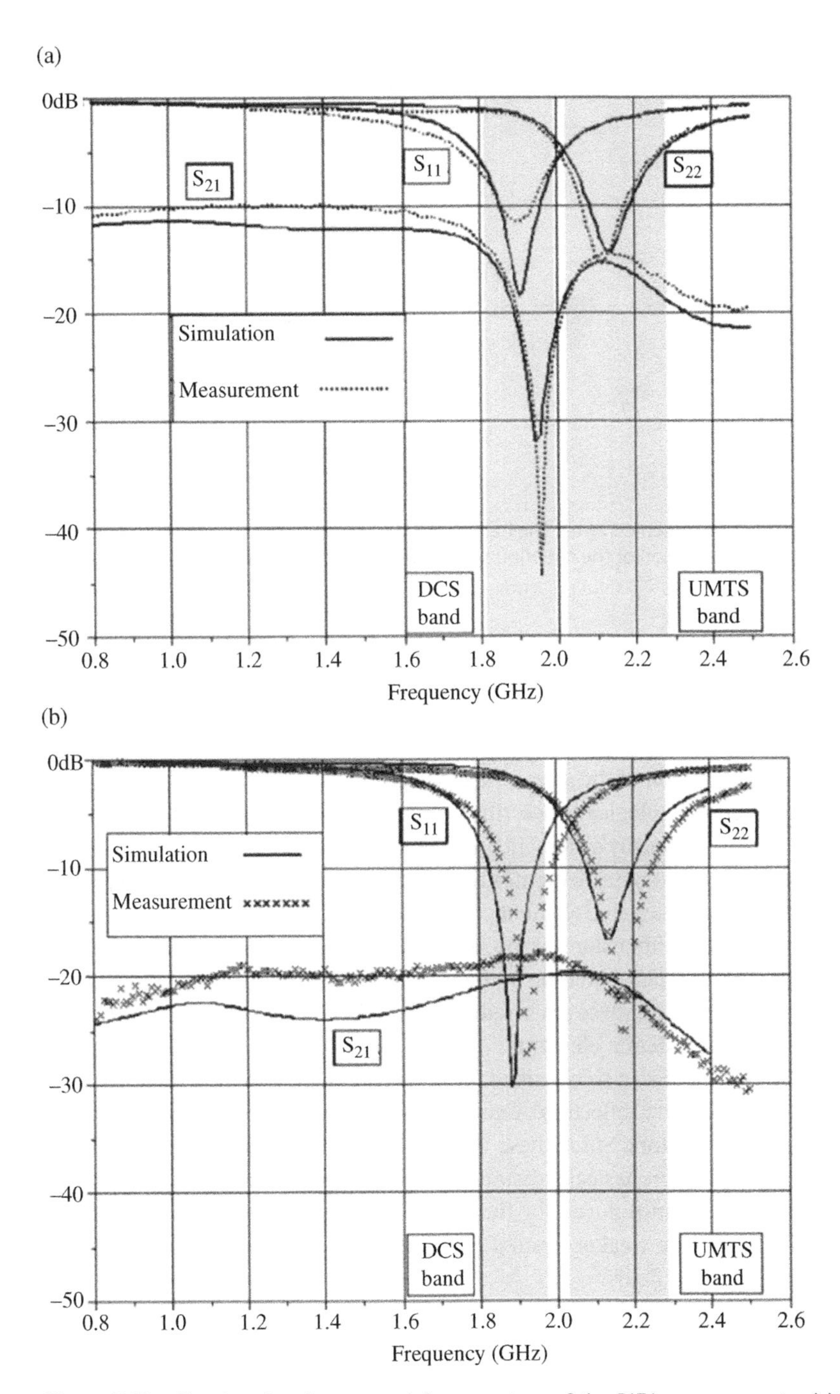

Figure 3.12 Simulated and measured *S*-parameters of the PIFA arrangements. (a) The feeding strips are parallel and linked by a suspended microstrip line. (b) The shorting strips are parallel and linked by a suspended microstrip line. In both cases, the line has the dimensions W = 0.5 mm, L = 18 mm. *Source:* From [16] / with permission of IEEE.

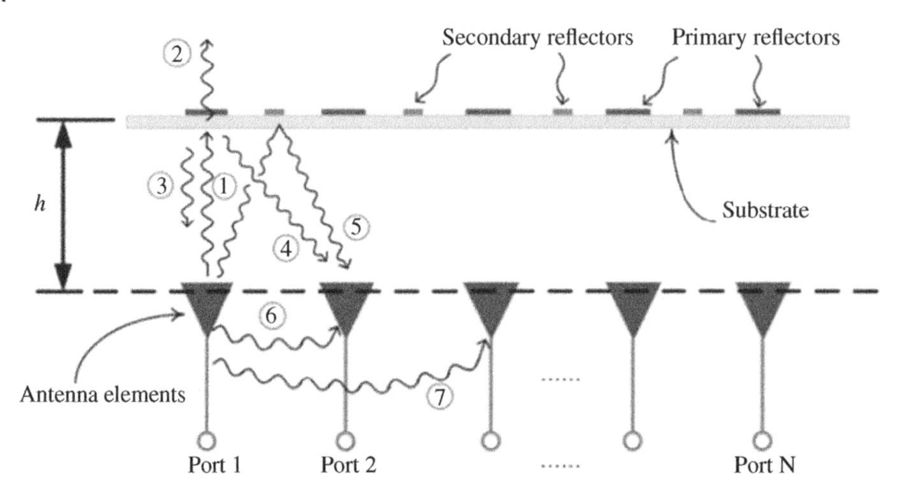

Figure 3.13 Schematic of an AADS-augmented array. The height of the AADS above its antenna elements, *h*, is optimized to achieve the desired cancelation of the coupling space waves. *Source*: From [24] / with permission of IEEE.

the far field. Since the primary objective of using an AADS is to reduce the mutual coupling between two adjacent antenna elements, it does not have to intervene with the mutual couplings among the nonadjacent elements since the latter are assumed to be very weak. Since the AADS is physically located in the reactive region of the array, the description, "reflected wave," is not accurate per se. It is only used to phenomenologically describe the waves scattered by the elements of the AADS into the directions received by array's elements. Furthermore, Figure 3.13 illustrates that a second signal pathway between the antenna elements arises from the presence of the AADS [24]. It arises from the fields scattered by the parasitic elements. These scattered waves can be controlled by subtly designing the pattern and the dimensions of the metal elements in the AADS and its height, *h*, above the array elements. With proper matching of the intensity, but with opposite phase of the incoming signals generated by the array, these scattered waves can be tailored to significantly cancel the coupling between adjacent antenna elements.

The reflecting elements in the AADS can be classified into two types based on their functionality: the primary reflectors and the secondary reflectors. A major portion of the waves scattered from the AADS arises from the primary reflectors. Since these waves provide a significant amount of the space-wave coupling, their polarization basically dictates the polarization of that coupling mechanism. The secondary reflectors are introduced for finte-tuning, i.e., they are designed to create minor scattered waves to mitigate the weaker mutual coupling that arises from other scattering pathways such as the cross-polarized fields.

Consider the 2-D dual-polarized 2×2 planar dipole array presented in Figure 3.14a that is integrated with an AADS. The system was designed to operate in the frequency band from 3.3 to 3.8 GHz [24]. The horizontal and vertical center-to-center distances between two antenna elements noted in Figure 3.14b are $D1 = 45$ mm and $D2 = 60$ mm, respectively. The shapes of the major and minor reflectors are illustrated in Figure 3.14c. The polarization directions associated with the major reflectors are assigned element numbers in Figure 3.14d for discussion purposes. The measured *S*-parameters of the ports feeding the array are presented in Figure 3.15.

As shown in Figure 3.15a, the simulated and measured return losses of the array with the AADS present are 15 dB or better at both port 1 and port 2 across the entire operational band from 3.3 to

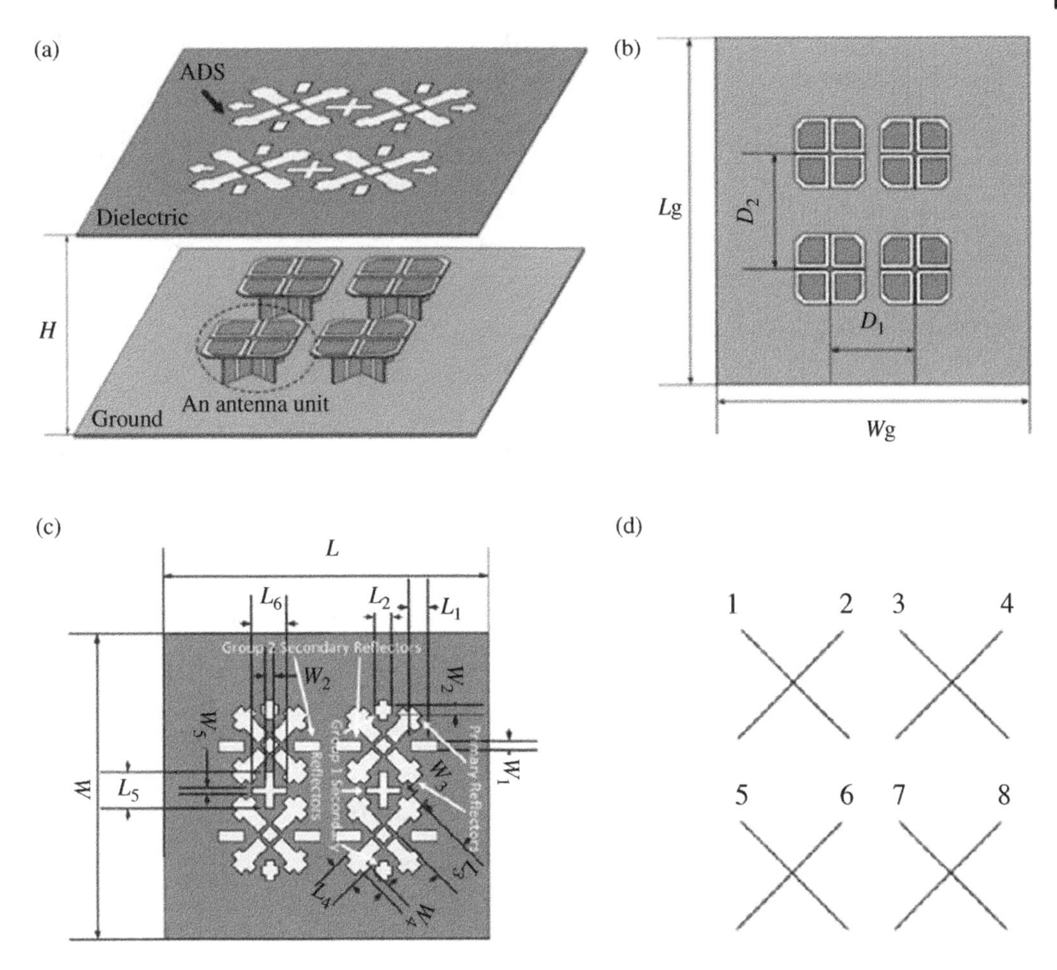

Figure 3.14 A 2 × 2 dual-polarized dipole array. (a) Perspective view of the array with the AADS above it. (b) Top view of the array. (c) Metal reflection patches on the AADS. (d) Numbers assigned to the antenna elements for discussion purposes. *Source*: From [24] / with permission of IEEE.

3.8 GHz. The mutual coupling levels between two adjacent elements having the same polarization in the horizontal and vertical directions are represented by $|S_{13}|$ and $|S_{15}|$, respectively, in Figure 3.15b. The presence of the AADS clearly reduces the $|S_{13}|$ values from about −14 to −25 dB and lower. On the other hand, the $|S_{13}|$ values are reduced slightly from −26 to −28 dB and lower. As shown in Figure 3.15c, the coupling between the two cross-polarized elements in the same unit, i.e., the $|S_{12}|$ values, is also improved to below −30 dB from −25 dB without the AADS. It is difficult to control the cross-polarization coupling levels between two adjacent array elements, namely the $|S_{14}|$ and $|S_{23}|$ values, when they are close to each other. This issue arises because that coupling mechanism is influenced significantly by the capacitance formed between the two closest ends of these radiators. Figures 3.15c and 3.15d show that these coupling levels are reduced, respectively, from −23 and −25 dB to −25 and −30 dB and lower when the AADS is present.

Note that, as assumed for the AADS design, the mutual coupling levels between the two distant coaxial and collinear radiators, namely $|S_{17}|$, and between the two yet further away co-polarized radiators, namely $|S_{28}|$, are inherently lower. Consequently, they truly are very minor contributions

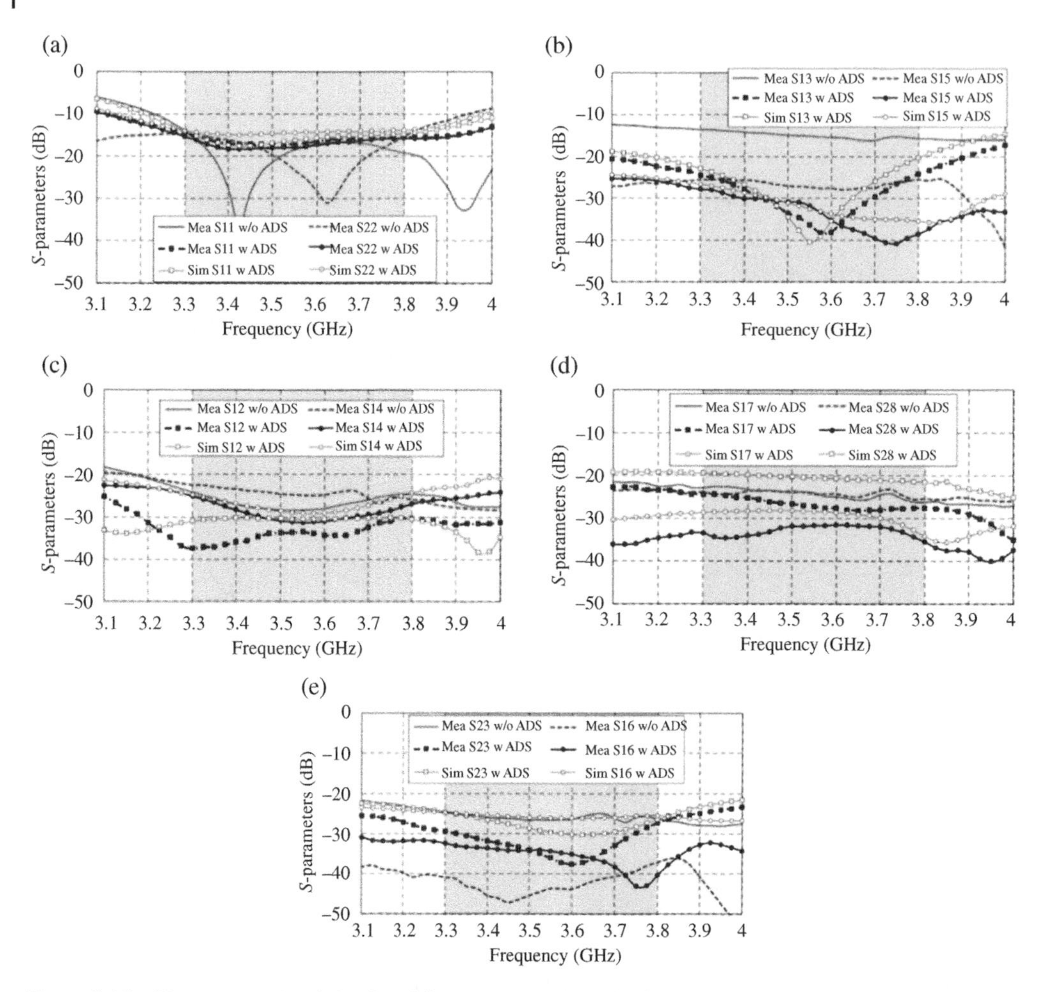

Figure 3.15 The measured and simulated *S*-parameters of the 2 × 2 array with and without the AADS. (a) $|S_{11}|$ and $|S_{22}|$. (b) $|S_{13}|$ and $|S_{15}|$. (c) $|S_{12}|$ and $|S_{14}|$. (d) $|S_{17}|$ and $|S_{28}|$. (e) $|S_{23}|$ and $|S_{16}|$. *Source*: From [24] / with permission of IEEE.

to the mutual coupling levels and no specific considerations were needed to deal with them. Note that the mutual coupling between elements 1 and 6 is the weakest due to their cross-polarized configuration and the large separation distance between them. Nevertheless, Figure 3.15e shows that the coupling level, $|S_{16}|$, decreased from −30 to −40 dB when the AADS was present. It is noted that because of its ability to efficiently reduce the coupling levels, the AADS method is attractive because it also enhances the quality of the array's radiation patterns.

3.5 Metamaterial Structures

The constitutive parameters of naturally occurring materials are generally characterized by their complex electrical permittivity and magnetic permeability. The former is generally in response to an applied electric field, while the latter is in response to a magnetic field. These permittivity and permeability values are both positive at microwave frequencies and, hence, these materials can be described as being double-positive (DPS) [26].

The propagation constant γ for waves in the material is calculated with the effective ε and μ of the material as:

$$\gamma = jk = j\omega\sqrt{\mu\varepsilon} = \alpha + j\beta \tag{3.3}$$

If k is a real (imaginary) number, the electromagnetic wave is propagating (evanescent). However, if either ε or μ has a negative value, k is imaginary and the attenuation constant α is nonzero. Artificial media, i.e., metamaterials, can be designed to realize large α values and, hence, strong attenuation appears in a prescribed band of frequencies. These so-called single-negative (SNG) metamaterials [26] are suitable for application between antenna elements to enhance their isolation characteristics [27–36].

Two types of resonant meta-structures were introduced for mutual coupling reduction in [33]. They were the grounded capacitively loaded loops (GCLLs) and the π-shaped elements. They are illustrated in Figures 3.16 and 3.17. The configuration of one GCLL unit cell is provided in Figure 3.16a together with its geometric parameters. The capacitively loaded loop (CLL) structure is oriented orthogonal to and is connected directly to the ground plane. The metallic traces of the CLL structure are supported on an Arlon AD450™ substrate with relative permittivity: $\varepsilon_r = 4.5$, and loss tangent: $\tan\delta = 0.0035$. The thickness of the substrate is 0.508 mm. Figure 3.16b presents the simulation model. It indicates the electromagnetic environment imposed on the GCLL element, i.e., it is illuminated by a plane wave propagating along the x-axis with its electric (E-) field being parallel to the z-axis and its magnetic (H-) field being parallel to the y-axis.

Details of the ANSYS high frequency structure simulator (HFSS) model used to simulate its performance are as follows. Perfect electric conducting (PEC) and perfect magnetic conducting (PMC) boundary conditions were imposed in the z- and y-directions, respectively. Two excitation ports were assigned in the x-direction. The simulated reflection and transmission properties obtained from this GCLL meta-structure model are shown in Figure 3.18a. Excellent isolation performance was exhibited; the peak isolation value is above 30 dB at the resonance frequency, 3.135 GHz. The presence of the meander lines not only makes the GCLL unit cell electrically smaller (10.29% reduction), but also provides more freedom to adjust its resonance frequency range. As shown in

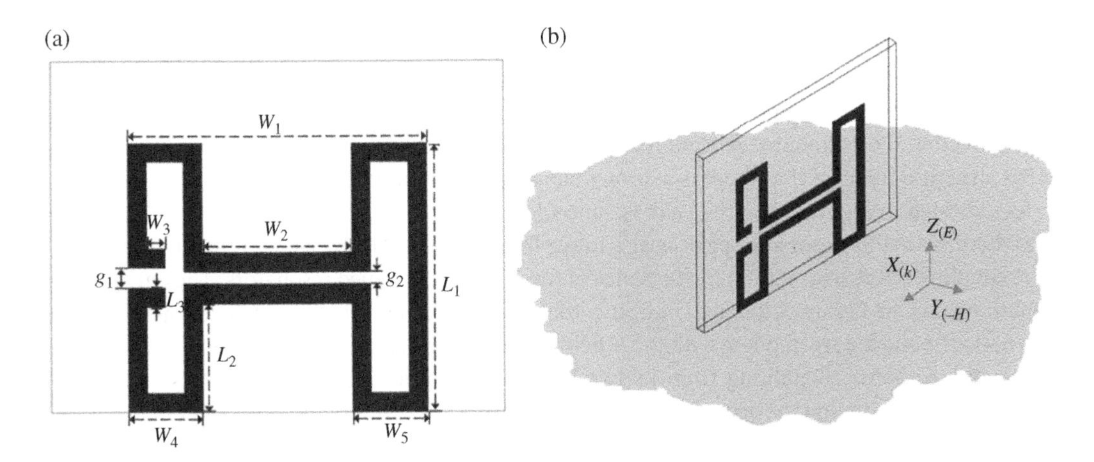

Figure 3.16 The GCLL meta-structure configuration. (a) The physical geometry with its defining parameters. (b) The unit cell simulation model. *Source*: From [33] / with permission of IEEE.

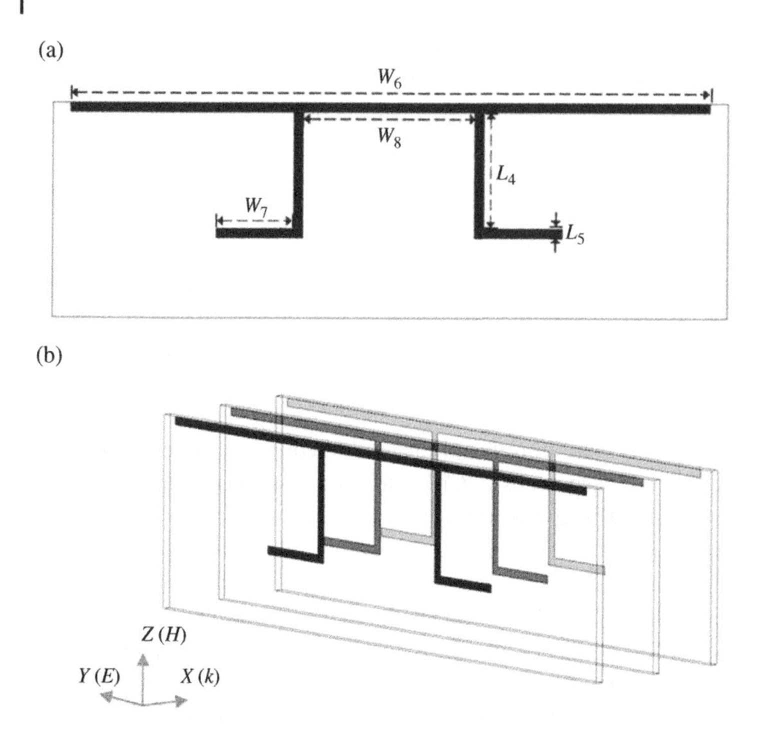

Figure 3.17 The π-shaped meta-structure configuration. (a) The physical geometry of the unit cell with its defining parameters. (b) The multiple unit cell simulation model. *Source:* From [33] / with permission of IEEE.

Figure 3.18b, the resonance frequency shifts from 2.75 to 3.37 GHz as the meander line length changes from 3.0 to 6.0 mm. Moreover, its effective medium parameters were retrieved from the simulated S-parameters of Figure 3.18a and are given in Figure 3.19.

As is illustrated in Figure 3.17a, the π-shaped resonator is printed on an Arlon AD450 substrate with the same thickness as that of the GCLL resonator in Figure 3.16. The corresponding simulation model and the electromagnetic environment it sees are shown in Figure 3.17b. In the decoupling configuration, three π-shaped unit cells are introduced. They are illuminated by a plane wave propagating along the x-axis with its E-field parallel to the y-axis and its H-field parallel to the z-axis. The HFSS simulation model had PEC and PMC boundary conditions imposed again on the simulation space in the y- and z-directions, respectively. Two excitation ports are assigned again in the x-direction. The simulated reflection and transmission properties of the resulting bulk metamaterial structure indicate that there is a strong bandgap behavior around 3.14 GHz.

As indicated in Figure 3.20a, the peak isolation level at the resonance frequency is ~45 dB. Note that there is another resonance peak near a lower frequency, 2.83 GHz, which is due to the capacitive coupling effect arising from the presence of the multiple π-shaped resonators. The current distribution inset in Figure 3.20a also indicates that the surface currents on each resonator at the resonance frequency are in phase with the E-field of the exciting plane wave and, hence, also exhibit an electric response. Benefiting from its two-leg configuration, one can shift its resonance frequency by only changing the distance (W_8) between its two legs. The results of the parametric study of W_8 are summarized in Figure 3.20b. They indicate that one can shift the resonance frequency in a large frequency interval: 2.6–3.2 GHz, simply by varying the distance W_8. The effective permittivity and permeability of this bulk metamaterial are also retrieved from the simulated S-parameters and

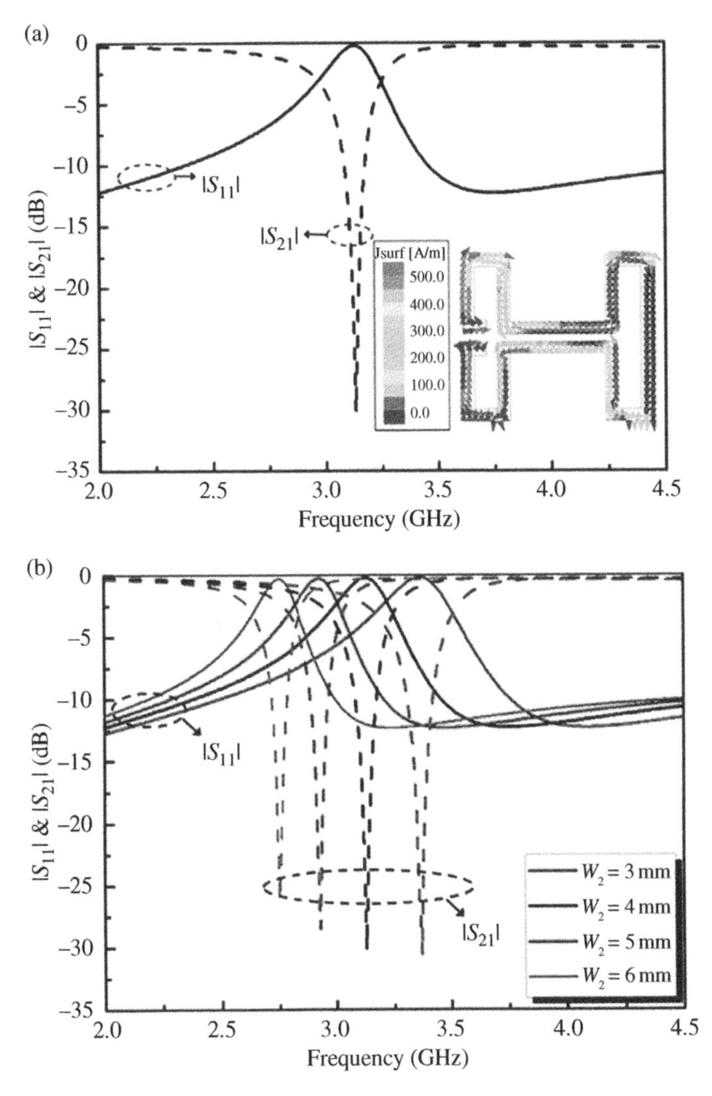

Figure 3.18 GCLL meta-structure simulation results. (a) Predicted *S*-parameters of the configuration in Figure 3.16b. (b) The results of the parametric study of the meander line length, W_2. *Source*: From [33] / with permission of IEEE.

are presented in Figure 3.21. The electric SNG nature of the π-shaped resonators in the frequency band of interest is clear.

Diagrams representing dual-polarized arrays with an arbitrary and a rectangular two-dimensional lattice arrangement are shown in Figure 3.22. The cross shapes, which are composed of *x*-directed (red) and *y*-directed (black) short line segments, represent each dual-polarized antenna element in the array. These elements are composed of a pair of linearly polarized radiators, one (red) in the *x*-direction and the other (black) in the *y*- direction. The parameters d_x and d_y are, respectively, the inter-element-spacing values along the *x*- and *y*-axes. The offset distance along the *y*-axis is *d*. When *d* decreases from a certain value to zero, the array then becomes a common rectangular array such as the one shown in Figure 3.22b. Therefore, arrays with arbitrary lattice arrangements can be formed simply by varying the offset distance *d*.

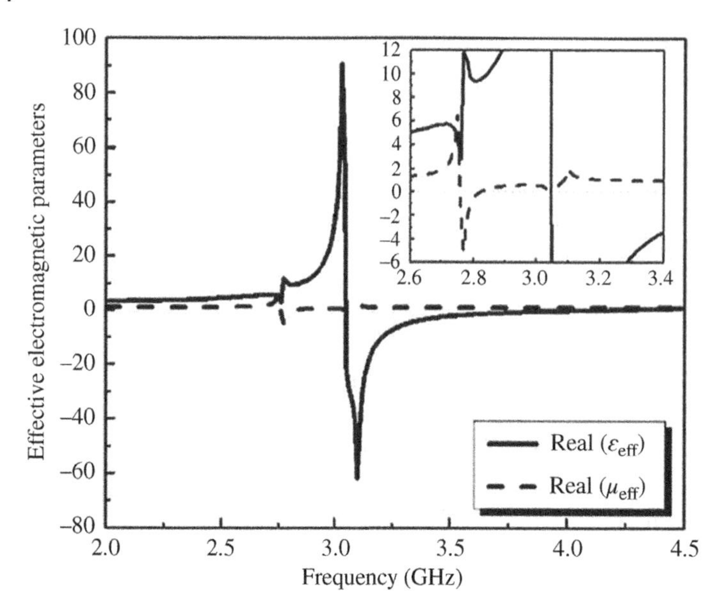

Figure 3.19 The retrieved effective medium parameters of the π-shaped element unit cell shown in Figure 3.18a. *Source*: From [33] / with permission of IEEE.

Figure 3.23 shows the rectangular array configuration composed of dual-polarized elements rotated by the angle α in comparison to those shown in Figure 3.22b. By varying α, a dual-polarized array with any specified polarization orientation can be achieved.

Figure 3.24 depicts a two-element array consisting of two dual-polarized elements, No. 1 and No. 2, with four ports numbered "1," "2," "3," and "4." The decoupling meta-structures, i.e., the seven resonant GCLLs and the three resonant π-shaped elements, are loaded halfway between the two radiating elements. As shown in Figure 3.24c, the GCLLs are incorporated into the region just underneath the patch layer, the middle region (in blue), and are connected to the parasitic ground into which the slot elements are etched. The three π-shaped elements are placed vertically on the top side of the GCLLs. The two-element array is configured with $d = 7.5$ mm and $\alpha = 0°$. This lattice configuration corresponds to a closely spaced triangular array with inter-element spacings of 43.42 mm and 32.81 mm along its x- and y-axes, respectively.

The simulated S-parameter results of the optimized array with and without the decoupling meta-structures are given in Figures 3.25 and 3.26, respectively. The simulated reflection coefficients of all of the ports, as well as the port isolation levels, are shown in Figures 3.25a and 3.26a, respectively. It is clear that both patch elements exhibit good impedance matching in the range from 3.3 to 3.6 GHz, where the reflection coefficient <-10 dB. They also exhibit excellent isolation levels for the cross-pol ports, i.e., they are as high as 35 dB for both elements No. 1 ($|S_{12}|$) and No. 2 ($|S_{34}|$). This outcome demonstrates that the presence of the decoupling meta-structures has little effect on the S-parameters of each antenna element. The port isolation levels between elements No. 1 and No. 2 are revealed by their co-polarization (co-pol) ($|S_{13}|$ and $|S_{24}|$) and cross-polarization (cross-pol) ($|S_{14}|$ and $|S_{23}|$) port levels for the cases with and without the decoupling meta-structures. They are shown in Figures 3.25b and 3.26b, respectively, as functions of the source frequency.

It is readily observed that the port isolation of the same polarization has been significantly improved over the entire operational band. In particular, $|S_{13}|$ ($|S_{24}|$) is decreased from -18.67

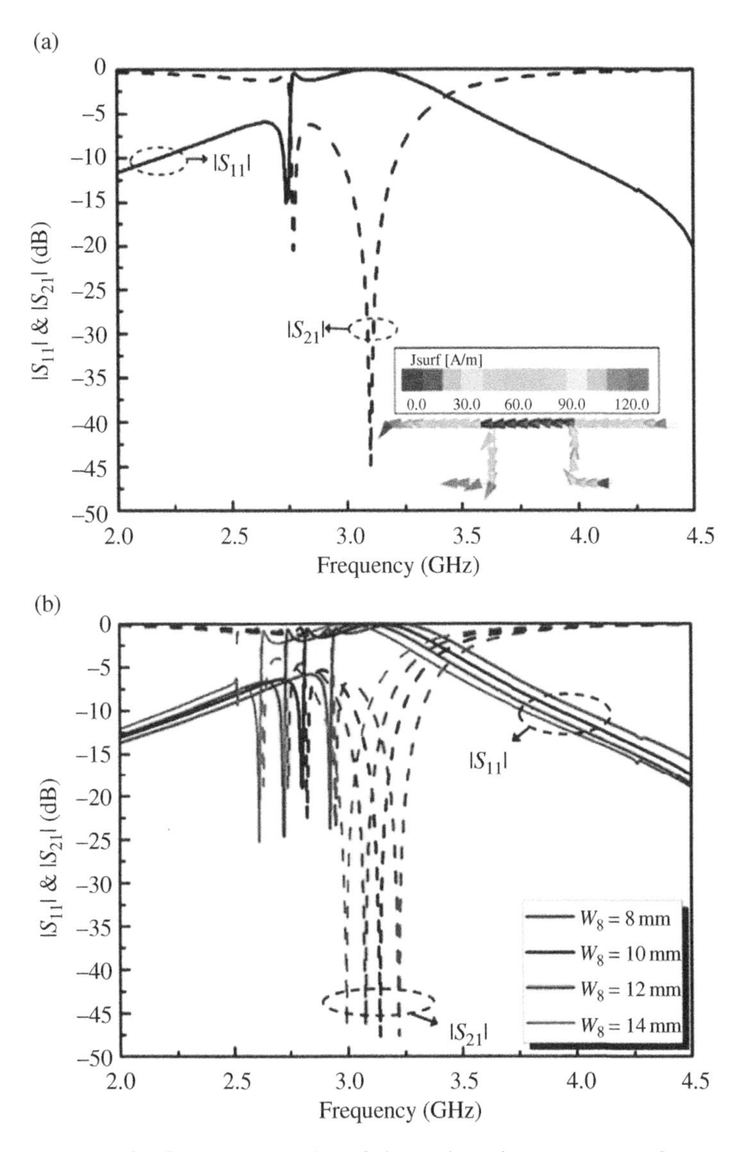

Figure 3.20 Simulation results of the π-shaped resonator. (a) S-parameters of the bulk metamaterial configuration in Figure 3.17b. (b) The results of the parametric study of the distance, W_8, between the two legs of each unit cell. *Source*: From [33] / with permission of IEEE.

dB (−18.36 dB) to −32.19 dB (−25.42 dB), witnessing a reduction in the coupling levels of ~13.52 dB (~7.06 dB). Taking the differences between the values in Figures 3.25 and 3.26, it is determined that the relative fractional bandwidth, where the isolation between both polarizations has been enhanced by up to 5 dB, is about 23.37%. Moreover, the cross-pol port isolation levels between elements No. 1 and No. 2 ($|S_{23}|$ and $|S_{14}|$) remain very low and almost unchanged.

In order to investigate the mutual coupling reduction effects of the identified meta-structures in a dual-polarized array defined by an arbitrary lattice array with an arbitrary orientation of its elements, a set of meta-structure-loaded two-element arrays having different offset distances d and rotation angles α was analyzed.

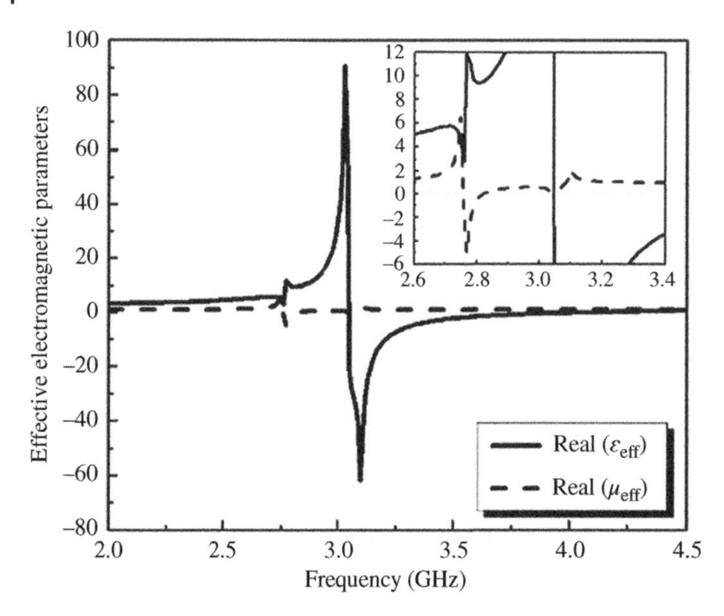

Figure 3.21 The retrieved effective medium parameters of the π-shaped element unit cell shown in Figure 3.20. *Source*: From [33] / with permission of IEEE.

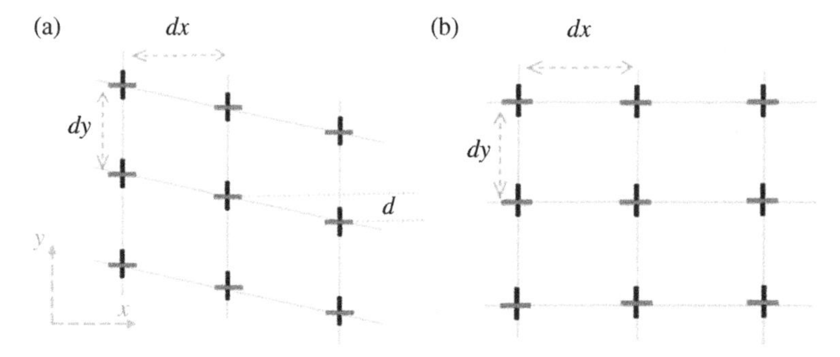

Figure 3.22 Diagram of the dual-polarized arrays with (a) arbitrary and (b) rectangular lattice arrangements. *Source*: From [33] / with permission of IEEE.

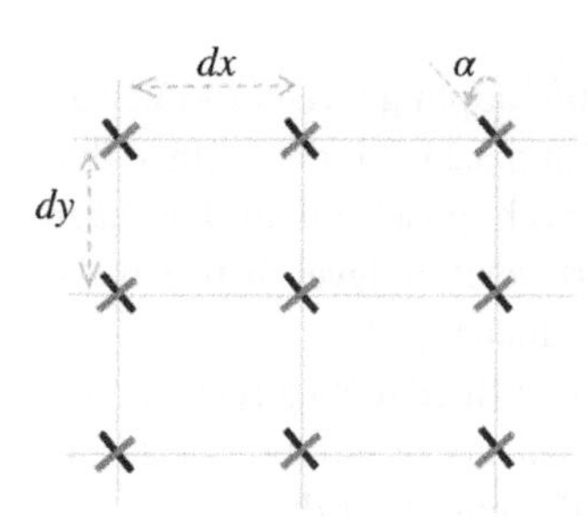

Figure 3.23 Diagram of a dual-polarized array with an arbitrary polarization orientation angle α. *Source*: From [33] / with permission of IEEE.

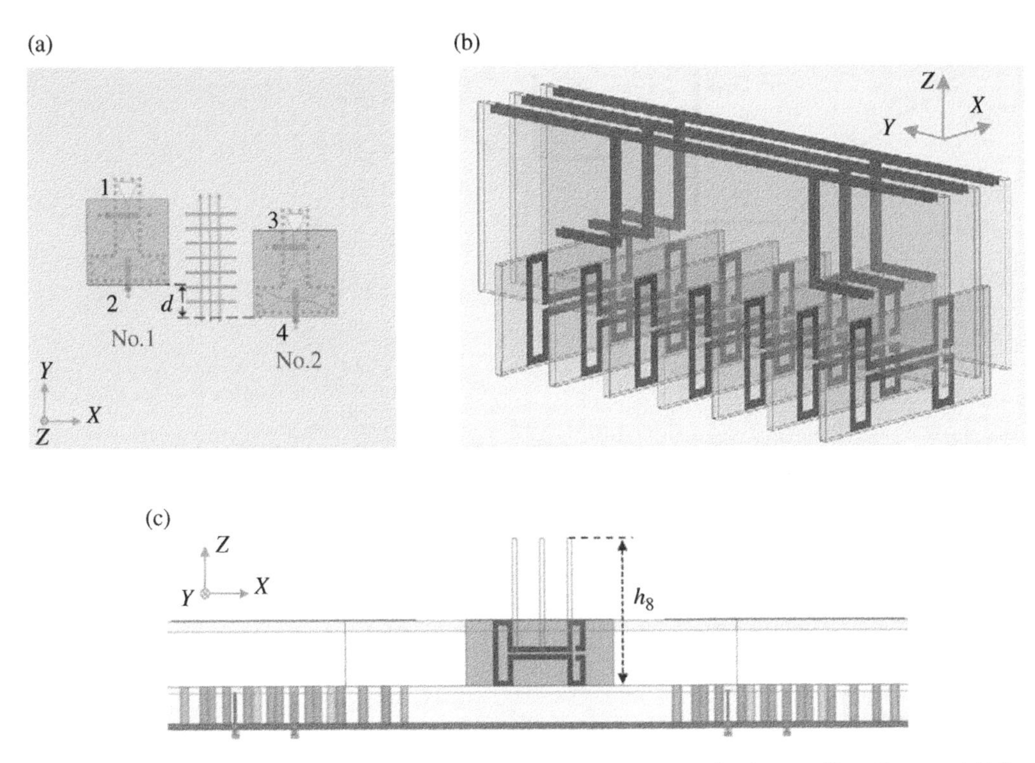

Figure 3.24 Two-element array loaded with decoupling meta-structures having an offset distance *d*. (a) Top view of the array. (b) 3-D zoom-in view of the meta-structures. (c) Side view of the array in the presence of the decoupling meta-structures. *Source*: From [33] / with permission of IEEE.

Figure 3.27a plots the port isolation levels of the co-pol fields ($|S_{13}|$ and $|S_{24}|$) between the two radiating elements as functions of *d* for the cases with and without the decoupling meta-structures. Since the port coupling levels of the cross-polarization (cross-pol) fields ($|S_{14}|$ and $|S_{23}|$) are far below those of the co-pol, only the mutual coupling suppression effects between the co-polarization (co-pol) ports are examined here. All of the isolation levels given in Figure 3.27a are the maximum value for each *d* across the entire frequency band. It is immediately apparent that without the decoupling meta-structures, the isolation levels between the two co-polarized ports decline only very slightly with an increase of *d* and generally remain above −20 dB. In contrast, after loading the array with the decoupling meta-structures, the port isolation of the co-pol fields has been significantly improved for all values of *d*. In particular, $|S_{13}|$ ($|S_{24}|$) shows a minimum reduction of 9.71 dB (7.01 dB), yielding all coupling levels below −25 dB. Figure 3.27b plots the port isolation levels between the co-pol ports of the two-element array as functions of the rotation angle α for the cases with and without the decoupling meta-structures. One finds the isolation curves, $|S_{13}|$ and $|S_{24}|$, are symmetrical with respect to the $\alpha = 45°$ case. This occurs because of the orthogonal polarization property of the two co-pol ports. It is obvious that the port isolation level between them remains above −20 dB as α increases if the decoupling meta-structures are not present. In contrast, when they are present, the co-pol port isolation levels are dramatically improved across the entire range of α values. In particular, both $|S_{13}|$ and $|S_{24}|$ show a minimum reduction of 6.48 dB and all the coupling levels are again below −25 dB. Therefore, it has been clearly demonstrated that these decoupling meta-structures are an effective means to reduce the mutual coupling between the elements of a dual-pol patch array in any lattice configuration or with any orientation of its LP elements.

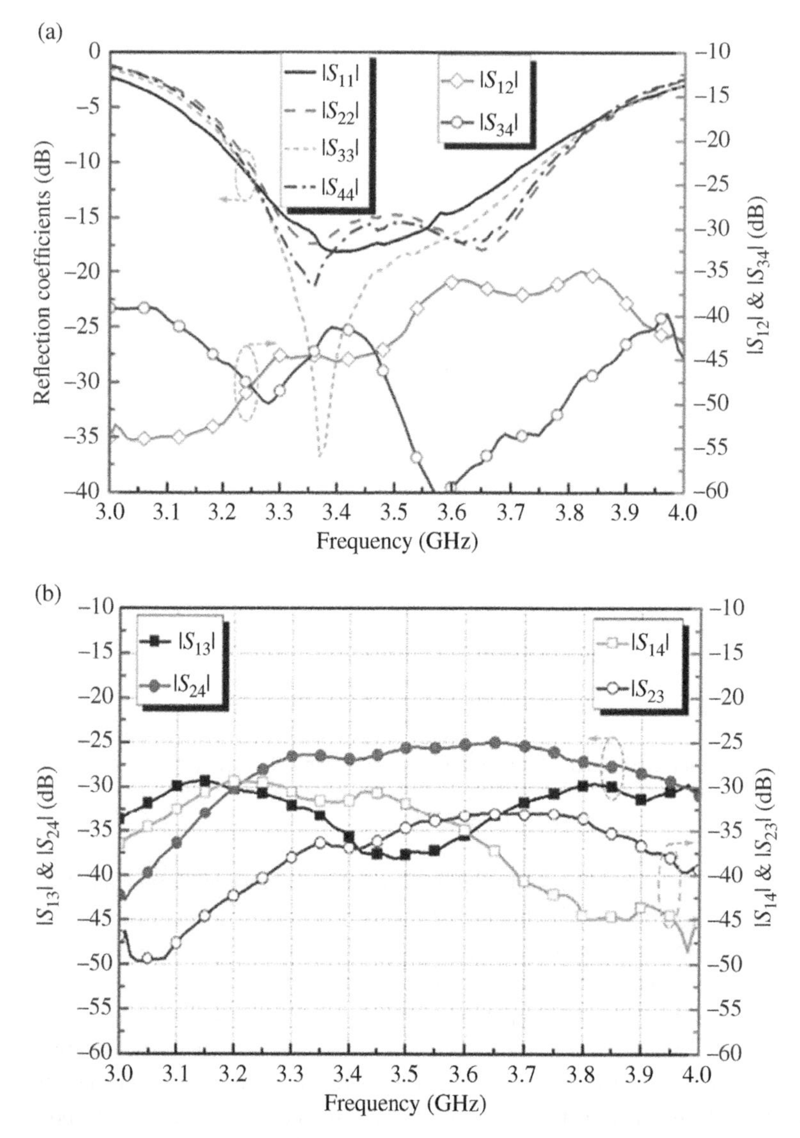

Figure 3.25 Simulated *S*-parameters for the array with the decoupling meta-structures present as functions of the excitation frequency. (a) Reflection coefficients and port isolation levels for each antenna element. (b) Port isolation levels between the two antenna elements. *Source*: From [33] / with permission of IEEE.

3.6 Parasitic Resonators

Loading parasitic elements between the antenna elements of an array is a very classical method for mutual coupling reduction [37–44]. A simplified model of using parasitic resonators to reduce the mutual coupling that takes into account only the primary interactions from the active element is illustrated in Figure 3.28 [39]. Consider first the two closely spaced antenna elements shown in Figure 3.28a. Because of their close proximity to each other, their mutual coupling will be strong.

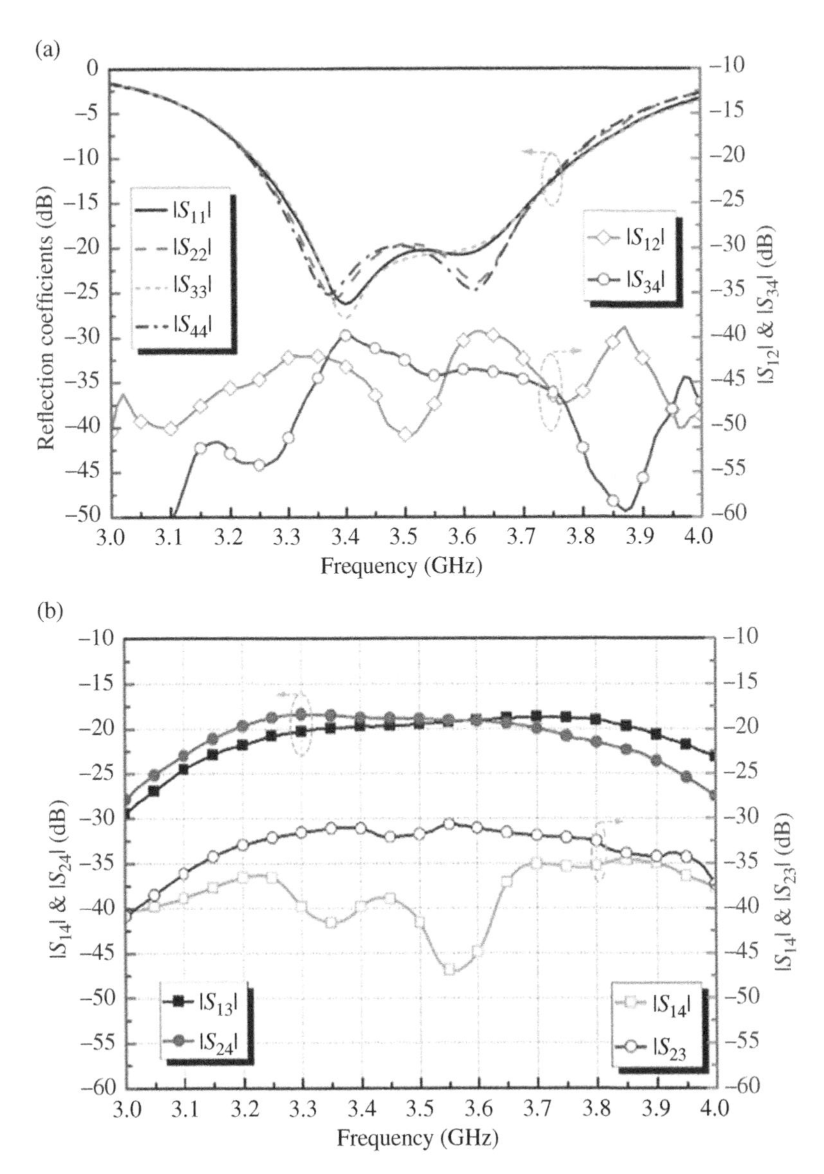

Figure 3.26 Simulated *S*-parameters for the array without the decoupling meta-structures present as functions of the excitation frequency. (a) Reflection coefficients and port isolation for each antenna element. (b) Port isolation levels between the two antenna elements. *Source*: From [33] / with permission of IEEE.

Element 1 is excited by a current $\boldsymbol{I_0}$. A current $\boldsymbol{aI_0}$ is then induced by the mutual coupling on element 2, where $\boldsymbol{a}$ represents the coupling coefficient. Then, as illustrated in Figure 3.28b, parasitic elements are added between the original two elements to establish an additional coupling path. This additional coupling path arises from the fields scattered by the parasitic elements and, hence, in a sense, it is a double-coupling one, i.e., it is a path in which the current in Element 1 is first space-wave coupled to the parasitic elements and then the scattered fields they generate are secondly space-wave coupled to Element 2. We assume that there are N parasitic elements because one

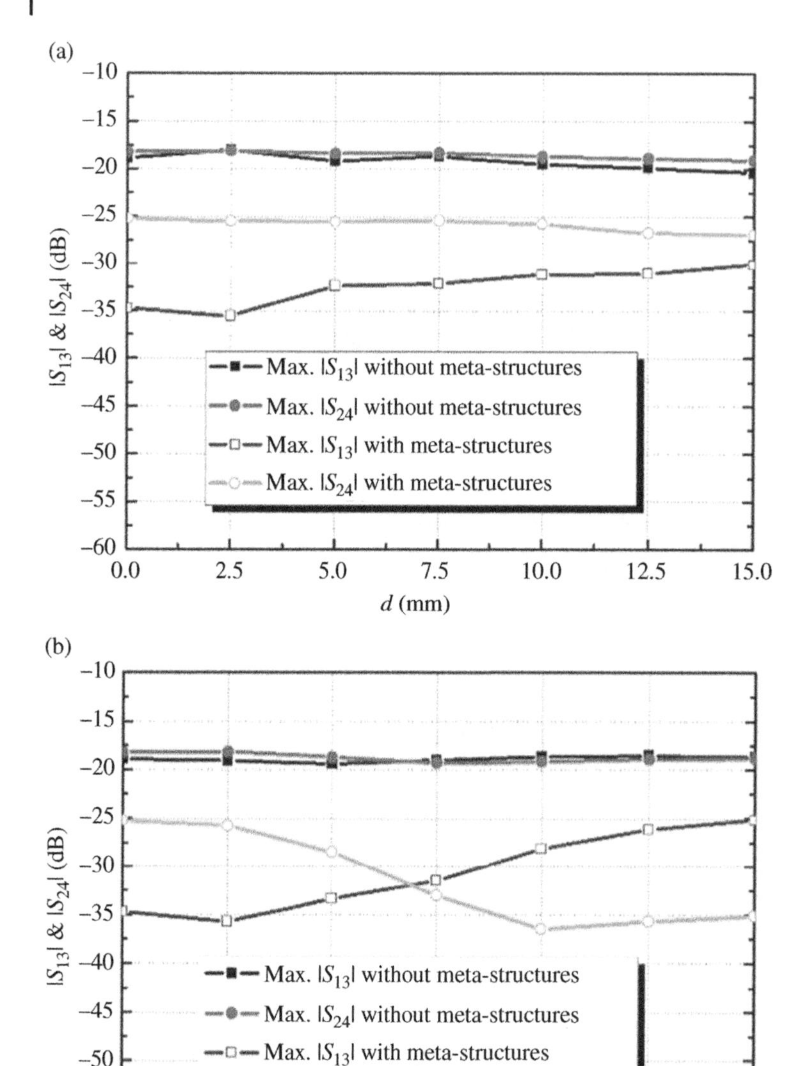

Figure 3.27 Simulated *S*-parameters as functions of the (a) offset distance *d*, and (b) rotation angle α. *Source*: From [33] / with permission of IEEE.

or more of them may be needed to attain the desired mitigation of the coupling. Therefore, the currents induced on them are given by the expressions:

$$\begin{cases} \boldsymbol{b_1 I_0}, & \text{for Parasitic Element 1} \\ \vdots & \\ \boldsymbol{b_N I_0}, & \text{for Parasitic Element } N \end{cases} \tag{3.4}$$

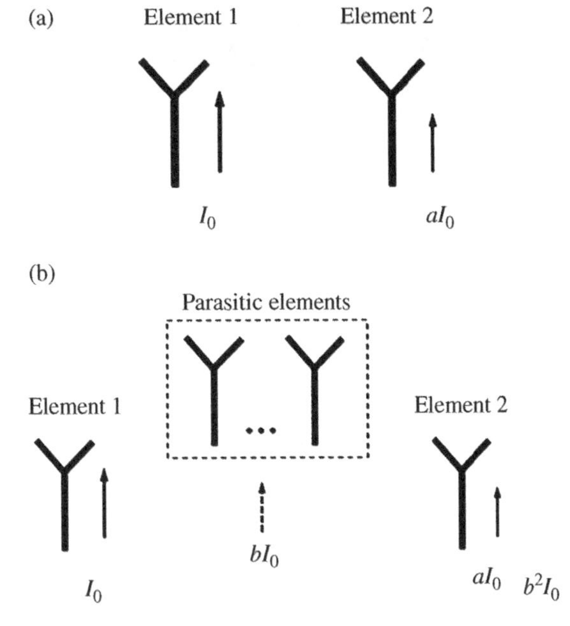

Figure 3.28 Simplified model of using parasitic elements to reduce the mutual coupling. (a) Two closely spaced antenna elements, element 1 being the driven one. (b) Parasitic elements introduced between the original ones. *Source:* From [39] / with permission of IEEE.

where $\boldsymbol{b_1}, \ldots, \boldsymbol{b_N}$ are the corresponding coupling coefficients. A coupling coefficient, $\boldsymbol{b}$, is introduced to represent the corresponding average induced current as $\boldsymbol{bI_0}$, which is convenient for considerations of the overall decoupling performance. Assuming that the set of parasitic elements is arranged symmetrically and the medium is reciprocal, a coupling current of amplitude $\boldsymbol{b^2I_0}$ is induced on Element 2 by this double-coupling path.

The mutual coupling can be tuned to be close to zero by properly designing the overall configuration of the array and parasitic elements. In particular, one can design the parasitic elements to make the two coupling coefficients $\boldsymbol{a}$ and $\boldsymbol{b}$ achieve the relation:

$$aI_0 + b^2I_0 = 0 \tag{3.5}$$

This means that the parasitic elements would induce currents through the double path which would be opposite to those created by the original path to reduce the mutual coupling level.

This approach was successfully employed in the dual-slot array shown in Figure 3.29. Two parasitic monopoles were introduced into the array to reduce the mutual coupling [39]. The array and the parasitic elements are etched and printed on an FR4 substrate board (95 mm × 60 mm). The substrate has a 0.8 mm thickness and a relative permittivity of 4.4. The array includes two symmetric slot elements, each of which is fed by a 50-Ω microstrip line whose end is shorted to ground with a via. The ground plane is printed on the bottom layer of the substrate. The parasitic monopoles were chosen to be metal strips with the same width as the feeding lines. This greatly simplified the design. A single rectangular portion was added at the one corner of each parasitic monopole to provide a means to adjust the impedance-matching level.

A plot of the simulated scattering parameters is given in Figure 3.30. These results are compared with the simulated performance of the array in the absence of the two parasitic monopoles. It was found that the mutual coupling level was greatly improved by adding the two parasitic monopoles. In particular, the $|S_{21}|$ values decreased from −8 to −20 dB. On the other hand, the operating band only shifted a small amount to lower frequencies while the bandwidth increased slightly.

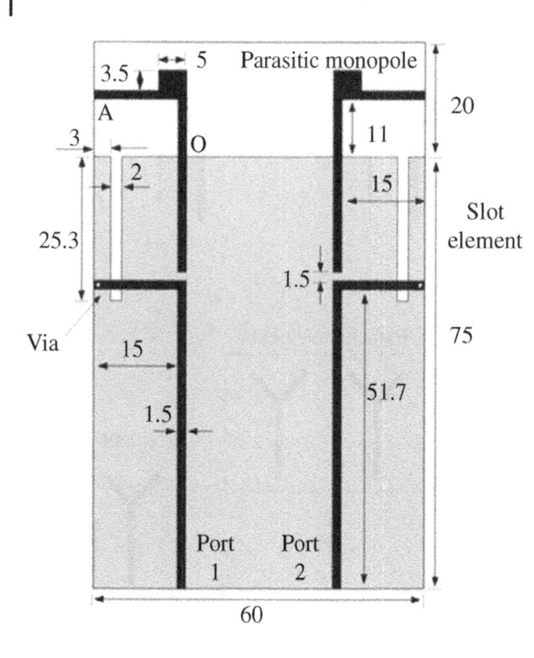

Figure 3.29 Geometry of the dual-slot-element antenna. Its dimensions are given in mm. The bold (black) line segments are the strips on the top surface of the substrate, which is the large rectangular (grey) region beneath them. The ground plane, the largest rectangular region, is beneath the substrate. *Source*: From [39] / with permission of IEEE.

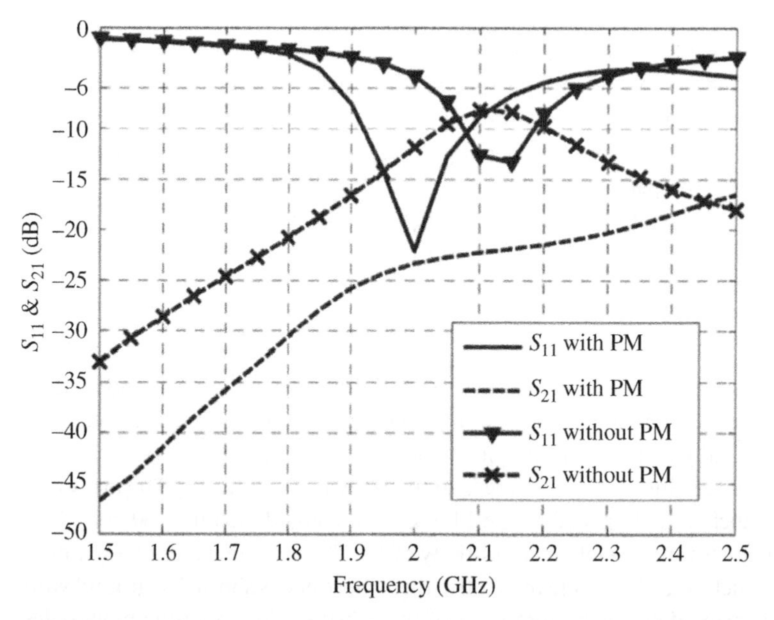

Figure 3.30 Simulated *S*-parameters as functions of the source frequency. The label "PM" parasitic-monopole. *Source*: From [39] / with permission of IEEE.

An analogous parasitic decoupling method was proposed in [44] for dual-polarized and circularly polarized (CP) high-density arrays. Resonant parasitic elements were employed to reduce the mutual coupling. Figure 3.31 shows a two-element dual-polarized array which is formed by two aperture-coupled dual-polarized antenna elements and one hybrid decoupling structure. The center-to-center distance between the two elements (No. 1 and No. 2) is 0.5 λ_0, where λ_0 indicates the

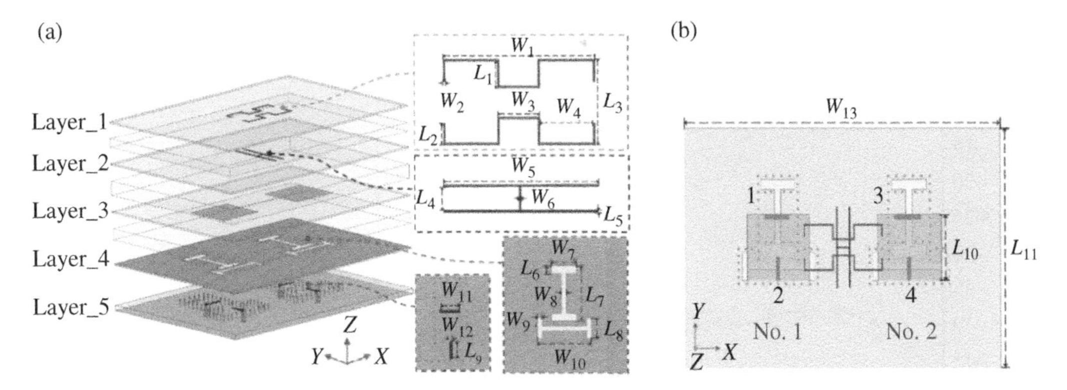

Figure 3.31 Two-element dual-polarized array loaded with the hybrid decoupling structure. (a) 3-D view of the array with its layers detached. (b) Top view of array. *Source*: From [44] / with permission of IEEE.

operational wavelength in the free space corresponding to the frequency f_0 at the lower bound of its operational bandwidth. The hybrid decoupling structure is shown in Figure 3.31a. It is comprised of one *H*-shaped strip structure and two meander lines that are placed above the array on separate dielectric layers and halfway between the two antenna elements. The hybrid decoupling structure has the advantages of being a planar configuration and being easy to fabricate.

The corresponding simulated reflection coefficients and port isolation levels for the two-element array with and without the planar hybrid decoupling structure are given in Figure 3.32. As shown in Figure 3.32a, the working band of the array (reflection coefficients <-10 dB) in both cases covers the range of 2.4–2.7 GHz. Moreover, the mutual couplings between cross-pol ports for antenna elements No. 1 ($|S_{12}|$) and No. 2 ($|S_{34}|$) are kept at very low levels, all being below -35 dB. These results demonstrate that the planar hybrid decoupling structure has little effect on the *S*-parameters of either antenna element.

The simulated isolation levels for both the co-pol ports ($|S_{13}|$ and $|S_{24}|$) and the cross-pol ($|S_{14}|$ and $|S_{23}|$) ports in the cases of the two-element array with and without the planar hybrid decoupling structure are given in Figure 3.32b. By comparing the results in the two cases, it is apparent that the isolation levels between the co-pol ports of the two-element array were improved significantly by loading it with the planar hybrid decoupling structure. In particular, the maximum value of $|S_{13}|$ ($|S_{24}|$) over the entire operational band is decreased from -14.02 (-17.20) to -25.79 dB (-26.44 dB) and thus exhibits a ~11.77 dB (~9.24 dB) improvement. Moreover, the isolation levels between the cross-pol ports ($|S_{23}|$ and $|S_{14}|$) are kept at very low levels in both cases when the array was loaded with the planar hybrid decoupling structure. The results show that a properly designed planar hybrid structure yields an excellent decoupling behavior for the compact dual-polarized array.

The individual contributions of the meander lines and the H element to the overall mitigation of the coupling were investigated as well. The surface current distributions on the traces of the two-element array loaded with only the two meander lines are shown in Figure 3.33. When ports 1 and 3 are excited with the same amplitude and phase, the two antenna elements of the array are equivalent to being oriented along their *E*-plane. The resulting field behaviors are thus dominated by electric field coupling between ports 1 and 3. It is clear that very strong out-of-phase currents are induced on the surfaces of the meander lines. Consequently, they offset the coupling currents to a large extent and yield significantly improved isolation levels between the two ports.

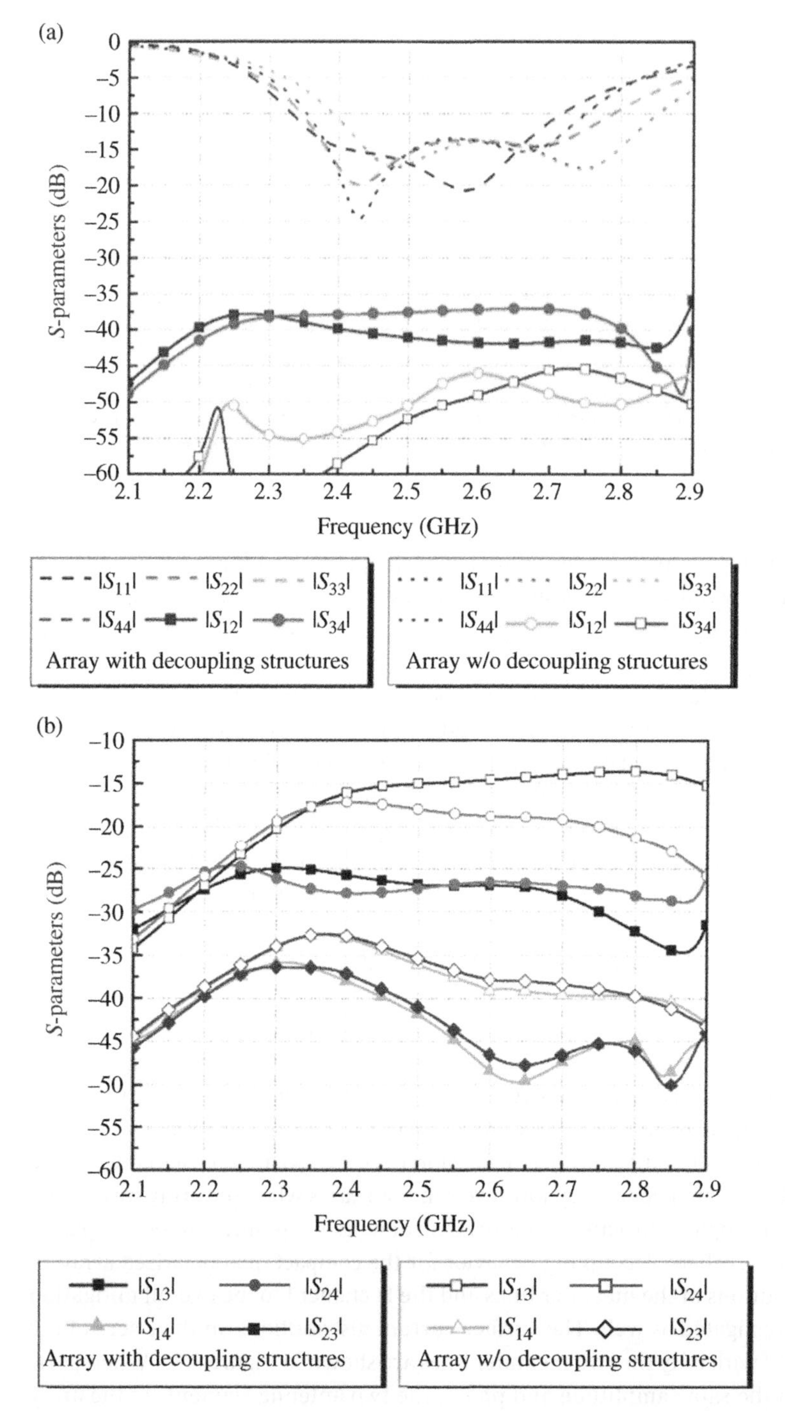

Figure 3.32 Simulated reflection coefficients and port isolation levels for the dual-polarized array with and without the planar hybrid decoupling structure as functions of the source frequency. (a) $|S_{12}|$, $|S_{34}|$, and reflection coefficient values. (b) $|S_{13}|$, $|S_{14}|$, $|S_{23}|$, and $|S_{24}|$ values. *Source*: From [44] / with permission of IEEE.

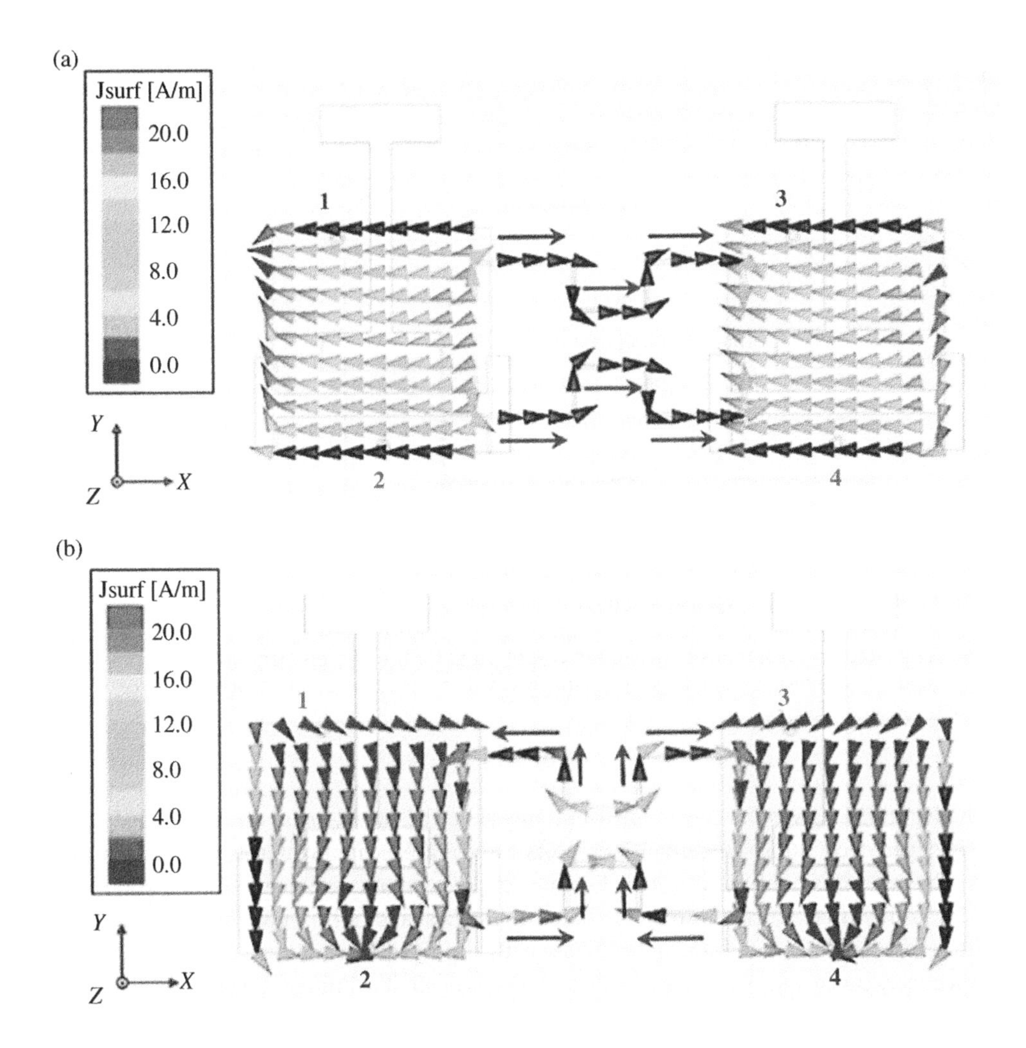

Figure 3.33 Surface current distributions of the two-element array loaded only with the two meander lines. (a) Only ports 1 and 3 are excited. (b) Only ports 2 and 4 are excited. The current polarizations on the meander lines are highlighted by the (blue) arrows. *Source*: From [44] / with permission of IEEE.

As shown in Figure 3.34, the mutual coupling level between ports 1 and 3 is reduced from −14.02 to −25.06 dB, a reduction of 11.04 dB. Note that the meander lines effectively increase the electrical length of the structure while maintaining its compact physical size for applications in high-density arrays. In contrast, when ports 2 and 4 are excited with the same amplitude and phase, the two antenna elements are equivalent to being oriented along the *H*-plane. Strong surface currents are excited on the meander lines. However, they are symmetrical with respect to their centers and are out-of-phase but with almost equal amplitude. This current behavior results in the net cancelation of their effects on the array. Thus, the meander lines are advantageous because they have a weak effect on the coupling between ports 2 and 4. As shown in Figure 3.34, the simulated $|S_{24}|$ value demonstrates little fluctuation as compared to the array without the decoupling structure.

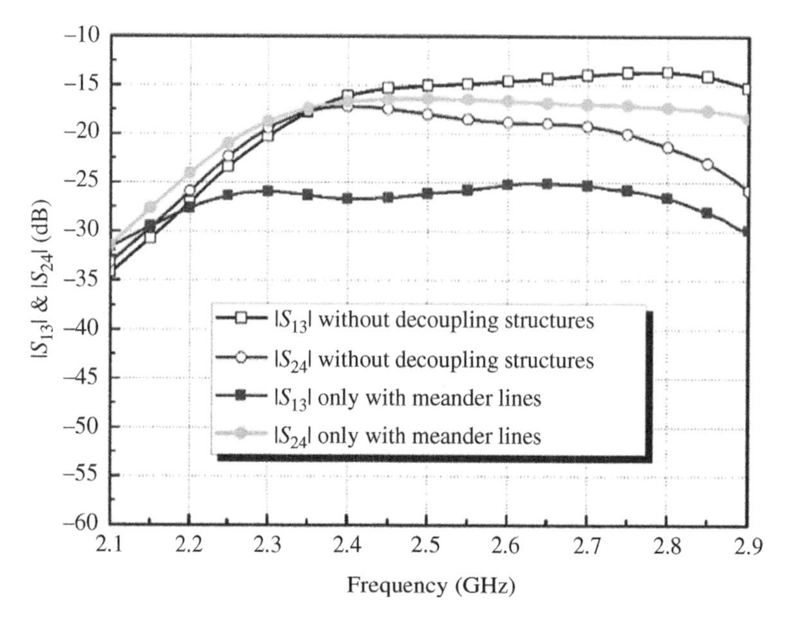

Figure 3.34 Simulated port isolations between the co-pol ports for the two-element array with only the decoupling meander lines present and the two-element array without the decoupling structure. *Source*: From [44] / with permission of IEEE.

Similarly, Figure 3.35 gives the current distributions when the two-element array is loaded only with the *H*-shaped decoupling structure. When ports 1 and 3 are excited with the same amplitude and phase, the system is equivalent to two *E*-plane-coupled antenna elements. As illustrated in Figure 3.35a, strong surface currents are then excited on the *H*-shaped strip. However, they are symmetric with respect to their centers and out-of-phase with almost equal amplitude. As a result, these currents effectively cancel out. Thus, the *H*-shaped strip has negligible influence on the mutual coupling levels between ports 1 and 3, i.e., as shown in Figure 3.36, the simulated $|S_{13}|$ values remain almost the same as those obtained from the array without the decoupling structure. On the other hand, when ports 2 and 4 are excited with the same amplitude and phase, the system is equivalent to two *H*-plane-coupled antenna elements. A strong out-of-phase current on the H-shaped strip is induced, exhibiting a dipole resonance. Thus, its scattered field induces currents that cancel out the original coupling currents and thus leads to a significant enhancement of the isolation level between ports 2 and 4. In fact, the $|S_{24}|$ values are reduced from the maximum −17.20 to −25.61 dB across the entire operating band. This result verifies the effectiveness of the H-shaped strip for mitigating the mutual coupling effects present in high-density arrays. Furthermore, it is now clear that the presence of both components of the hybrid decoupling structure is necessary to provide the overall decoupling improvements demonstrated in Figure 3.32 for all of the ports.

In order to further explore its potential for a wider range of array applications, the proposed hybrid decoupling structure is applied to the high-density two-element CP array shown in Figure 3.37 to demonstrate its insensitivity to various polarization states. The CP elements are placed with a center-to-center separation of $0.5\,\lambda_0$. The proposed decoupling structure is placed halfway between the two CP elements, similar to the dual-polarized array configuration in Figure 3.31. The simulated S-parameter results of the optimized CP array with and without the planar hybrid structure are presented in Figure 3.38. They clearly show that both systems exhibit good impedance

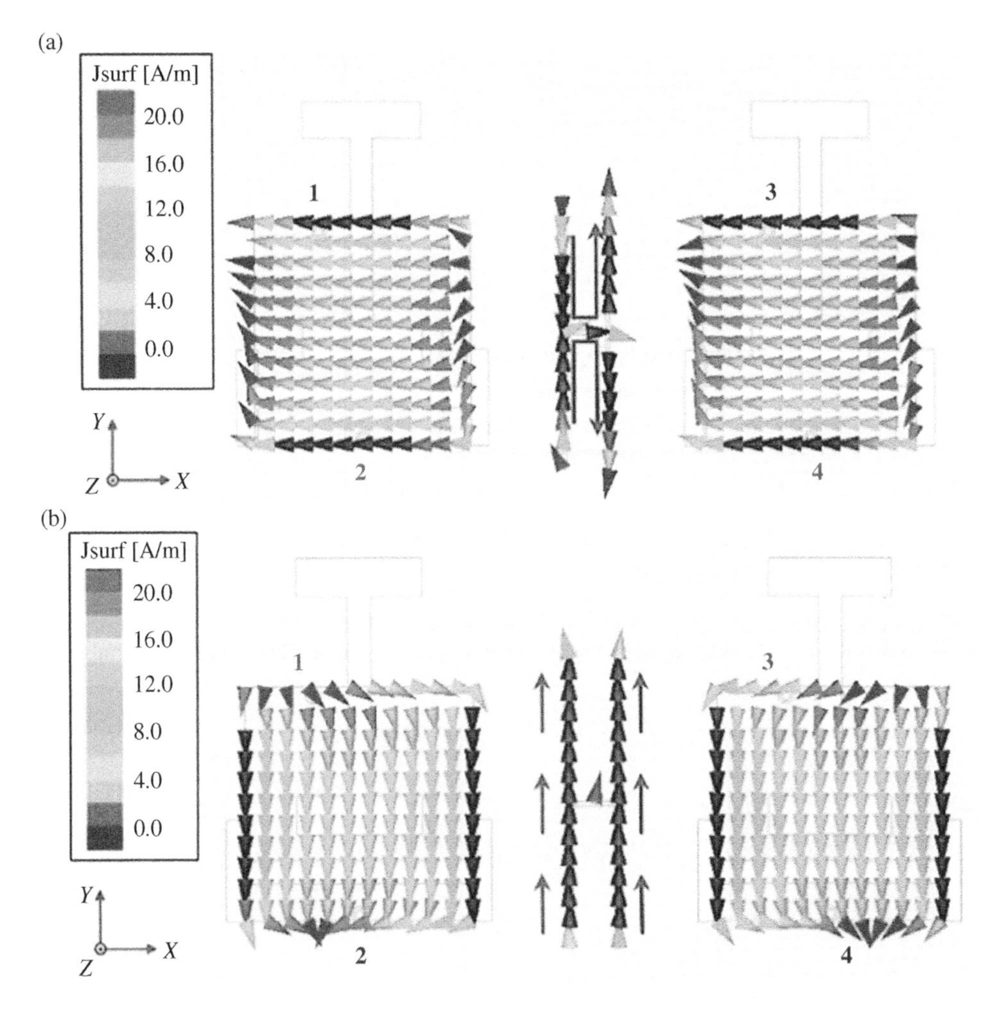

Figure 3.35 Surface current distributions of the two-element array loaded only with the H-shaped strip. (a) Only ports 1 and 3 are excited. (b) Only ports 2 and 4 are excited. (The current directions on the *H*-shaped strip are highlighted by the (red) arrows.) *Source*: From [44] / with permission of IEEE.

matching. The operating band (reflection coefficients <−10 dB) covers the range from 2.4 to 2.7 GHz. Significant enhancement of the isolation level between the two CP elements results from the presence of the planar hybrid decoupling structure. Notably, $|S_{12}|$ decreases from −16.5 to −29.2 dB, which is a ~12.7 dB reduction. The simulated AR values for the arrays with and without the planar hybrid decoupling structure are given in Figure 3.39. The array augmented with the hybrid decoupler maintains the original excellent CP characteristics (AR < 3 dB) over the entire operating band.

The surface current distributions of the two-element CP array loaded with the planar hybrid structure are shown in Figure 3.40. When ports 1 and 2 are excited with the same amplitude and phase, one observes that strong currents are also induced on the surface of the decoupling structure. These currents and their indicated orientations show that they are strongly induced on both the meander line and *H* elements over an entire period of the source frequency. However, notice that at the fixed phase $\pi/4$, only the meander lines contribute to the mutual coupling

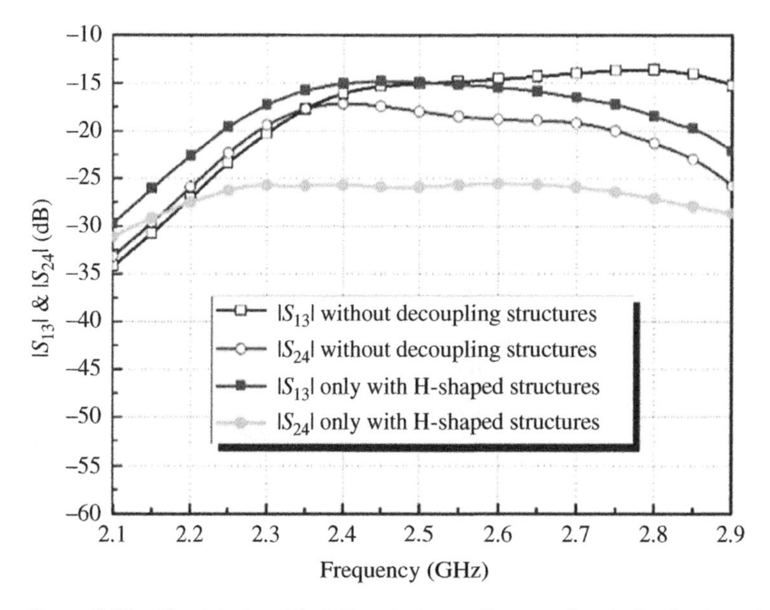

Figure 3.36 Simulated port isolations between the co-pol ports for the two-element array with only the *H*-shaped strip and the two-element array without any decoupling structure. *Source*: From [44] / with permission of IEEE.

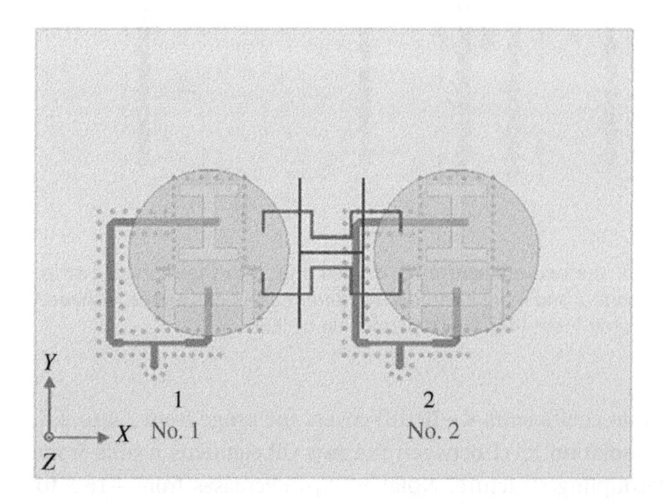

Figure 3.37 Two-element CP array loaded with the planar hybrid decoupling structure. *Source*: From [44] / with permission of IEEE.

reduction. This behavior is illustrated in Figure 3.33. At the fixed phase $\pi/2$, only the *H*-shaped strip contributes to the mutual reduction. This behavior is illustrated in Figure 3.35. Both the *H*-shaped strip and the meander lines could be combined to reduce the mutual coupling. This phenomenon indicates that the decoupling of the CP array is in fact the consequence of the collaboration of the two types of decoupling structures, which empowered the resulting hybrid decoupling structure to possess the unique advantage of polarization insensitivity.

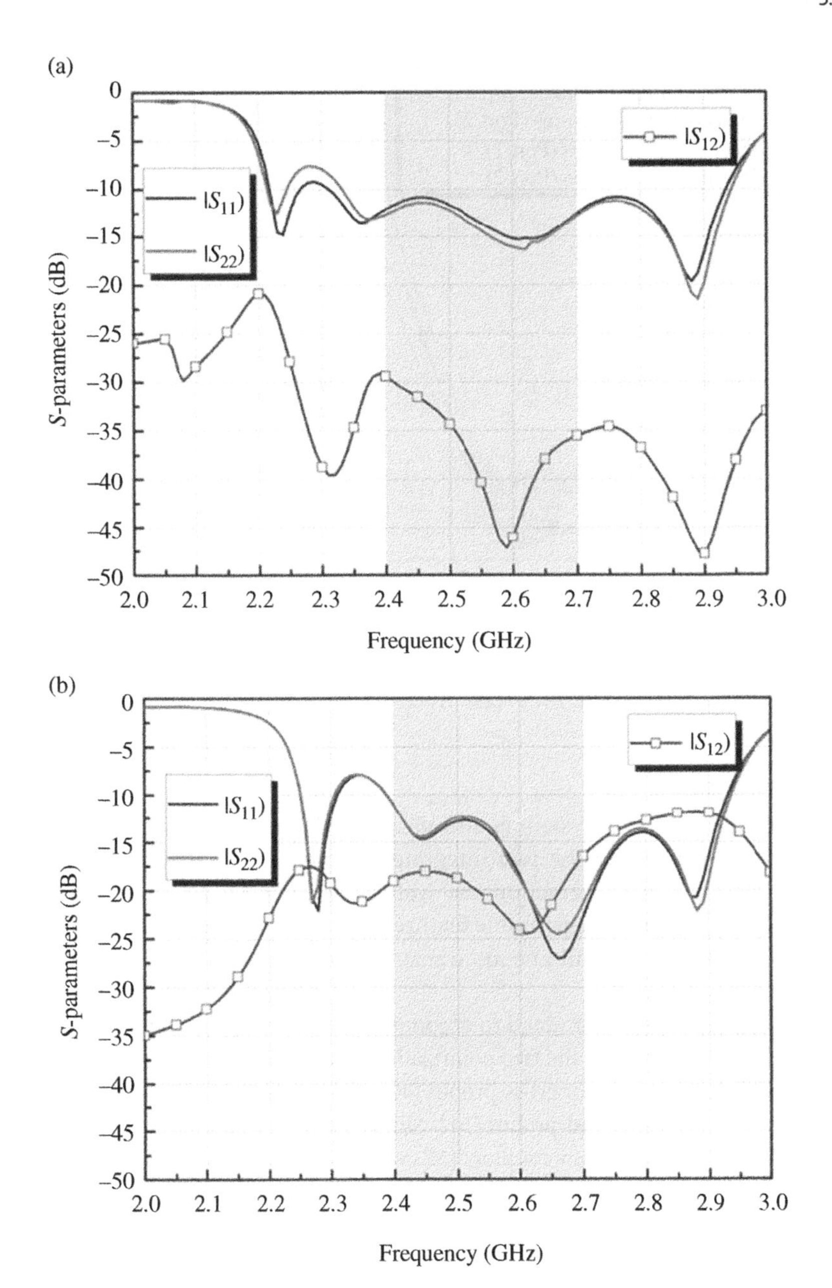

Figure 3.38 Simulated reflection coefficients and port isolation levels of the CP array as functions of the source frequency. (a) With the planar hybrid decoupling structure. (b) Without the planar hybrid decoupling structure. *Source*: From [44] / with permission of IEEE.

3.7 Polarization Decoupling

Another form of decoupling occurs naturally from the electromagnetics theory point-of-view, i.e., two orthogonal polarization states. Consider an antenna with two ports that radiate orthogonal polarization states, one associated with each port. High port isolation between

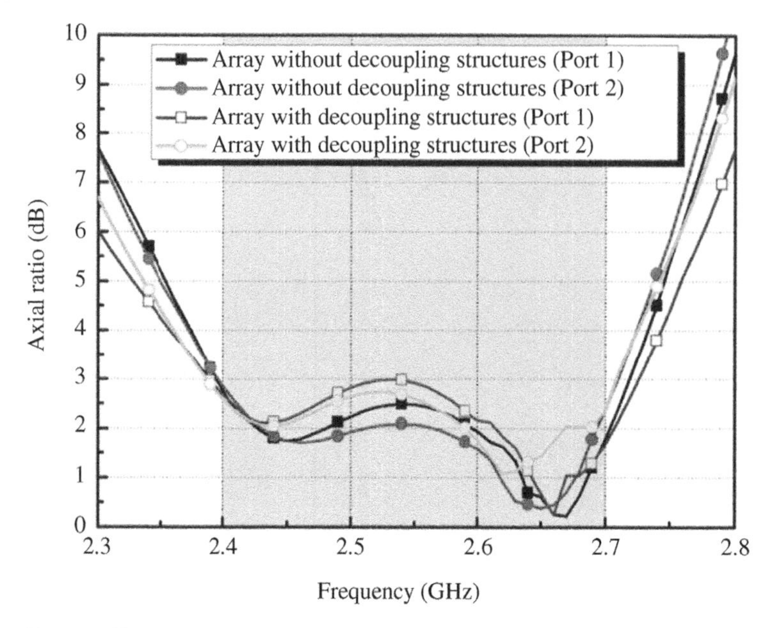

Figure 3.39 Simulated AR values as functions of the source frequency for the two-element CP array with and without the decoupling structures. *Source*: From [44] / with permission of IEEE.

each element in a two-port dual-polarized system is an indispensable feature. However, it is difficult to achieve high isolation between the two integrated elements in practice because of proximity current and surface wave interactions as well as space-wave co- and cross-polarization effects. Nevertheless, it is an indispensable feature for many applications requiring more functionality in small footprints. Simultaneous transmit and receive (STAR) systems are practical examples.

Numerous effective structures have been examined to accomplish polarization decoupling, i.e., high isolation between the elements achieving the two polarization states. Examples include aperture-coupled feed networks [45, 46], meander or cross-probes [47, 48], coupled feed networks [49, 50], and hybrid feed networks of apertures and probes [51]. Similarly, high isolation in dual-CP designs has been reported. These include hybrid couplers [52–54], even-odd mode feed networks [55], loading modified slot feeding structure below a meta-surface [56], strategic placements of inverted L-shaped grounded strips [57], and decoupling networks [58].

Achieving polarization decoupling between collocated antennas is particularly difficult when the overall volume of the system is required to be small. Properly codesigned feed and distinctly different radiating elements achieved high two-port isolation in [59]. Clever codesigned arrangements of both the driven and near-field resonant parasitic (NFRP) radiating elements have led to demonstrated high isolation between the ports in two-LP [60], two-CP [61], and two-function [62] electrically small Huygens dipole antennas. The latter seamlessly integrated together a wireless power transfer (WPT) element with one LP state and a communications element with the orthogonal LP state with very effective polarization decoupling.

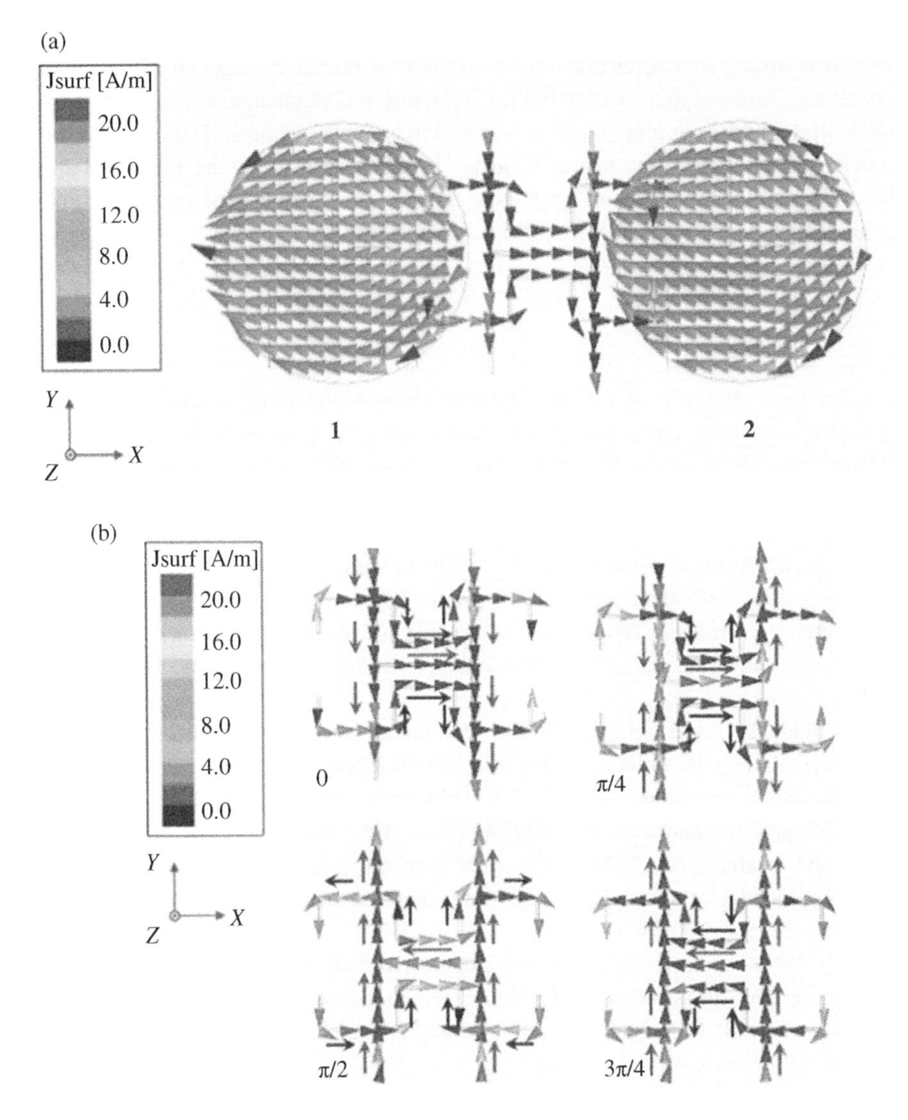

Figure 3.40 Surface current distributions. (a) Two-element CP array loaded with the planar hybrid decoupling structure. (b) The planar hybrid decoupling structure alone for different excitation phases. The current orientations are highlighted with arrows on the meander lines (blue) and the *H*-shaped strip (red). *Source*: From [44] / with permission of IEEE.

3.8 Conclusions

With the desire for high-density arrays for 5G and beyond wireless ecosystems, the need to develop simpler yet very effective decoupling structures remains. A large variety of examples that have demonstrated their decoupling effectiveness were reviewed in this chapter. There is not one method to date that is vastly superior to all the others. Consequently, decoupling structures for general and densely packed arrays will continue to be a prime area for active research efforts. It is clear that one must tailor the decoupling structures to the types of array being developed and to their radiation characteristics. The decoupling structures must be able to handle either the associated surface or space-wave coupling mechanisms and usually both. Computational electromagnetics software is extremely useful for studying the coupling mechanisms and to analyze simple and complex decoupling structures to improve the performance of an array. With modern tools, the simulated and measured results are generally in very good agreement. Hence, they provide the means to thoroughly optimize an array's performance before fabrication and testing.

References

1. Sievenpiper, D., Zhang, L., Broas, R. et al. (1999). High-impedance electromagnetic surfaces with a forbidden frequency band. *IEEE Trans. Microwave Theory Tech.* **47** (11): 2059–2074.
2. Yang, F. and Rahmat-Samii, Y. (2003). Microstrip antennas integrated with electromagnetic bandgap (EBG) structures: a low mutual coupling design for array applications. *IEEE Trans. Antennas Propag.* **51** (10): 2936–2946.
3. Farahani, H.S., Veysi, M., Kamyab, M., and Tadjalli, A. (2010). Mutual coupling reduction in patch antenna arrays using a UC-EBG superstrate. *IEEE Antennas Wirel. Propag. Lett.* **9**: 57–59.
4. Expósito-Domínguez, G., Fernández-Gonzalez, J.-M., Padilla, P., and Sierra-Castañer, M. (2012). Mutual coupling reduction using EBG in steering antennas. *IEEE Antennas Wirel. Propag. Lett.* **11**: 1265–1268.
5. Al-Hasan, M.J., Denidni, T.A., and Sebak, A.R. (2015). Millimeter-wave compact EBG structure for mutual coupling reduction applications. *IEEE Trans. Antennas Propag.* **63** (2): 823–828.
6. Yang, X., Liu, Y., Xu, Y.-X., and Gong, S.-X. (2017). Isolation enhancement in patch antenna array with fractal UC-EBG structure and cross slot. *IEEE Antennas Wirel. Propag. Lett.* **16**: 2175–2178.
7. Mohamadzade, B. and Afsahi, M. (2017). Mutual coupling reduction and gain enhancement in patch array antenna using a planar compact electromagnetic bandgap structure. *IET Microw. Antennas Propag.* **11** (12): 1719–1725.
8. Tan, X., Wang, W., Wu, Y. et al. (2019). Enhancing isolation in dual-band meander-line multiple antenna by employing split EBG structure. *IEEE Trans. Antennas Propag.* **67** (4): 2769–2774.
9. Wong, H., Lau, K.-L., and Luk, K.-M. (2004). Design of dual-polarized L-probe patch antenna arrays with high isolation. *IEEE Trans. Antennas Propag.* **52** (1): 45–52.
10. Chiu, C.-Y., Cheng, C.-H., Murch, R.D., and Rowell, C.R. (2007). Reduction of mutual coupling between closely-packed antenna elements. *IEEE Trans. Antennas Propag.* **55** (6): 1732–1738.
11. Xiao, S., Tang, M.-C., Bai, Y.-Y. et al. (2011). Mutual coupling suppression in microstrip array using defected ground structure. *IET Microw. Antennas Propag.* **5** (12): 1488–1494.
12. Tang, M.-C., Xiao, S., Li, C.-J. et al. (2011). Scan blindness elimination using composite defected ground structures and edge-coupled split ring resonator. *Appl. Comput. Electromagn. Soc. J.* **26** (7): 572–583.
13. Zhang, S., Lau, B.K., Tan, Y. et al. (2012). Mutual coupling reduction of two PIFAs with a T-shape slot impedance transformer for MIMO mobile terminals. *IEEE Trans. Antennas Propag.* **60** (3): 1521–1531.
14. Wei, K., Li, J.-Y., Wang, L. et al. (2016). Mutual coupling reduction by novel fractal defected ground structure bandgap filter. *IEEE Trans. Antennas Propag.* **64** (10): 4328–4335.
15. Hwangbo, S., Yang, H.Y., and Yoon, Y.-K. (2017). Mutual coupling reduction using micromachined complementary meander-line slots for a patch array antenna. *IEEE Antennas Wirel. Propag. Lett.* **16**: 1667–1670.

16. Diallo, A., Luxey, C., Thuc, P. et al. (2006). Study and reduction of the mutual coupling between two mobile phone PIFAs operating in the DCS1800 and UMTS bands. *IEEE Trans. Antennas Propag.* **54** (11): 3063–3074.
17. Chebihi, A., Luxey, C., Diallo, A. et al. (2008). A novel isolation technique for closely spaced PIFAs for UMTS mobile phones. *IEEE Antennas Wirel. Propag. Lett.* **7**: 665–668.
18. Ling, X. and Li, R. (2011). A novel dual-band MIMO antenna array with low mutual coupling for portable wireless devices. *IEEE Antennas Wirel. Propag. Lett.* **10**: 1039–1042.
19. Li, L., Huo, F., Jia, Z., and Han, W. (2013). Dual zeroth-order resonance antennas with low mutual coupling for MIMO communications. *IEEE Antennas Wirel. Propag. Lett.* **12**: 1692–1695.
20. Ban, Y.-L., Chen, Z.-X., Chen, Z. et al. (2014). Decoupled hepta-band antenna array for WWAN/LTE smartphone applications. *IEEE Antennas Wirel. Propag. Lett.* **13**: 999–1002.
21. Wang, Y. and Du, Z. (2014). A wideband printed dual-antenna with three neutralization lines for mobile terminals. *IEEE Trans. Antennas Propag.* **62** (3): 1495–1500.
22. Zhang, S. and Pedersen, G.F. (2016). Mutual coupling reduction for UWB MIMO antennas with a wideband neutralization line. *IEEE Antennas Wirel. Propag. Lett.* **15**: 166–169.
23. Amjadi, S.M. and Sarabandi, K. (2016). Mutual coupling mitigation in broadband multiple-antenna communication systems using feedforward technique. *IEEE Trans. Antennas Propag.* **64** (5): 1642–1652.
24. Wu, K.-L., Wei, C., Mei, X., and Zhang, Z.-Y. (2017). Array-antenna decoupling surface. *IEEE Trans. Antennas Propag.* **65** (12): 6728–6738.
25. Niu, Z., Zhang, H., Chen, Q., and Zhong, T. (2019). Isolation enhancement in closely coupled dual-band MIMO patch antennas. *IEEE Antennas Wirel. Propag. Lett.* **18** (8): 1686–1690.
26. Engheta, N. and Ziolkowski, R.W. (eds.) (2006). *Metamaterials: Physics and Engineering Explorations*. New York: Wiley.
27. Tang, M.-C., Xiao, S.-Q., Guan, J. et al. (2010). Composite metamaterial enabled excellent performance of microstrip antenna array. *Chin. Phys. B* **19** (7): 074214.
28. Tang, M.-C., Xiao, S., Deng, T. et al. (2010). Novel folded single split ring resonator and its application to eliminate scan blindness in infinite phased array. *2010 International Symposium on Signals, Systems and Electronics* (*ISSSE2010*), Nanjing China, 17–20 September 2010.
29. Bait-Suwailam, M.M., Boybay, M., and Ramahi, O.M. (2010). Electromagnetic coupling reduction in high-profile monopole antennas using single-negative magnetic metamaterials for MIMO applications. *IEEE Trans. Antennas Propag.* **58** (9): 2894–2902.
30. Tang, M.-C., Xiao, S., Wang, B. et al. (2011). Improved performance of a microstrip phased array using broadband and ultra-low-loss metamaterial slabs. *IEEE Antennas Propag. Mag.* **53** (6): 31–41.
31. Yang, X.M., Liu, X.G., Zhou, X.Y., and Cui, T.J. (2012). Reduction of mutual coupling between closely packed patch antennas using waveguided metamaterials. *IEEE Antennas Wirel. Propag. Lett.* **11**: 389–391.
32. Qamar, Z., Naeem, U., Khan, S.A. et al. (2016). Mutual coupling reduction for high-performance densely packed patch antenna arrays on finite substrate. *IEEE Trans. Antennas Propag.* **64** (5): 1653–1660.
33. Tang, M.-C., Chen, Z., Wang, H. et al. (2017). Mutual coupling reduction using meta-structures for wideband, dual-polarized, and high-density patch arrays. *IEEE Trans. Antennas Propag.* **65** (8): 3986–3998.
34. Jafargholi, A., Jafargholi, A., and Choi, J.H. (2019). Mutual coupling reduction in an array of patch antennas using CLL metamaterial superstrate for MIMO applications. *IEEE Trans. Antennas Propag.* **67** (1): 179–189.

35. Yang, F.-m., Peng, L., Liao, X. et al. (2019). Coupling reduction for a wideband circularly polarized conformal array antenna with a single-negative structure. *IEEE Antennas Wirel. Propag. Lett.* **18** (5): 991–995.
36. Ghosh, J., Mitra, D., and Das, S. (2019). Mutual coupling reduction of slot antenna array by controlling surface wave propagation. *IEEE Trans. Antennas Propag.* **67** (2): 1352–1357.
37. Tang, M.-C., Xiao, S., Deng, T., and Wang, B. (2010). Parasitic patch of the same dimensions enabled excellent performance of microstrip antenna array. *Appl. Comput. Electromagn. Soc. J.* **25** (10): 862–866.
38. Lau, B.K. and Andersen, J.B. (2012). Simple and efficient decoupling of compact arrays with parasitic scatterers. *IEEE Trans. Antennas Propag.* **60** (2): 464–472.
39. Li, Z., Du, Z., Takahashi, M. et al. (2012). Reducing mutual coupling of MIMO antennas with parasitic elements for mobile terminals. *IEEE Trans. Antennas Propag.* **60** (2): 473–481.
40. Vishvaksenan, K.S., Mithra, K., Kalaiarasan, R., and Raj, K.S. (2017). Mutual coupling reduction in microstrip patch antenna arrays using parallel coupled-line resonators. *IEEE Antennas Wirel. Propag. Lett.* **16**: 2146–2149.
41. Sun, X. and Cao, M.Y. (2017). Low mutual coupling antenna array for WLAN application. *Electron. Lett.* **53** (6): 368–370.
42. Xun, J.H., Shi, L.F., Liu, W.R. et al. (2017). Compact dual-band decoupling structure for improving mutual coupling of closely placed PIFAs. *IEEE Antennas Wirel. Propag. Lett.* **16**: 1985–1989.
43. Dhevi, B.L., Vishvaksenan, K.S., and Rajakani, K. (2018). Isolation enhancement in dual-band microstrip antenna array using asymmetric loop resonator. *IEEE Antennas Wirel. Propag. Lett.* **17**: 238–241.
44. Chen, Z., Li, M., Liu, G. et al. (2019). Isolation enhancement for wideband, circularly/dual-polarized, high-density patch arrays using planar parasitic resonators. *IEEE Access* **7**: 112249–112257.
45. Barba, M. (2008). A high-isolation, wideband and dual-linear polarization patch antenna. *IEEE Trans. Antennas Propag.* **56** (5): 1472–1476.
46. Chiou, T.-W. and Wong, K.-L. (2002). Broad-band dual-polarized single microstrip patch antenna with high isolation and low cross polarization. *IEEE Trans. Antennas Propag.* **50** (3): 399–401.
47. Lai, H.W. and Luk, K.M. (2007). Dual polarized patch antenna fed by meandering probes. *IEEE Trans. Antennas Propag.* **55** (9): 2625–2627.
48. Bao, Z., Nie, Z., and Zong, X. (2014). A novel broadband dual-polarization antenna utilizing strong mutual coupling. *IEEE Trans. Antennas Propag.* **62** (1): 450–454.
49. Wong, H., Lau, K.L., and Luk, K.M. (2004). Design of dual-polarized L-probe patch antenna arrays with high isolation. *IEEE Trans. Antennas Propag.* **52** (1): 45–52.
50. Mak, K.-M., Lai, H.-W, and Luk, K.-M. (2012). Wideband dual polarized antenna fed by L- and M-probe. *Proceedings of Asia-Pacific Microwave Conference*, Kaohsiung, Taiwan, 4–7 December 2012, pp. 1058–1060.
51. Sim, C.Y.D., Chang, C.C., and Row, J.S. (2009). Dual-feed dual-polarized patch antenna with low cross polarization and high isolation. *IEEE Trans. Antennas Propag.* **57** (10): 3405–3409.
52. Ferreira, R., Joubert, J., and Odendaal, J.W. (2017). A compact dual-circularly polarized cavity-backed ring-slot antenna. *IEEE Trans. Antennas Propag.* **65** (1): 364–368.
53. Yang, Y.-H., Sun, B.-H., and Guo, J.-L. (2019). A low-cost, single-layer, dual circularly polarized antenna for millimeter-wave applications. *IEEE Antennas Wirel. Propag. Lett.* **18**: 651–655.
54. Wang, A., Yang, L., Zhang, Y. et al. (2019). A novel planar dual circularly polarized endfire antenna. *IEEE Access* **7**: 64297–64302.
55. Narbudowicz, A., Bao, X., and Ammann, M.J. (2013). Dual circularly-polarized patch antenna using even and odd feed-line modes. *IEEE Trans. Antennas Propag.* **61** (9): 4828–4831.

56. Liu, S., Yang, D., and Pan, J. (2019). A low-profile broadband dual-circularly-polarized metasurface antenna. *IEEE Antennas Wirel. Propag. Lett.* **18**: 1395–1399.
57. Saini, R.K. and Dwari, S. (2016). A broadband dual circularly polarized square slot antenna. *IEEE Trans. Antennas Propag.* **64** (1): 290–294.
58. Lu, L., Jiao, Y.-C., Liang, W., and Zhang, H. (2016). A novel low-profile dual circularly polarized dielectric resonator antenna. *IEEE Trans. Antennas Propag.* **64** (9): 4078–4083.
59. Yetisir, E., Chen, C., and Volakis, J.L. (2016). Wideband low profile multiport antenna with omnidirectional pattern and high isolation. *IEEE Trans. Antennas Propag.* **64** (9): 3777–3786.
60. Tang, M.-C., Wu, Z., Shi, T. et al. (2018). Dual-linearly-polarized, electrically small, low-profile, broadside radiating, Huygens dipole antenna. *IEEE Trans. Antennas Propag.* **66** (8): 3877–3885.
61. Wu, Z., Tang, M.-C., Shi, T., and Ziolkowski, R.W. (2021). Two-port, dual-circularly polarized, low-profile broadside-radiating electrically small Huygens dipole antenna. *IEEE Trans. Antennas Propag.* **69** (1): 514–519 https://doi.org/10.1109/TAP.2020.2999747.
62. Lin, W. and Ziolkowski, R.W. (2019). Electrically small Huygens antenna-based fully-integrated wireless power transfer and communication system. *IEEE Access* **7**: 39762–39769.

4

De-scattering Methods for Coexistent Antenna Arrays

As more and more 4G, 5G, and 6G systems will be supported by a common wireless communications platform, antenna real estate on it will become more and more scarce. Consequently, the number of antenna elements on any platform, being either a mobile terminal or base station, will continue to increase. Correspondingly, the spacing between antenna elements servicing either the same bands or different bands will become smaller. This smaller distance typically results in two problems, namely strong mutual coupling and distortion of the radiation patterns of the elements due to mutual scattering. In Chapter 3, we discussed various methods to achieve effective decoupling between antenna elements operating in the same band with the aim to reduce mutual coupling. In this chapter, we present a number of approaches for de-scattering with the aim to maintain the integrity of element radiation patterns in multiband arrays.

Before diving into the details, we would like to simply emphasize the following facts. Assume two antenna elements are placed close to each other. When excited, the fields they radiate will induce currents on each other. Two effects then arise. First, they will no longer be well matched to their sources even if they were well matched as standalone radiators. Second, the radiation patterns of the two elements will become significantly different from those when they were standalone radiators. These effects are particularly serious in multiple-input-multiple-output (MIMO) systems where different antenna elements may be dedicated to different signal transmissions/receptions, and the correlations between the signals radiated by adjacent antennas need to be kept to a minimum. While one can solve the basic coupling problem by using the various decoupling methods described in Chapter 3, most decoupling methods reported to date only address impedance-matching issues. They do not fundamentally address the problem of pattern distortion. On the other hand, if one can somehow remove the scattering associated with the second antenna element when the first one is excited, and vice versa, then the presence of an adjacent antenna would not contribute any detrimental scattering effects to the array's performance. In this chapter, we shall present a number of such "de-scattering" techniques. Since the mobile communications industry has rather stringent requirements on the radiation patterns produced by base station antennas in order to provide adequate coverage, these de-scattering methods are expected to become invaluable for future antenna array designs.

4.1 De-scattering vs. Decoupling in Coexistent Antenna Arrays

It is well known that electromagnetic interference exists among antennas placed in close proximity. This electromagnetic interference causes high correlations among the signals received by the different antennas. It introduces unacceptable distortions of their radiation patterns and degradation

Advanced Antenna Array Engineering for 6G and Beyond Wireless Communications, First Edition.
Y. Jay Guo and Richard W. Ziolkowski.

of their impedance-matching performance. The net result is a complete deterioration of the overall performance of a communication network. Negative effects include a reduction of the data rate, channel capacity, and coverage [1]. Reducing unwanted electromagnetic interference is one of the most significant challenges faced by antenna designers. To better address this scientific and engineering issue, one needs to recognize that there are two different types of electromagnetic inferences involved and each needs to be addressed systematically but differently.

1) Definitions of antenna mutual coupling and scattering

The interference effects between two antennas can be classified into two categories based on the fundamental physical mechanisms producing them, i.e., the **mutual coupling** and **scattering**. Consider Figure 4.1. Assume that Antenna 1 is well-designed, lossless, and radiates a symmetric pattern. When it is radiating by itself (left subplot), the majority of the input power ($P_{1\text{fwd}}$) is radiated ($P_{1\text{rad}}$) and only a small portion is reflected back ($P_{1\text{ref}}$) toward the source, i.e., the impedance mismatch is small. However, when another antenna (Antenna 2) is placed in close proximity to it (right subplot), these two antennas are coupled. On one hand, a portion of the electromagnetic power radiated by Antenna 1 will be coupled to Antenna 2 (P_c) via a space wave and currents on the ground plane if one is present. A portion of P_c will then pass into the feed port of Antenna 2 ($P_{2\text{ref}}$) and lead to mutual coupling between the two antennas. As discussed in Chapter 3, antenna decoupling refers to the techniques/approaches introduced to reduce this transfer of power between two antenna elements, i.e., to minimize $P_{2\text{ref}}/P_{1\text{fwd}}$.

On the other hand, a portion of the coupled power (P_c) will be reradiated by Antenna 2 ($P_{2\text{rad}}$), i.e., it will scatter the field incident on it. The fields radiated by both Antenna 1 and Antenna 2 then combine and generally lead to an overall distorted radiation pattern and reduced antenna gain, as depicted in Figure 4.1 (right). In the same manner, a portion of the field scattered by Antenna 2 will be recaptured by Antenna 1. Some of this power will propagate into the feed port of Antenna 1, contributing to the reflected power ($P'_{1\text{ref}}$), i.e., it will change the input impedance and thus degrade the impedance match of Antenna 1 to its source. **De-scattering** refers to the process that attempts to restore the radiation pattern and input impedance of Antenna 1 to their condition when Antenna 2 was not present.

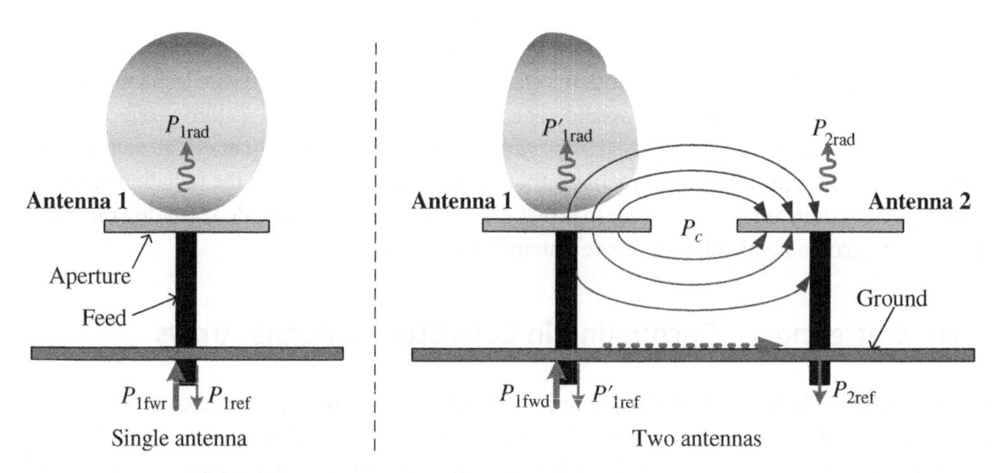

Figure 4.1 The schematic power flow when exciting Antenna 1 in two cases. (Left) Antenna 1 stands alone. (Right) Antenna 1 is placed in close proximity to Antenna 2.

2) Why is de-scattering important?

It is now clear that both mutual coupling and scattering naturally exist in any antenna system employing more than one antenna element. However, in comparison to the long-standing and intensive interest on antenna decoupling, research on antenna de-scattering has been limited and the problem has not been addressed well. The main reason is that the main scattering effect, i.e. distortion of the radiation pattern by adjacent antenna elements, was not a big issue in the early days of wireless systems and networks. As shown in Figure 4.2a, early communication systems employed arrays of one basic form, i.e., an antenna array consisting of periodic distribution of one type of radiator. All of its elements worked together simultaneously to obtain a desired radiation pattern, which is mainly determined by the array factor and the element pattern. Any distortion in the radiation pattern of an individual element due to scattering was easily compensated by adjusting the weighting factors of all the elements in the array. When multiple arrays were needed for network coverage, they were generally spatially isolated from each other. Thus, the scattering effects between the different arrays were quite weak. In contrast, modern wireless platforms contain many antenna elements and multiple antenna arrays operating in the presence of significantly complex host systems. These changes have arisen, for instance, due to economic and environmental pressures associated with cellular operators having extreme difficulties in acquiring new antenna sites. Basically, mobile communication technologies have been forced to collocate antennas in the same antenna panels. Their densely packed panels have been enabled by the continuous pursuit of antenna miniaturization and multi-functionalities enabled, for example, by reconfiguration.

Figure 4.2b shows a schematic diagram of a typical 3G/4G base station antenna array currently deployed in Australia and overseas. It hosts two kinds of arrays in an interleaved configuration within one cover (radome). One array operates at a lower frequency band, e.g., 698–960 MHz; the other at a higher frequency band, e.g., 1.71–2.69 GHz. The two arrays work independently; each array needs to have a stable and symmetric radiation pattern across its operational band. The close proximity of the low-band (LB) and high-band (HB) antenna elements in this system can cause severe, unacceptable scattering effects. Note that both the LB and HB antennas will impose

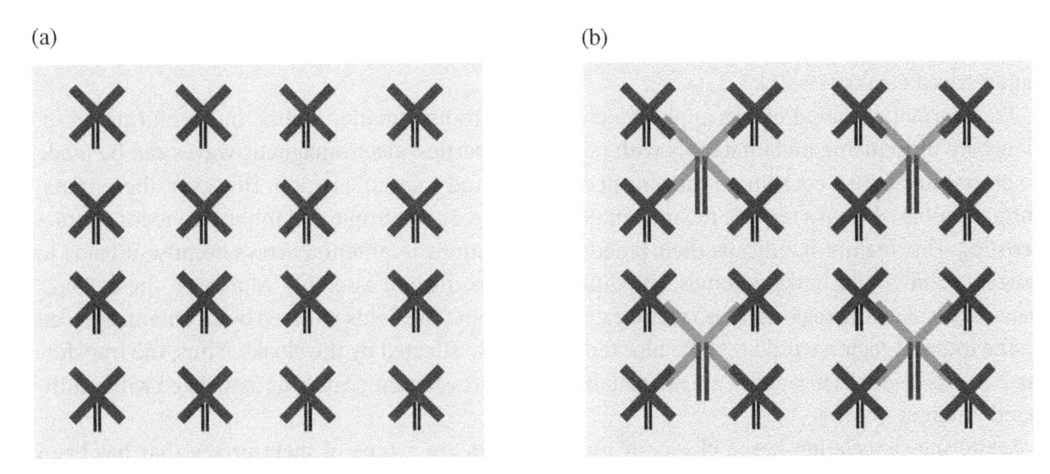

Figure 4.2 The schematic configuration of Antenna Array 1 in two cases. (a) Antenna Array 1 stands alone. (b) Antenna Array 1 (high band system) is closely packed with Antenna Array 2 (low band system).

detrimental scattering effects on each other. Owing to their size difference, the influence of the LB antennas on the HB antennas is relatively stronger while the HB antennas' influence on the LB antennas is relatively weaker. This type of dual-band base station system needs de-scattering techniques to empower its collocated antenna arrays to achieve satisfactory performance characteristics.

3) Antenna de-scattering methods

As discussed in Chapter 3, many technologies are available to reduce mutual coupling, including decoupling networks [2], parasitic elements [3], neutralization lines [4], defected ground structures [5], and metamaterial structures [6]. However, these techniques only aim to reduce the transfer of power between the antenna elements, and most of them do not have enough ability to mitigate the radiation pattern distortion due to the scattering. To suppress this unwanted scattering, one needs to make the scatterers "invisible." To endow invisibility to the scatterers, different cloaking methods based on transformation optics [7, 8] and scattering cancelation [9, 10] have been demonstrated theoretically and experimentally recently. They are able to suppress the scattering within a limited bandwidth.

There have also been intense research efforts to develop filtering antennas, i.e., filtennas, which intrinsically combine filters with antennas and thus have both radiating and filtering functions. Although most of the filtering antennas have focused on the decoupling between antennas [11, 12], a few choke-based techniques have been reported [13–15] that are analogous to the filtenna concepts and that are able to suppress the crossband scattering.

These currently available de-scattering techniques will be discussed in detail in the following sections.

4.2 Mantle Cloak De-scattering

The use of electromagnetic metamaterials, i.e., artificial materials with custom-designed properties, has stimulated much research enthusiasm in recent years. De-scattering in a multi-antenna platform scenario could be attained by introducing an "invisibility" cloak around each radiating element to minimize the overall electromagnetic scattering. Among the various cloaking methods, there are two major categories: the transformation-based cloaks [16–18] and the scattering-cancelation-based cloaks [19–21].

Transformation-based cloaks apply the concepts of transformation optics and conformal mappings. By introducing metamaterials with tailored properties, electromagnetic waves can be made to propagate along a coordinate-transformed path around a specific region. However, these transformation-based cloaks require the metamaterials to have anisotropic and inhomogeneous characteristics. This feature handicaps their practical applications to antenna arrays because it leads to narrow bandwidths, bulky volumes, and difficult processing and assembly. Moreover, these cloaks generally isolate the regions internal and external to them. The fields radiated by an antenna placed in the internal region would then be blocked or severely affected by the cloaks. Thus, the transformation-based cloaks in general are not suitable for suppressing the scattering associated with multiband antenna systems.

Scattering-cancelation-based cloaks or mantle cloaks are a type of metasurface that has been introduced specifically to reduce scattering from an antenna. The electromagnetic mantle cloak and its scattering cancelation theory were introduced in 2009 [22]. As shown in Figure 4.3, the basic mantle cloak can be thought to be a cover of a cylindrical or spherical object that is composed of a periodic set of sub-wavelength unit cells [23]. By optimizing the pattern and size of these unit cells,

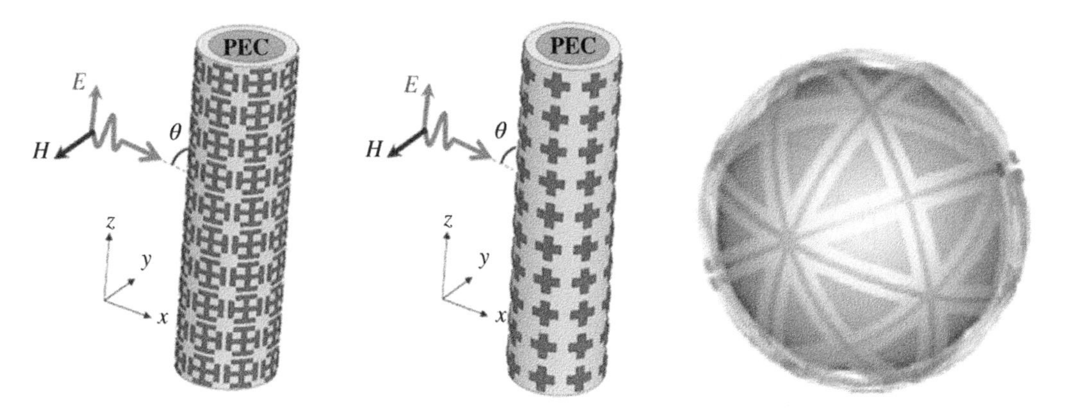

Figure 4.3 Illustrations of cylindrical and spherical mantle cloaks.

the resulting metasurface has tunable effective average surface impedance. The currents induced on the metasurface elements are designed to radiate an "antiphase" scattering field that causes a destructive interference with field scattered by the target itself, thus reducing the visibility of the entire structure. Moreover, when they are applied to a radiator, the cloaks can be adjusted to create a pass band that minimizes their effect on the fields radiated into or out from the internal area. Nevertheless, the low- or negative-index metamaterials required, for example, for a plasmonic cloak are also difficult to realize, again particularly for wide multiband operation.

The mantle cloak concept was first applied to eliminate the scattering from a pair of dipole antennas in 2012 [24]. The two dipoles were designed to resonate at 3.07 and 3.33 GHz, respectively. Consider the results shared in Figures 4.4 and 4.5. It can be seen in Figures 4.4a–d and 4.5a that the presence of a second dipole at a very close distance, 0.1 λ at 3 GHz, to another one significantly affects the radiation patterns and input impedances of each antenna. When mantle cloaks with appropriately designed metal strips are loaded around the dipole antennas, the radiation patterns and input impedances of the antennas are restored to the isolated radiation characteristics as observed in Figures 4.4a–d and 4.5b. Hence, the desired de-scattering of the antennas is achieved.

A related multilayer metasurface-based mantle cloak was applied to a dual-band monopole antenna in [25] as shown in Figure 4.6. It led to the desired de-scattering over a wider frequency band. There have been additional papers [26, 27] that have used mantle cloaks around antennas to reduce their scattering effects. It should be noted that all the reported antennas are traditional, simple dipole, or monopole antennas composed only of metal rods or strips. Moreover, all the associated mantle cloaks conformed to cylinders placed around them.

The mantle cloak is based on nonresonant characteristics lending it to broader bandwidths than resonant-based metamaterial cloaks. Consequently, it has better robustness in the microwave frequency range. Moreover, it is realized with an ultrathin patterned conducting surface and, hence, it has the advantages of being very low profile and lightweight. However, all of the metasurface-based mantle cloaks to date have been developed for cylindrical surfaces. Furthermore, traditional printed circuit board (PCB) technology to process the requisite metasurfaces may introduce large errors in the resulting system. New processing technologies such as additive manufacturing 3D printing may prove beneficial in this regard.

Metamaterial research is only two decades old and the associated cloaking technologies are even younger. The currently available cloaking techniques suffer from several practical limitations, e.g., large sizes, narrow bandwidths, complex fabrication procedures, and they have only been

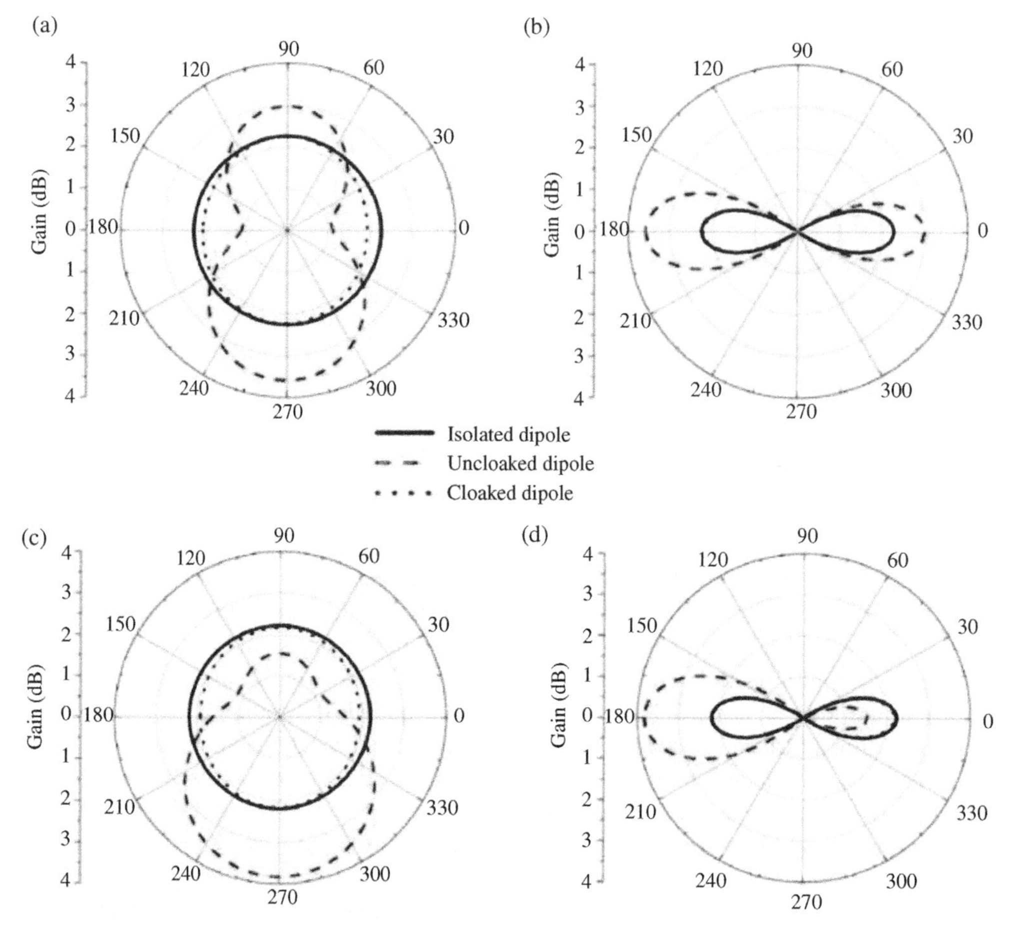

Figure 4.4 The (a) *H*-plane and (b) *E*-plane gain patterns of the first antenna at 3.298 GHz when it is isolated, and is uncloaked and cloaked in the presence of a nearby second one. The gain patterns in (c) and (d) show the same quantities for the second antenna at 2.970 GHz. *Source*: From [24] / with permission of IEEE.

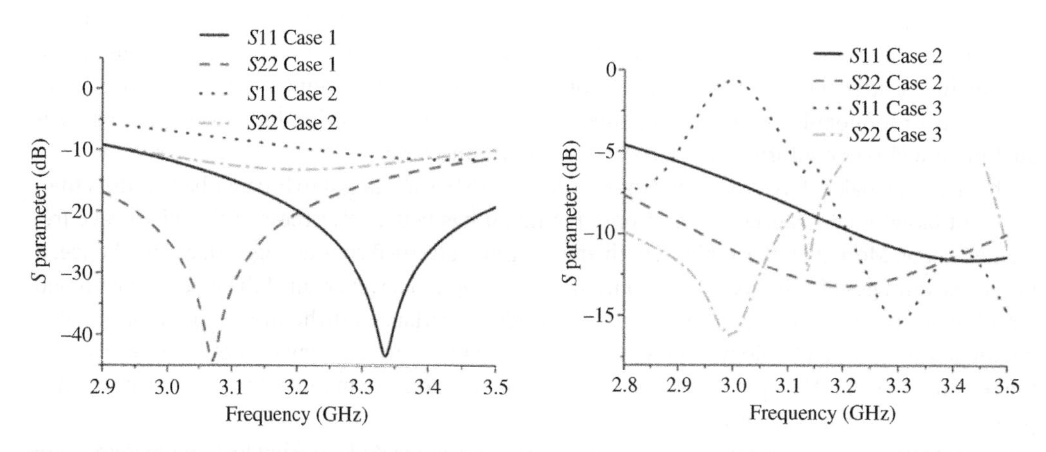

Figure 4.5 Simulated reflection coefficients of the two dipoles in three cases. Case 1: Two dipoles standing alone. Case 2: Two dipoles closely spaced without cloaks. Case 3: Two dipoles closely spaced with cloaks. *Source*: From [24] / with permission of IEEE.

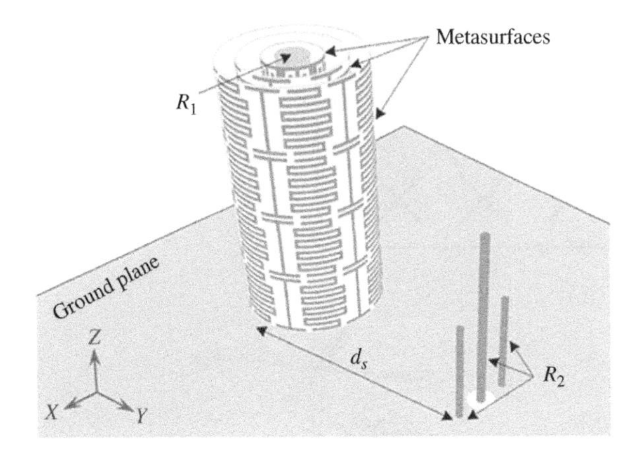

Figure 4.6 Multilayer metasurface-based mantle cloak for monopole antenna. *Source*: From [25] / with permission of John Wiley & Sons.

demonstrated on radiators with simple shapes such as conventional cylindrical dipoles/monopoles. More research efforts will be required before it may prove useful in practical, densely packed, complicated multiband antenna array systems.

4.3 Lumped-Choke De-scattering

A de-scattering technique based on lumped chokes was developed in [13]. Chokes were integrated directly into the radiating elements to suppress the unwanted scattering. Its application to a multiband 3G/4G base station antenna (BSA) array serves as a typical example to facilitate the explanation of its operating principles.

1) Demonstration of the effects of scattering

Figure 4.7a shows a common arrangement [28] of the HB and LB dual-polarized elements in a multiband 3G/4G BSA. One LB column is located midway between the two HB columns. Due to the proximity of the HB and LB elements and the electrically large dimension of the LB element at the HB wavelengths, the HB radiator induces currents on the LB element. As a result, the LB element radiates an unwanted HB signal. This scattered signal causes a major distortion of the HB radiation pattern. Although the reverse effect of the distortion of the LB pattern due to the presence of the HB elements can also occur, it generally only affects a relatively narrow band around the LB resonance.

Consider the dual-band array shown in Figure 4.7a. The configuration of each section of the array is illustrated in Figures 4.7b and 4.7c. A strip-shaped cross-dipole is used as the unaltered LB element, and square-shaped cross-dipoles are used as the HB elements. Baluns are used to provide balanced feeding and impedance matching for these elements. The two HB columns form two HB subarrays; they are fed from independent wideband phase-shifters. The latter are represented as power dividers for modeling purposes. The HB elements with the same polarizations in one column are excited simultaneously. The LB element is fed separately by inputs to the two baluns. The parameters for the array arrangement are marked in Figure 4.7b. Those parameters were chosen to achieve good MIMO performance while keeping the array applications compact.

Note that current BSAs usually need the LB array to operate from 0.69 to 0.96 GHz, and the HB array to operate from 1.71 to 2.69 GHz. However, the arrays considered here were designed to operate at relatively narrower bands, i.e., the LB band ranges from 0.82 to 1.0 GHz and the HB band

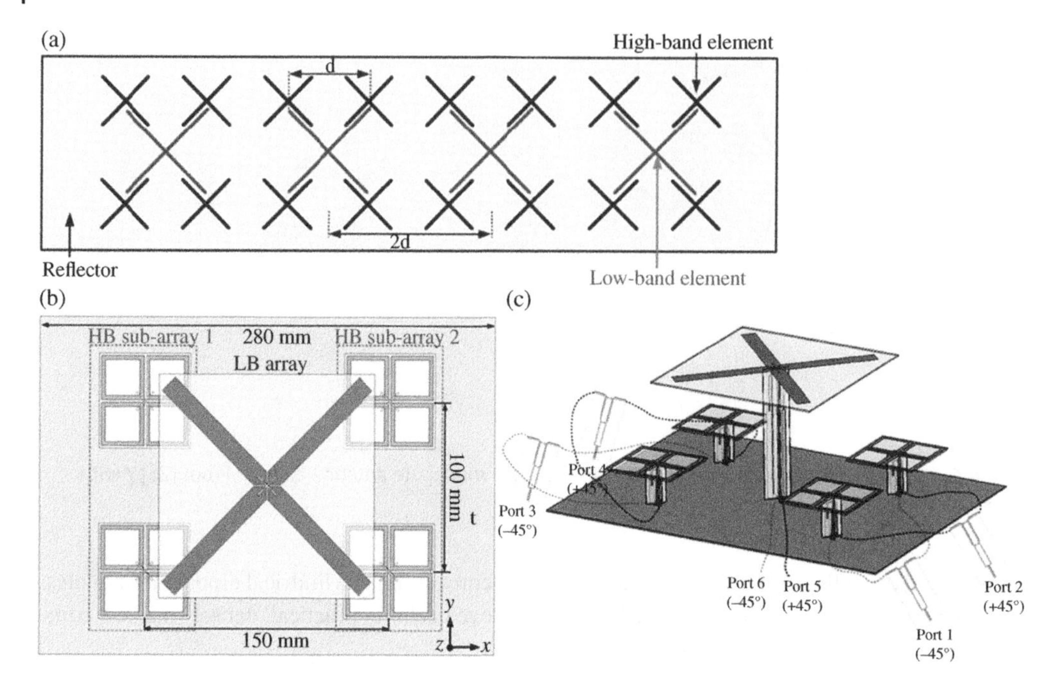

Figure 4.7 3G/4G dual-band dual-polarized base station antenna array. (a) Schematic top view of entire array. *Source:* (a) Based on [14] / IEEE. (b) Top and (c) perspective views of one section of the array. *Source*: (b and c) From [14] / with permission of IEEE.

ranges from 1.71 to 2.28 GHz. This choice acknowledges the fact that the realized lumped-choke de-scattering technique is still limited and cannot yet cover the full 3G/4G band.

The *S*-parameters for the LB and HB elements when they are working alone (no mutual coupling and scattering) are shown in Figure 4.8. On the other hand, the simulated current distributions on one array section with the conventional, unaltered LB element are shown at 1.7 GHz in Figure 4.9. It is seen that nontrivial currents are induced on the LB arms and they are in the same direction as the HB-driven currents. These induced currents re-radiate and deteriorate the HB radiation pattern. The radiation patterns in the horizontal plane (*xz*-plane) with and without the LB element are shown in Figure 4.10 at 1.7, 1.8, and 1.9 GHz. Without the LB element, the radiation pattern of the HB array is naturally symmetric, with its main beam pointing in the boresight direction. When the LB element is present, the HB radiation pattern deteriorates, i.e., the pattern is no longer symmetric and the main beam splits or shifts away from boresight. The worst-case outcome, as shown in Figure 4.10a, has the main lobe direction of the pattern tilted to 19º. As a consequence, the HB array in the presence of the LB elements has severely deteriorated radiation patterns. Thus, it cannot provide the required coverage, which leads to signal loss in particular areas. Actually, what is even worse is that the pattern distortions take place across a wider bandwidth and the resulting patterns are irregular as the frequency changes. This is generally unacceptable performance to cellular operators. Therefore, it is very desirable to suppress the scattering and restore the HB pattern.

2) Equivalent circuit and physical realization of a choke

An effective way to reduce the induced HB currents on the LB arms and, hence, minimize the destructive LB-induced scattering is to insert chokes periodically into the LB dipole arms. The chokes are designed to block the HB currents, but with minimum impact on the LB currents. These chokes should present an open circuit to the HB currents and a short circuit to the LB currents. They can be

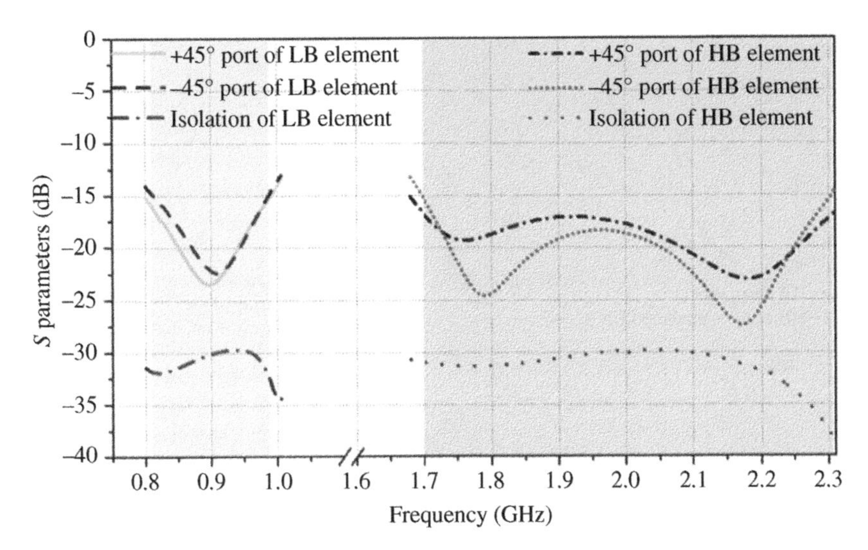

Figure 4.8 Simulated *S*-parameters of the individual LB and HB antenna elements when working alone, i.e. there is no mutual coupling or scattering.

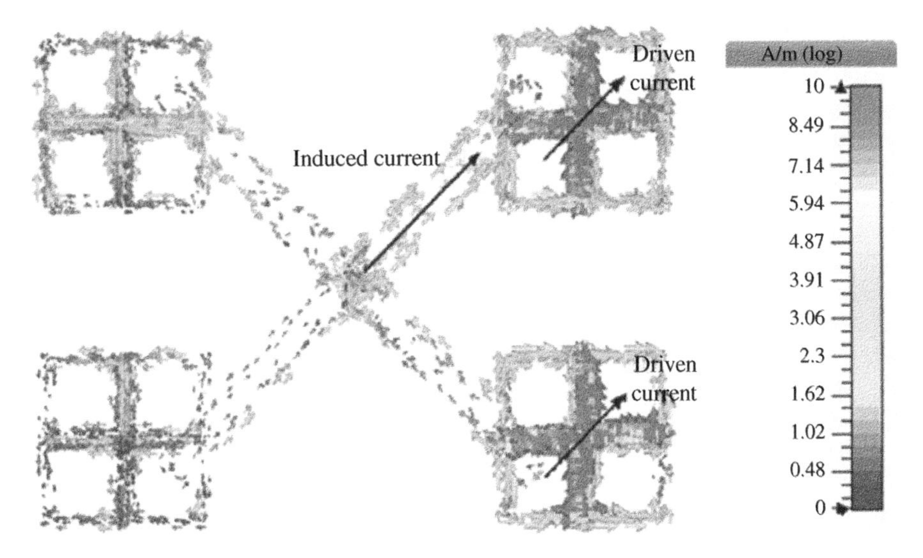

Figure 4.9 Simulated current distributions on a single section of the array at 1.7 GHz when the right column HB subarray is excited. *Source*: From [14] / with permission of IEEE.

modeled as the simple equivalent circuit shown in Figure 4.11a. It consists of a parallel resonator defined by L_1 and C_1 for the HB, and a series resonator for the LB attained with two additional capacitances C_2. The resonance frequencies of both these resonators are obtained from the relations:

$$j2\pi f_h C_1 + \frac{1}{j2\pi f_h L_1} = 0 \tag{4.1}$$

$$\frac{2}{j2\pi f_l C_2} + \frac{1}{j2\pi f_l C_1 + \frac{1}{j2\pi f_l L_1}} = 0 \tag{4.2}$$

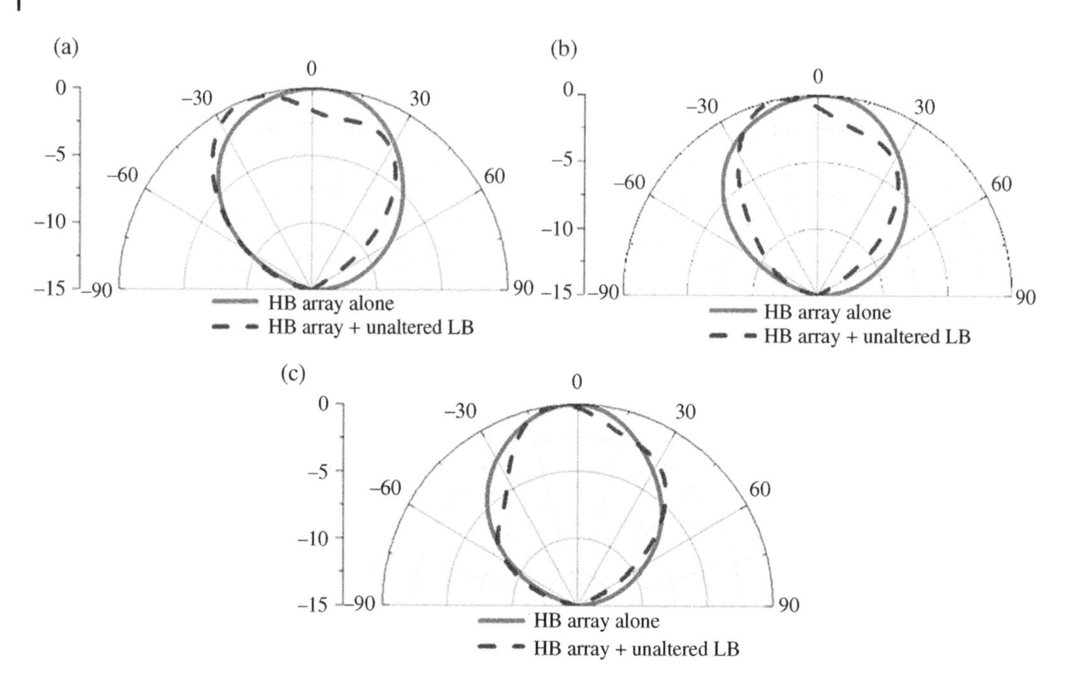

Figure 4.10 Radiation patterns of the HB sub-array without (solid line, black) and with (dashed line, red) the unaltered LB antenna element at (a) 1.7 GHz, (b) 1.8 GHz, and (c) 1.9 GHz.

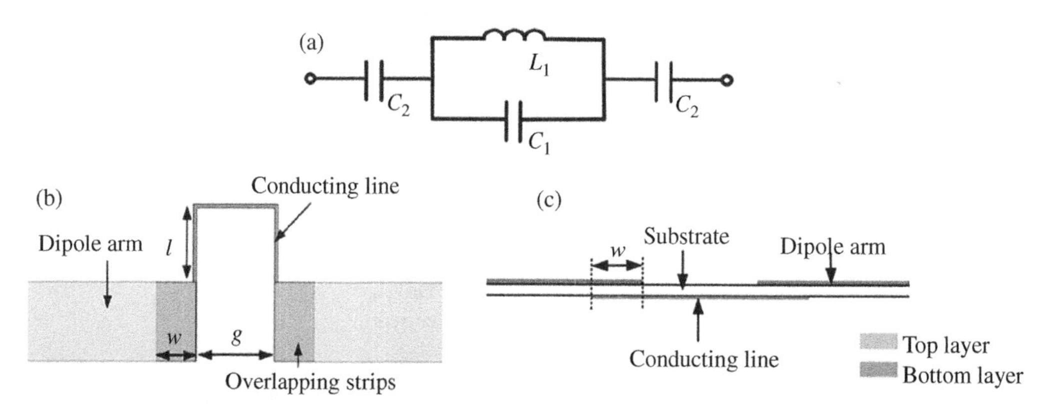

Figure 4.11 De-scattering choke design. (a) Equivalent circuit of the choke. (b) Top and (c) side views of the realized choke using conducting lines. *Source*: From [14] / with permission of IEEE.

where f_h is the open-circuit frequency in the HB and f_l is the short-circuit frequency in the LB. By suitably choosing L_1, C_1, and C_2, one can attain the desired open-circuit and short-circuit frequency points in the HB and LB, respectively.

The choke element can be realized with the conducting strip structure shown in Figures 4.11b and 4.11c. It is a two-layer structure with a dipole arm placed on the top layer and a conducting line placed on the bottom layer. The thin conducting line provides L_1; the gap in the dipole arm provides C_1; the overlap of the conducting strips and the dipole arms located on the two sides of the substrate provide C_2. Therefore, the thin inductive line and the gap capacitance control the open circuit frequency in the HB; and the overlapping strips control the short-circuit frequency in the LB. The equivalent circuit of this choke realization guides its optimization, which must ensure two important performance characteristics, i.e., the choke suppresses the current in the HB and creates a pass band in the LB. Finally, one needs to determine the number of chokes inserted in the dipole arms.

3) Optimization of the choke

Step 1 – Maximizing the choke's HB de-scattering performance

The choke is first optimized to maximize its ability to suppress induced currents in the HB around 2.0 GHz. The scattering shows the strongest effect on the HB antennas at this frequency according to the simulation results. To assess the scattering suppression ability of the choke, two models are built and compared, i.e., one with the choke and one without it. These models are shown in Figure 4.12. Both are around $\lambda/2$ in length at 2.0 GHz. This resonant wavelength includes the effects of the strip width and the dielectric constant of the substrate. Both models are shown in Figure 4.11. Model 1 is the strip without modification; Model 2 has a choke embedded in the strip.

Recall that the open circuit performance in the HB is determined by L_1 and C_1. These values are determined by the length of the conducting line $(2l + g)$ and the gap width (g). Therefore, adjusting g and l can effectively tune the open-circuit frequency.

To assess the de-scattering ability of a choke, the two models with and without the choke are illuminated by a plane wave with its E-field oriented parallel to the length of the strip. The maximum induced HB currents flowing on the strips are then monitored. Note that the HB currents and radiation response can be very complicated, i.e., it can have different responses depending on the polarization and angle of incidence of the plane wave. Nevertheless, the choke's ability to suppress the HB current is an intrinsic property that is only related to its dimensions and the frequency of the HB waves. If the choke shows good ability in suppressing the induced currents when exposed to an ideal wave at a specific frequency, then it will always have essentially the same HB suppression ability at the same frequency when it is exposed to waves with different incident angles and polarizations and placed in the near-field range of the HB antenna.

Figures 4.13a and 4.13b illustrate the magnitudes of the induced HB currents obtained from Model 1 and Model 2 with different values of g and l when they are illuminated by an ideal, normally incident plane wave. The Model 1 currents show a noticeable amount of HB contribution induced on the strip with the peak occurring near 2.0 GHz. In comparison, Model 2 exhibits much less HB current being induced on the strips. This outcome represents the scattering suppression ability of the choke. Moreover, it is observed from Figure 4.13a that increasing l with g fixed makes

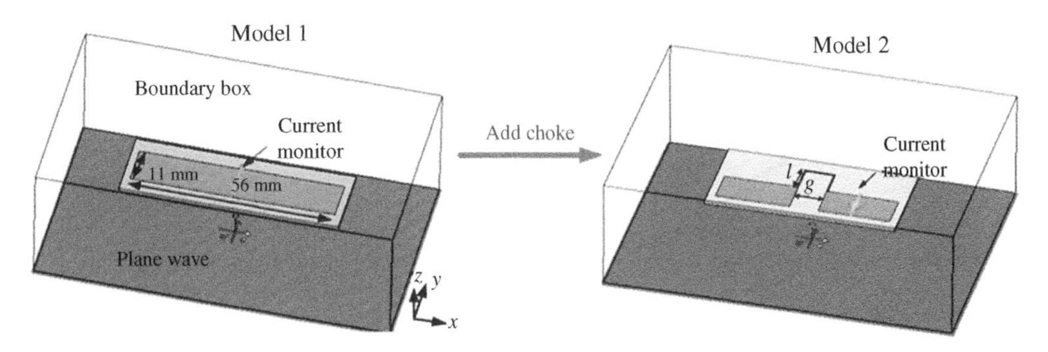

Figure 4.12 Two models used to assess the scattering suppression ability of the choke. Model 1: A strip whose length is around $\lambda/2$ at 2.0 GHz (middle frequency at the HB) and whose width is 11 mm. Model 2: The strip in Model 1 is cut in its middle and is then bridged with an inductive line.

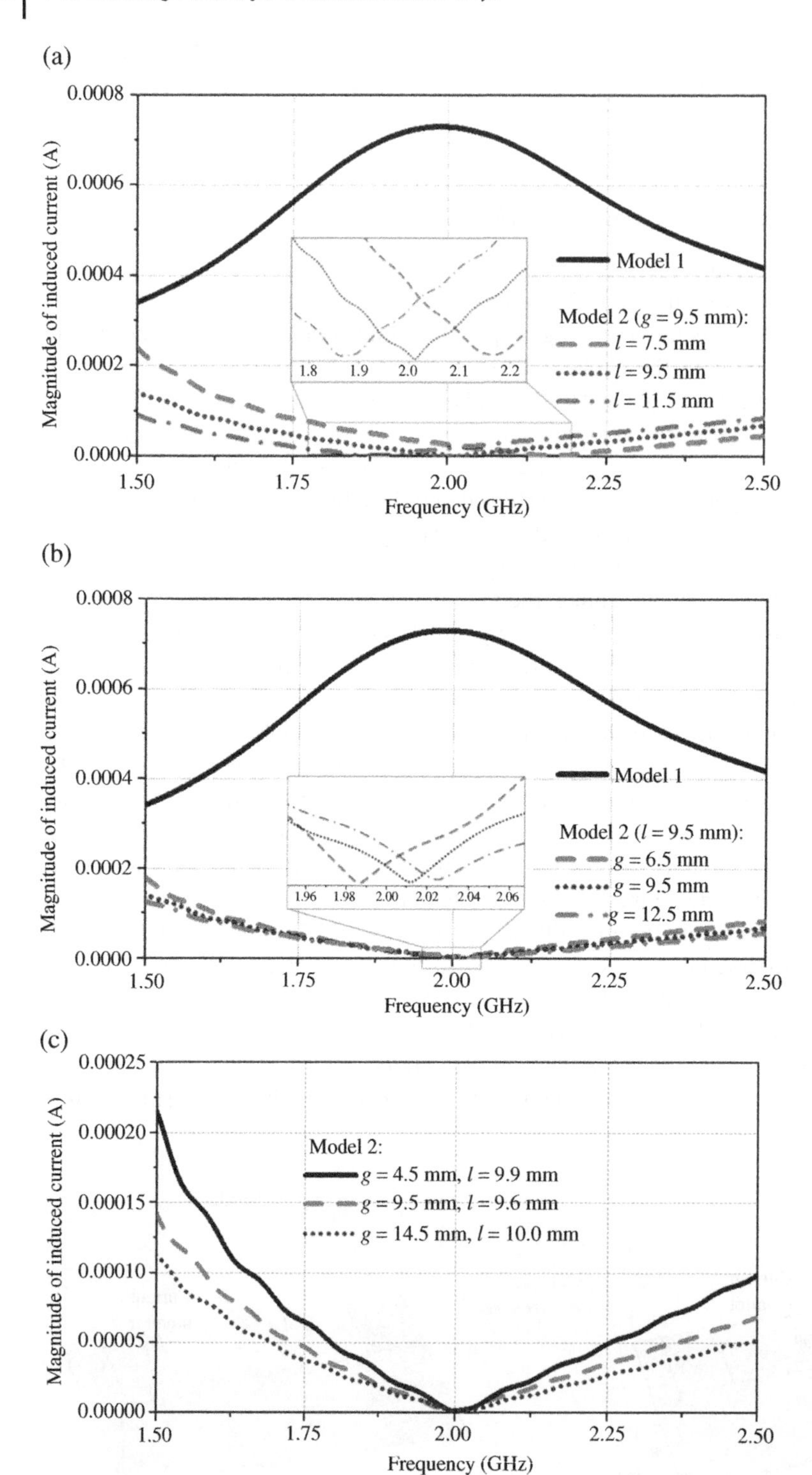

Figure 4.13 Magnitude of the induced HB currents obtained with Model 1 and Model 2. (a) Comparison for different values of l. (b) Comparison for different values of g. (c) Comparison for different combinations of $\{g, l\}$ to attain the open circuit frequency point at 2.0 GHz.

the minimum induced current point shifts toward a lower frequency. This feature results because increasing l increases L_1 and, hence, makes f_h lower. Similarly, as indicated in Figure 4.13b, increasing g with l fixed reduces C_1 and moves the open-circuit point slightly to a higher frequency.

Different combinations of g and l can be chosen to achieve the open-circuit condition at 2.0 GHz. The maximum magnitudes of the currents flowing on the strip as functions of the source frequency for some suitable combinations {g, l} are plotted in Figure 4.13c. These results demonstrate that a combination with larger g provides current suppression across a wider bandwidth. This behavior occurs because increasing g increases $Z_c = \sqrt{L_1/C_1}$ and thus widens the bandwidth of the scattering suppression.

Step 2 – Minimizing the choke's influence on the LB performance

After determining {g, l} to attain the maximum de-scattering ability over the HB, the next step is to optimize the choke to minimize its influence over the LB around 0.89 GHz, the middle frequency of the LB. The two models with and without the choke shown in Figure 4.14 were developed to assess the LB performance. Both have the same length, around $\lambda/2$ at 0.89 GHz, where $\lambda/2$ is the resonant length taking into account the effects of the strip width and the dielectric constant of the substrate. Model 3 is the strip without modification; Model 4 is the choked strip. The overlap width w determines C_2 in the choke model. The short-circuit frequency point is tuned by adjusting it.

These models are also illuminated with ideal, normally incident LB plane waves, and the induced LB currents on the strips are monitored. The results are illustrated in Figure 4.15a. One observes that the induced currents in both models are similar. This outcome demonstrates that the choke has limited influence on the LB performance. Note that the magnitude of the induced currents in Model 2 depends on the value of w and that the frequency of the current maximum corresponds to the short-circuit frequency f_l. Increasing w increases C_2 and thus moves the short-circuit point to a lower frequency as shown in Figure 4.15a.

For the different combinations of {g, l} that achieve the maximum de-scattering outcome at 2.0 GHz, different values of w are required to also realize a minimal influence on the currents at 0.89 GHz. Figure 4.15b gives the induced current results for Model 3 for different optimized combinations of {g, l, w}. It has been found that the combinations with larger g yield narrower

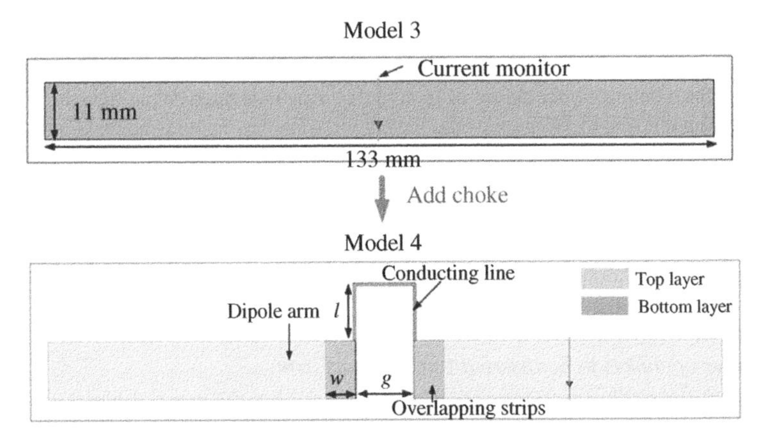

Figure 4.14 Two models used to assess the influence of the choke on the LB performance. The strip length in both models is around λ/2 at 0.89 GHz, the middle frequency of the LB, and a width of 11.0 mm. Model 3 is without the choke. Model 4 modifies the strip in Model 3 by cutting out its middle and bridging the resulting gap with a choke.

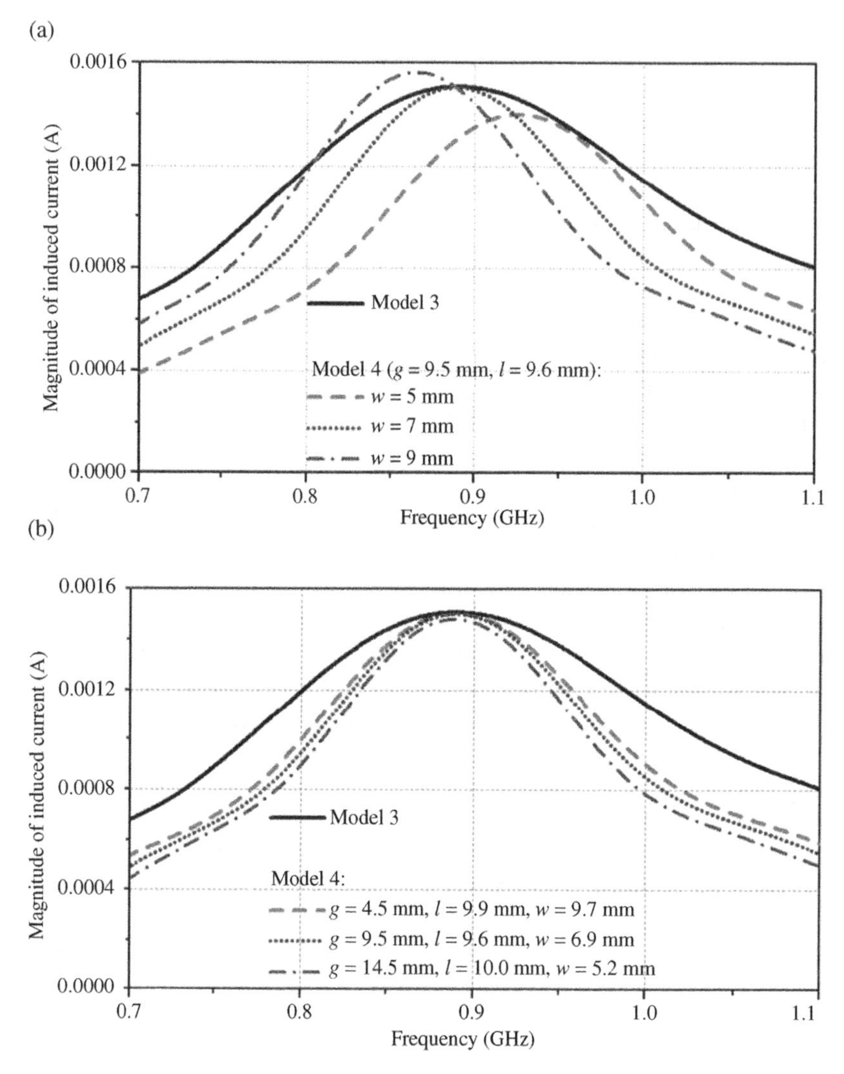

Figure 4.15 Comparison of the induced LB current magnitudes obtained from Model 3 and Model 4. (a) Results for different values of *w*. (b) Results for different combinations of {*g*, *l*, *w*} that attain the open circuit frequency point at 2.0 GHz and the short-circuit point at 0.89 GHz.

bandwidths in the LB since they require a smaller value of C_2. Therefore, one must be aware that there exists a trade-off between the LB passband and HB stopband performance when optimizing the dimensions {g, *l*, *w*}.

Step 3 – Determining the number of chokes to be inserted into a dipole arm
The next step is to properly insert the chokes into the dipole arms. Generally, the more chokes one inserts, the better the de-scattering performance at the HB. On the other hand, the radiation performance at the LB worsens with more chokes because they inevitably introduce losses into the radiator. Thus, it is necessary to carefully determine the minimum number of chokes required in the LB antenna to attain the desired overall LB and HB performances.

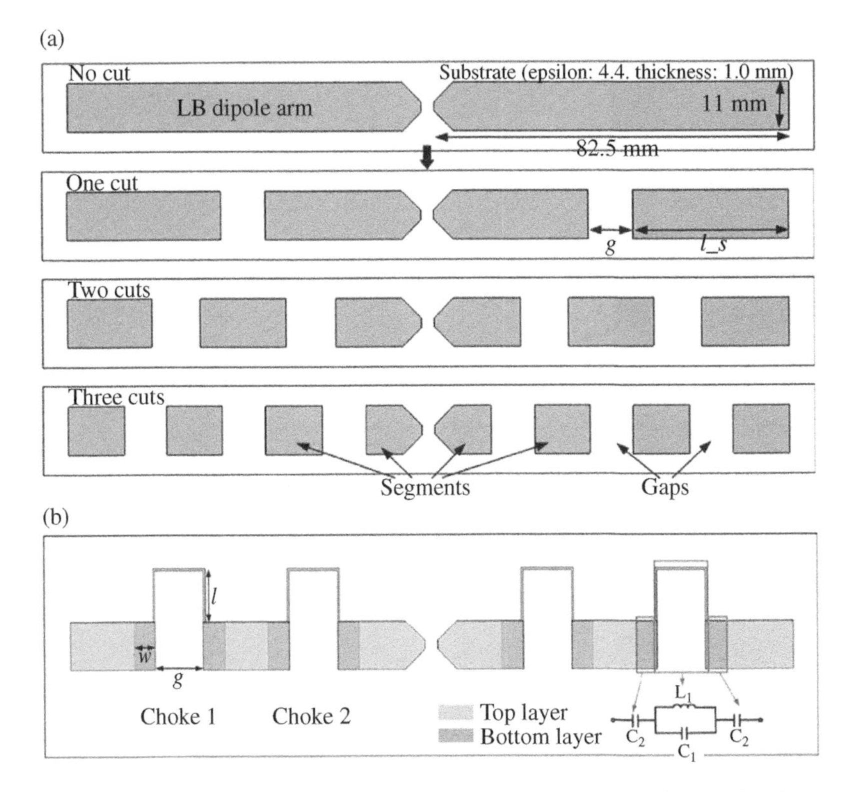

Figure 4.16 The introduction of chokes into the LP dipole arms. (a) Subdividing the arm with different numbers of cuts. (b) Top view of an LB dipole arm with chokes inserted into it.

As explained, the LB dipoles act as scatterers for the HB fields since they are electrically large for the HB wavelengths. If the dipole arm is divided into several shorter segments that are electrically small in the HB as shown in Figure 4.16a, then it naturally realizes a minimum scattering of the HB fields. Subsequently, the chokes must connect together these shorter segments in order to support the LB current. One effective scheme is depicted in Figure 4.16b.

The length of each segment, l_s, that assures suitably low HB currents on it is determined by simulating the scattering of the ideal, normally incident plane wave from an LB arm with segments of differing lengths. The model used for this purpose is shown in Figure 4.17a. The induced HB currents were monitored; the results are shown in Figure 4.17b. As expected, the shorter the segment, the lower is the induced current. Considering the overall performance, the LB dipole arm was broken into six small segments with lengths around 20 mm. This choice was found to guarantee a minimum scattering from the currents induced by the fields radiated across the HB band. The gaps between the segments have a width of 11 mm to fit the chokes. Simulations confirm that the HB radiation pattern is almost unchanged with this LB choke configuration.

Step 4 – Fine-tuning of the chokes

The final step is to re-optimize the chokes inserted in the LB dipole arms. The optimized choked LB radiators are shown in Figure 4.18. They remain arranged in the cross-dipole configuration to realize the required dual-polarization performance. The optimized design parameters are listed in Table 4.1. Only two chokes per arm were required to attain the desired performance. The substrate

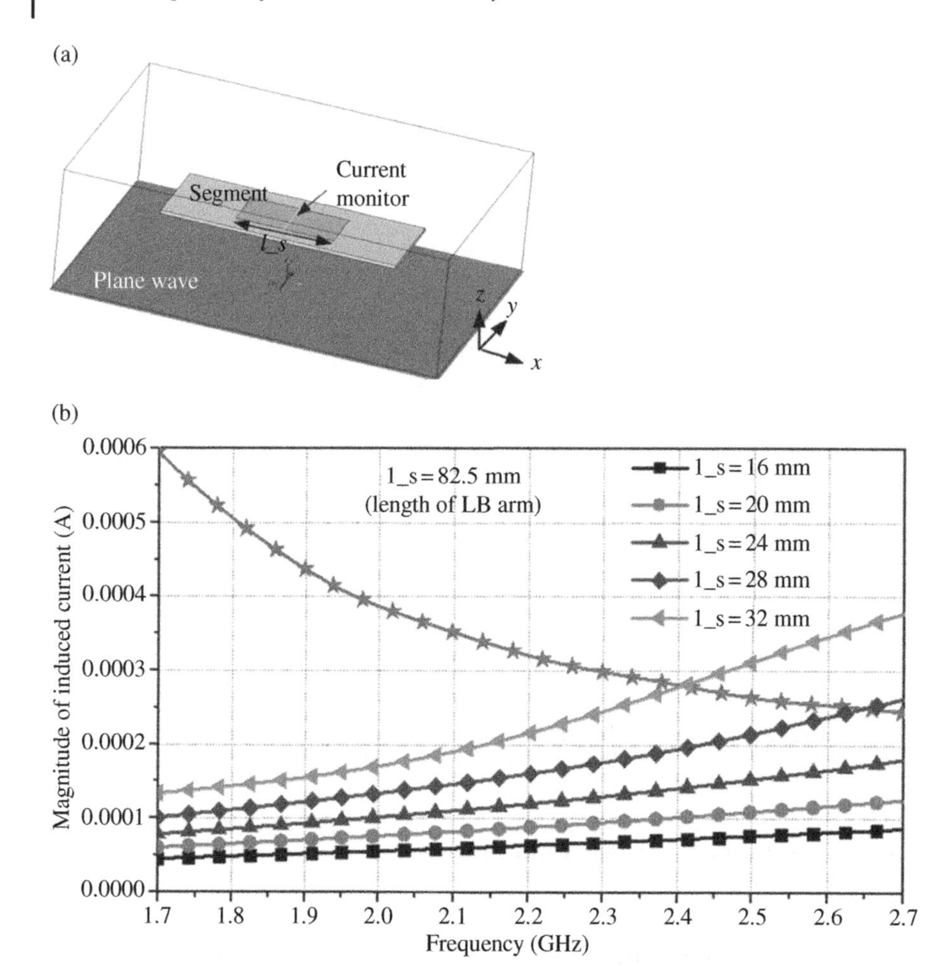

Figure 4.17 Determination of the length of the segments in the LB arms. (a) Simulation model used. (b) Maximum induced HB current on an individual segment for different segment lengths *l_s*.

supporting the radiator has a dielectric constant of 4.4, loss tangent of 0.0025, and a 1.0 mm thickness.

To achieve the desired HB choking performance across a wide band, one of the two chokes in each LB arm was tuned to a slightly different frequency. Because a higher magnitude of the LB current exists near the center of the radiator, the chokes near it were tuned to have the better LB pass performance. The chokes near the ends of the arms have a stronger HB suppression performance because they are closer to the HB radiators where the magnitudes of the HB currents induced on the LB arms are relatively larger.

Step 5 – Impedance matching of the choked LB antenna

As the realized chokes are not ideal and can only approximate a short circuit over a limited range of LB frequencies, they change the impedance properties of the LB element. This change makes the impedance-matching task much more difficult. Nevertheless, the impedance matching can be achieved across a relatively narrower bandwidth by baluns designed following the guidelines given in [29, 30].

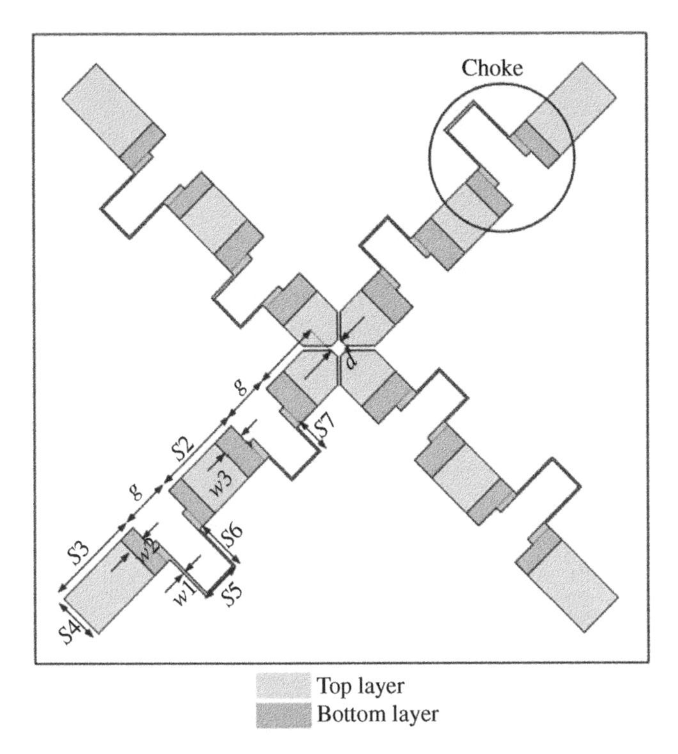

Figure 4.18 Configuration of the optimized choked LB radiator. *Source*: From [14] / with permission of IEEE.

Table 4.1 Optimized parameters of the choked antenna.

Parameters	s1	s2	s3	s4	s5	s6
Values (mm)	16.75	20.5	22.25	11	9.5	11
Parameters	s7	D	g	w1	w2	w3
Values (mm)	7	3	11.5	0.5	4	5.5

Two baluns are orthogonally arranged to feed the two pairs of choked dipoles. The configurations of the choked LB element and the baluns are shown in Figure 4.19. The baluns are printed on both sides of another piece of the same substrate, but whose thickness is 1.5 mm. The dimensions of the baluns are listed in Table 4.2. The simulation results are discussed below.

4) Performance Evaluation of the Choke

The choke's de-scattering ability

To demonstrate the de-scattering ability of the choked LB radiator, the electric field distribution in the x-z plane near the radiators for three cases: (i) HB array only; (ii) HB array with unaltered LB radiator; and (iii) HB array with choked LB were simulated. These results are compared in Figure 4.20. It is clear that when the unaltered LB radiator is present, it imposes strong scattering

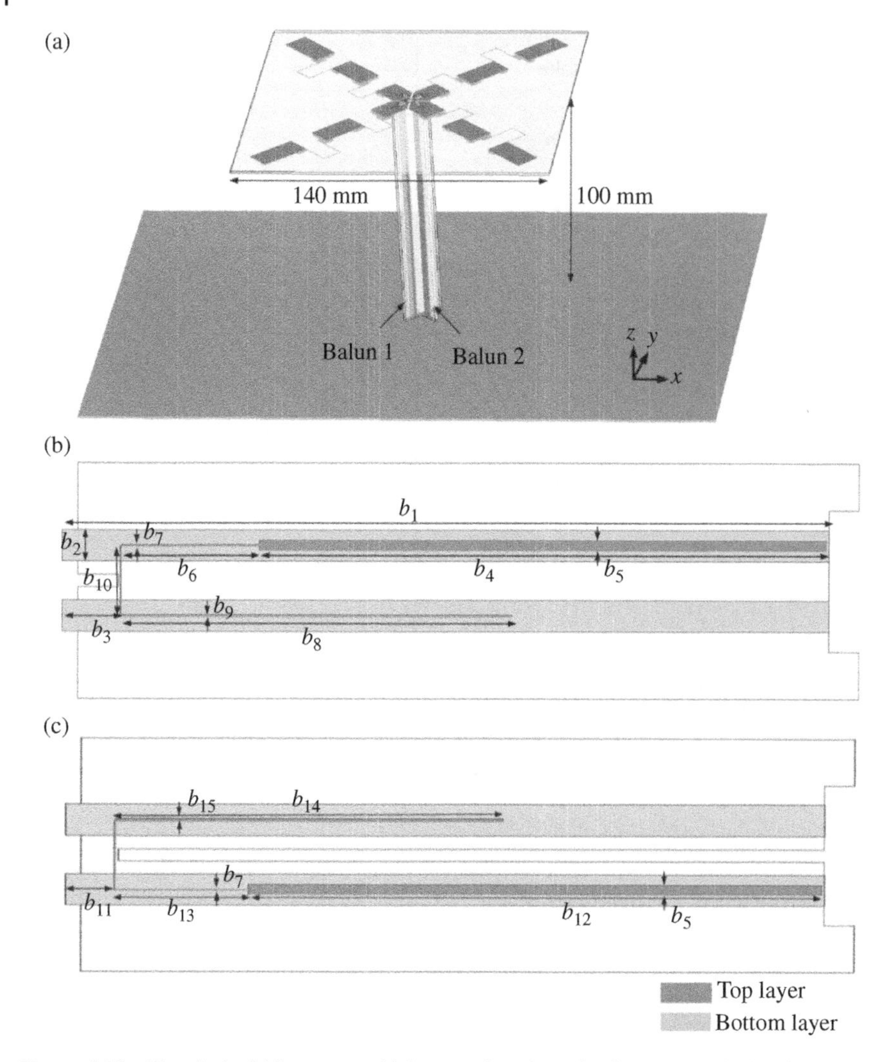

Figure 4.19 The choked LB antenna. (a) Perspective view. Configuration of (b) balun 1 and (c) balun 2.

Table 4.2 Optimized parameters of the baluns for the choked antenna.

Parameters	*b*1	*b*2	*b*3	*b*4	*b*5	*b*6	*b*7	*b*8
Values (mm)	101	4	7.6	72.98	1.4	20	0.2	50
Parameters	*b*9	*b*10	*b*11	*b*12	*b*13	*b*14	*b*15	
Values (mm)	0.2	9	6.475	74.08	20	50	0.25	

effects on the HB radiation. Moreover, it interferes strongly with the HB-radiated electric fields to a large extent, especially at the low frequencies. In comparison, the choked LB radiator shows a minimum scattering effect on the HB radiation and the radiated electric fields at the sample frequencies are very similar to those of the case without the LB radiator.

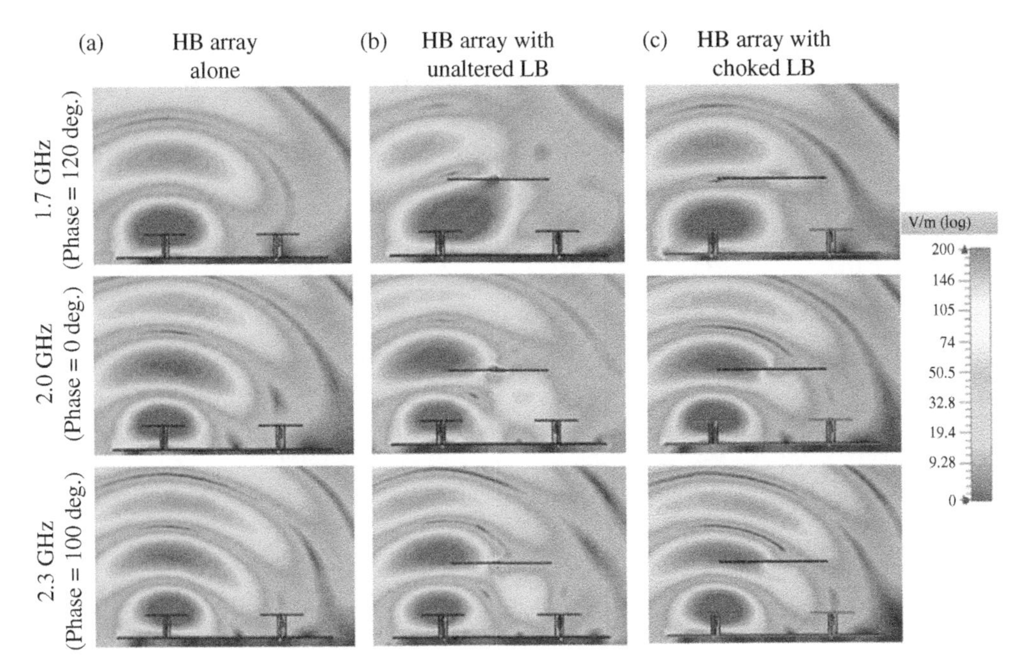

Figure 4.20 The electric field distribution in the *xz*-plane at 1.7, 2.0, and 2.3 GHz for three configurations: (i) HB array only, (ii) HB array with unaltered LB radiators, and (iii) HB array with choked LB radiators.

The radiation patterns in the *x-z* plane for cases (i) and (iii) are compared in Figure 4.21. It is observed that the HB radiation patterns in the presence of the choked LB radiators are almost the same as those of HB array alone. Thus, the effectiveness of the chokes in reducing the HB pattern distortion across the band is demonstrated.

The choke's effect on the LB performance

The input impedance of the LB antenna changes once the chokes are implemented. Unfortunately, the impedance matching of the choked LB antenna is nontrivial. An unaltered cross-dipole radiator can be successfully matched across the entire required 3G/4G band from 690 to 960 MHz [28, 29]. However, the choked LB antenna can only be matched across a much narrower bandwidth from 820 to 960 MHz. Figure 4.22 compares the *S*-parameters of the unaltered and choked LB radiators. These results show, except for the narrower bandwidth, that the choked LB antenna does not show any other deficiencies in terms of the scattering coefficients. The LB antenna is well matched over this narrower bandwidth.

The normalized radiation patterns of the unaltered LB radiator and the choked LB radiator are compared in Figure 4.23. The patterns in these two cases are almost identical, showing that the choked LB arms do not affect the shape of the radiation pattern. Nevertheless, the chokes do introduce some losses and, hence, do reduce the gain of the LB radiator. The choked LB antenna shows approximately a 1 dB gain loss on average when compared to the unaltered antenna.

5) Results and discussion

A prototype of the dual-band dual-polarized interleaved array section was fabricated and tested. Photographs of the fabricated section are shown in Figure 4.24. Note that each of the two HB

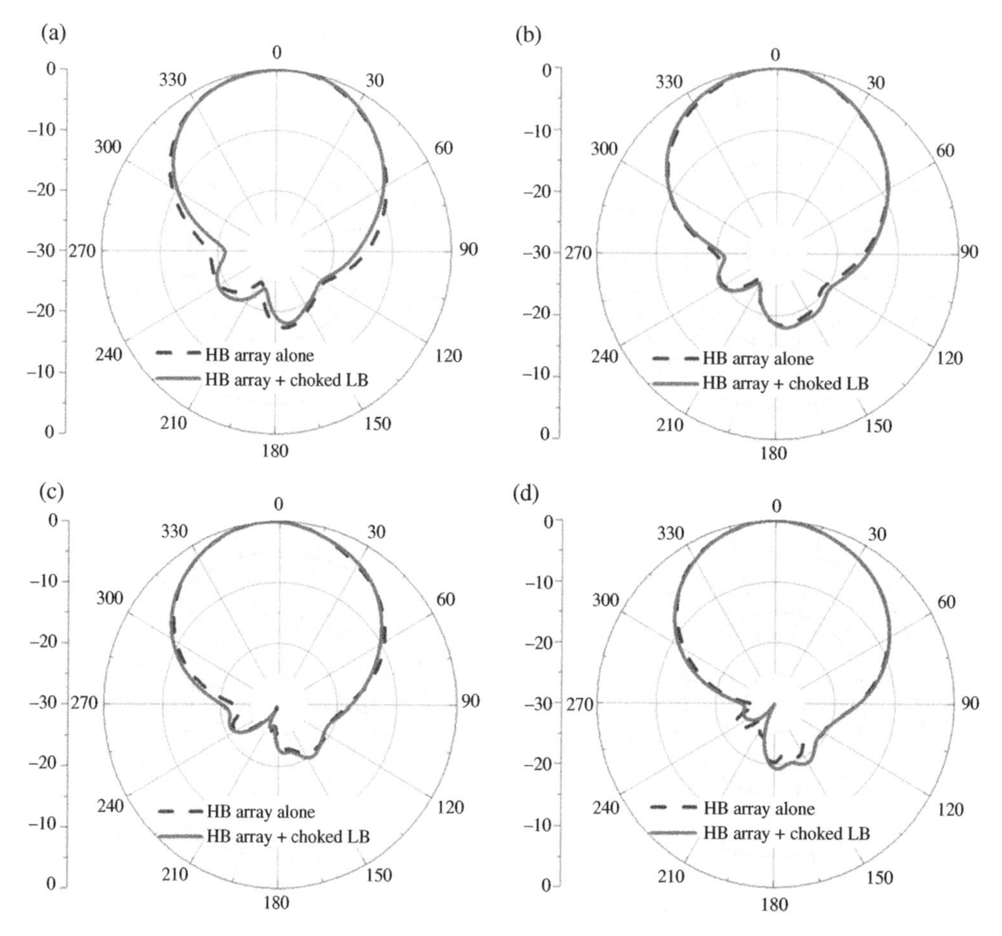

Figure 4.21 Comparison of the HB radiation patterns when only the HB array is present and when both the HB array and the choked LB radiator are present at (a) 1.7, (b) 1.9, (c) 2.1, and (d) 2.3 GHz.

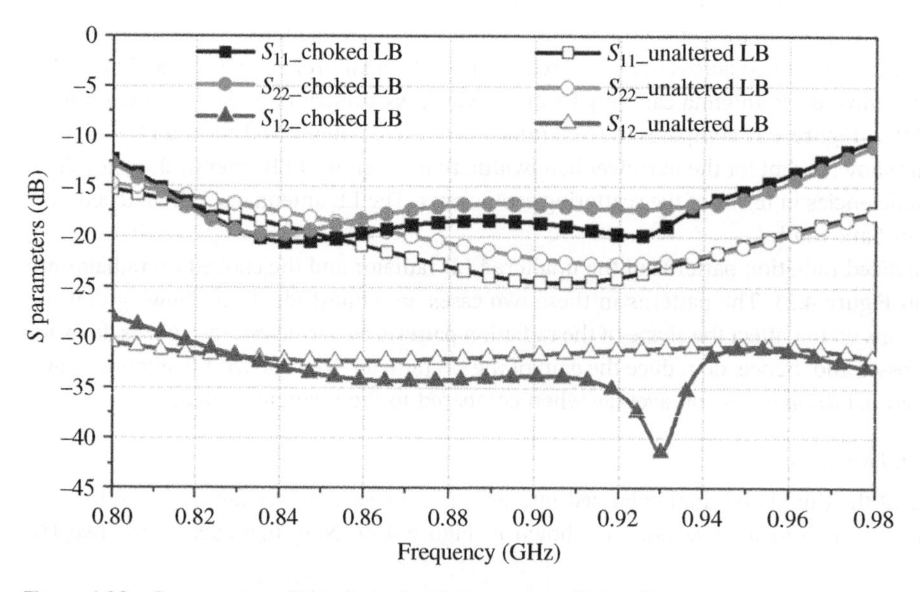

Figure 4.22 S-parameters of the choked and the unaltered LB radiator.

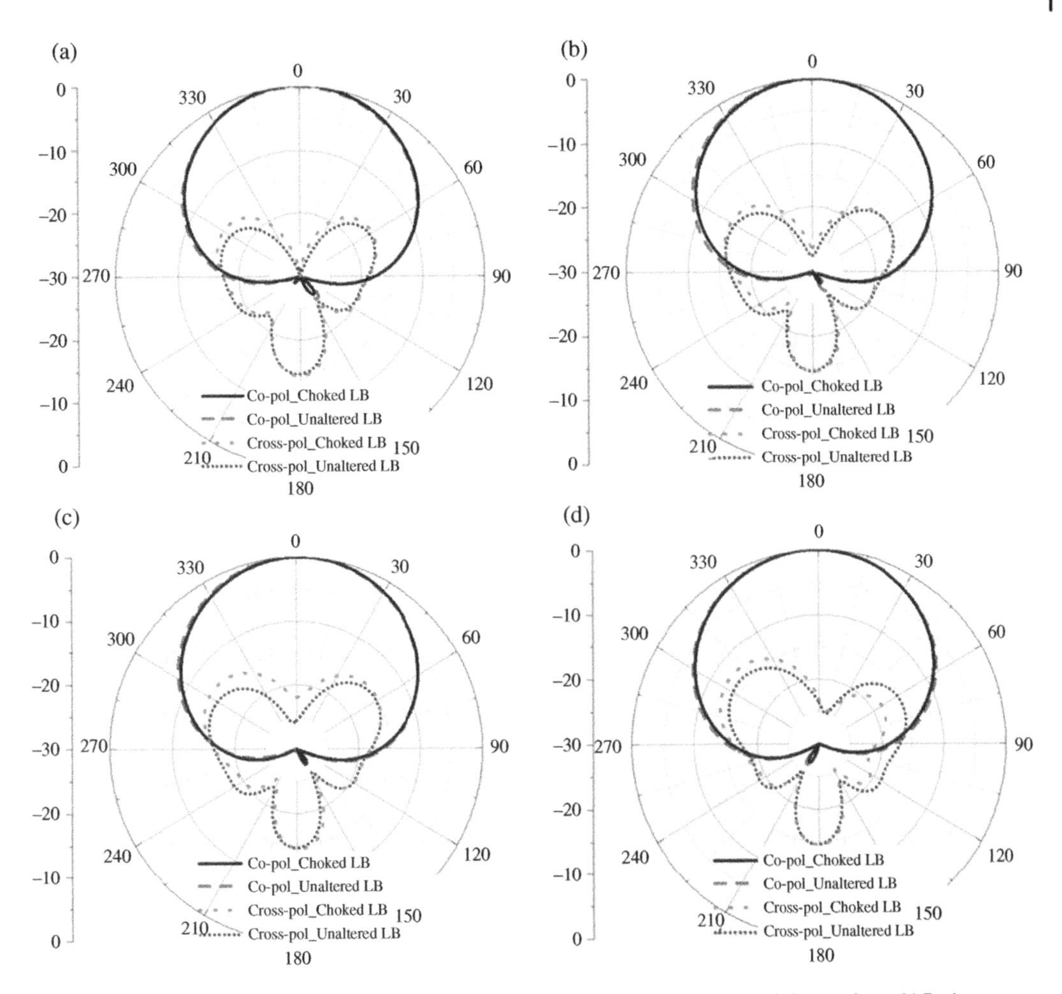

Figure 4.23 Comparison of the radiation patterns of the choked LB element and the unaltered LB element at (a) 0.82, (b) 0.88, (c) 0.92, and (d) 0.96 GHz.

dipoles in one column is fed by a 1-to-2 power divider as an HB subarray. There are four power dividers employed to feed the two HB subarrays for their two polarizations. The input ports of the power dividers to feed the left subarray are listed as Ports 1 and 2 for the 45°- and −45°-polarizations. The input ports of the power dividers to feed the right subarray are listed as Ports 3 and 4 for the two polarizations. The input ports of the LB antenna for the two polarizations are Ports 5 and 6. Detailed information on the ports can be found in Figure 4.7c.

The simulated and measured reflection coefficients at the HB and LB ports are given in Figures 4.25a and 4.25b, respectively. The measured results agree well with the simulated ones. The HB antennas have a good matching ($|S_{11}|/|S_{22}| < -15$ dB) across a band of 28.6% from 1.71 to 2.28 GHz. For the choked LB antenna, the measured bandwidth is 19.7% from 0.82 to 1.0 GHz with $|S_{55}|/|S_{66}| < -10$ dB. We believe that the chokes implemented on the LB antenna will impose some negative effects on the impedance matching of the LB antenna, but have limited effects on that of the HB antennas.

(a)

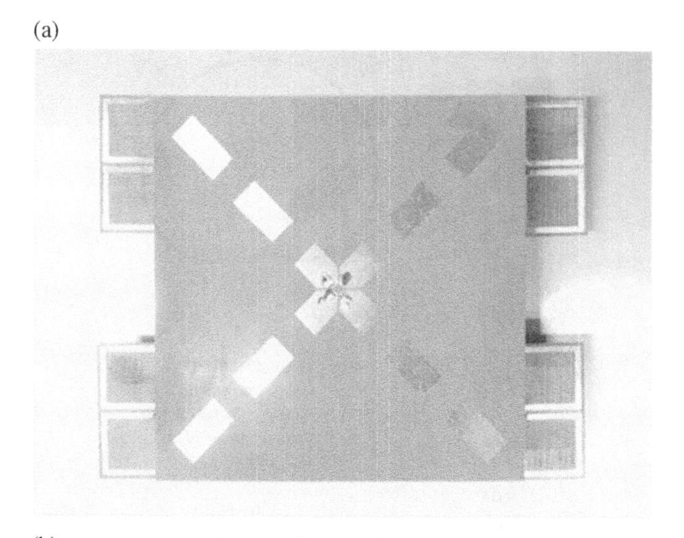

(b)

Figure 4.24 Fabricated prototype of the dual-band dual-polarized interleaved array section. (a) Top view. (b) Side view.

The simulated and measured radiation patterns for one HB subarray at 1.7, 1.9, and 2.2 GHz are shown in Figure 4.26 for its two polarizations. The simulated and measured HPBWs and realized gains of the HB subarray are displayed in Figure 4.27. Since the HB subarray on the left and right are necessarily the same, only the results of the subarray on the left are presented. The measured results agree well with the simulated ones. They confirm that the HPBWs vary within $65^\circ \pm 5^\circ$, the XPDs at the boresight are >16 dB, and the realized gains vary from 10 to 12 dBi for the two polarization states. The radiation performance of the HB subarrays in the presence of the choked LB antenna is very similar to those of the case without the LB antenna. This outcome demonstrates the fact that the chokes employed in the LB antenna have minimal effects on the HB radiation characteristics despite the close proximity of both the LB and HB antennas.

The simulated and measured radiation patterns of the two polarization states of the LB antenna at 0.82, 0.88, and 0.96 GHz are shown in Figure 4.28. The simulated and measured HPBWs and

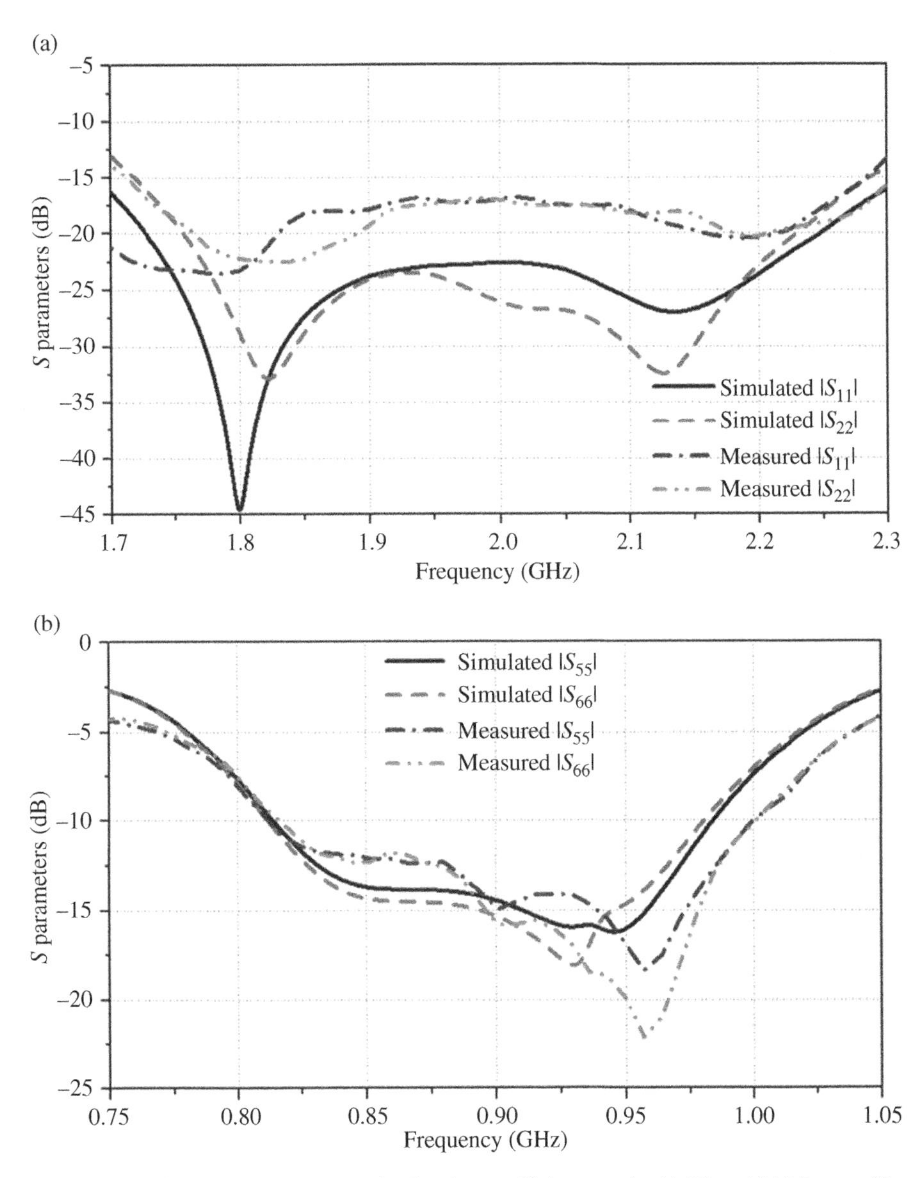

Figure 4.25 Simulated and measured reflection coefficients at the (a) HB and (b) LB ports. The numbering of the ports is indicated in Figure 4.7c.

realized gains of the LB antenna are displayed in Figure 4.29. Again, the simulated and measured results agree well with each other. The measured HPBWs vary within 69.5° ± 4°, the XPDs at the boresight are >20 dB, and the realized gains vary from 6 to 7 dBi for both the two polarization states. It is noted that the measured realized gains of the LB antenna are lower than those of conventional LB antennas. This feature is attributed to two reasons. One is the lossy feed cables used in the experiment, and the other is the loss resulting from the chokes. In practical engineering situations, both of them can be reduced to negligible levels.

A method of suppressing crossband scattering by employing lumped chokes on LB elements was presented in this section. The technique was analyzed and demonstrated in the context of a

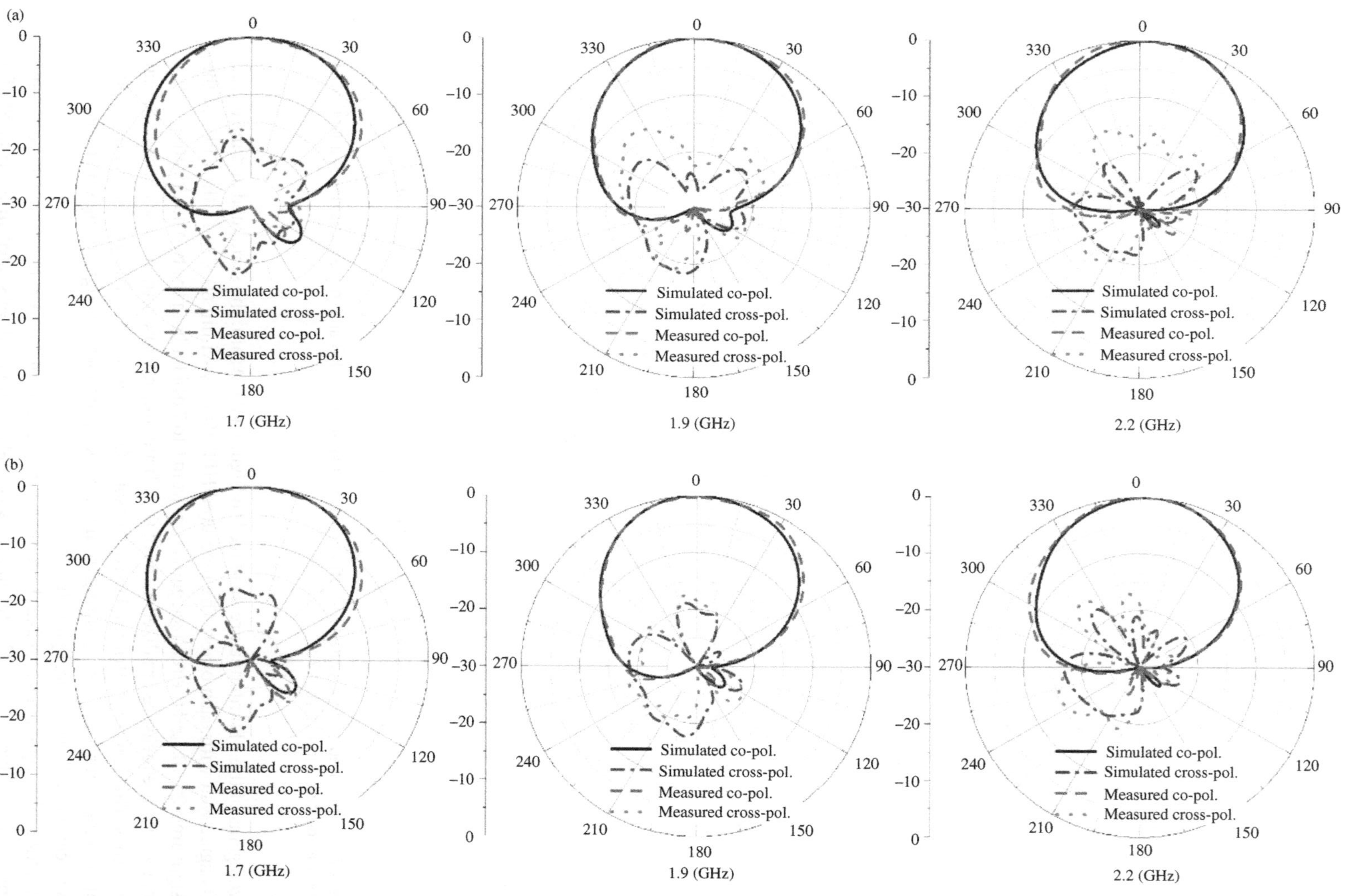

Figure 4.26 Simulated and measured radiation patterns of the left column HB subarray when (a) port 1 and (b) port 2 are excited at 1.7, 1.9, and 2.2 GHz.

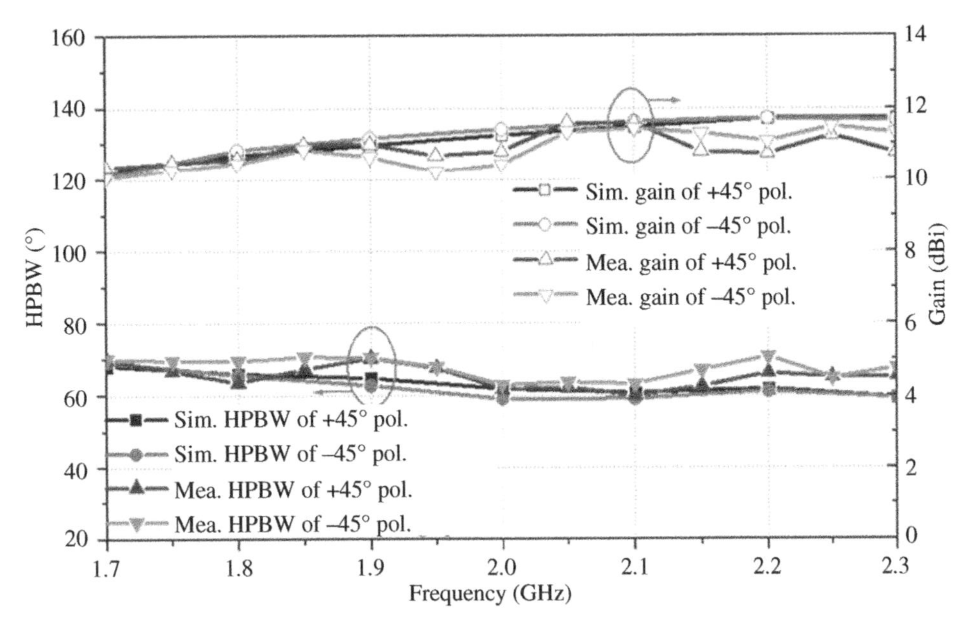

Figure 4.27 Simulated and measured HPBW and realized gains of the left column HB subarray when port 1 is excited (45° polarization state) and port 2 is excited (–45° polarization state), respectively.

dual-band dual-polarized interleaved base station array configuration that covers the bands: 0.82 to 1.0 GHz and 1.71 to 2.28 GHz. The simulated and measured results both demonstrate that choking the LB element largely restores the HB radiation pattern. The chokes have some minor effects on the LB impedance characteristics. Nevertheless, satisfactory performance can be obtained with suitable choice of choke impedances and optimized feed networks. The realized prototype array section is compact and has stable radiation patterns in both the high and low bands. Similar choking methods can be adopted to solve crossband scattering issues in other multiband antenna systems.

4.4 Distributed-Choke De-scattering

Several lumped-choke styles have been considered. For instance, stripline chokes [13] and coaxial chokes [31] have been implemented on the LB radiator and have proved to be effective in suppressing the HB currents. However, as noted, the impedance matching becomes difficult and a narrower-than-desired bandwidth occurs. To overcome this issue, a distributed choke structure, i.e. a spiral arm, version of the LB radiator has proved effective. It has been applied successfully to achieve a dual-band, dual-polarization interleaved 4G and 5G base station antenna [15].

1) Configuration of the distributed spiral choke

The distributed spiral choke configuration is shown in Figure 4.30. A close-up view of the spiral arm is illustrated in Figure 4.30a. Its implementation in an interleaved 4G and 5G base station antenna is shown in Figure 4.30b. As with the discontinuous design, the distributed choke has to demonstrate a high-stop low-pass characteristic when it is inserted into the LB radiator. These properties suppress the HB scattering and maintain the LB performance.

The spiral structure, which is similar to the helix coil, can be represented as a parallel LC circuit. It acts as an open circuit at its HB resonant frequency. When the LB frequency is much lower than

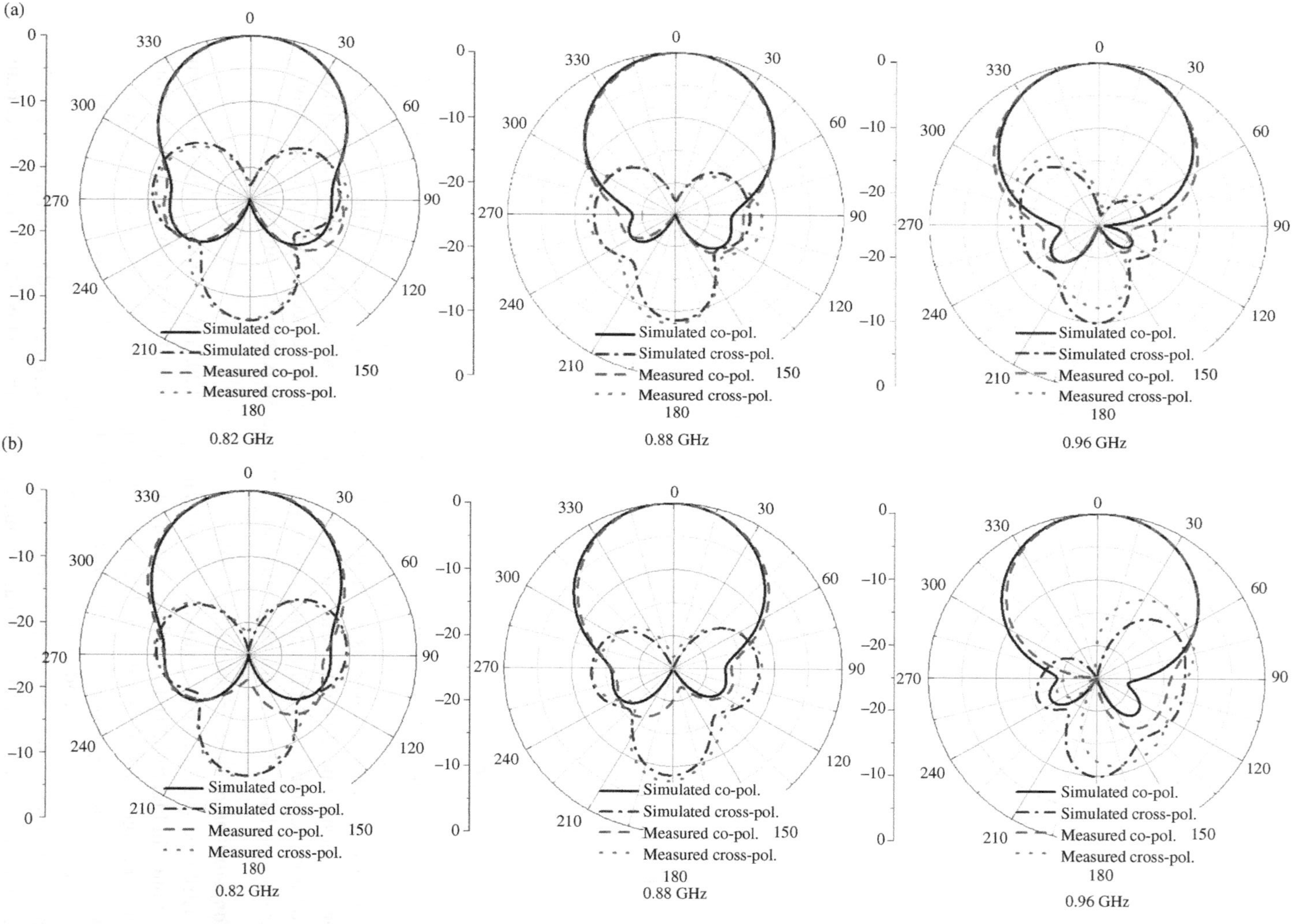

Figure 4.28 Simulated and measured radiation patterns of the LB antenna when (a) port 5 and (b) port 6 are excited at 0.82, 0.88, and 0.96 GHz.

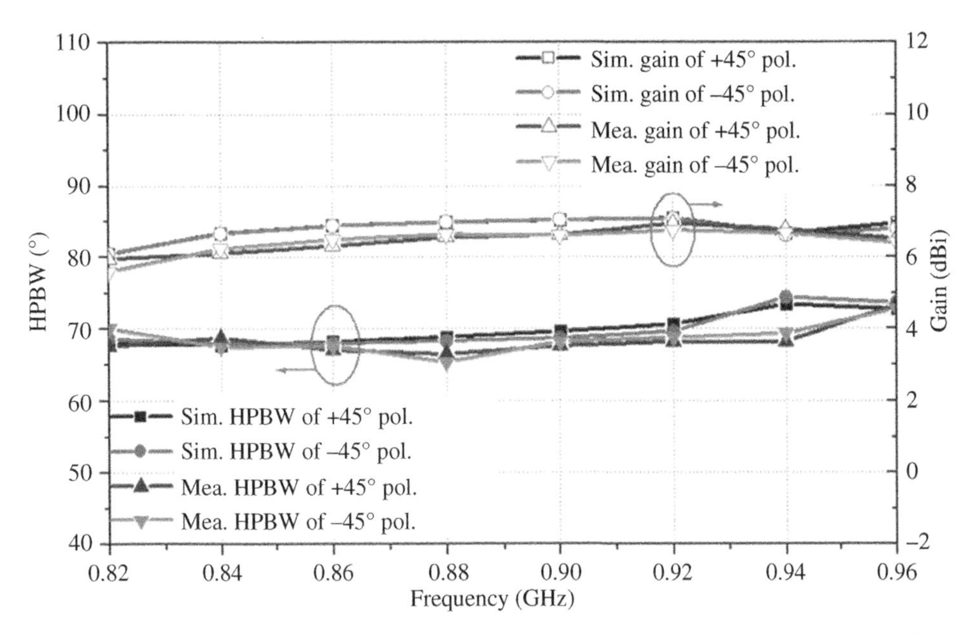

Figure 4.29 Simulated and measured HPBW and realized gains of the LB antenna when port 5 (45°-polarization state) and port 6 (−45°-polarization state) are excited.

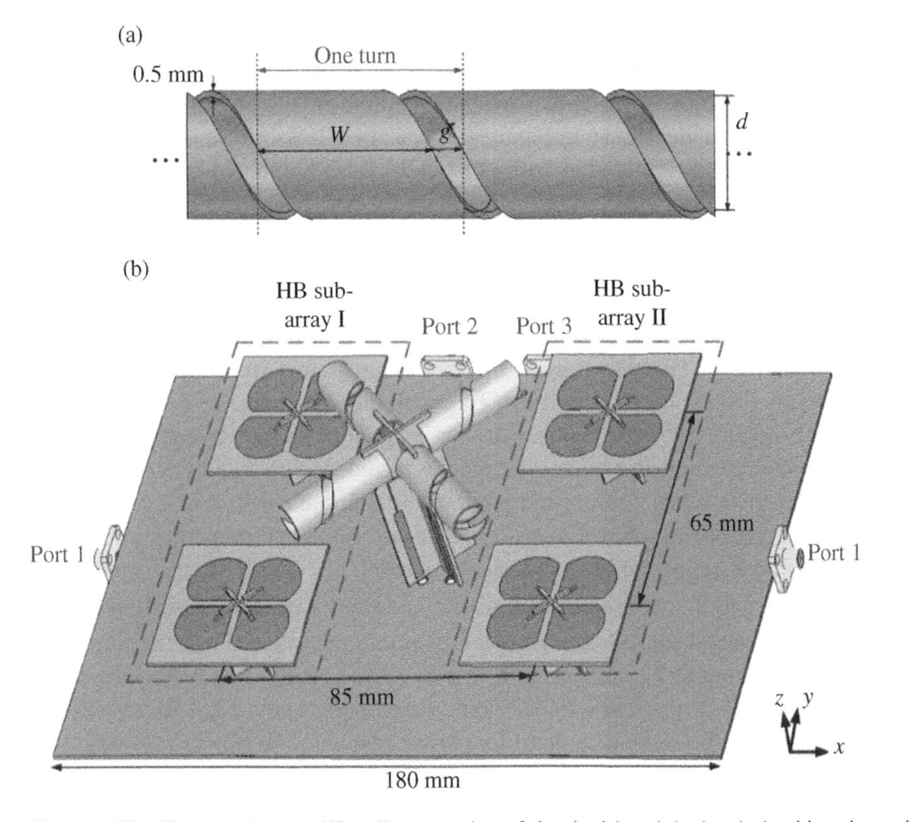

Figure 4.30 The spiral-based HB radiator version of the dual-band dual-polarized interleaved array section. (a) Design parameters of the spiral arm. (b) Single subsection of the interleaved dual-band 4G/5G BSA. *Source:* From [15] / with permission of IEEE.

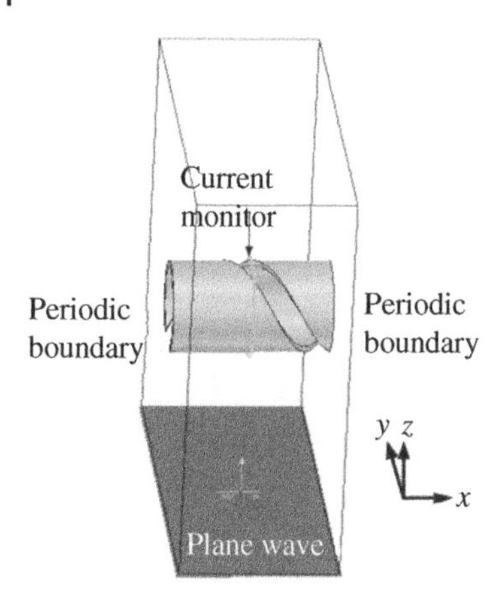

Figure 4.31 Simulation model to design the spiral choke. It represents an infinitely long spiral structure illuminated by the ideal, normally incident, plane wave. *Source*: From [15] / with permission of IEEE.

that resonance frequency, the spiral has low inductive impedance and high shunt capacitive impedance. Compared with a conventional cylindrical conductor, the spiral structure only introduces a small additional inductance in the LB. Therefore, it can be used directly as the LB radiator since its intrinsic low-pass high-stop property not only suppresses the unwanted HB currents induced on it, but also maintains the LB radiation performance.

2) Analysis of infinitely long spiral

To study the ability of the spiral to suppress the HB currents, its resonance frequency point must be determined. The currents induced by an incident HB electric field are minimized at this frequency. Consequently, a properly designed spiral choke will then minimize the scattering effects. To achieve an optimal design, the model shown in Figure 4.31 was used. It is a single cell of the spiral with periodic boundary conditions imposed at both of its ends. It is illuminated by a plane wave whose electric field is oriented along the axis of the spiral, the *x*-axis. The currents induced on it were monitored. The periodic boundary was set to simulate an infinitely long spiral to eliminate the effects of the end discontinuities associated with a finite version.

There are three important parameters used to tune the choke's resonance frequency. As noted in Figure 4.30a, they are the distance between neighboring turns g, the inner diameter of the spiral *d*, and the width of the strip *w*. The thickness of the conductor here is set to be 0.5 mm. The magnitudes of the currents flowing on the spiral structure with different values of g, *d*, and *w* are monitored and plotted in Figure 4.32. For comparison, the magnitude of the currents flowing on a cylindrical conductor with the same diameter as the spiral is included in the figures as a reference.

The subplots show that there exists a minimum point of the induced currents at a certain frequency for each combination of the spiral dimensions. The corresponding frequency points are approximately the resonance points of the equivalent parallel LC circuit of the spiral, i.e. $f_{\text{res}} = 1/\left(2\pi\sqrt{LC}\right)$, where L and C are the series inductance and shunt capacitance per unit length of the spiral. As shown in Figure 4.32a, a larger g moves the minimum induced current point to a higher frequency, corresponding to f_{res} moving to a higher value as well. This behavior occurs because increasing *g* decreases the capacitance and, hence, increases f_{res}. As shown in Figure 4.32b, a larger *d* moves the resonance point to a lower frequency, as it increases both the

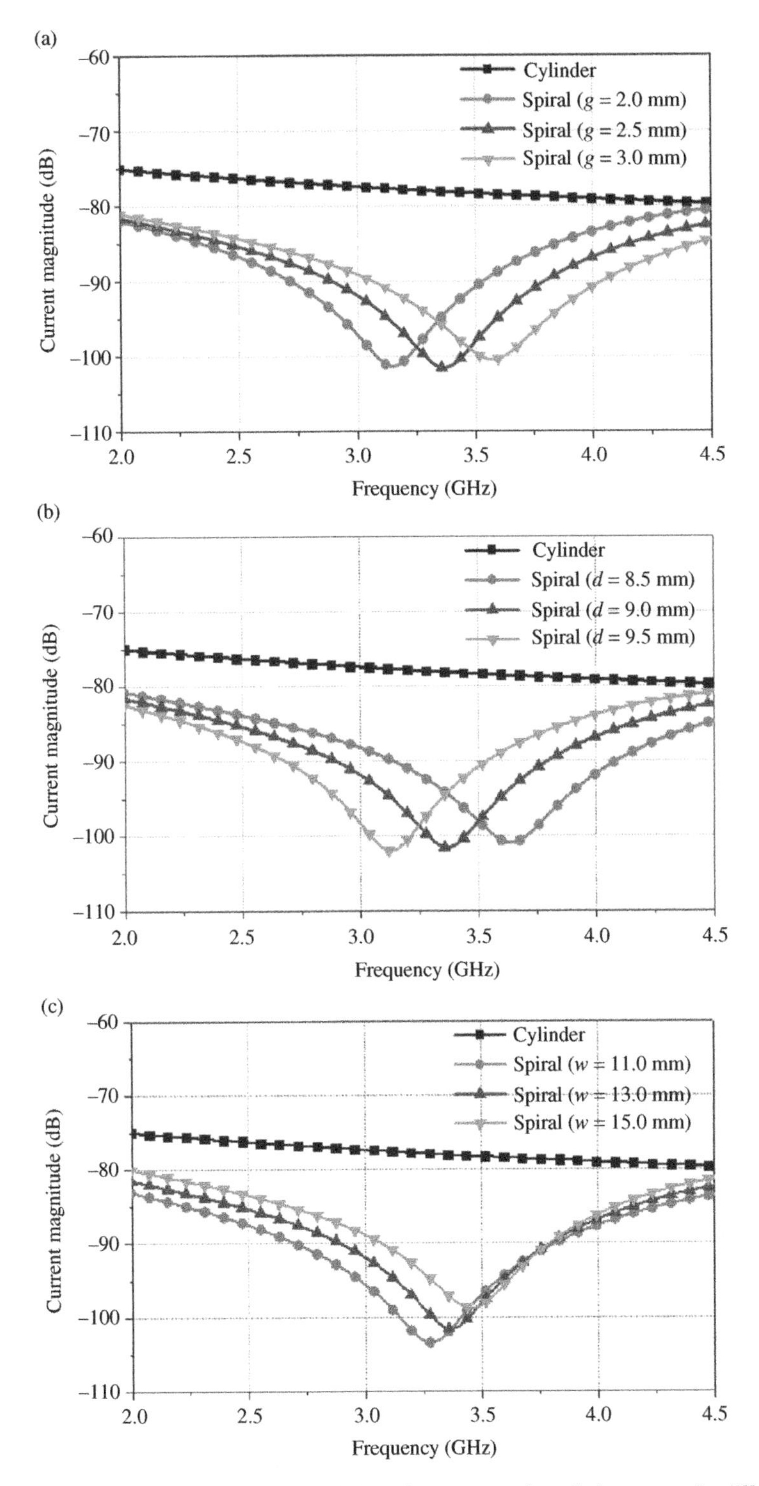

Figure 4.32 Magnitudes of the induced currents on the spiral structure for different values of (a) gap g, (b) inner diameter d, and (c) width of strip w.

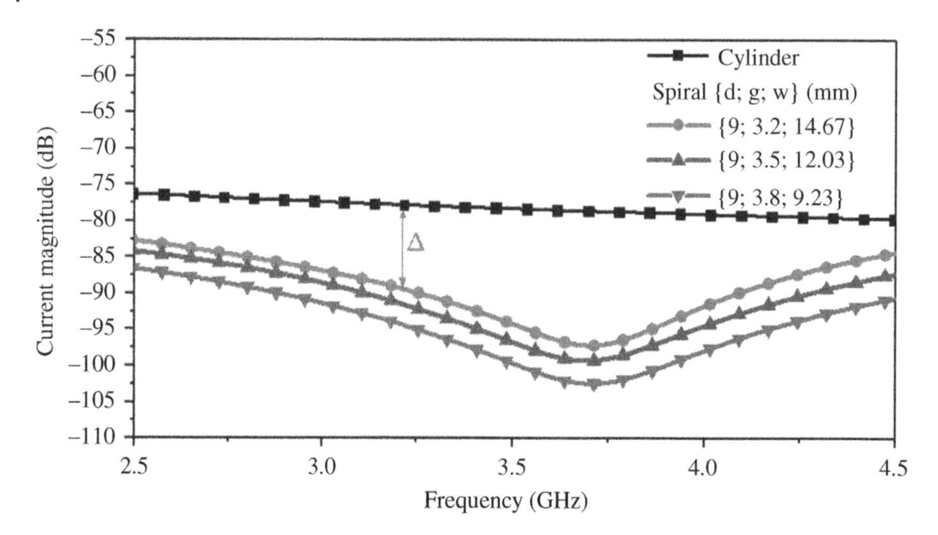

Figure 4.33 Magnitudes of the currents induced on the spiral structure with different combinations of its dimensions, but all of which are resonating at 3.7 GHz.

Table 4.3 SSBW of spirals with different combinations of dimensions.

{*d*, *g*, *w*} (mm)	SSBW (%)
{9.0; 3.2; 14.67}	30.5
{9.0; 3.5; 12.03}	40.9
{9.0; 3.8; 9.23}	61.9

inductance and capacitance of the spiral. As observed from Figure 4.32c, a larger w moves the resonance point to a higher frequency, as it decreases the inductance of the spiral.

To quantitatively assess the bandwidth of a choke's ability to efficiently suppress the scattering currents, the scattering suppression bandwidth (SSBW) is defined as the bandwidth in which $\Delta >$ 10 dB, where Δ is the difference in dB between the currents induced on a structure with and without chokes being present. In the current case, the structures are the spiral and the corresponding cylindrical rod. While several different spirals have the same resonance point at 3.7 GHz, their different dimensions lead to different induced currents on them. The selected cases and the induced currents on them are plotted in Figure 4.33. The currents induced on the unmodified cylinder are also simulated and plotted for comparison. The difference Δ is labeled in the figure. The SSBW values for the spirals with different combinations of their design parameters are calculated and listed in Table 4.3. The results show that a spiral with a larger g and smaller w has a wider SSBW value. The spiral with a smaller w corresponds to a smaller capacitance and a larger inductance per unit length, which leads to a larger L/C ratio. This result indicates the fact that the distributed spiral choke has a similar property to that of the lumped chokes, i.e. the larger the L/C ratio, the broader the SSBW.

3) Analysis of a finite long spiral

The periodic boundary conditions applied to the unit cell model were suited for simulating an infinitely long spiral to theoretically study its resonance. However, the length of the spiral used for an antenna arm is finite in practice. Its optimal length is determined by the operating frequency of the LB antennas. We have conducted numerous parameter studies to optimize all of the dimensions of the finite length spiral structure.

Figure 4.34 shows the simulation setup. A spiral of 29 mm in length is placed in free space and centred in the simulation box with open (radiation) boundaries on all six faces. It is illuminated by an ideal, normally incident plane wave with its electric field along the axis of the spiral. The current induced on the spiral is monitored. The three design parameters g, *d*, and *w* were varied. The results are shown in Figures 4.35a–c. The corresponding results for the finite-length cylindrical rod

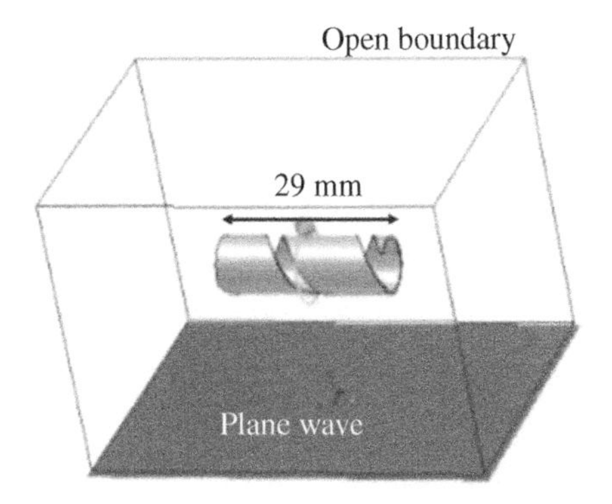

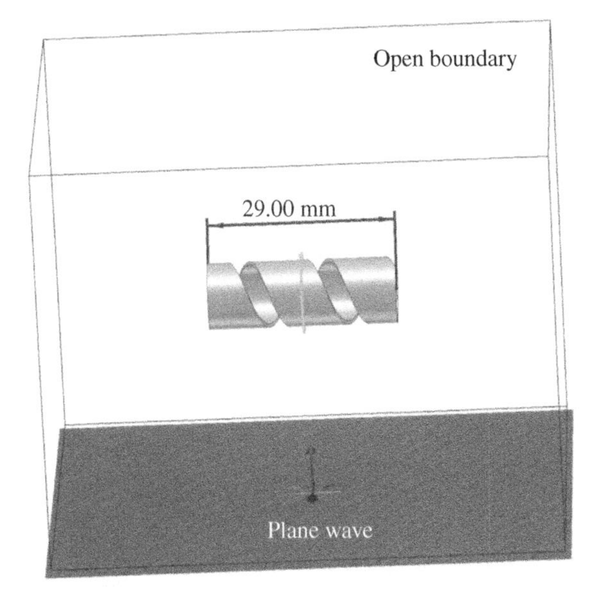

Figure 4.34 Simulation model of the finite spiral structure illuminated by the ideal, normally incident plane wave.

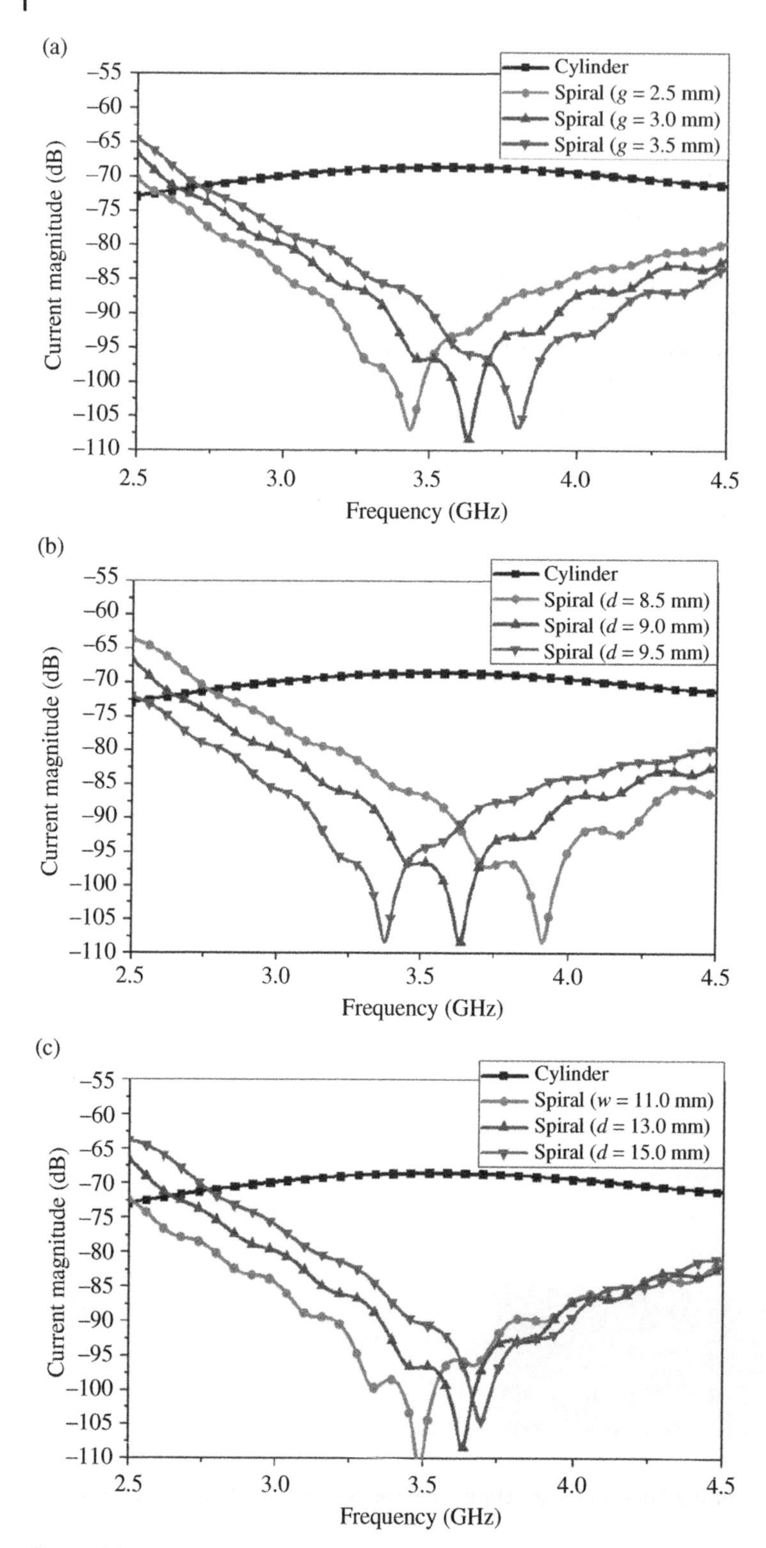

Figure 4.35 Magnitudes of the induced currents on the finite-length spiral structure with different values of (a) gap g, (b) inner diameter d, and (c) width of strip w. (d) Magnitudes of the induced currents on the finite-length spiral structure resonating at 3.7 GHz with different combinations of dimensions.

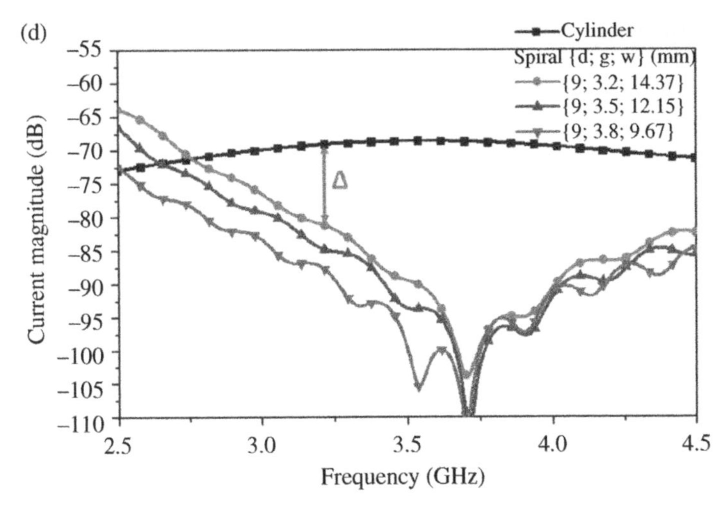

Figure 4.35 (Continued)

are also shared for comparison purposes. Note that the ripples shown in the curves are due to the discontinuities at the edges of the spiral structure.

Compared with the results shown in Figure 4.32, the finite-length case has similar outcomes. A larger g, a larger w, or a smaller d moves the minimum induced current point to a higher frequency and, hence, the resonance point of the spiral moves to a higher frequency. Also, note that the positions of the resonance points are almost identical for both the finite- and infinite-length cases. Figure 4.35d shows the case with the same resonance points but different combination of dimensions. As with the infinite-length case results, the finite-length spiral with a larger g and a smaller w has a wider scattering suppression bandwidth.

According to previous studies, the lumped choke [13, 31] and the distributed choke (for both the finite and infinite cases) [15] share similar properties in regards to suppressing the HB scattering. The chokes suppress the induced HB current and exhibit the strongest suppression at the resonance frequency f_{res}. Moreover, the SSBW of the chokes remains related to the ratio of L/C. The larger the ratio, the wider is the SSBW. These aspects of the design can be used as general guidelines when designing de-scattering chokes in the future.

4) Performance Evaluation of the Spiral Choke

To evaluate its performance, the spiral structure is implemented as the LB radiator in a compact 4G and 5G dual-band antenna array to suppress the crossband scattering. This system is shown in Figure 4.36. Both the LB and HB radiators have cross-dipole arrangements and yield the required dual-polarized radiation for base station applications. The LB and HB antennas are designed to operate from 1.7 to 2.3 GHz, and from 3.3 to 3.7 GHz for the 4G and 5G operations, respectively. The parameters of the spirals in this design were optimized to have the resonance point at 3.7 GHz since it was observed that the pattern distortion is the most severe at this frequency. An additional constraint on the parameters is that the value of d must be selected to allow matching of the LB element across its required bandwidth. The optimized dimension of the spiral was determined to be $d = 9.0$ mm, $w = 14.1$ mm, and $g = 3.23$ mm.

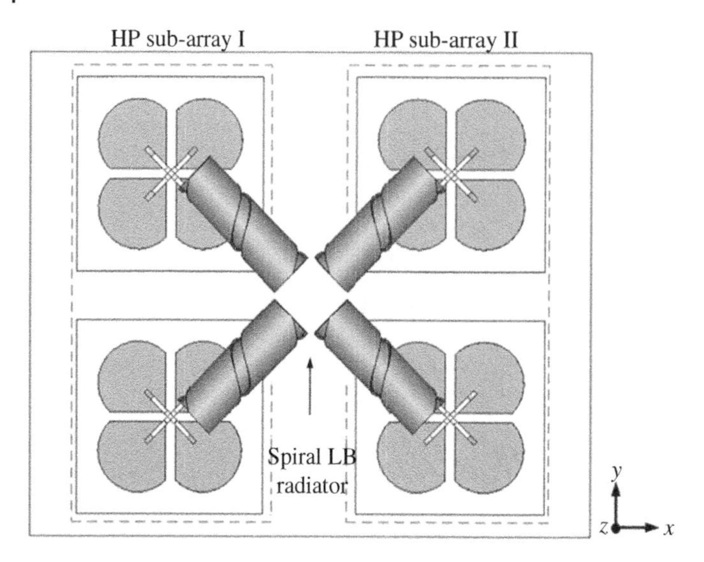

Figure 4.36 Top view of the interleaved 4G/5G BSA array with the spiral LB radiator. *Source*: From [15] / with permission of IEEE.

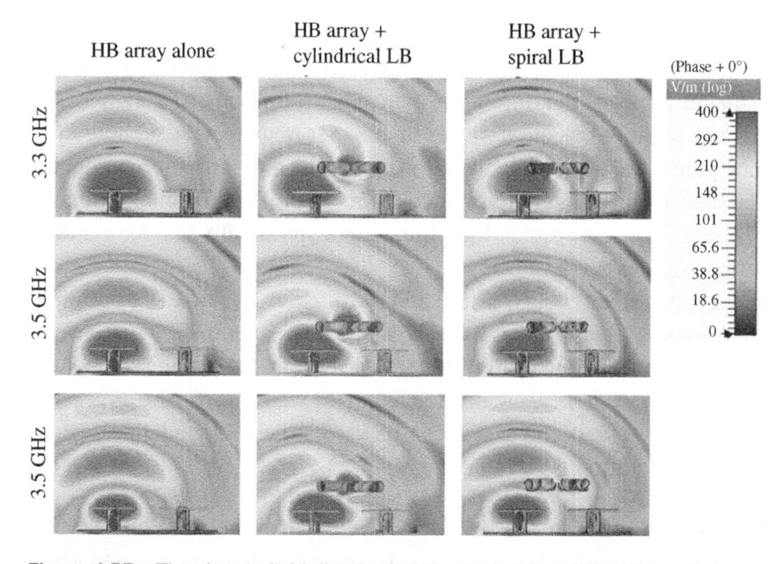

Figure 4.37 The electric field distributions in the *xz*-plane when the radiators consist of (i) only the HB array, (ii) the HB array with cylindrical LB radiators, and (iii) the HB array with the spiral LB radiators. *Source*: From [15] / with permission of IEEE.

To examine the spiral's ability in suppressing the HB scattering, the electric field distributions and the radiation patterns when one of the HB subarrays is excited were investigated. Three cases are again compared, i.e., the (i) HB array alone, (ii) HB array with cylindrical LB arms, and (iii) HB array with spiral LB arms. The LB feed networks were not considered and, hence, any impact they may have is not reflected in the results. This study only addressed the scattering caused by the HB currents on the LB elements. Figure 4.37 shows plots of the electric field distributions in the *x-z* plane for the three cases. Compared with those when only the HB array exists, the cylindrical LB radiators block the HB electric field to a large extent, especially near 3.7 GHz. The spiral LB radiators have much less impact on the electric field distributions across the entire HB. The resulting horizontal patterns in all three cases from 3.3 to 3.7 GHz are shown in Figure 4.38. The deterioration of the HB radiation pattern in the presence of the cylindrical LB radiators was largely eliminated using the spiral LB arms. These results clearly demonstrate the effectiveness of the spiral structure in reducing the HB pattern distortion.

Having demonstrated the effectiveness of the spiral structure in suppressing the crossband scattering, the spiral LB radiators were re-matched to their source using baluns. The baluns are designed following the guidelines given for the impedance-matching network given in [29, 30]. Their design takes into account the change in the dipole feed point impedance caused by the spiral structure. The substrate is FR4 with a dielectric constant of 4.4 and a thickness of 1.0 mm. Two baluns are orthogonally arranged to feed the two pairs of spiral arms. The final configuration of the spiral LB antenna is shown in Figure 4.39. Figure 4.40 shows the matching results for this

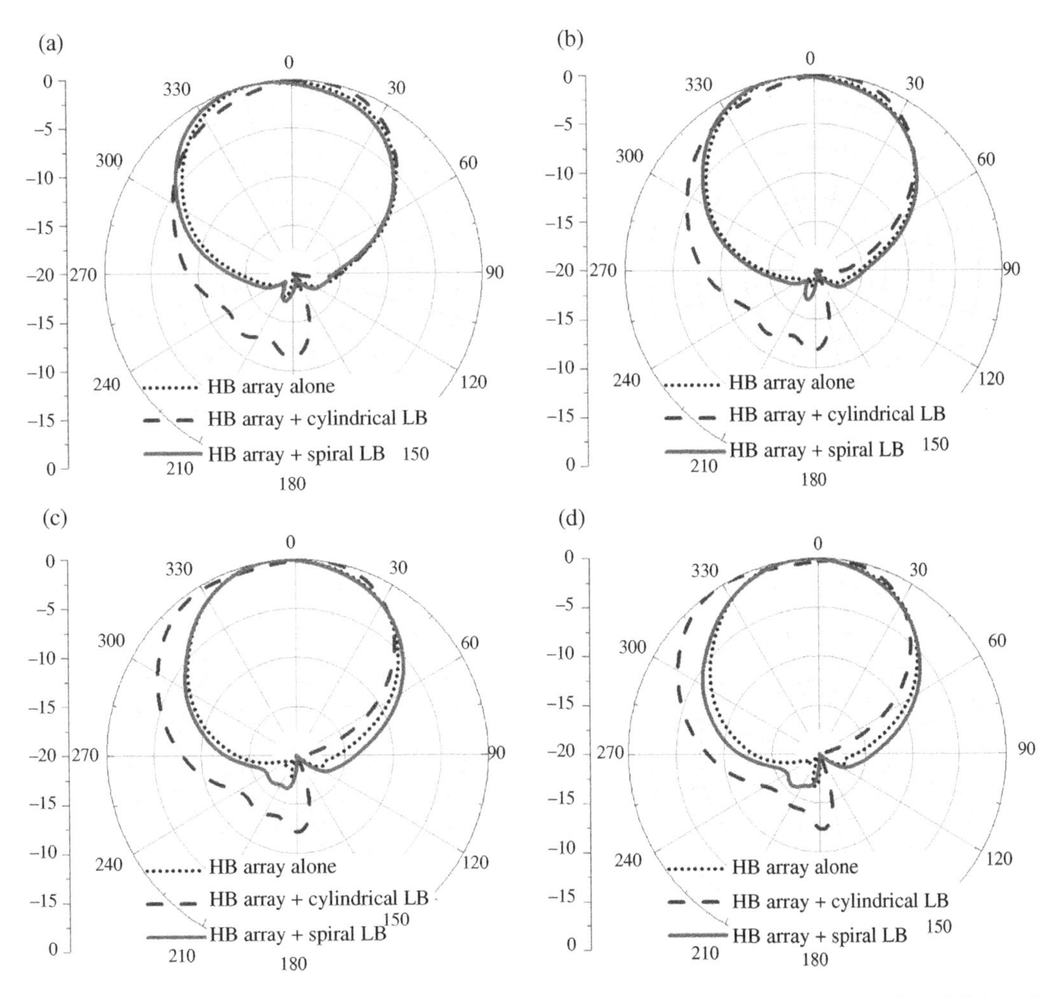

Figure 4.38 Comparison of the HB radiation pattern in the vertical *xz*-plane at (a) 3.3, (b) 3.4, (c) 3.6, and (d) 3.7 GHz for the three cases: (i) HB array only, (ii) HB array with cylindrical LB radiators, and (iii) HB array with spiral LB radiators. *Source:* From [15] / with permission of IEEE.

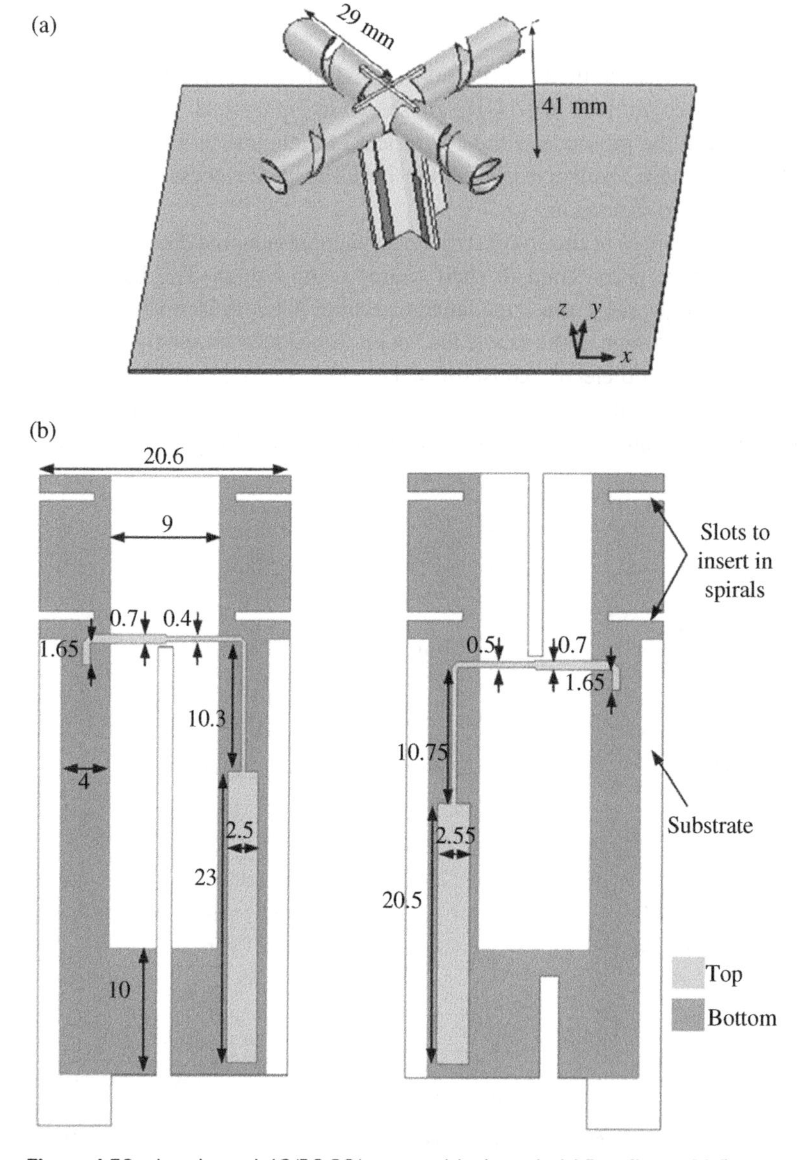

Figure 4.39 Interleaved 4G/5G BSA array with the spiral LB radiator. (a) Perspective view of the spiral LB element. (b) Configuration of the two baluns driving the spiral LB element. *Source*: From [15] / with permission of IEEE.

choked element. The antenna is matched with its reflection coefficients being less than −15 dB from 1.66 to 2.22 GHz. The simulated radiation patterns of the spiral arm LB element and the unaltered strip arms are shown in Figure 4.41 at 1.7, 1.9, and 2.2 GHz. The radiation patterns are very stable across the matched band and the cross-polarization level is less than −20 dB at boresight. All the simulated results demonstrate that the spiral element performs well across the entire LB range.

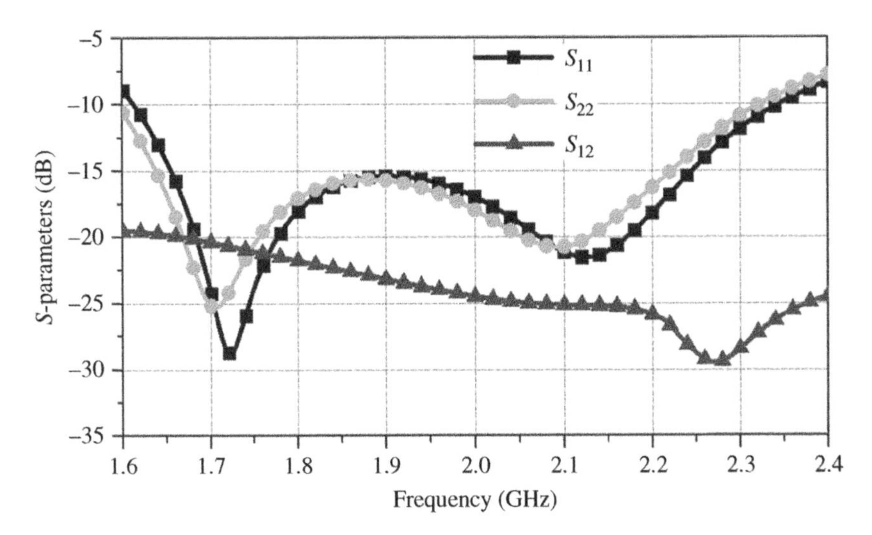

Figure 4.40 Simulated S-parameters of the spiral LB antenna in the interleaved 4G/5G BSA array.

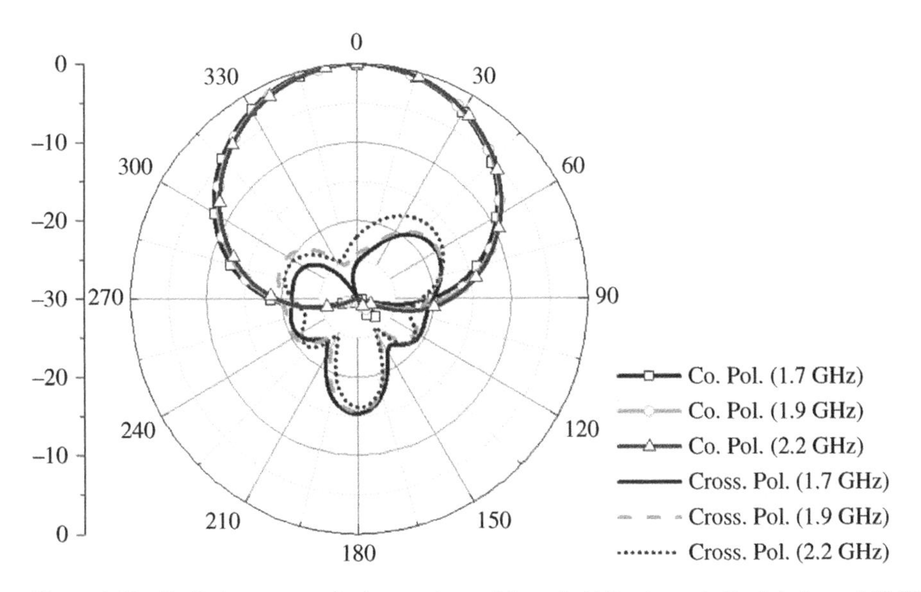

Figure 4.41 Radiation patterns in the x-z plane of the spiral LB antenna in the interleaved 4G/5G BSA array at 1.7, 1.9, and 2.2 GHz.

5) Results and Discussion

The final configuration of the dual-band 4G/5G BSA section is shown in Figures 4.42a and 4.42b. Power dividers printed on the back of the reflector are used to feed the two HB antennas in each column together as a subarray. The distance between the two HB subarrays is set as 1.0 λ at 3.5 GHz to guarantee enough isolation and to meet the MIMO requirements. As Figure 4.42b indicates, ports 1 and 3 feed the +45° polarization state, and ports 2 and 4 feed the −45° polarization state of the two HB subarrays. Ports 5 and 6 excite the LB antenna's +45°- and −45°-polarization states, respectively.

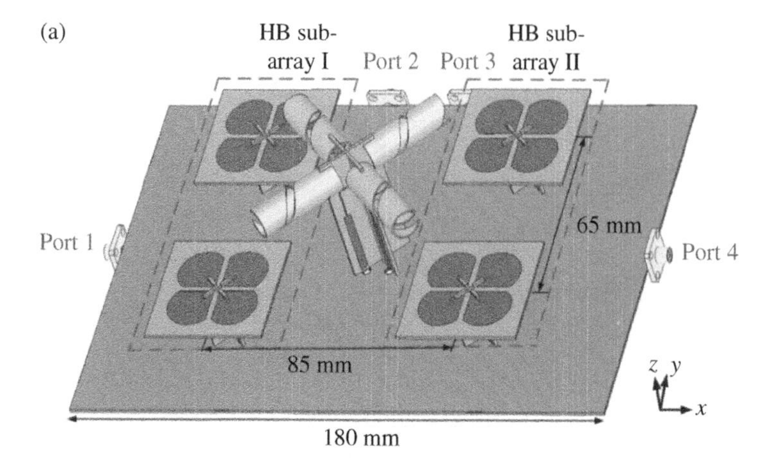

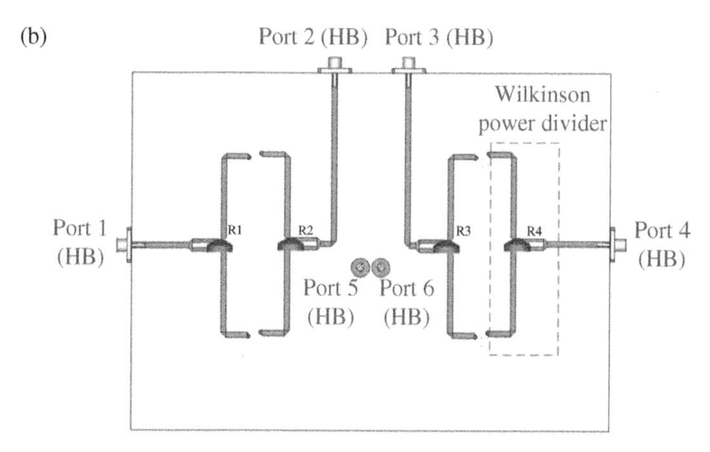

Figure 4.42 The interleaved 4G/5G BSA array with the spiral LB radiator. (a) Simulation model. (b) Feed network. *Source*: From [15] / with permission of IEEE.

The dual-band base station antenna array section that incorporates the spiral LB antenna was fabricated and tested. The array is shown in Figure 4.43. Note that the spirals were obtained with laser cutting technology. Figures 4.44a and 4.44b give the simulated and measured results, respectively, of the reflection coefficients at the HB ports. The measured bandwidth is 11.4% from 3.3 to 3.7 GHz with reflection coefficients <−15 dB. Figures 4.44c and 4.44d show the simulated and measured results, respectively, of the transmission coefficients between the HB ports. The measured results demonstrated that good isolations, >20 dB, were achieved. The simulated and measured reflection and transmission coefficients at the LB ports are plotted in Figure 4.45. The measured results are similar to the simulated ones. The LB antenna has a bandwidth of 28.3% from 1.7 to 2.26 GHz with reflection coefficient magnitudes < −15 dB. Compared with the LB antenna with lumped chokes [13], the distributed-choked spiral LB antenna has a much wider bandwidth.

The simulated and measured HB radiation patterns are shown in Figure 4.46. The +45°- and −45°-polarization state radiation patterns for subarray I are presented in Figures 4.46a and 4.46b,

(a)

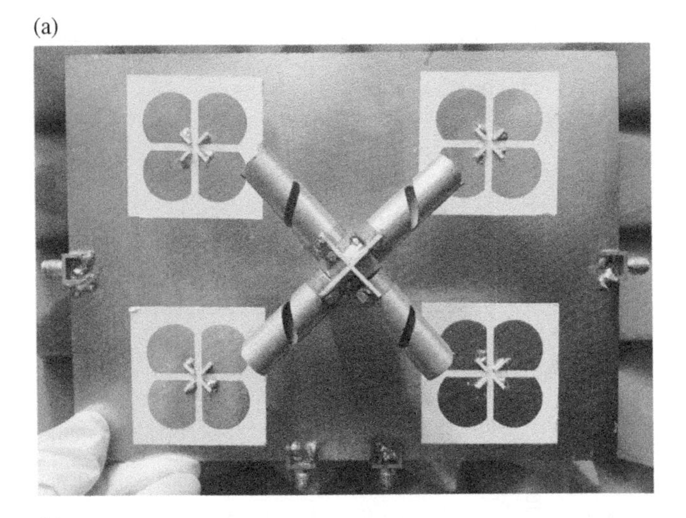

(b)

Figure 4.43 Fabricated prototype of the interleaved 4G/5G BSA array with the spiral LB radiator. (a) Front view. (b) Side view.

respectively. The radiation patterns for subarray II are very similar to those of subarray I. Consequently, they are not presented here. These results demonstrate that the measured and simulated radiation patterns are quite similar. They are stable across the operating band. The simulated and measured cross-polarization levels are < -20 dB and < -15 dB, respectively, in the boresight direction. Figure 4.47 shows the HPBWs and realized gains of the two HB subarrays for both polarization states. Again, the radiation patterns are stable across the operating band. The HPBWs vary within $66.5° \pm 3.5°$. The measured gain is around 11 dBi, which is only 0.5 dBi less than the simulated value. This difference is attributed to the losses in the SMA terminators and cables used for the measurements.

The simulated and measured +45°- and −45°-polarization state radiation patterns in the *x-z* plane of the LB antenna are plotted in Figures 4.48a and 4.48b, respectively. The simulated and measured patterns agree well with each other; the cross-polarization levels are < -17 dB across the band in the boresight direction. The horizontal HPBW and gain of the LB element are plotted in Figure 4.49. The radiation patterns have consistent HPBWs of $65° \pm 5°$. The simulated and

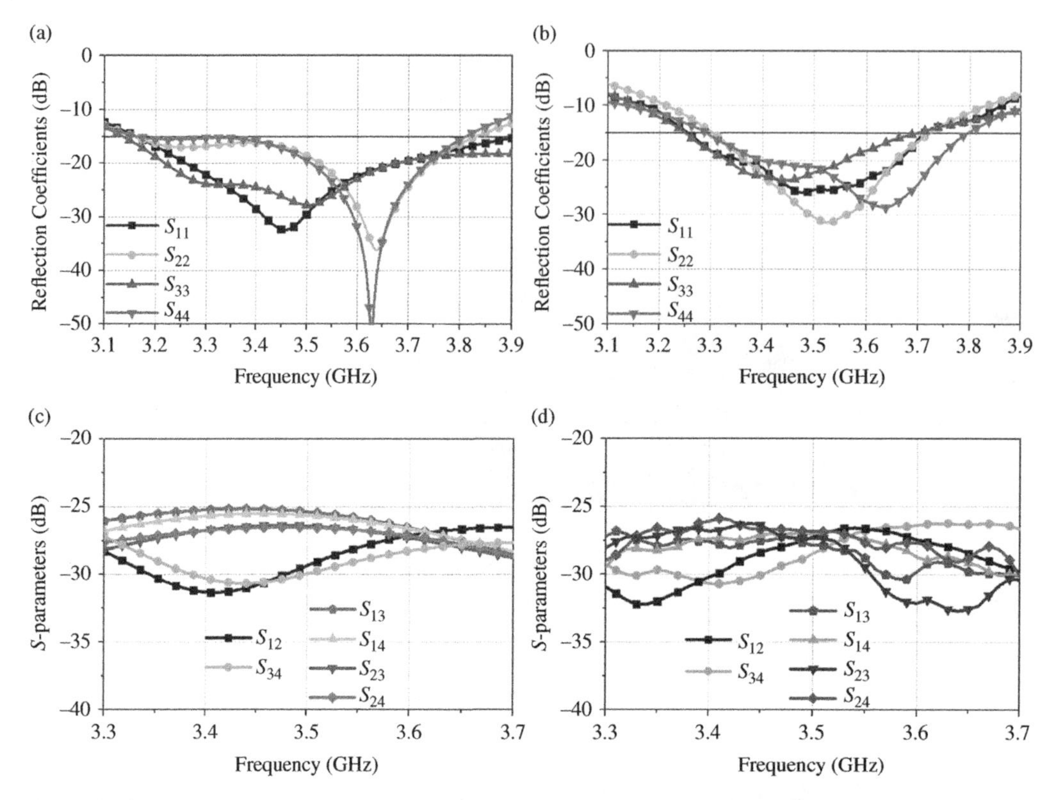

Figure 4.44 Magnitudes of the (a) simulated and (b) measured reflection coefficients at the HB ports. Magnitudes of the (c) simulated and (d) measured transmission coefficients between the HB ports.

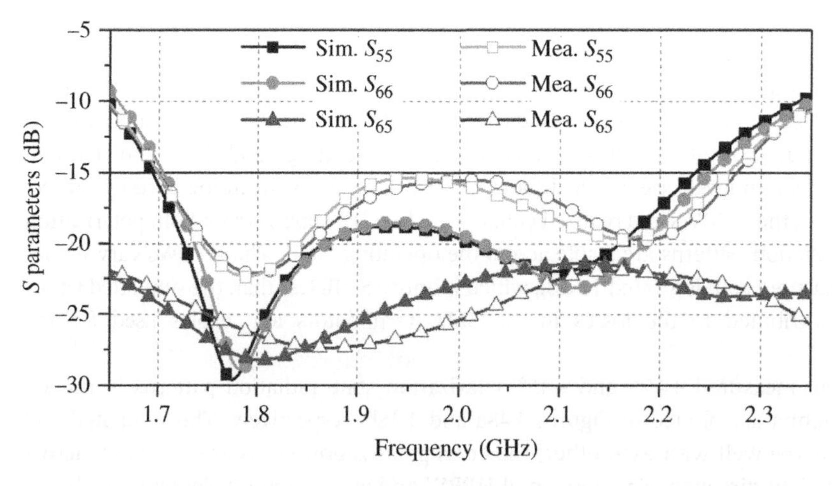

Figure 4.45 Magnitudes of the simulated and measured reflection and transmission coefficients at the LB ports. *Source*: From [15] / with permission of IEEE.

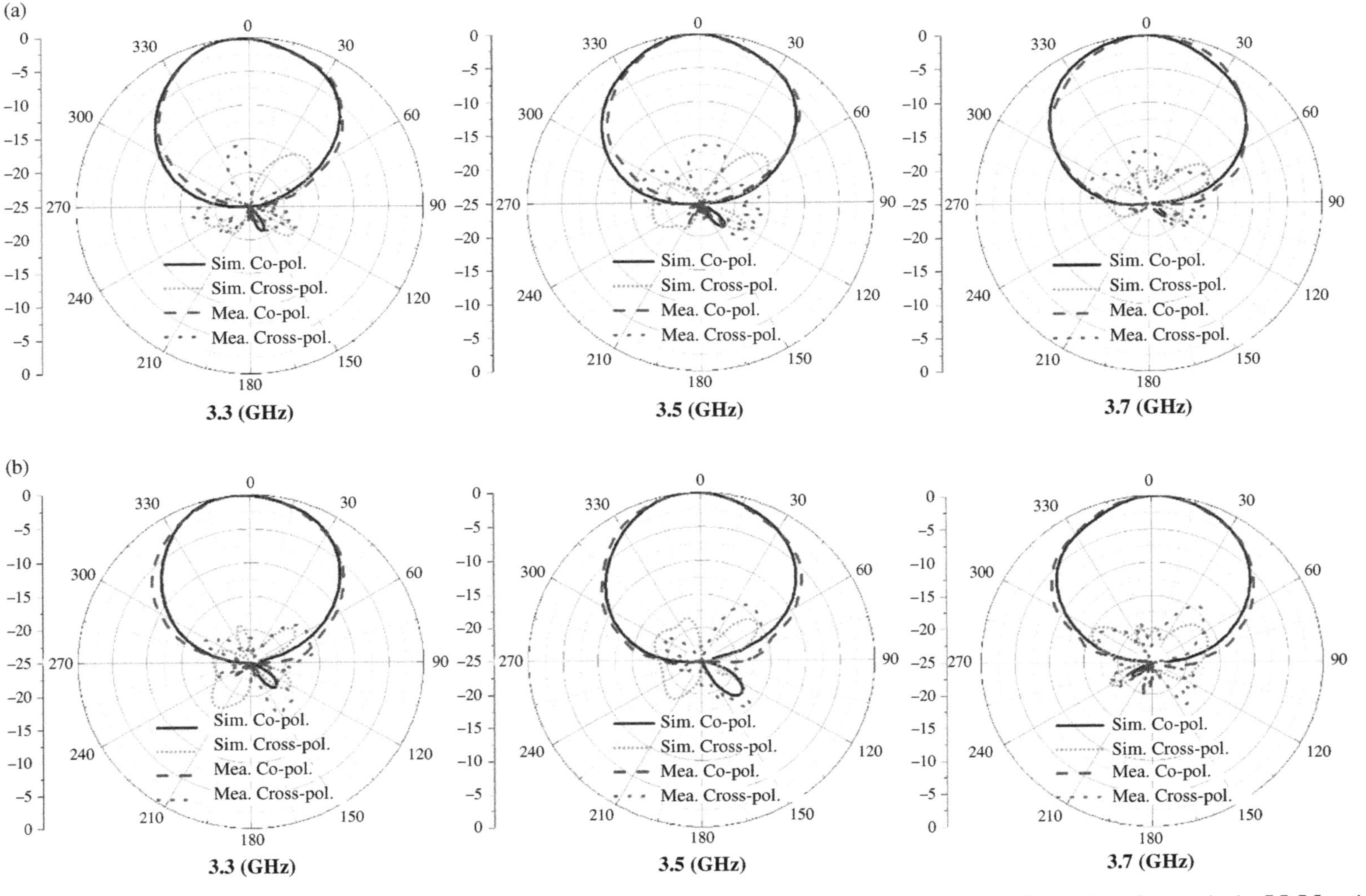

Figure 4.46 Simulated and measured radiation patterns in the *x-z* plane of the HB subarray in the left column when (a) port 1 and (b) port 2 are excited at 3.3, 3.5, and 3.7 GHz.

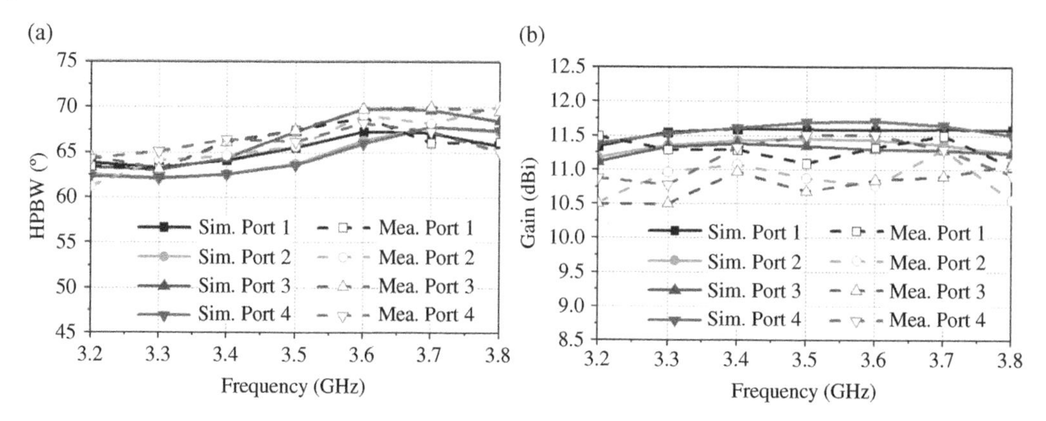

Figure 4.47 Simulated and measured results of the HB subarrays. (a) HPBW in the *x-z* plane. (b) Realized gain.

measured gain is around 8 and 7 dBi, respectively. The difference is caused by the losses in coaxial cable, the spiral-shaped metal and the substrate in the feed network.

All of the presented results clearly demonstrate that the distributed-choked spiral LB antenna has a minimum scattering effect on the HB antennas and that a good radiation pattern is maintained across its operating band. Compared to the lumped-choked LB antenna presented in the previous section, the distributed-choked antenna is easier to match and has a wider bandwidth. However, both the lumped-choked and distributed-choked antennas suffer from slightly reduced gain values when they are compared with conventional cross-dipoles.

4.5 Mitigating the Effect of HB Antennas on LB Antennas

While there have been many efforts to reduce the scattering issues associated with radiators in complicated multiband antenna array systems, they have focused mainly on reducing the LB radiators' scattering on the HB performance. This strategy results from the recognition that the LB radiators have larger sizes and are naturally big scatterers in relation to the shorter wavelength HB radiation. On the other hand, several HB radiators are placed under the LB radiators in multiband antenna arrays as illustrated in Figure 4.50. Since the HB radiators are connected to the ground, they form a set of elements that resemble a mushroom structure and cause scattering effects to the LB radiation as well. This scattering is especially significant when the LB radiator works over a wide band. Although the scattering effects on the LB elements due to HB radiators are weaker, the best performance would be attained if all scattering events were addressed properly with de-scattering techniques.

It has been found that there is a significant amount of current induced on the HB elements when the LB element is excited [32]. This current would re-radiate and deform the original LB antenna radiation. Although this effect tends to happen within a fraction of the LB antenna band, the LB patterns at these frequencies can be significantly widened and the realized gains dramatically reduced. One method to mitigate the effect is to introduce some inductive or capacitive loading to the HB antenna. The net result is to move the scattering effect outside of the LB band [32].

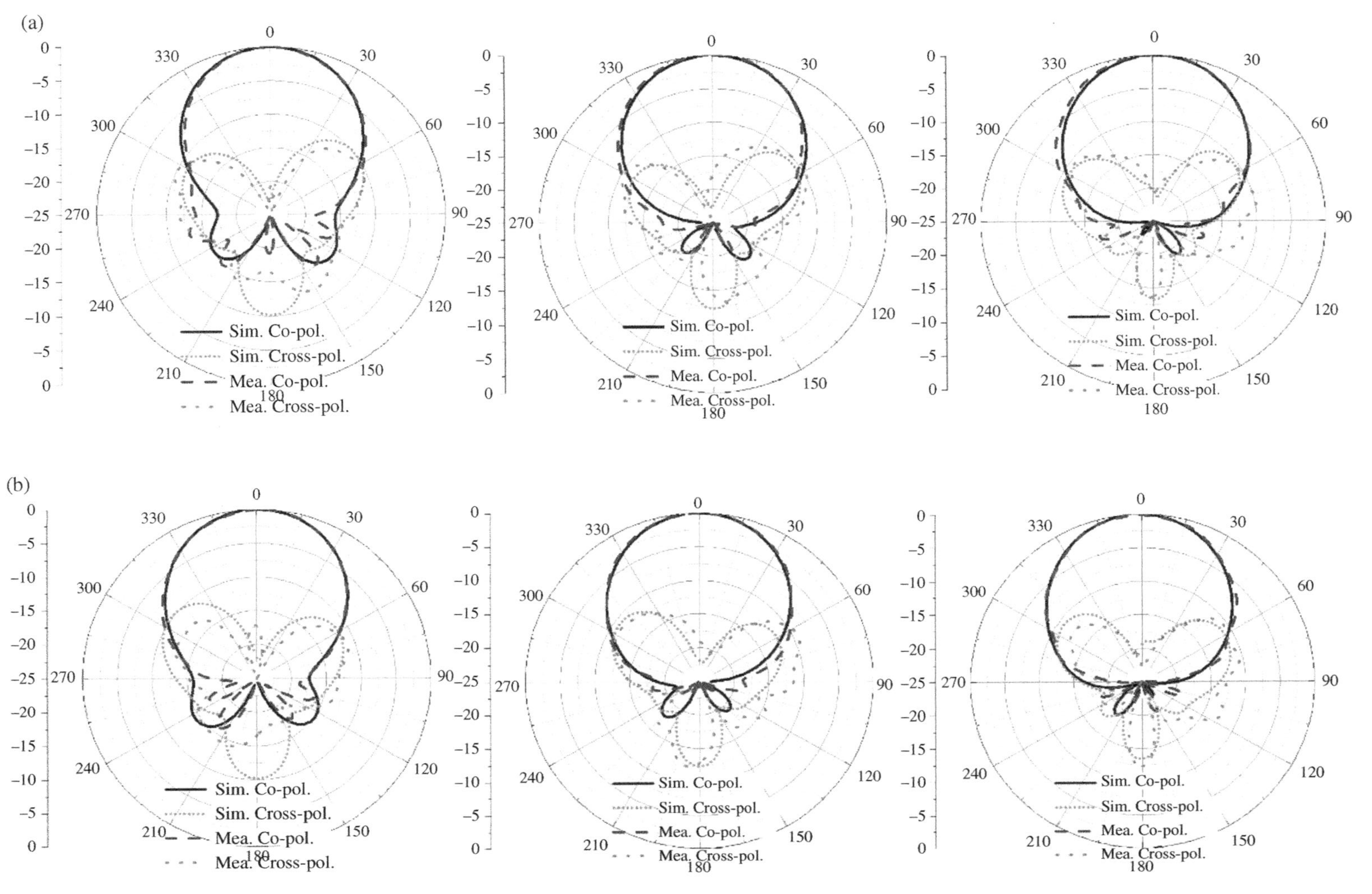

Figure 4.48 Simulated and measured radiation patterns in the *x-z* plane of the LB antenna when (a) port 5 and (b) port 6 are excited at 1.7, 1.9, and 2.2 GHz.

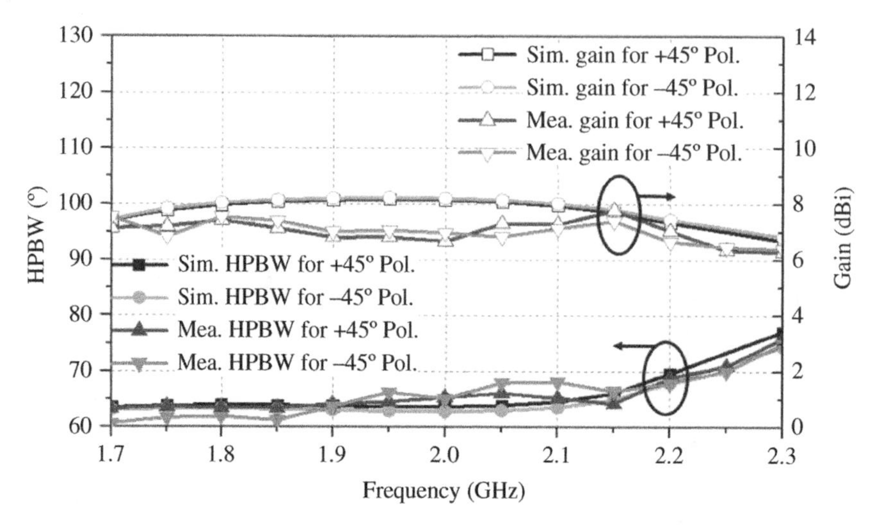

Figure 4.49 LB antenna's simulated and measured HPBW and gain values. *Source:* From [15] / with permission of IEEE.

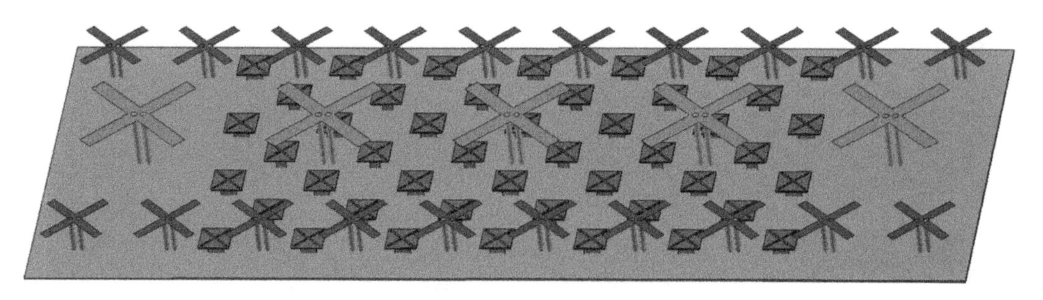

Figure 4.50 Tri-band 3G/4G/5G BSA.

4.6 Conclusions

This chapter described successful de-scattering strategies and efforts applied to multiband base station arrays. Lumped chokes were designed and implemented in a representative 3G/4G BSA. Distributed spiral chokes were designed and implemented in an interleaved 4G/5G BSA array. The HB crossband scattering was essentially eliminated in both cases. It was illustrated that the distributed choke approach led to a significantly wider bandwidth. The simulation and experimental results of the prototype dual-band arrays with de-scattering elements incorporated in them demonstrated that the distortion of the HB pattern caused by LB scattering was largely eliminated. Finally, some thoughts were shared for future research into de-scattering technologies.

References

1. Tse, D. and Viswanath, P. (2010). *Fundamentals of Wireless Communication*, 120. Cambridge, United Kingdom: Cambridge University Press.

2. Xia, R., Qu, S., Li, P. et al. (2015). An efficient decoupling feeding network for microstrip antenna array. *IEEE Antennas Wirel. Propag. Lett.* **14**: 871–874.
3. Lau, B.K. and Andersen, J.B. (2012). Simple and efficient decoupling of compact arrays with parasitic scatterers. *IEEE Trans. Antennas Propag.* **60** (2): 464–472.
4. Zhang, S. and Pedersen, G.F. (2016). Mutual coupling reduction for UWB MIMO antennas with a wideband neutralization line. *IEEE Antennas Wirel. Propag. Lett.* **15**: 166–169.
5. Niu, Z., Zhang, H., Chen, Q., and Zhong, T. (2019). Isolation enhancement for 1×3 closely spaced E-plane patch antenna array using defect ground structure and metal-vias. *IEEE Access* **7**: 119375–119383.
6. Liu, F., Guo, J., Zhao, L. et al. (2019). A meta-surface decoupling method for two linear polarized antenna array in sub-6 GHz base station applications. *IEEE Access* **7**: 2759–2768.
7. Leonhardt, U. (2006). Optical conformal mapping. *Science* **312** (5781): 1777–1780.
8. Kwon, D. and Werner, D.H. (2010). Transformation electromagnetics: an overview of the theory and applications. *IEEE Antennas Propag. Mag.* **52** (1): 24–46.
9. Chen, P.Y., Soric, J., and Alu, A. (2012). Invisibility and cloaking based on scattering cancellation. *Adv. Opt. Mat.* **24**: OP281–OP304.
10. Alu, A. and Engheta, N. (2005). Achieving transparency with plasmonic and metamaterial coatings. *Phys. Rev. E* **72** (1): 016623.
11. Zhang, X.Y., Xue, D., Ye, L.-H. et al. (2017). Compact dual-band dual-polarized interleaved two-beam array with stable radiation pattern based on filtering elements. *IEEE Trans. Antennas Propag.* **65** (9): 4566–4575.
12. Zhang, Y., Zhang, X.Y., Ye, L., and Pan, Y. (2016). Dual-band base station array using filtering antenna elements for mutual coupling suppression. *IEEE Trans. Antennas Propag.* **64** (8): 3423–3430.
13. Sun, H.-H., Ding, C., Zhu, H., and Guo, Y.J. (2019). Suppression of cross-band scattering in multiband antenna arrays. *IEEE Trans. Antennas Propag.* **67** (4): 2379–2389.
14. Ding, C, Sun, H.-H, and Guo, Y.J. (2019). Enabling the co-existence of multiband antenna arrays. *Proceedings of 9th International Symposium on Phased Array Systems and Technology (ISPAST)*, Waltham, MA, USA, October 2019, pp. 1–4.
15. Sun, H.H., Jones, B., Ding, C. et al. (2020). Scattering suppression in a 4G and 5G base station antenna array using spiral chokes. *IEEE Antennas Wireless Propag. Lett.* **19** (10): 1818–1822.
16. Schurig, D., Mock, J.J., Justice, B.J. et al. (2006). Metamaterial electromagnetic cloak at microwave frequencies. *Science* **314** (5801): 977–980.
17. Landy, N. and Smith, D. (2013). A full-parameter unidirectional metamaterial cloak for microwaves. *Nat. Mater.* **12**: 25–28.
18. Chen, X., Luo, Y., Zhang, J. et al. (2011). Macroscopic invisibility cloaking of visible light. *Nat. Commun.* **2**: 176.
19. Xu, S., Cheng, X., Xi, S. et al. (2012). Experimental demonstration of a free-space cylindrical cloak without superluminal propagation. *Phys. Rev. Lett.* **109**: 223903.
20. Jiang, Z.H. and Werner, D.H. (2013). Exploiting metasurface anisotropy for achieving near-perfect low-profile cloaks beyond the quasi-static limit. *J. Phys. D. Appl. Phys.* **46** (50): 505306.
21. Selvanayagam, M. and Eleftheriades, G.V. (2013). Experimental demonstration of active electromagnetic cloaking. *Phys. Rev. X* **3** (4): 041011.
22. Alù, A. (2009). Mantle cloak: Invisibility induced by a surface. *Phys. Rev. B* **80** (24): 245115.
23. Chen, P., Monticone, F., and Alù, A. (2011). Suppressing the electromagnetic scattering with an helical mantle cloak. *IEEE Antennas Wirel. Propag. Lett.* **10**: 1598–1601.
24. Monti, A., Soric, J., Alù, A. et al. (2012). Overcoming mutual blockage between neighboring dipole antennas using a low-profile patterned metasurface. *IEEE Antennas Wirel. Propag. Lett.* **11**: 1414–1417.

25. Jiang, Z.H., Sieber, P.E., Kang, L., and Werner, D.H. (2016). Restoring intrinsic properties of electromagnetic radiators using ultralightweight integrated metasurface cloaks. *Adv. Funct. Mater.* **26** (18): 2986–2986.

26. Monti, A., Soric, J., Barbuto, M. et al. (2016). Mantle cloaking for co-site radio-frequency antennas. *Appl. Phys. Lett.* **108** (11): 113502.

27. Vellucci, S., Monti, A., Barbuto, M. et al. (2017). Satellite applications of electromagnetic cloaking. *IEEE Trans. Antennas Propag.* **65** (9): 4931–4934.

28. Ding, C., Sun, H.-H., Ziolkowski, R.W., and Guo, Y.J. (2017). Simplified tightly-coupled cross-dipole arrangement for base station applications. *IEEE Access* **5**: 27491–27503.

29. Ding, C., Jones, B., Guo, Y.J., and Qin, P.-Y. (2017). Wideband matching of full-wavelength dipole with reflector for base station. *IEEE Antennas Wirel. Propag. Lett.* **65** (10): 5571–5576.

30. Ding, C., Sun, H.-H., Zhu, H., and Guo, Y.J. (2020). Achieving wider bandwidth with full-wavelength dipoles (FWDs) for 5G base stations. *IEEE Trans. Antennas Propag.* **68** (2): 1119–1127.

31. Shang, C., Jones, B.B., and Isik, O. (2012). Dual-band interspersed cellular basestation antennas. US Patent 9,570,804, filed 24 December 2012 and issued 3 July 2014.

32. Sun, H.-H., Jones, B., Jay Guo, Y., and Hui Lee, Y. (2020). Suppression of cross-band scattering in interleaved dual-band cellular base-station antenna arrays. *IEEE Access* **8**: 222486–222495.

5

Differential-Fed Antenna Arrays

Owing to their numerous advantages including high immunity to environmental noise and electromagnetic interference, high level of suppression of even-order harmonics, and very low level of cross-polarization in their radiation patterns, differential-fed antennas have attracted a lot of recent interest in the antennas research community. In contrast to conventional single feed antennas, differential-fed antennas have two feed ports with a 180° phase difference between them. This provides more flexibility in manipulating the current distribution on the antenna and the feed, and, therefore, potentially better antenna performance.

As shown in Figure 5.1, a typical RF transmitter or receiver is composed of multiple balanced devices. These include differential filters, amplifiers, mixers, and antennas. Differential antennas can be directly connected to these balanced devices without needing a balun, which avoids complicated feeding structures. Most papers published to date in this area have focused on differential-fed antenna elements. To fully exploit the benefits of differential-fed antennas, however, one needs to develop differential-fed antenna arrays to meet the requirements of current 5G and evolving 6G and beyond wireless communication networks that demand high gain and scanning beams. Whilst the design of differential-fed antenna elements is a challenging task itself, the design of the differential feeding network (DFN) for antenna arrays, multi-beam antenna arrays in particular, is even more challenging.

One of the most critical issues to be addressed in DFNs is the common-mode (CM) suppression. Although baluns can convert unbalanced signals to balanced ones, they also introduce CM currents which degrade the antenna pattern to some extent. Differential power dividers (PDs), which are core components in DFNs, can divide one signal into two signals and transform an unbalanced signal into a balanced one, or vice versa. To guarantee an antenna array with excellent performance characteristics, differential PDs must have high CM signal rejection and low mixed-mode conversion.

In this chapter, we discuss a number of major issues in the design of differential-fed antenna arrays. These include the design of differential PDs, differential beamforming networks (BFNs), differential Butler matrices (BMs), linearly polarized differential-fed antenna arrays, circularly polarized differential-fed antenna arrays, and differential-fed multi-beam antenna arrays.

5.1 Differential Systems

Differential circuits and systems are widely used in communication systems. From the perspective of signal processing, differential signals have high immunity to environmental noises. Thus, they lead to a high signal-to-noise ratio (SNR) in the receiver. The use of differential signals also can

Advanced Antenna Array Engineering for 6G and Beyond Wireless Communications, First Edition.
Y. Jay Guo and Richard W. Ziolkowski.

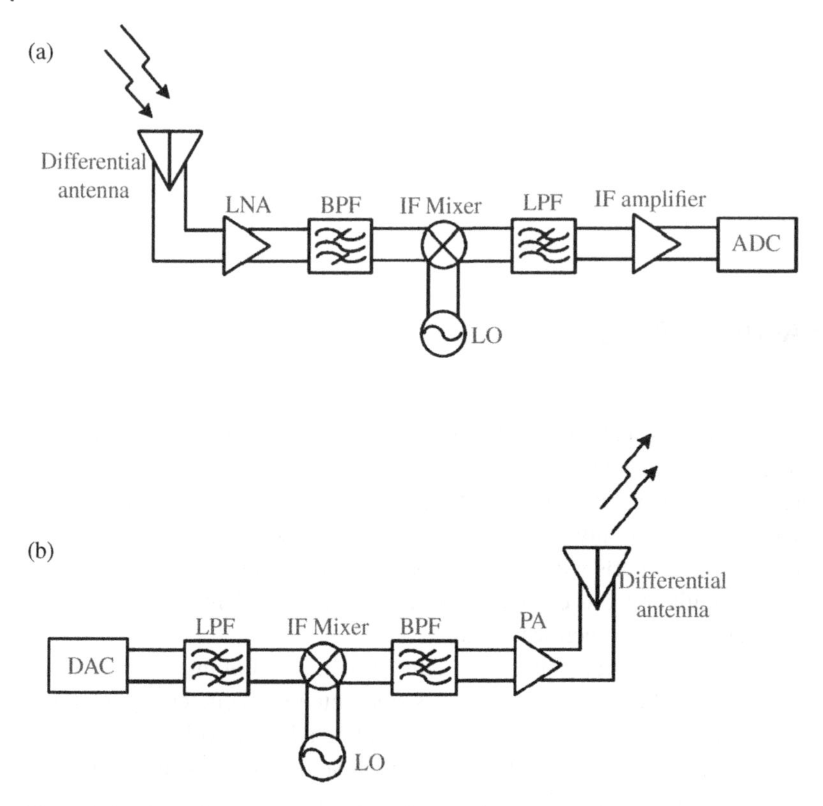

Figure 5.1 Typical configuration of a differential RF (a) receiver, and (b) transmitter.

reduce the coupling between the circuits and the antennas. This advantage arises from the fact that most of the capacitive mutual coupling occurs in the form of CM interactions.

Differential systems have two major benefits: one is noise immunity and the other is even-mode cancelation. Figure 5.2 shows a block diagram of a single-ended and a differential amplifier. In comparison with the single-ended one which only supports one mixed signal v_{in}, the differential device can transmit differential signals, i.e., a pair of balanced signals v_{in}^{+} and v_{in}^{-}. These two balanced signals have the same amplitude, but have a 180° phase difference.

Figure 5.3 shows how a differential device can suppress the environmental noise during the transmission process of a signal. In Figure 5.3a, a mixed signal v_{in} (as illustrated in blue) with noise (indicated in red) enters a single-ended amplifier. The noise in the output signal v_{out} is then amplified by the same amount that of the original signal. This is clearly an undesirable outcome during the transmission of signals. On the other hand, differential devices can effectively suppress the noise, e.g., they can suppress environmental noise. As illustrated in Figure 5.3b, both of the input signals, v_{in}^{+} and v_{in}^{-}, in a differential amplifier experience the same noise (as indicated in red). Furthermore, the noise is again amplified in both of the output signals, v_{out}^{+} and v_{out}^{-}. However, because of the phase difference, the composite output signal has no noise component, i.e., the differential device has an inherent noise cancelation property. Similarly, differential devices can suppress even-order harmonics because the even-mode components of v_{out}^{+} and v_{out}^{-} cancel each other out when the composite output signal is recovered. Only the odd-mode parts are maintained.

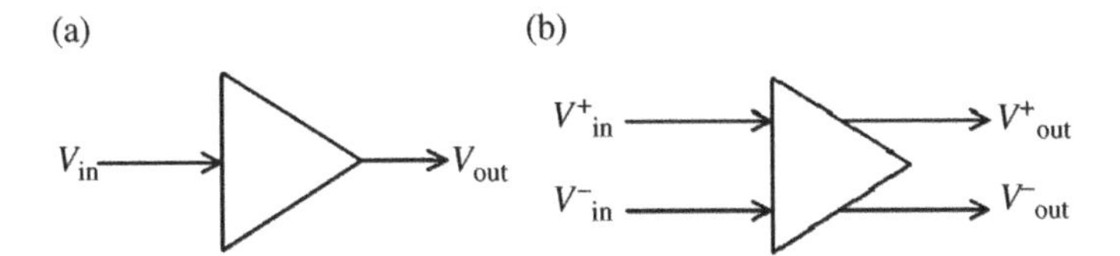

Figure 5.2 Block diagram of (a) single-ended, and (b) differential circuitry.

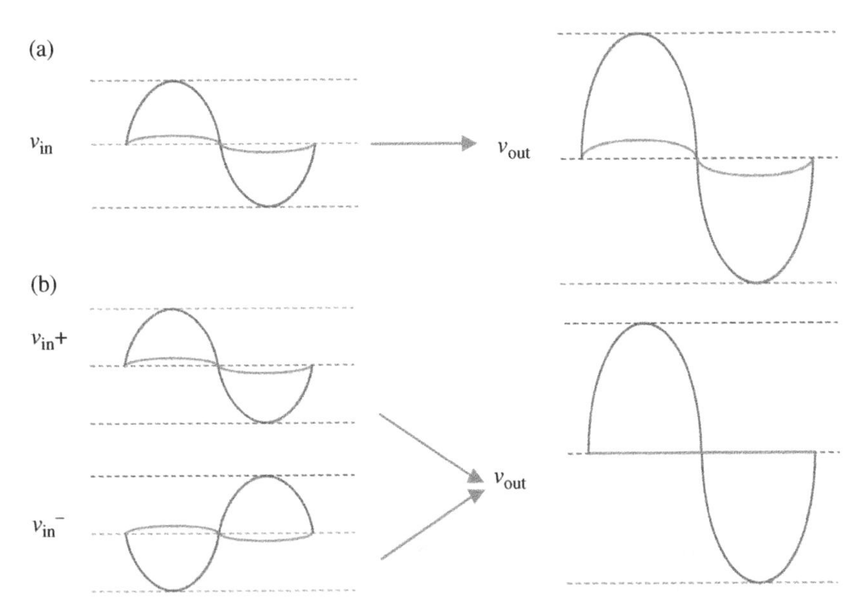

Figure 5.3 Block diagram of (a) mixed-mode, and (b) differential signals.

5.2 Differential-Fed Antenna Elements

Conventional antennas typically have only a single feed in the form of a coaxial cable through a connector or a microstrip transmission line. In contrast, differential-fed antennas have two out-of-phase ports, each being connected to a transmission line [1, 2]. As noted, they exhibit a number of distinct advantages over single feed antennas. They can be connected directly to differential amplifiers without any extra connecting or matching circuit. Differential-fed antennas can have higher gain, more symmetric patterns, lower cross-polarization (X-pol) level, and less sensitivity to the environment due to their ability to suppress CM signals. In order to examine the advantages of differential-fed antennas, two types of wideband antennas that radiate linear polarized (LP) and circular polarized (CP) fields are discussed using both a single feed and a differential feed in the following.

Microstrip patch antennas are of great importance and are widely used in microwave systems. They have many advantages including simple planar configurations, easy fabrication, low manufacturing costs, low profiles, mechanical robustness, and easy integration with microwave integrated circuits. Patch antennas can support both linear and circular polarization. Traditional rectangular microstrip patch antennas, as shown in Figure 5.4a, are printed on a single printed

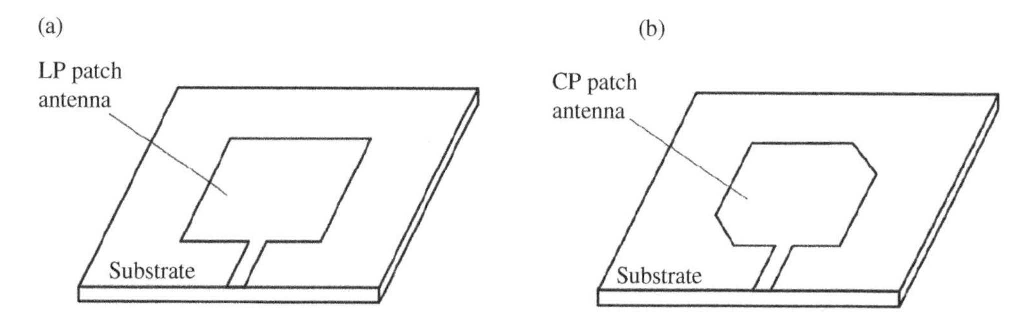

Figure 5.4 Traditional microstrip patch antennas. (a) Linear polarization. (b) Circular polarization. *Source:* From [3], Stutzman and Thiele (2012) and [4], Garg et al. (2001).

circuit board with a conductor on the backside as the common ground. The type of LP patch antenna depicted usually has a relatively narrow bandwidth (around 5%) with a maximum gain of about 5 dBi. By trimming the corners of a square patch, as shown in Figure 5.4b, two orthogonal modes can be excited simultaneously, thus leading to the generation of CP radiation [3, 4]. Similar to the LP patch antenna, the CP patch antenna also suffers from a narrow bandwidth. The illustrated LP and CP antennas have single feeds. In the following subsections, we will discuss wideband versions of them to demonstrate the advantages of the corresponding differential-fed antennas.

5.2.1 Linearly Polarized Differential Antennas

One way to enhance the bandwidth of a rectangular patch antenna is to generate one more resonance to produce a dual-mode response in the band using another substrate with a larger ground. Such a stacked substrate configuration is shown in Figure 5.5a. A patch antenna is printed on the smaller upper substrate that is supported by using four nylon posts. A second substrate with a larger size is placed below the upper substrate. The ground plane is printed on the backside of the lower substrate. The length and width of the patch antenna are denoted as a and b, respectively. Figure 5.5b shows the backside of the upper substrate of the single feed antenna. There is a circular patch with radius r printed on the backside of the substrate. Its position is the location of the feed point of the patch antenna. The inner conductor of the coax-feed is extended via an air hole through the ground plane and the lower substrate and is connected to the circular patch on the bottom layer of the upper substrate. The outer conductor of the coax is connected to the ground. Figure 5.5c shows the side view of the structure.

The size of the small patch is adjusted to achieve optimum matching. This configuration generates two resonances within the desired band, which, in turn, results in a wide operating bandwidth. The distance between the upper and lower substrates is h. In this design, both substrates have the same dielectric constant of 4.4 and 1.0 mm thickness.

The wideband patch antenna in Figures 5.5b and 5.5c has an unbalanced single feed. It is readily modified into a differential-fed version by employing a pair of balanced ports. In the differential case shown in Figure 5.5d, one more circular patch is added at the symmetric position on the backside of the upper substrate. Figure 5.5e shows the side view of the differential-fed antenna. The two cables that constitute its pair of differential ports (port + and port −) are clearly seen. The distance between these two coaxial feeds is c, which includes the radius of the conductors.

The performance of both of these wideband antenna designs was simulated with the full-wave electromagnetic modeling environment HFSS (high-frequency structure simulator). The center

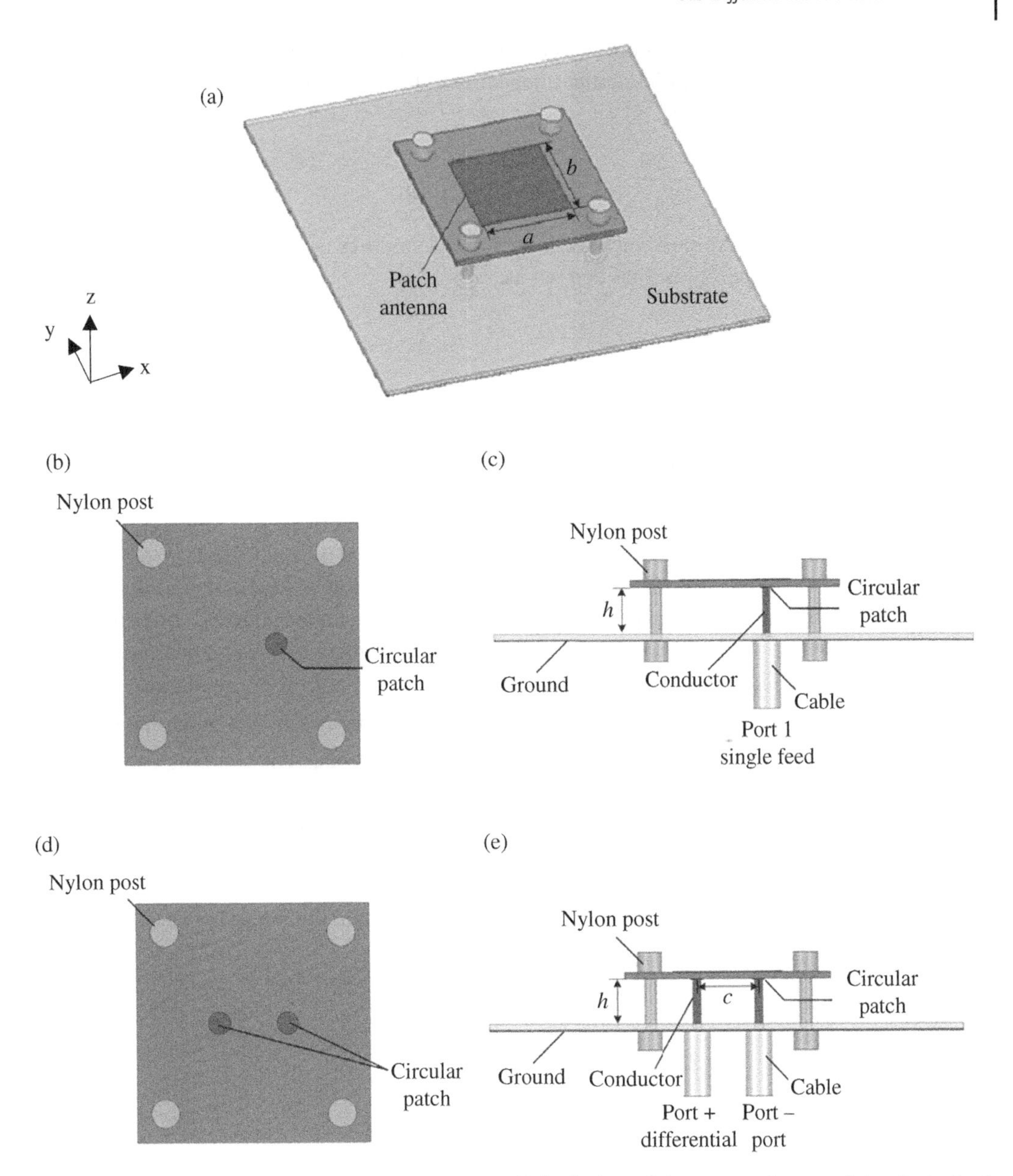

Figure 5.5 Wideband patch antennas. (a) 3D model. (b) Back view of the upper substrate of the single feed version. (c) Side view of the single feed version. (d) Back view of the upper substrate of the differential-fed version. (e) Side view of the differential-fed version.

frequency was selected to be 5 GHz. The dimensions were: $a = b = 17.4$ mm, $c = 9.7$ mm, $d = 42.0$ mm, and $h = 7.0$ mm. It is noted for comparison purposes that the same parameter values were used for both the single feed and the differential-fed antenna.

The mixed-mode S-parameters for the differential-fed antenna can be expressed using single port ones by evaluating the quantity:

$$S_{dd11} = S_{11} - S_{12} \tag{5.1}$$

where S_{11} and S_{12} refer to the reflection coefficient at port + and the transmission coefficient between port + and port –. The differential impedance of the balanced port of a symmetric differential-fed antenna is given by [1]:

$$Z_{dd} = 2Z_o \frac{1 - S_{11}^2 + S_{21}^2 - 2S_{21}}{(1 - S_{11})^2 - S_{21}^2} \tag{5.2}$$

Since the corresponding two-port network is reciprocal, one has $S_{11} = S_{22}$ and $S_{12} = S_{21}$. The relation between S_{dd11} and Z_{dd} is then derived as:

$$S_{dd11} = \frac{Z_{dd} - 2Z_o}{Z_{dd} + 2Z_o} \tag{5.3}$$

Thus, the characteristic impedance of the differential-fed antenna is $2Z_o$, whereas it is Z_o for the single feed version.

The equivalent circuit of the wideband patch antenna is shown in Figure 5.6a. It is composed of a serial *RLC* network: R_1, L_1, and C_1 and a parallel *RLC* network: R_2, L_2, and C_2. In the serial *RLC* network, L_1 represents the inductance produced by connecting the input port and the upper substrate; C_1 represents the capacitance between the circular patch and the main patch of the antenna; and R_1 is the inner resistance. In the parallel *RLC* network, L_2 and C_2 represent the resonance generated by the patch antenna and R_2 is the inner resistance. The input impedance Z_{in} should be equal to Z_o for the single feed antenna and $2Z_o$ for the differential-fed antenna. A resonance occurs when either $j\omega L_1 + \frac{1}{j\omega C_1} = 0$ or $j\omega C_2 + \frac{1}{j\omega L_2} = 0$. Consequently, the angular frequencies of these two resonances are $\omega_1 = \frac{1}{\sqrt{L_1 C_1}}$ and $\omega_2 = \frac{1}{\sqrt{L_2 C_2}}$. It is noted that ω_1 is generated by the patch mode of the patch antenna. On the other hand, ω_2 is produced by the electromagnetic coupling between the conductor and the patch.

The simulated reflection coefficients of the single feed and differential-fed antennas are shown in Figure 5.6b. Dual-mode responses are realized in both cases; they yield similar bandwidths. Moreover, the bandwidth is significantly enlarged in comparison to a traditional patch antenna with one resonance. It is observed in Figure 5.6b that there is a slight difference between the $|S_{dd11}|$

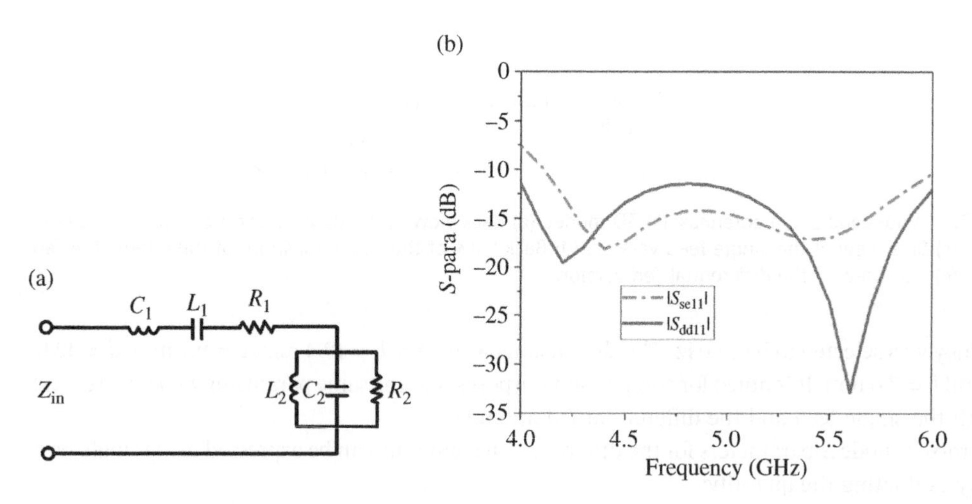

Figure 5.6 Wideband patch antennas. (a) Equivalent circuit. (b) Reflection coefficients of the single feed and the differential-fed antennas.

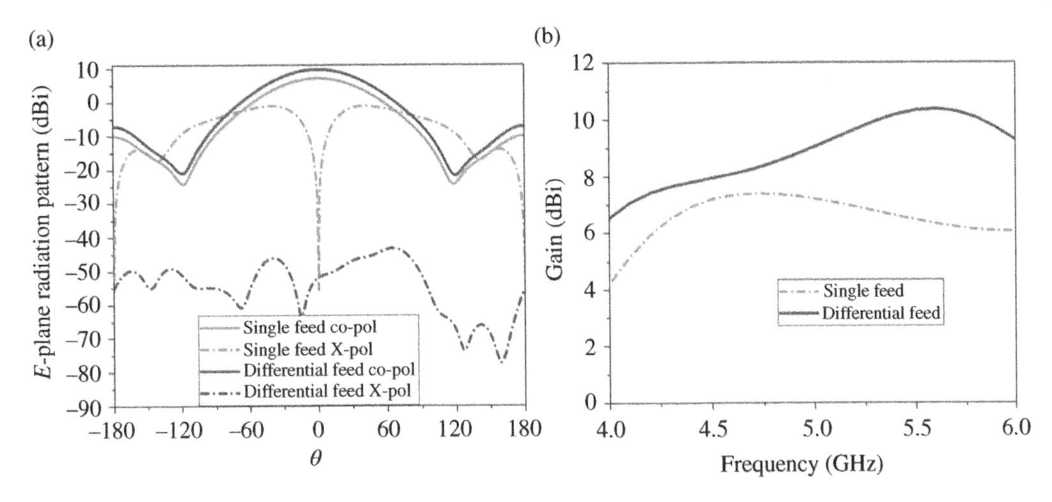

Figure 5.7 Simulated performance characteristics of the single feed and the differential-fed LP antennas. (a) *E*-plane radiation patterns. (b) Maximum realized gain values.

and $|S_{se11}|$ values. This behavior arises from the different current and electrical field distributions of the two antennas. The resonances occur at 4.4 and 5.4 GHz for the single feed antenna; they occur at 4.2 and 5.6 GHz for the differential-fed version. The coupling coefficient between the two resonant modes of the differential-fed antenna is stronger in comparison to the single-ended version. This feature of the differential-fed antenna is also attributed to the fact that its excitations are balanced.

It is known that differential-fed antennas have much better radiation performance characteristics in comparison to single feed ones. These include more stable and symmetric patterns, higher gain, and better polarization purity [2]. Figure 5.7a compares the simulated E-plane radiation patterns of the single feed and the differential-fed antennas given in Figure 5.5. The co-polarization (co-pol) pattern shapes are quite similar, including the positions of their nulls. Their front-to-back ratios are also similar. On the other hand, the cross-polarization patterns of the two antennas are significantly different. There are only two nulls in the pattern of the single feed antenna at $\theta = 90°$ and $\theta = 180°$, whereas the X-pol pattern is at very low level at all θ for the differential-fed version. In fact, the overall X-pol level of the differential-fed antenna is below −50 dB, while it is only −15 dB for the single feed version. This huge difference is attributed to the balanced excitations associated with the differential port which prohibit the propagation of the CM signals.

The difference between the two E-plane patterns also indicates that the differential-fed antenna can realize a higher gain than the single feed one. Figure 5.7b shows the maximum gain of these two antennas across a wide frequency range. It is seen that the maximum gain of the differential-fed antenna is 6.5–10.2 dBi from 4 to 6 GHz, whereas the maximum gain of the single feed antenna over the same frequency range is only 4.2–7 dBi. These results also mean that the differential-fed antenna has an aperture efficiency that is higher than the single feed version for the same basic overall geometry.

Differential excitations can also produce more stable and symmetric radiation patterns because the subsequent current distributions on the radiators are more balanced. Figures 5.8a and 5.8b show the simulated H-plane radiation patterns of the single feed and the differential-fed antenna, respectively, 4.5, 5.0, and 5.5 GHz. Apart from the difference in the maximum gain that was noted above, the shape of the H-plane patterns of the single feed antenna at these frequencies varies a lot.

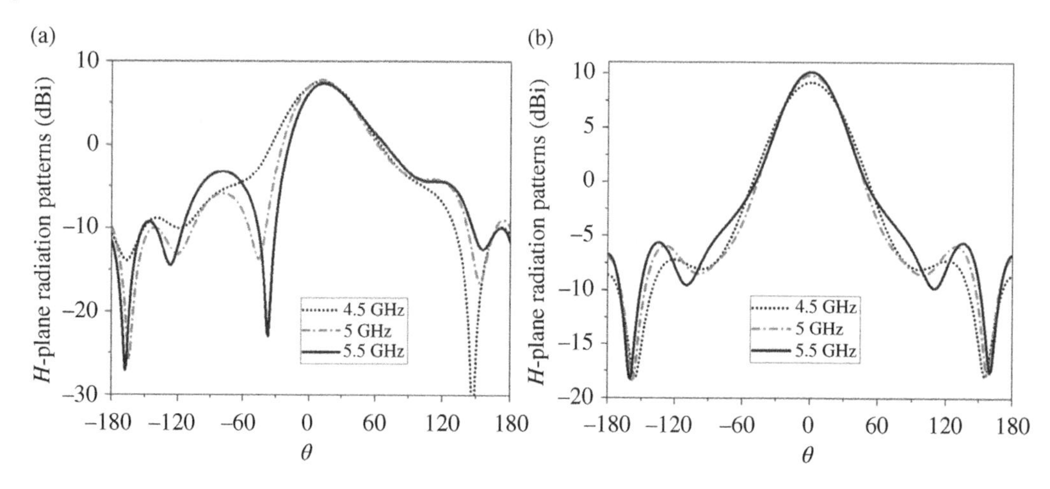

Figure 5.8 Simulated H-plane radiation patterns of the LP patch antenna at 4.5, 5.0, and 5.5 GHz. (a) Single feed version. (b) Differential-fed version.

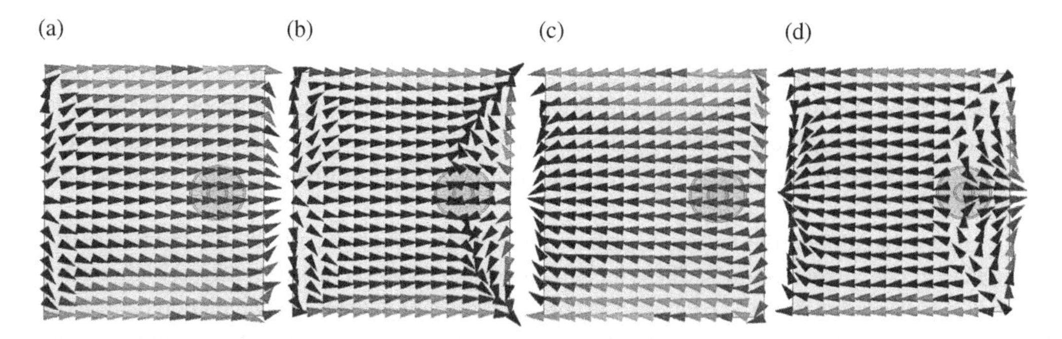

Figure 5.9 Simulated current distributions on the single feed LP patch antenna at 5.0 GHz at specific times in one period T. (a) $t = 0$. (b) $t = 1/4\ T$. (c) $t = 2/4\ T$. (d) $t = 3/4\ T$.

Moreover, they have some distortions as observed in Figure 5.8a. In particular, these H-plane patterns are asymmetric when comparing values at different $|\theta|$ and their nulls are located at different observation angles. Furthermore, the patterns are not stable, i.e., the 3-dB beamwidth varies and the maximum radiation direction moves away from boresight as the frequency increases. Such variations in the radiation patterns are undesirable for most wireless communications systems.

On the other hand, the H-plane patterns of the differential-fed antenna are perfectly symmetric. They attain the same gain and null levels at the symmetric observation angles $|\theta|$ and their 3-dB beamwidth is the same at different frequencies. The maximum radiation direction is also stable at the boresight. These features are very desirable for applications. All of these results manifest the importance and the advantages of using a differential feed for LP antennas. The differences in their performance characteristics are attributed to the differences in the current distributions on their radiating elements.

As shown in Figures 5.9a–d, the current distributions on the single feed antenna are not symmetric about the vertical centerline. It is much stronger in the vicinity of the feed point and is much weaker on the other side of the conductor. Furthermore, one sees that there are strong vertical components in the current near the feed point which contribute to the cross-polarization in the

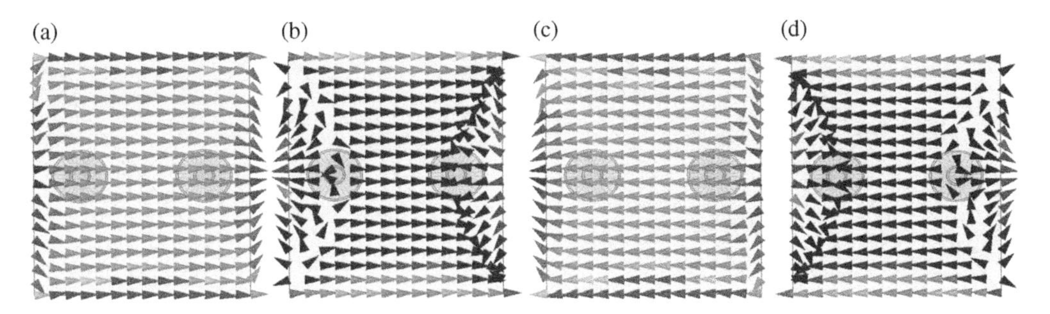

Figure 5.10 Simulated current distributions on the differential-fed LP patch antenna at 5.0 GHz at specific times in one period *T*. (a) *t* = 0. (b) *t* = 1/4 *T*. (c) *t* = 2/4 *T*. (d) *t* = 3/4 *T*.

radiation pattern. Admittedly, these asymmetries can be alleviated if a different antenna configuration, such as a smaller rectangular patch, is used.

In contrast, since the differential excitation leads to balanced currents, the electric field lines are symmetric. This behavior is illustrated in Figures 5.10a–d. The current density on the conductor is essentially symmetric about the vertical centerline. This kind of current distribution leads to stable and symmetric radiation patterns that have almost no distortion when the excitation frequency changes. More importantly, there is no CM component in the balanced excitations. Thus, environmental noise and electromagnetic interference from the RF transceivers are eliminated. Consequently, the CM suppression capability is considered to be a design priority for wireless communication antenna arrays as was discussed in Section 5.1.

5.2.2 Circularly Polarized Differential Antennas

Differential feeds can also be adopted in CP antennas for improving their radiation performance. Similar to the LP case, a larger substrate backed with a ground plane is used to enhance the bandwidth of CP antennas. The differential-fed CP antenna, as shown in Figure 5.11a, also has its radiating patches printed on the top layer of the upper substrate. The main patch is a rectangle over the center of the substrate that is oriented at 45° with respect to both edges of the substrate and has two 45°–45° right triangle pieces symmetrically cut from each end. There are also four rectangles sequentially rotated around the edges of the square substrate with some offset from the center points of each edge [5]. As shown in Figure 5.11b, the upper substrate is supported by two cylindrical conductors which are the center conductors of the two coax-feed cables. These conductors are connected to two small patches on the backside of the upper substrate. The distance between the two substrates is h. Figures 5.11c and 5.11d show the top and bottom views of the upper substrate. Both substrates again have a dielectric constant of 4.4 and 1.0 mm thickness. Note that this antenna can also be excited with a single feed. This alternate configuration is implemented by removing port 1+ (or port 1−) and exciting the remaining port with an unbalanced signal. The optimized design dimensions (in millimeters) are: $a = 7.5$, $b = 17.4$, $c = 6.6$, $d = 12.3$, $e = 32.9$, $g = 9.7$, $h = 7.0$, $t = 5.5$, $s = 1.15$, and $r = 1.0$.

Compared with the single feed CP antenna shown in Figure 5.4b, this double-substrate design has a much larger bandwidth for similar reasons as those for the LP design. The $|S_{11}|$ values of this CP antenna also exhibit a double-tuning response, i.e., two resonances appear within the band. One resonance is generated by the main patch on the substrate. The other one is produced by the conductors and the small circular patches. Its response can be modeled as an equivalent series *LC* circuit.

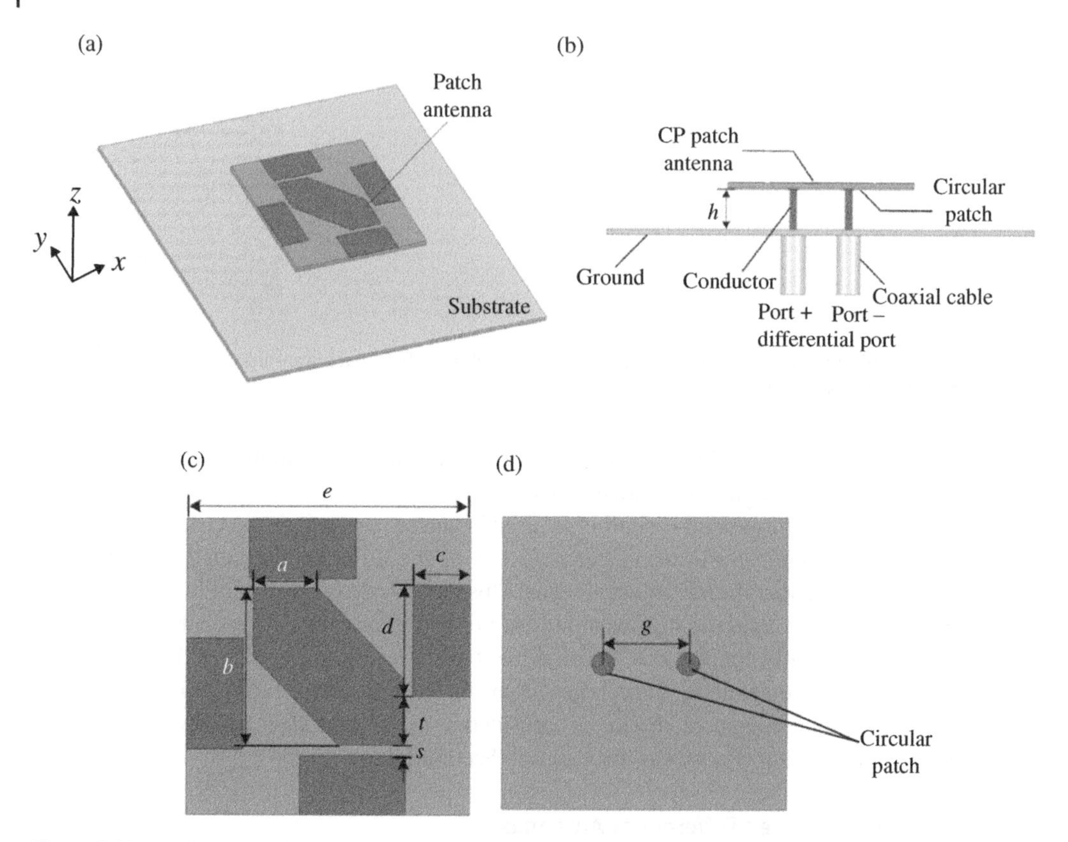

Figure 5.11 Wideband differential-fed CP patch antenna design. (a) 3D isometric view. (b) Side view. (c) Top view of the upper substrate. (d) Bottom view of the upper substrate. *Source*: Modified from [5] / IEEE.

HFSS simulations of both the single feed and the differential-fed CP antennas were conducted. The corner truncations and the rotation of the four rectangular edge patches indicated in Figure 5.11c give this antenna a left-handed CP (LHCP) response. The corresponding right-handed CP (RHCP) version is attained simply by flipping the center patch with respect to the vertical centerline and the rotation sequence of the rectangular edge patches.

Figure 5.12a shows the radiation patterns of the single feed and differential-fed versions of the CP patch antenna in the *x0z*-plane, which is orthogonal to the patch along the vertical centerline. It is observed that there is an obvious difference between the maximum gain of the two antennas. The maximum realized gain of the differential antenna is 7.8–9.0 dBi from 4.5 to 5.5 GHz, whereas it is only 5.3–6.0 dBi in the single feed case. The differential-fed values average around 2.8 dBi larger than the single feed ones. The differential-fed antenna also has more symmetrical LHCP and RHCP patterns in comparison to the single feed ones. The average RHCP X-pol level is much lower in the LCHP differential-fed case as well.

The significant difference in the maximum realized gain between the two antennas is due to undesirable current distributions on the single feed patch. They are asymmetric and cause unwanted distortions of the radiation patterns that lower its overall gain. To illustrate these effects, Figure 5.13 shows the simulated radiation patterns of the single feed and differential-fed CP antennas in the *y0z*-plane. The LHCP radiation patterns of the single feed antenna shown in Figure 5.13a

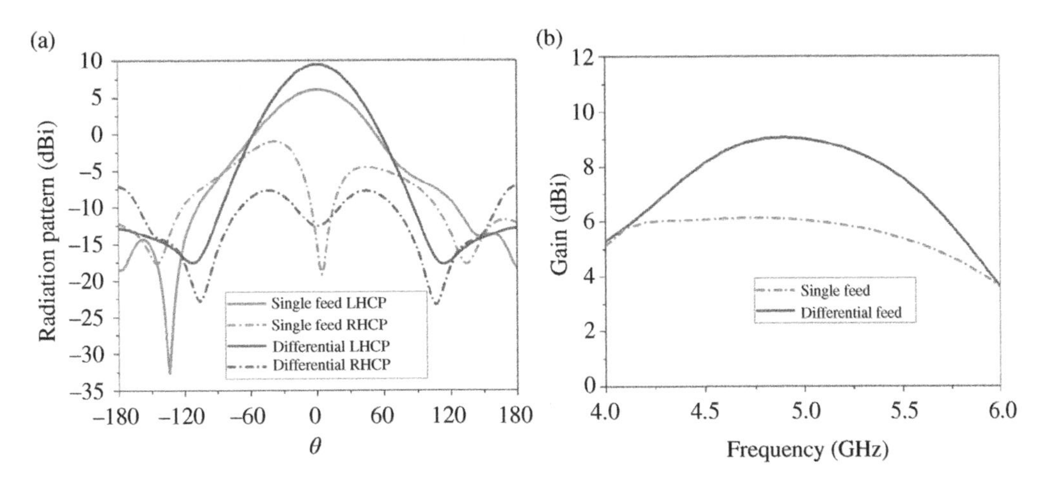

Figure 5.12 Simulated performance characteristics of the single feed and differential-fed CP antennas at 5.0 GHz. (a) xOz-plane radiation patterns. (b) Maximum realized gain values.

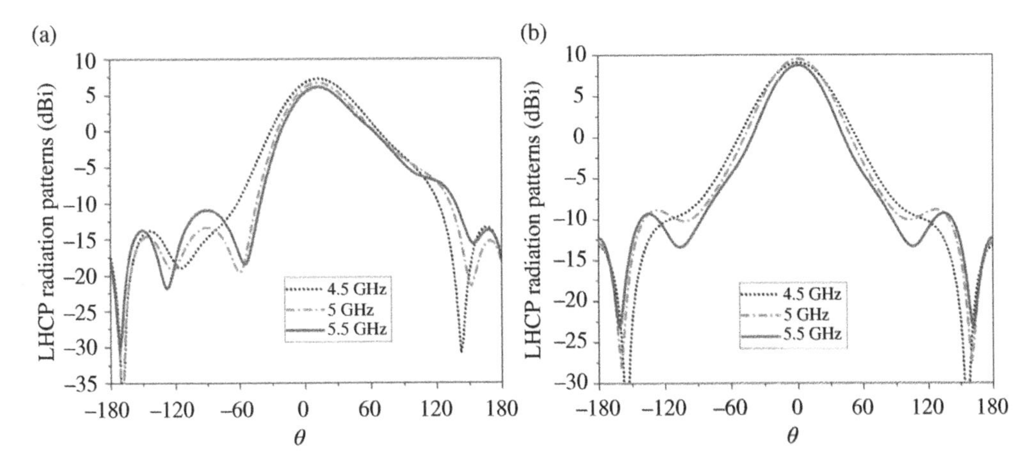

Figure 5.13 Simulated radiation patterns of the CP patch antenna in the *yOz*-plane at 4.5, 5.0, and 5.5 GHz. (a) Single feed version. (b) Differential-fed version.

are asymmetrical. Moreover, they are also not consistent at different frequencies. The nulls are not located at the same values of $|\theta|$ and the maximum radiation direction is not along boresight.

On the other hand, the radiation patterns of the differential-fed antenna in Figure 5.13b are perfectly symmetric in the same vertical orthogonal plane at 4.5, 5.0, and 5.5 GHz. Moreover, they exhibit a constant beamwidth and the positions of their nulls appear at the same observation angle $|\theta|$. More importantly, the maximum gain direction is always located on boresight. As noted previously, this is crucial and desirable for wireless communication systems.

It is to be noted that differential-fed antennas can be designed to produce radiation patterns with very unique characteristics. By employing distributed micro-patches backed by a metal cavity and four feeding ports as shown in Figure 5.14, a very wide 3 dB beamwidth was achieved in [6]. An antenna with this feature is a very useful element for the design of phased arrays that require wide scanning ranges for sensing applications.

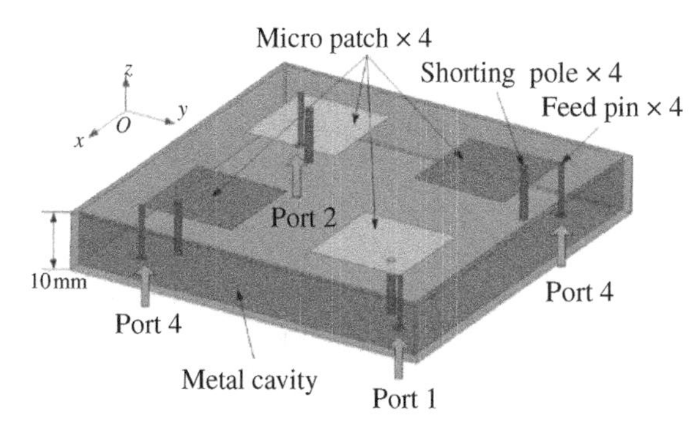

Figure 5.14 Illustration of a differential-fed distributed microstrip antenna backed by a cavity. *Source*: From [6] / with permission of IEEE.

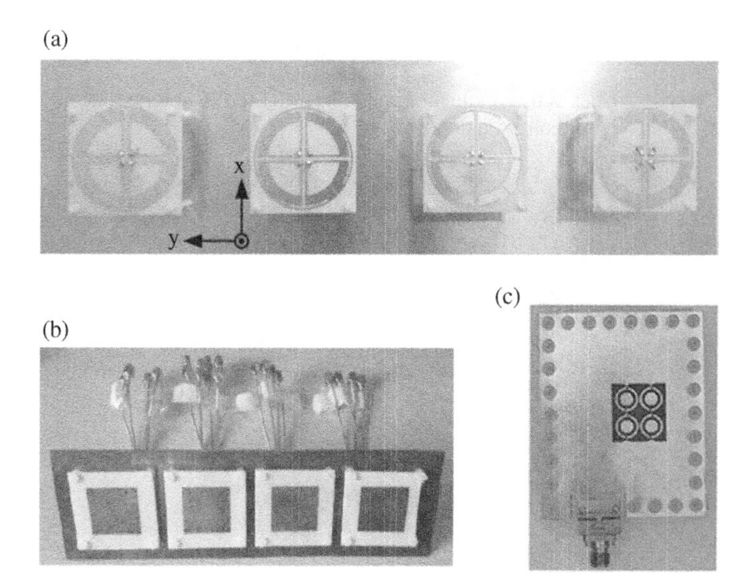

Figure 5.15 Some differential-fed antenna arrays reported in the literature. *Source*: (a) From [12] / with permission of IEEE, (b) From [13] / with permission of IEEE, (c) From [15] / with permission of IEEE.

5.3 Differential-Fed Antenna Arrays

Antenna arrays are of great importance to wireless microwave communication systems. They can provide higher gains, higher directivities, and more selective beamwidths in comparison to single antenna elements. Different polarizations are often used in different scenarios, i.e., linear polarization [7–9], dual-linear polarization [10–13], and left-handed/right-handed circular polarization [14, 15]. Large-scale differential arrays are favored in differential transceivers. However, their complicated feeding networks are a great obstacle that inhibits the wide use of differential-fed arrays.

Some differential-fed antenna arrays reported in the literature are shown in Figure 5.15. A wideband DFN was applied in [12] to feed a wideband dual-polarized planar array with high CM suppression. A low-cost differentially driven dual-polarized patch antenna array was presented in [13] to achieve high port isolation and low cross-polarization level. A differential feeding PD was designed in [15] to feed a CP differential-fed array. It was based on aperture coupling to achieve a stable 180° phase imbalance over a 20% operating fractional bandwidth at 30 GHz.

There are numerous challenges and difficulties in the development of suitable feeding networks for differential-fed antenna arrays. Traditional feeding networks are usually composed of multiple PDs cascaded with baluns which are quite bulky in size. Therefore, it is critically important to miniaturize the circuit size. Moreover, the CM currents need to be eliminated from the feeding network in order to obtain the desired enhanced radiation performance. This goal requires a minimum level of CM signal transmission and differential-mode-to-CM (DM-to-CM) conversions throughout the circuit.

A new type of balanced PD is introduced and discussed in the next subsection. It has a high level of CM signal suppression and wide bandwidth. It facilitates larger-scale feeding networks that can be established for both LP and CP linear arrays. Moreover, differential BFNs, such as BMs, can also be developed to produce multi-beams. Those will be introduced and described in the next section.

5.3.1 Balanced Power Dividers

Balanced PDs play important roles in wireless systems. They are particularly useful since the input signals are usually unbalanced and the input ports are single-ended in most cases. To design a feed network for differential-fed antenna arrays, a single-ended-to-balanced (SETB) PD is often needed to convert the unbalanced signals to balanced ones and to split the power equally with identical phase and amplitude.

An SETB PD works as a combination of a single-ended PD with two baluns, as indicated in Figure 5.16a. There are one single-ended input port and two balanced output ports. For integration and miniaturization purposes, the compact design shown in Figure 5.16b integrates the functions of the PD and the two baluns. It is more favorable than other designs because it has low loss and small size.

SETB PDs have been investigated in recent years, especially for use in microwave communication systems such as transceivers and feeding networks. Gysel- and Wilkinson-based PDs were described in [16, 17], respectively. A compact, low loss, and planar SETB PD was presented in [18] using folded coupled lines. A planar wideband SETB PD was shown in [19] with a zero for the CM transmission at the center frequency. Filtering characteristics were included later in the designs of SETB PDs as described in [20, 21]. The prototypes in [16–21] are shown in Figure 5.17.

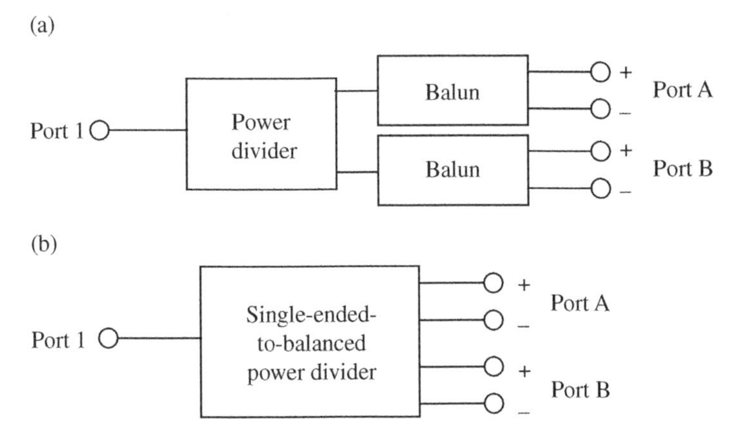

Figure 5.16 Diagrams representing power dividers having multiple cascaded components. (a) Traditional cascaded topology. (b) The developed single-ended-to-balanced (SETB) device.

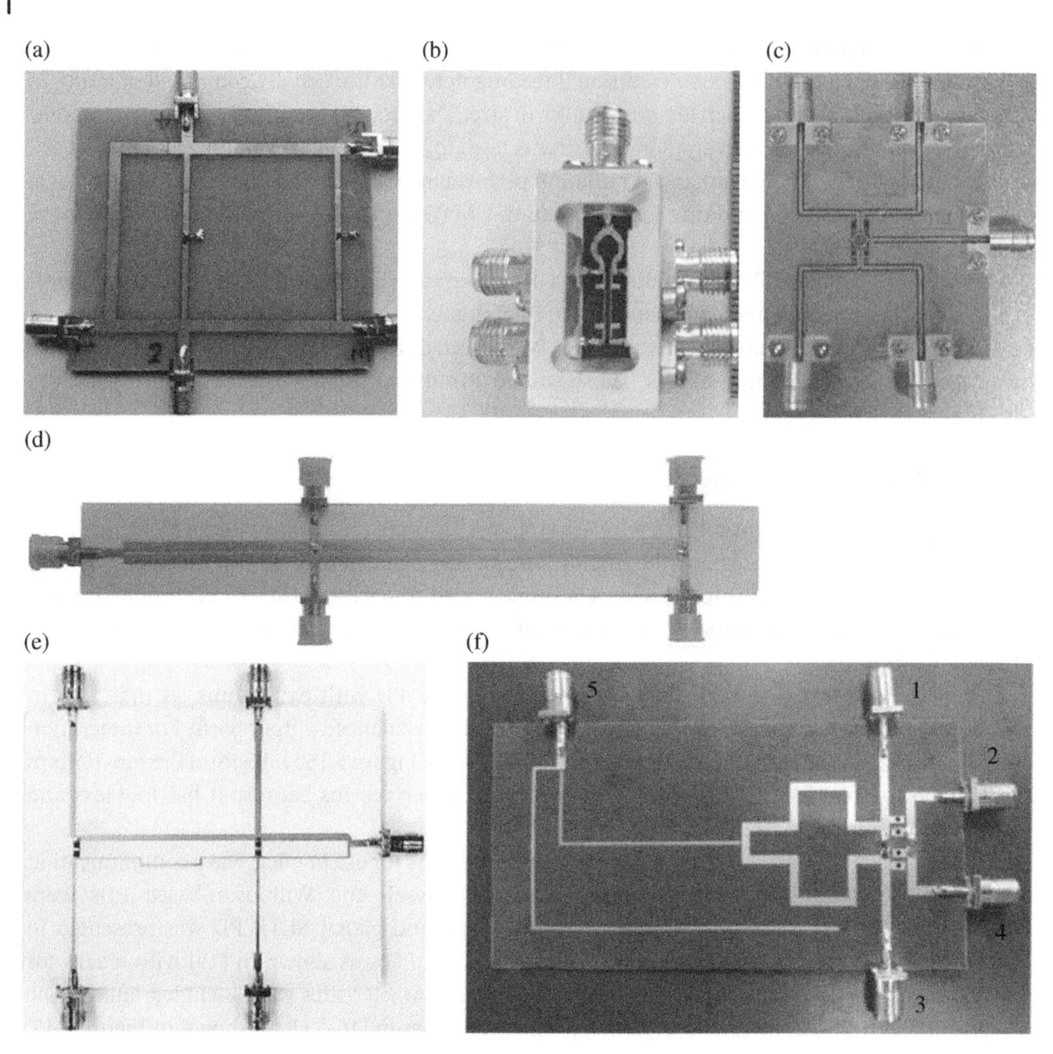

Figure 5.17 Some design examples of single-ended-to-balanced power dividers in the literature. *Source*: (a) From [16] / with permission of IEEE, (b) From [17] / with permission of John Wiley & Sons, (c) From [19] / with permission of IEEE, (d) From [20] / with permission of IEEE, (e) From [21] / with permission of IEEE, and (f) From [22] / with permission of IEEE.

The primary issue to be considered for a balanced component is to maximize the levels of CM transmission suppression and the DM-to-CM conversions. It arises because the CM suppression ability is very critical for a balanced device as it largely determines how much noise and electromagnetic interference can be rejected by it. Unfortunately, the CM suppression issue tends to be neglected in most existing designs. This shortcoming has occurred because most current designs employ 180° transmission lines to connect the two output ports. However, these 180° transmission lines are actually frequency-dependent. Although a wideband range of CM suppression was achieved in [21] by employing 180° phase inverters, there is still great potential to further improve its CM suppression level and frequency range.

Facing the challenges associated with the two concurrent issues of CM bandwidth and CM suppression levels, a new approach to the design of SETB PDs with high levels of CM transmission and DM-to-CM conversion suppression was developed [23]. It is presented using the slotline-to-microstrip transition depicted in Figure 5.18.

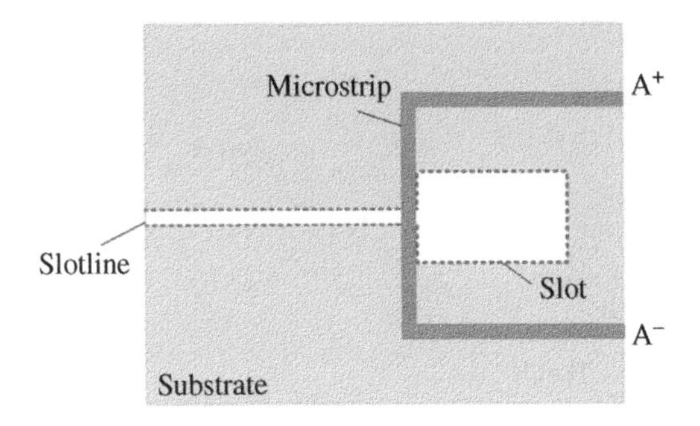

Figure 5.18 Layout of a slotline-to-balanced-microstrip line transition. *Source*: Based on [23] / IEEE.

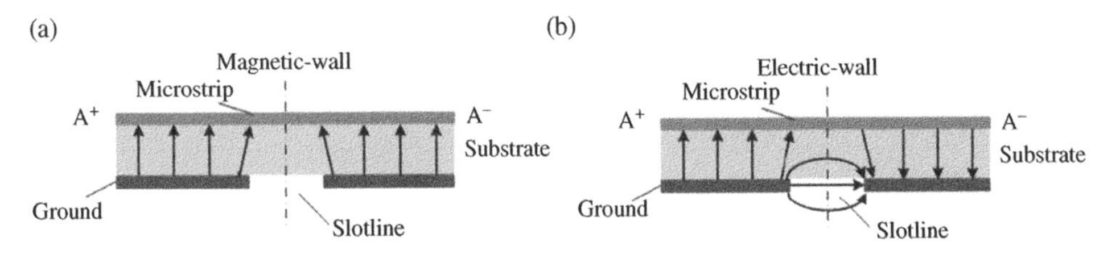

Figure 5.19 Side view of the electric fields in the substrate between the boundaries of the slotline and the microstrip line under (a) CM operation and (b) DM operation.

Our main target was to achieve more than 30 dB in-band CM rejection and an extremely wide CM rejection bandwidth. The developed design utilizes multimode slotline resonators and microstrip-to-slotline transitions to create a constant band. The CM signals are terminated in slotlines because they do not support the CM-excited electric fields. As depicted in Figure 5.18, the slotline ends with a short-ended stub that is etched in the ground plane. A microstrip line connecting port A+ and port A− is located across the slotline on the other side of the substrate in the perpendicular direction. Figures 5.19a and 5.19b show the side views of the slotline-to-microstrip transition and the associated electrical field lines produced under DM and CM operations.

When the balanced ports are under CM operation, the middle plane can be treated as a virtual electric wall. This is depicted in Figure 5.19a. The signals from the slotline are split into two parts with identical magnitude and opposite phase [22]. This configuration is attained because the middle plane of the slotline can also be regarded as a virtual electric wall. Thus, it converts via magnetic coupling of the electrical field behavior of the microstrip line into that of a slotline.

On the other hand, when the circuit is under DM operation, the slotline will no longer support the transmission of CM signals due to the field distribution shown in Figure 5.19b. The CM signals are terminated in the slotline and, hence, are totally reflected. Note that the termination property of the CM signals applies to all frequencies. Thus, the structure has the potential to achieve wideband CM suppression in the design.

Based on the slotline-to-microstrip transition, the developed SETB PD is shown in Figure 5.20a. The function of this DM device is to provide wideband transmission with a prescribed filtering response as well as extremely low levels of CM transmission and DM-to-CM conversion. A T-junction with a slotted stub and a resistor underneath the microstrip line are used to connect the input port to the two output ports while splitting the input power delivered to them equally. The electrical length of the slot stub is a quarter-wavelength. Thus, it is equivalent to an open circuit at the T-junction. Since the slotline structure is frequency-independent, the electrical fields will be

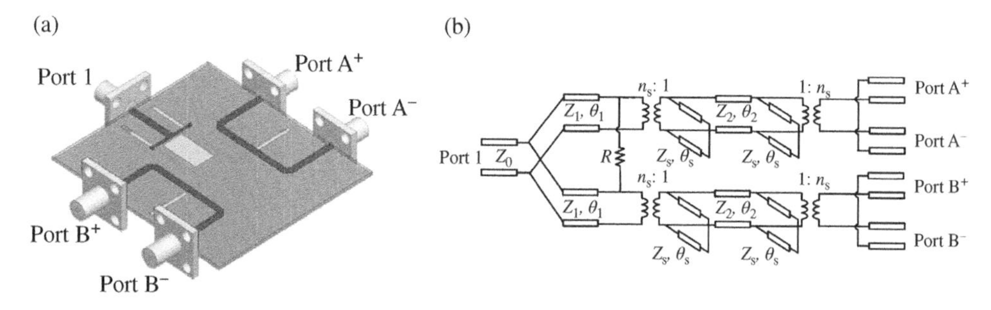

Figure 5.20 The developed SETB PD. (a) Layout. *Source:* (a) From [25] / with permission of IEEE. (b) Equivalent circuit. *Source:* (b) Based on [23] / IEEE.

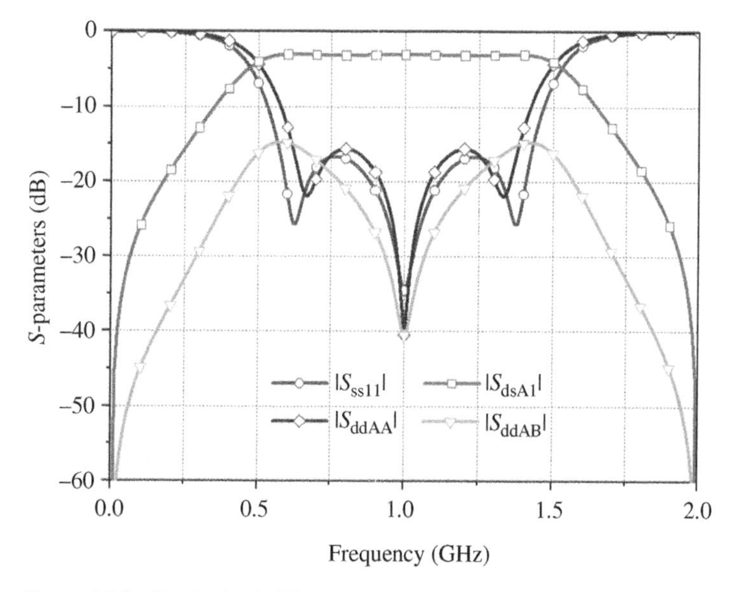

Figure 5.21 Synthesized differential-mode *S*-parameters of the SETB PD.

terminated. Consequently, it is possible to achieve a wide frequency range with high isolation between the two parts on both sides of the slotline. A wideband filtering response is also obtained in the DM transmission because multiple resonance poles are introduced within the operating frequency band. Moreover, wideband isolation and ideal port matching are realized between the two output ports. An equivalent circuit is shown in Figure 5.20b; it can assist the quantitative analysis of the structure.

The differential-mode transmission coefficients ($|S_{dsA1}|$ and $|S_{dsB1}|$) of the SETB PD from its single-ended input port to its two balanced output ports satisfy the half-power requirement that $|S_{dsA1}| = |S_{dsB1}| = 1/\sqrt{2}$. Moreover, the structure satisfies the requirement that the single-ended input port and the two balanced output ports are matched and isolated, i.e., that $S_{ss11} = S_{ddAA} = S_{ddBB} = S_{ddAB} = 0$. Figure 5.21 shows the synthesized differential-mode *S*-parameters of the ideal SETB PD using the equivalent circuit.

The slotline plays an important role in the design; it generates multiple resonant modes in the operating band. The $|S_{dsA1}|$ values indicate a constant, wide 3-dB operating band of around 80%.

Two transmission zeros are located at DC and at $2f_o$, providing a filtering response in $|S_{dsA1}|$. They produce the filtering response exhibited in the $|S_{dsA1}|$ values. Three resonance poles are present in the $|S_{ss11}|$ and $|S_{ddAA}|$ responses and a zero is located at f_o in the $|S_{ddAB}|$ response. The results indicate the PD has excellent matching and isolation characteristics.

On the other hand, high levels of CM rejection and low levels of DM-to-CM conversion in the device's transmission characteristics require that $S_{csA1} = S_{csB1} = S_{dcAA} = S_{dcBB} = 0$ at all frequencies. Traditional methods use 180° transmission lines to realize a transmission zero at the center frequency of the CM. This approach largely limits the CM suppression bandwidth. To extend the level of CM suppression and its operating bandwidth, the slotlines of the SETB PD are employed to replace those 180° transmission lines. Because the slotlines do not support the CM electric fields, the CM signals are terminated and totally reflected.

An SETB PD prototype was fabricated using monolithic PCB to verify its performance characteristics. Figure 5.22 shows the simulated and measured DM and CM performance. Inset photos of the fabricated prototype are included in Figure 5.22b. It is observed that the EM simulation and test results are very close to those obtained with the equivalent circuit. The measured DM transmission coefficient $|S_{dsA1}|$ has a fractional bandwidth of 79.6%, from 2.98 to 3.92 GHz, which was very close to the objective of 80%. The measured in-band insertion loss is around 1.2 to 2.5 dB in comparison to its simulated values of 0.6 to 2.0 dB. Wideband matching performance was realized for the input port 1 and the output ports A and B. The in-band reflection coefficients, $|S_{ss11}|$ and $|S_{ddAA}|$, are smaller than −17 dB and −12.5 dB, respectively. The isolation between the two differential ports is less than −15 dB across the whole band. The CM transmission coefficient $|S_{dsA1}|$ is less than −27 dB at all frequencies. The DM-to-CM conversion levels indicated by $|S_{dcAA}|$ and $|S_{dcAB}|$ is quite low, i.e., they are less than −25 dB and −28 dB at all frequencies - a 200% fractional bandwidth.

This measured CM and DM-to-CM conversion suppression levels of the SETB PD prototype were at least 27 and 25 dB, respectively, at all frequencies. They are much higher than those reported in the open literature. Moreover, the CM suppression range of this design achieved a 200% fractional bandwidth. The simulated and measured results verified that this design achieved a minimum level of CM transmission and DM-to-CM conversion at all frequencies. Consequently, it serves as a favorable choice for building feeding networks for wideband differential antenna arrays.

5.3.2 Differential-Fed Antenna Arrays Employing Balanced Power Dividers

Differential PDs are widely used in the feeding networks of differential antenna arrays to equally split the input power, convert mixed-mode signals to balanced ones, and deliver the balanced signals to the differential antenna elements of the array through transmission line connections. Arrays having 2^n elements are typically favored and widely used. This format is simply the most convenient one for building a simple feeding network. Figure 5.23 shows the configuration of a typical DFN. It consists of a traditional Wilkinson PD and two SETB PDs. The electromagnetic DFN model shown in Figure 5.24 was developed and simulated. It is based on the design shown in Figure 5.20a. There is one unbalanced port, Port 1, and four balanced ports: Port A, Port B, Port C, and Port D. Port 1 is connected to the signal source which delivers mixed-mode signals; the balanced ports are connected to the feed points of the elements in a four-element antenna array as depicted in Figure 5.23.

A four-element linear-polarized differential antenna array was built to verify the performance of the DFN. Its model is shown in Figure 5.25a. The same element depicted in Figure 5.5a was adopted for it. The distance between the center of any two adjacent array elements is d. The optimized

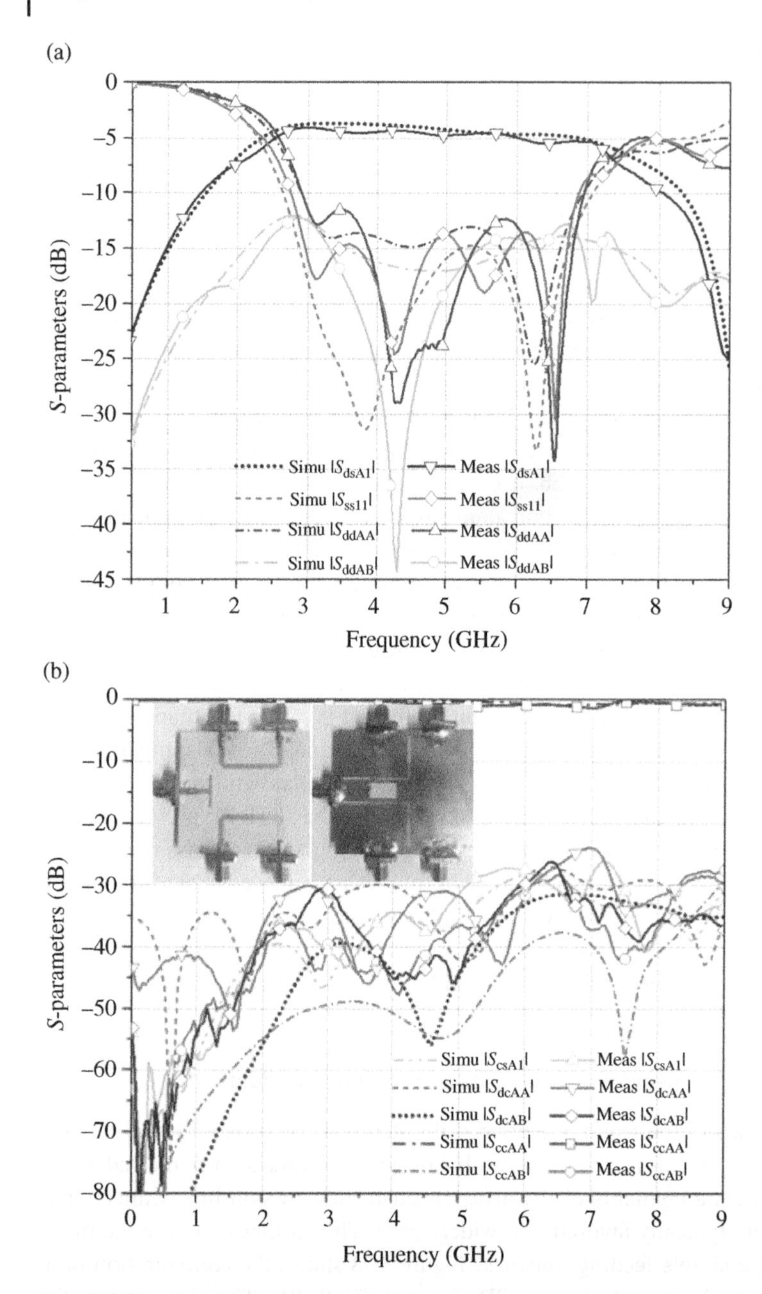

Figure 5.22 Simulated and measured *S*-parameters of the SETB PD prototype. (a) DM performance. (b) CM performance. *Source*: Based on [23] / IEEE.

design had $d = 42.0$ mm, which is 0.7λ at 5.0 GHz in free space. Four pairs of differential feeding points, as displayed in Figure 5.25b, were connected to Port A to Port D of the DFN.

Figure 5.26 shows the E-plane and H-plane radiation patterns of the four-element differential-fed array with the developed DFN. Note that differentially fed array patterns, similar to single feed array patterns, can be decomposed into a product of the element pattern and the array factor. The beamwidth in the E-plane is much narrower than H-plane due to the existence of multiple

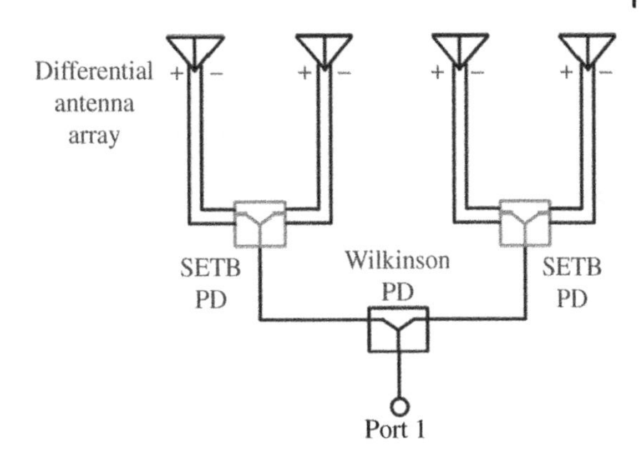

Figure 5.23 Configuration of a four-element array fed by a differential feeding network.

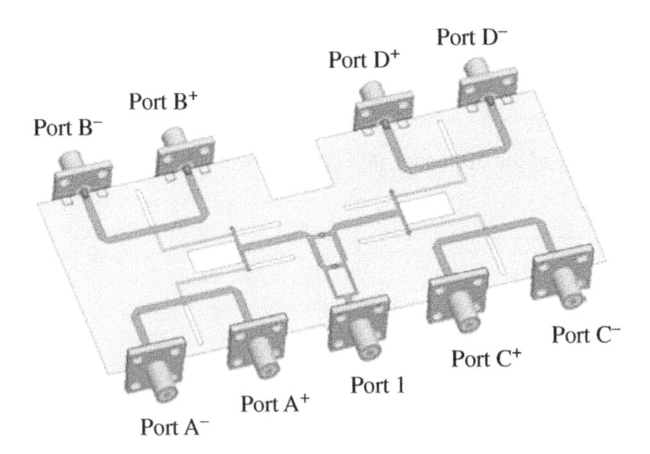

Figure 5.24 Electromagnetic model of the developed DFN based on the SETB PD model shown in Figure 5.20a. *Source*: Based on [25] / IEEE.

elements, i.e., a wider aperture length. The maximum gain of the array is around 14.0 dBi, and the sidelobe level is about −14.0 dB. The co-pol patterns exhibit premium symmetric results including the beam shapes, sidelobe levels, and positions of the nulls. All of these features are attributed to the symmetric current distribution facilitated through the use of the differential feeding sources, as illustrated in Figure 5.10. The X-pol level is also extremely low, less than −50 dB at all observation angles. This feature arises because the CM currents are eliminated using the differential signals associated with the DFN and the differential array. Such low X-pol levels are very advantageous for large-scale array applications. These differential array characteristics help prevent the negative impact of environmental noise and electromagnetic interference on communication channels.

Circular-polarized antennas can also be used in differential-fed antenna arrays [24]. There are two common configurations for CP arrays, one is a linear $1 \times n$ arrangement and the other one is an $n \times n$ format. The usefulness of either depends on the application scenario. Both configurations will be discussed as a 1×4 and a 2×2 array. Both types of differential CP arrays can be excited by a feeding network comprised of a 90° hybrid coupler and two SETB PDs. Typical configurations of both are shown, respectively, in Figures 5.27a and 5.27b.

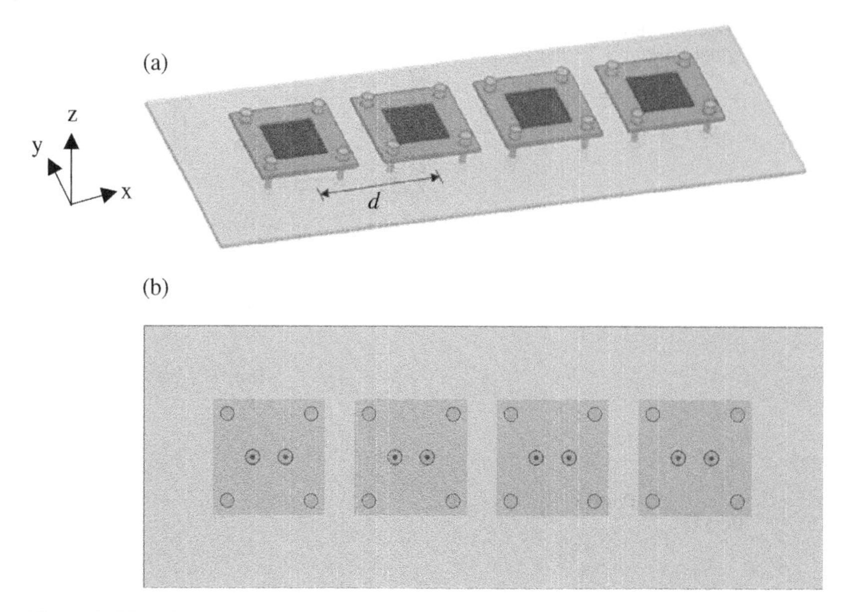

Figure 5.25 LP differential array. (a) Isometric view of the 3D model. (b) Bottom view of the model.

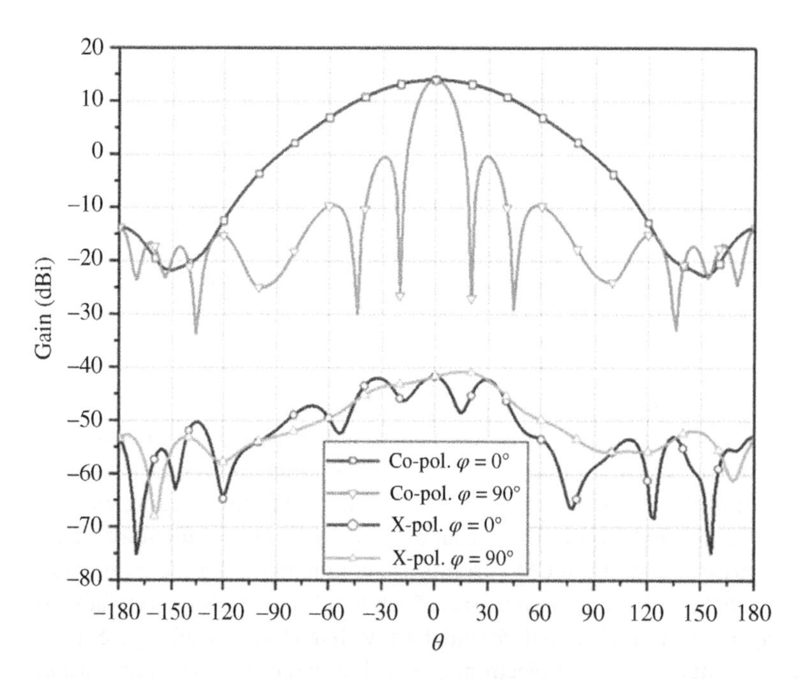

Figure 5.26 Radiation patterns of the four-element differential array fed by the developed DFN.

A four-element linear-configured CP array that adopted the CP antenna element in Figure 5.11 was built. The model of this 1×4 array is shown in Figure 5.28. Notice that each element is rotated clockwise to guarantee an LHCP phase distribution in the array. Compared with the traditional CP array using four single feed elements, this differential CP array exhibits much better performance in many aspects.

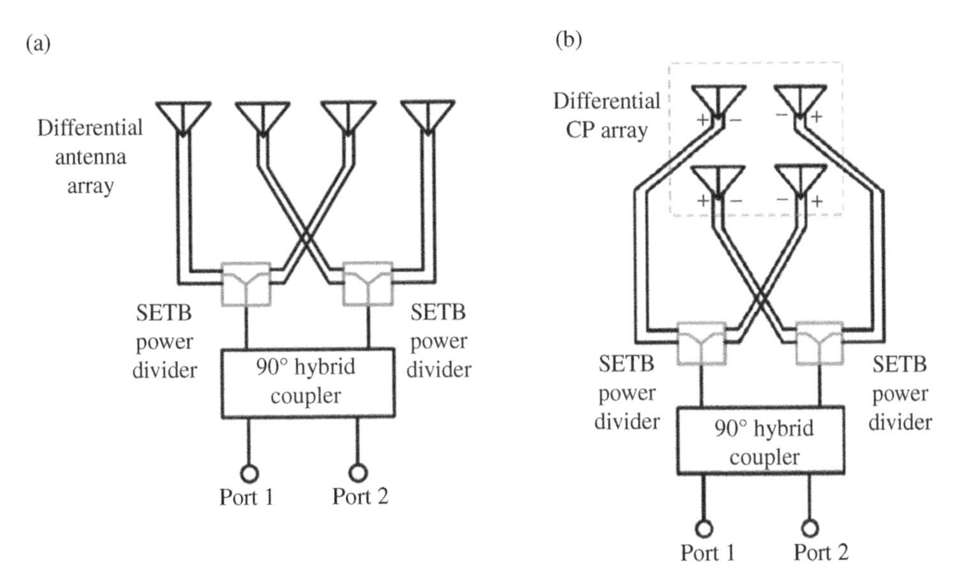

Figure 5.27 The two differential-fed antenna array configurations. (a) 1 × 4 array. *Source*: From [25] / with permission of IEEE. (b) 2 × 2 array.

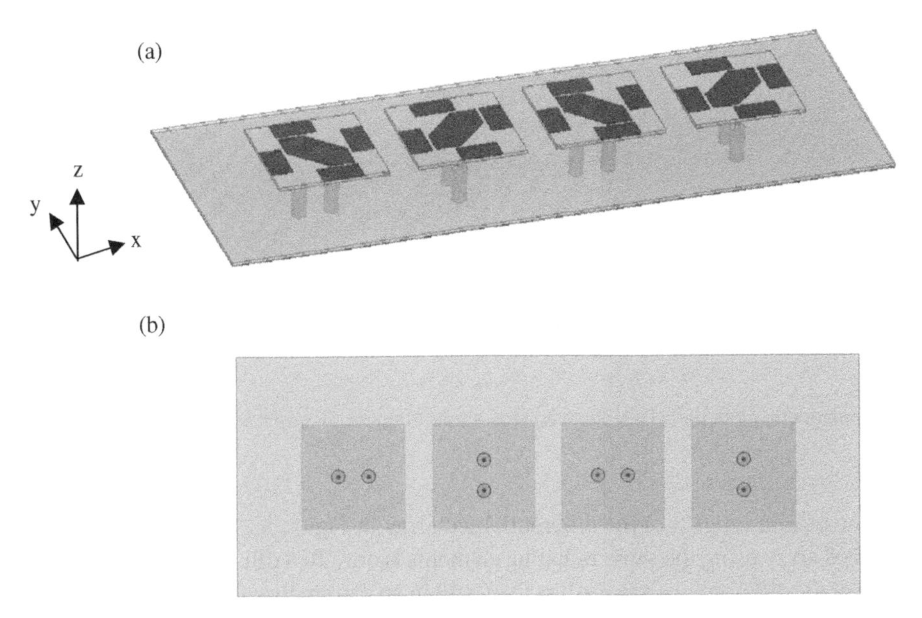

Figure 5.28 Circular-polarized 1 × 4 differential array: (a) 3D model. *Source:* (a) From [25] / with permission of IEEE. (b) Bottom view of the model. *Source*: (b) Based on [25] / IEEE.

In particular, the differential-fed CP array performance characteristics share the advantages over the corresponding single feed ones that were discussed for the LP versions. Their radiation patterns are stable and symmetric as characterized by their beamwidths, maximum radiation angles, and positions of nulls. The comparisons shown in Figure 5.29 illustrate that the differential-fed array has higher gains and better axial ratio (AR) performance when compared to the corresponding

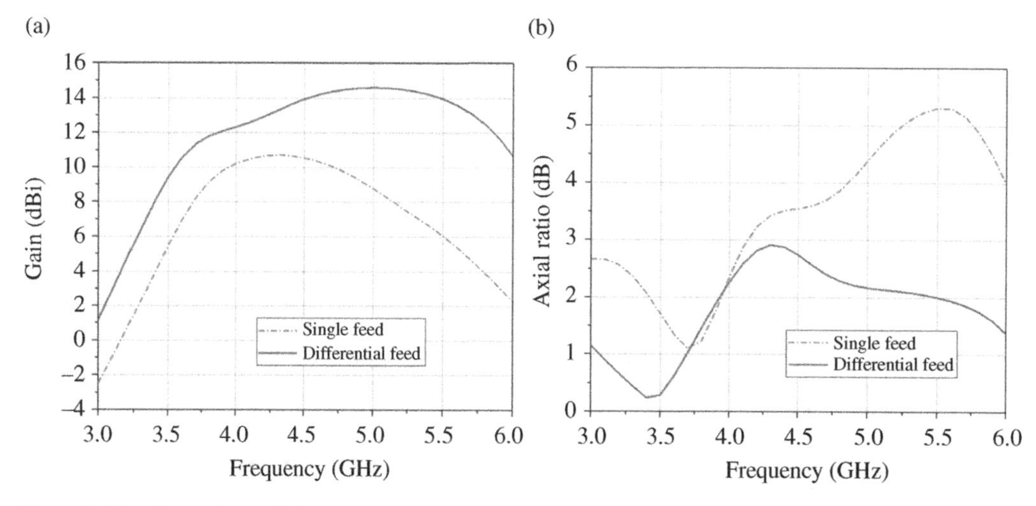

Figure 5.29 Comparison of the array performance of the single feed and differential-fed CP antenna arrays. (a) Realized gain. (b) Axial ratio.

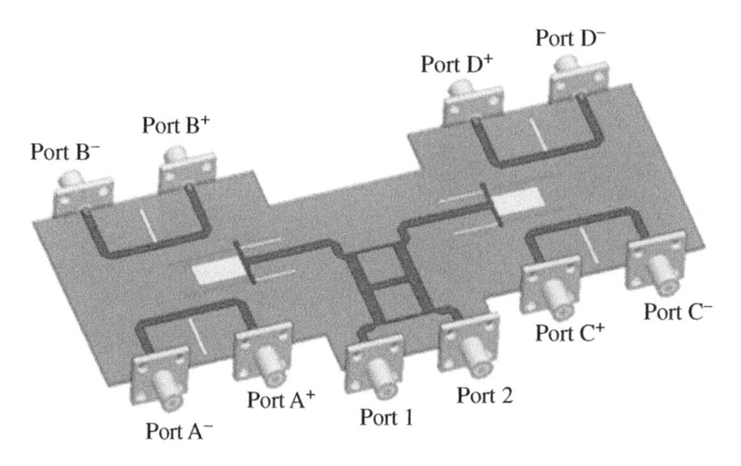

Figure 5.30 Electromagnetic model of a DFN that can feed a four-element CP array. *Source*: From [25] / with permission of IEEE.

single feed one. The maximum gain of the differential-fed array in Figure 5.29a is 14.6 dBi, while that of the single feed array using the same radiating elements is only 10.8 dBi. Figure 5.29b shows that the AR values of the differential array are less than 3 dB in an extremely wide bandwidth range from 3.0 to 6.0 GHz. The observed AR bandwidth of the single feed array is much narrower. It can be concluded that the performance of the differential-fed CP array is significantly better than the single feed one in all aspects.

Although premium performance characteristics can be realized using a differential feed, the main challenge lies in the design of high-performance DFNs, especially the ones that achieve a high level of CM suppression. The SETB PD described above serves as a good candidate for building these kinds of DFNs. Using the configuration shown in Figure 5.27a, a complete four-element DFN for a differential-fed CP array is illustrated in Figure 5.30. It consists of a 90° hybrid coupler and two balanced PDs.

When Port 1 is excited, balanced signals are produced at the four output ports with a phase increment of −90°. On the other hand, balanced signals will be produced at the four output ports with a phase increment of 90° when Port 2 is excited. To feed a four-element differential-fed CP array, the differential feeding points of Ant 1, 2, 3, and 4 in Figure 5.27a are connected to Ports A, C, B, and D of the DFN. It is noted that the order of the ports is important. Ports C and D should be opposite to Ports A and B to obtain a 180° phase difference between Ports C and A, and between Ports D and B. Therefore, the DFN can be used to feed an LHCP differential array by exciting port 1 and to feed an RHCP array by exciting port 2. Only the LHCP array design is discussed in this chapter. Its simulated performance characteristics have been verified with a measured prototype. The RHCP array version can be realized simply by using the same structure and feeding network, but exciting it via Port 2.

The prototypes of the DFN and the 1×4 CP array were fabricated and tested. Photos of the prototypes and the measurement configuration are shown in Figure 5.31. The radiation patterns of the array were measured in an anechoic chamber. The simulated and measured radiation patterns in the two principal planes at 4.5 GHz (center frequency) are compared in Figure 5.32. The 3-dB beamwidths in the plane along the length of the array, shown in Figure 5.32a and orthogonal to it, shown in Figure 5.32b, are 18° and 56°, respectively. The first sidelobe level shown in Figure 5.32a is 14.0 dB.

The simulated and measured peak realized gain and AR values over the frequency range from 3.5 to 5.5 GHz are presented in Figure 5.33. The measured peak realized gain values vary from 9.8 to 13.5 dBi. In comparison, the simulated values vary from 9.2 to 14 dBi. The 3-dB AR bandwidth is from 3.55 to 5.5 GHz, which is almost the same as the matching bandwidth. These measured results verify that the DFN provides an effective feeding approach for differential-fed arrays and facilitates their premium radiation performance.

For certain applications, identical patterns in all vertical planes are needed. This would require an $n \times n$ array configuration. The HFSS 2×2 CP array model shown in Figure 5.34 utilized the same radiating elements employed in the 1×4 version. Since each radiating element is an LHCP antenna, each one of them is rotated clockwise by 90° in the array.

To guarantee that the array emits LHCP fields, each element should emit LHCP fields and the phase sense of each element in the array should also be LHCP. This configuration requires a clockwise 90° discrete rotation from Ant 1 to Ant 4. Similarly, an RHCP array requires a 90° discrete anticlockwise rotation from Ant 1 to Ant 4. To provide the required 90° phase shift between the

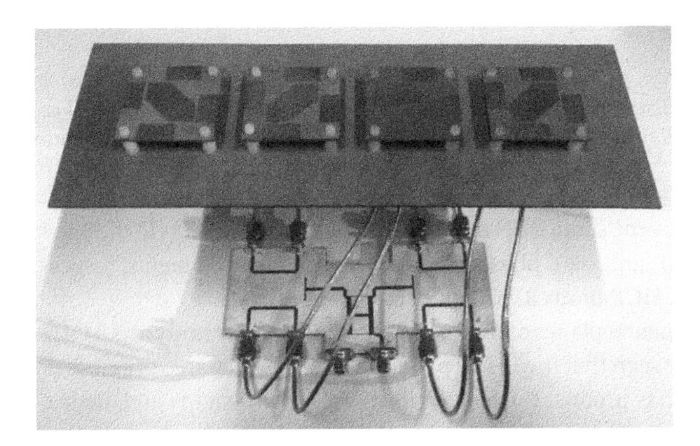

Figure 5.31 Prototype of the DFN connected to a differential-fed 1 × 4 CP array. *Source*: From [25] / with permission form IEEE.

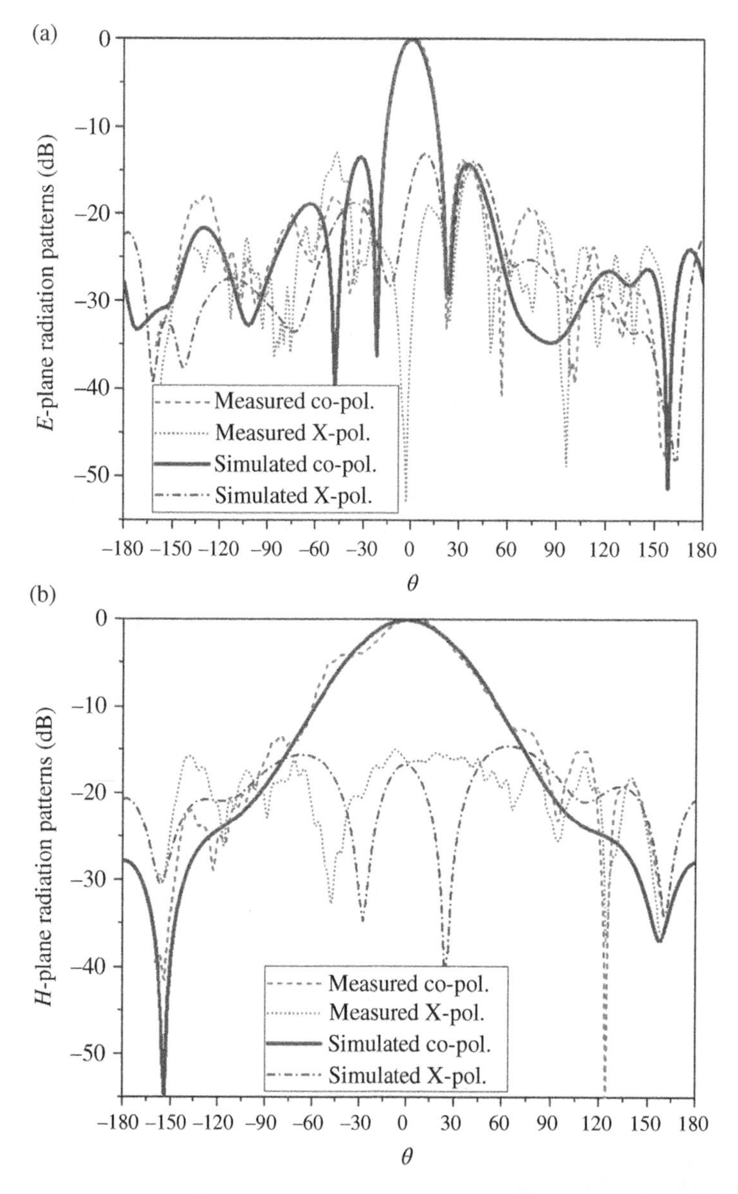

Figure 5.32 Simulated and measured radiation patterns of the 1 × 4 differential CP array fed by the DFN. (a) Plane along the length of the array. (b) Plane orthogonal to the length of the array. *Source*: From [25] / with permission of IEEE.

excitations to each of the array elements, the same DFN in Figure 5.29 was used. Similar to the 1 × 4 CP array, Port 1 and Port 2 provide +90° and −90° phase increments to the array elements, respectively. Thus, it can support either an LHCP or an RHCP outcome, respectively.

The radiation patterns in the two principal planes of the 2 × 2 differential-fed CP array at 5.0 GHz are displayed in Figure 5.35. It is clearly seen that the LHCP patterns in both planes are very close to each other, indicating that the design has produced a very symmetrical pattern shape and, hence, constant beamwidth. The maximum gain is around 14.5 dBi and the X-pol, RHCP level is less than −35 dB at boresight.

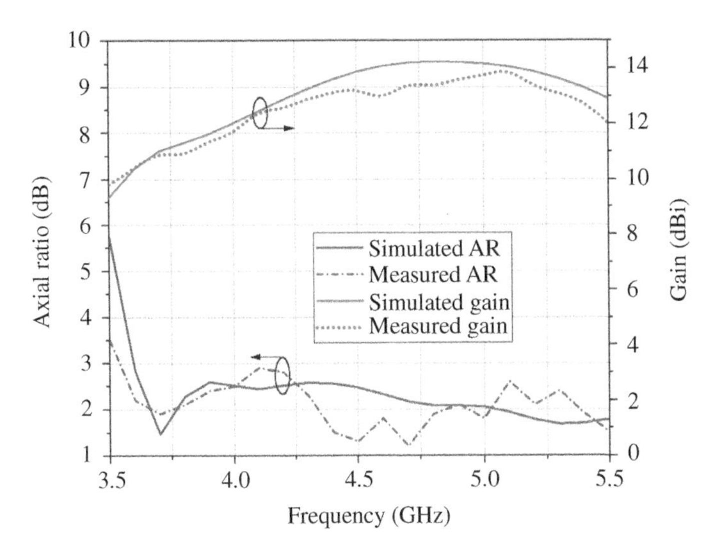

Figure 5.33 Simulated and measured peak realized gain and axial ratio values of the 1 × 4 differential CP array fed by the DFN. *Source*: Based on [25] / IEEE.

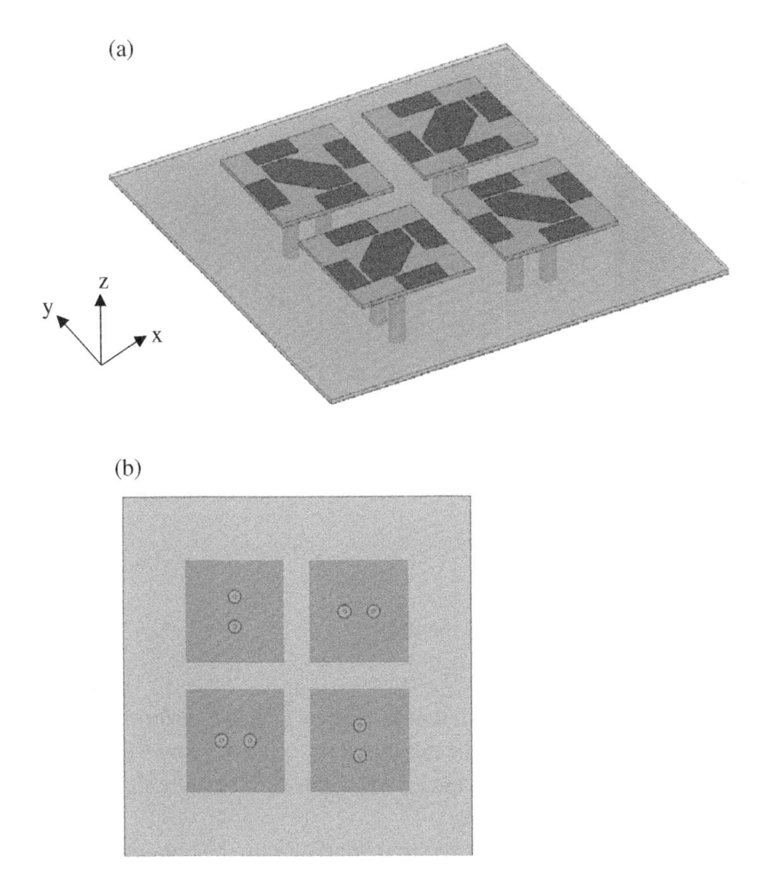

Figure 5.34 2 × 2 differential CP array fed by the DFN. (a) Isometric view of the 3D model. (b) Bottom view of the model.

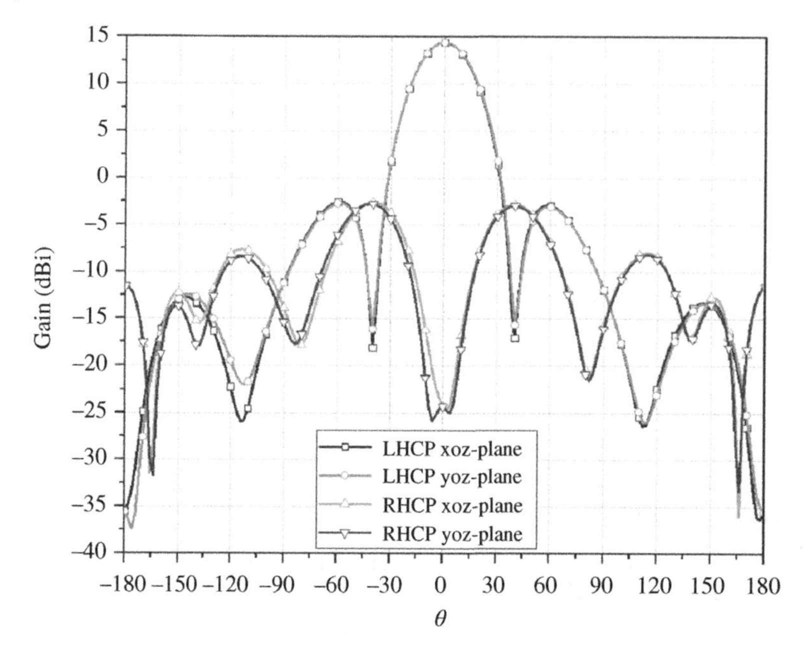

Figure 5.35 Radiation patterns of the 2 × 2 differential CP array fed by the DFN.

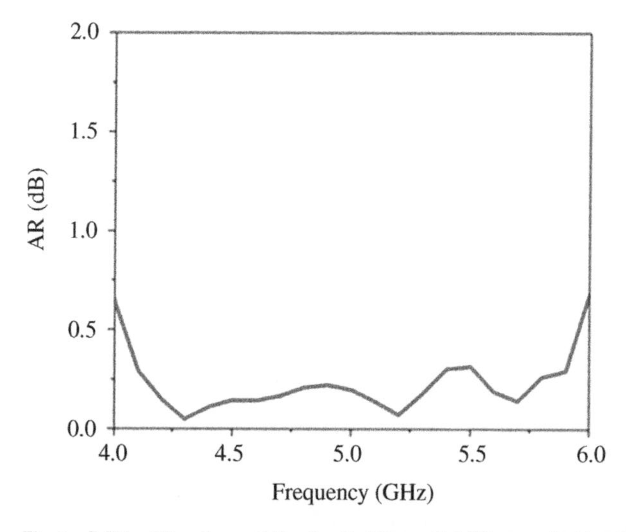

Figure 5.36 AR values of the 2 × 2 differential CP array fed by the DFN.

This 2 × 2 CP array has a quite wide operating bandwidth thanks to the wideband performance of its CP antenna elements and the DFN. Figure 5.36 also indicates that it has relatively low AR values and a wide AR bandwidth. The AR is less than 0.7 dB across an AR bandwidth of more than 40%. This performance is very desirable for CP antenna arrays.

It has been demonstrated above that excellent performance characteristics can be obtained from both LP and CP differential arrays by using the developed DFNs. Their exceptional performance is attributed to the enhanced CM suppression ability of the DFN and the individual radiating elements and the wide operating bandwidths of the balanced PDs. Given these results, it is of substantial interest for current 5G and evolving 6G and beyond applications to consider differential-fed arrays as multiple beam antenna solutions. This ability will be discussed in the next section.

5.4 Differential-Fed Multi-Beam Antennas

Multi-beam BFNs are popular microwave circuits to produce a number of antenna beams simultaneously. Among them, BMs are the most commonly used ones because they are both simple and lossless [25]. Multiple beams can be realized with differential antenna systems by introducing DFNs to drive them.

The traditional single-ended 2×4 BM, as shown in Figure 5.37a, is composed of a 90° hybrid coupler, two PDs, and two 180° phase shifters. The BFN is connected here to a 1×4 four-element array that can produce two beams when Port 1 or Port 2 is fed. A phase increment of +90° or −90° can be realized at each element. The result is the realization of two non-boresight beams in the E-plane.

Similarly, a differential 2×4 BM is needed to feed the corresponding 1×4 four-element differential array. The 2×4 differential BM arrangement reported in [23] is shown in Figure 5.37b. It is composed of a 90° hybrid coupler and two SETB PDs. The output ports of the SETB PDs are connected to the feeds of radiating elements. It is noted that although a 180° phase difference is still needed between any two nonadjacent elements, 180° phase shifters are no longer necessary with this approach. The 180° phase difference is realized by simply reversing the connection port order of the two differential elements. The BFN illustrated in Figure 5.37b is actually the same as the DFN for the CP arrays presented in Figure 5.28. The +90° and −90° phase increments associated with that BFN provide the means for beam steering. The feeding network given in Figure 5.30 was thus used to achieve a differential 2×4 BM.

Since the bandwidth of the requisite branch-line couplers is limited, the differential 2×4 BM design targeted only a 40% fractional bandwidth, from 4.0 to 6.0 GHz. Two SETB PDs were cascaded with the two output ports of the branch-line coupler. The BFN had two single-ended input ports,

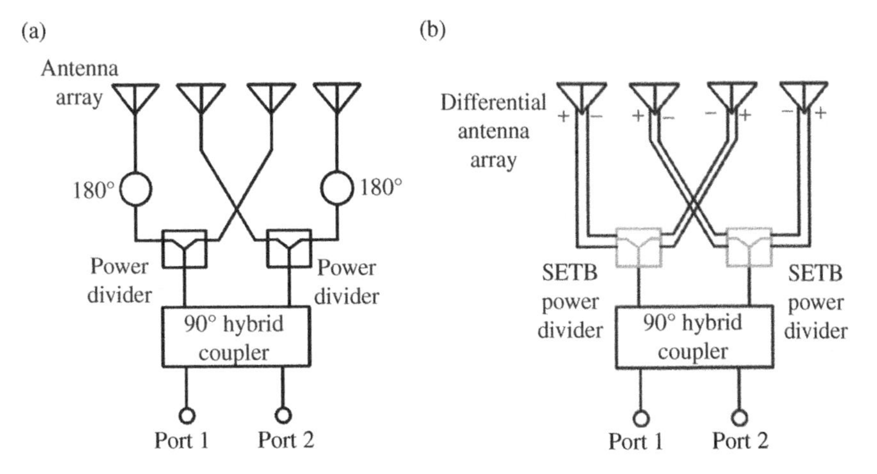

Figure 5.37 2 × 4 Butler matrix configurations. (a) Traditional single-ended Butler matrix. (b) Differential Butler matrix. *Source:* Based on [23] / IEEE.

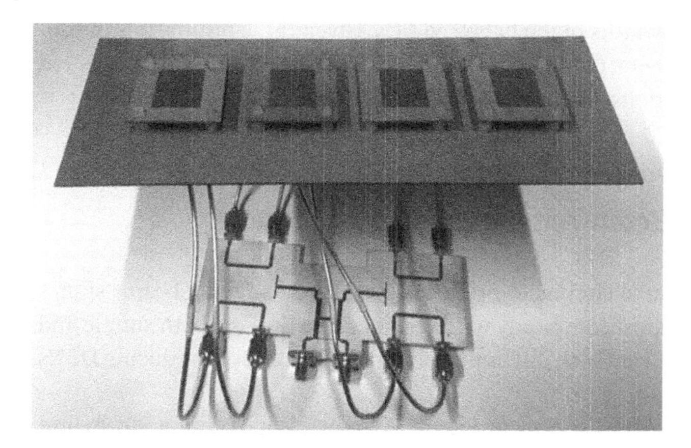

Figure 5.38 Prototype of the Butler matrix driving the multi-beam 1 × 4 differential-fed LP array. *Source*: From [25] / with permission form IEEE.

Port 1 and Port 2, and four differential output ports: Port A, Port B, Port C, and Port D. Ports A and B and Ports C and D are in-phase. On the other hand, there is a phase difference φ between Ports A and C, as well as between Ports B and C, Ports A and D, and Ports B and D. When Port 1 is fed, this phase difference is $\varphi = +90°$; when Port 2 is fed, $\varphi = -90°$.

To verify the simulated performance characteristics of this differential 2 × 4 BM, a four-element differential LP patch array driven with a BM DFN was fabricated and tested. Figure 5.38 shows the 1 × 4 differential-fed LP array, the associated beamforming network, and its test setup. The output ports of the BM are connected to the feed ports of the array. The order of these connections is as follows. Port A is connected to the first element, Port C is connected to the second one, Port B is connected to the third one, and Port D is connected to the fourth one.

It is noted that the connection order of the third and fourth elements is reversed from that of the first and second elements. This difference occurs because a 180° phase difference is needed for the third and fourth elements as illustrated in Figure 5.37b. When Port 1 is excited, the phase increment between the elements is +90°; while when Port 2 is excited, the phase augment between elements is −90°. This phase distribution results in multiple beams in the E-plane of this differential linear array.

The radiation patterns of the BM DFN-driven differential LP antenna array were obtained in the same anechoic chamber with the same test procedures. The simulated and measured E-plane patterns obtained at 5.0 GHz are shown in Figure 5.39. A good agreement between the simulated and measured patterns was obtained. Figure 5.39a shows the normalized co-pol radiation patterns. When Port 1 or Port 2 was excited, the main beam points to the −18.5° or 18.5° direction, respectively. The half-power beamwidth is around 19° and the gain is 12.5 dBi. It is clearly seen that multiple nulls are generated including one at boresight. Figure 5.39b shows E-plane X-pol patterns. The measured X-pol levels when the array is excited with Port 1 or Port 2 are below −32 dB. This is actually reasonably close to the simulated value of −40 dB. It is noted that the X-pol level of the differential array is much lower than the single-ended one. Similar patterns were obtained at other frequencies from 4.5 to 5.5 GHz, so they are not included here. The measured results fully demonstrate the utility of the BM DFN for antenna arrays and verify the efficacy of its SETB PDs.

The above differential BM can be further extended to feed larger-scale antenna arrays with 8, 16, or even more elements. Figure 5.40 shows the configuration of a typical 2 × 8 differential BM that feeds an eight-element array. It is composed of a 90° hybrid coupler, two single-ended PDs and four SETB

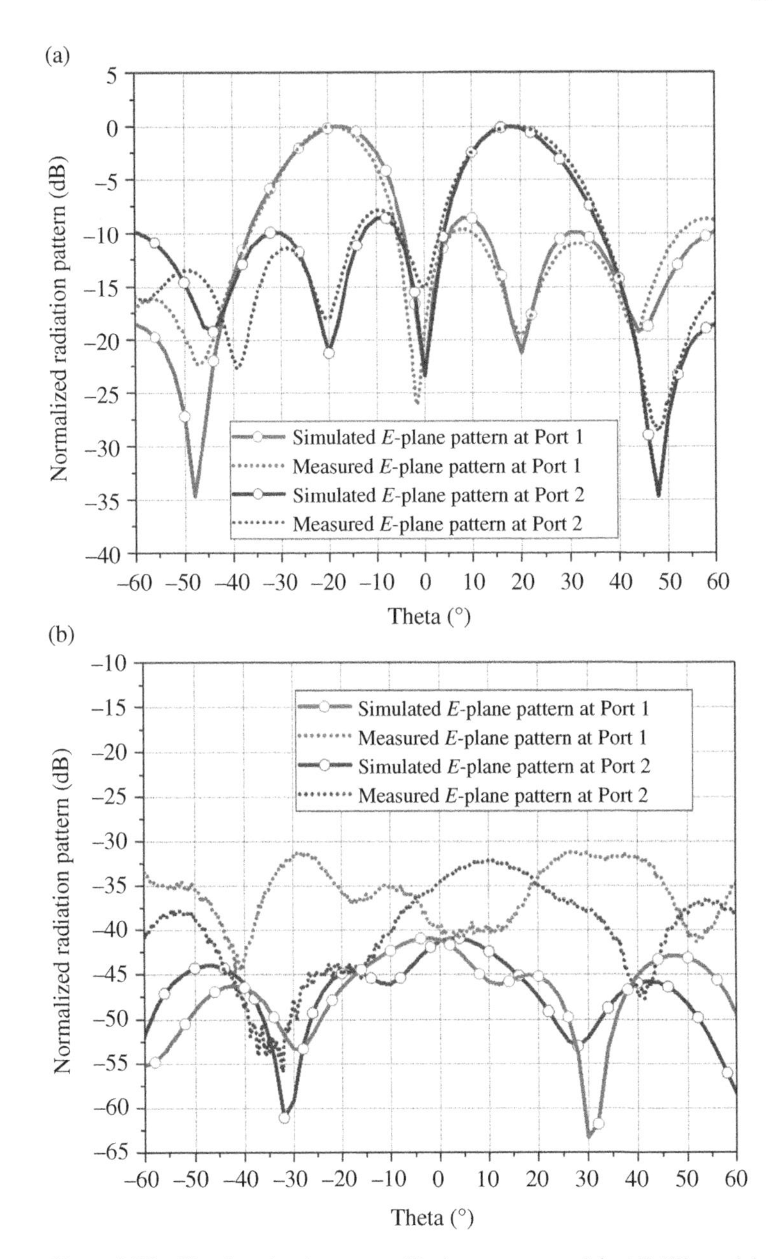

Figure 5.39 Simulated and measured E-plane patterns of the LP differential array fed by the differential 2 × 4 Butler matrix. (a) Co-polarization radiation patterns. (b) Cross-polarization radiation patterns. *Source*: Based on [23] / IEEE.

PDs. Since the phase increment is 90°, the outputs of the first SETB PD are connected to antennas 1 and 5. The remaining SETB PDs are connected to antennas 2 and 6, 3 and 7, and 4 and 8, respectively. Since there is 180° phase difference between antenna 1 and 3 as well as between 2 and 4, 5 and 7, and 6 and 8, the connecting order of the port + and port − is reversed at antennas 3, 4, 7, and 8.

The feeding network can be used to feed an eight-element differential array and attain multiple beam functions. Figure 5.41 shows the E-plane co-pol radiation pattern when port 1 *and* port 2 are

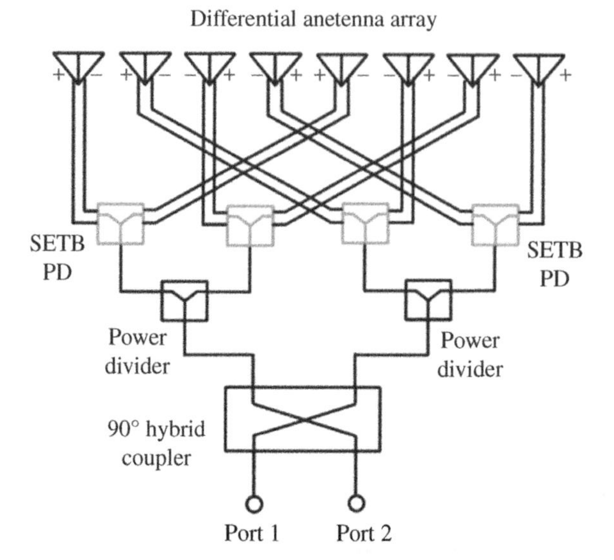

Figure 5.40 The extended 2 × 8 Butler matrix DFN configuration used to drive an eight-element LP differential antenna array. *Source*: Based on [23] / IEEE.

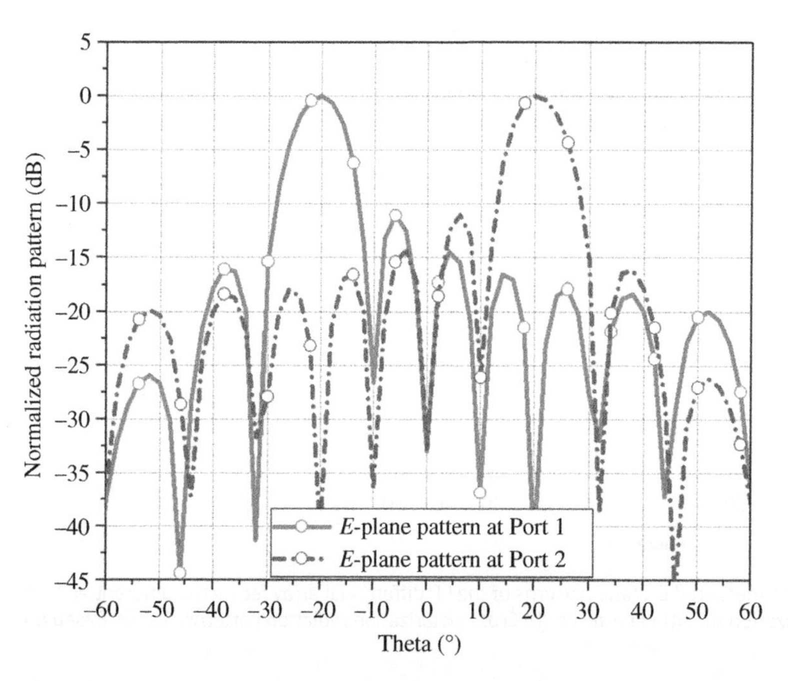

Figure 5.41 E-plane radiation patterns of the eight-element LP differential array fed by the extended 2 × 8 Butler matrix. *Source*: Based on [23] / IEEE.

excited. Two beams radiating at −20° and +20° are obtained. The peak realized gain is 15.7 dBi and the half-power beamwidth is around 10° for both beams. It is obvious that in comparison to the initial 2 × 4 BM design, the 2 × 8 one achieves better array performance, including higher gain, narrower beamwidth, better selectivity of the main lobe, and more nulls in the pattern. Both the 2 × 4

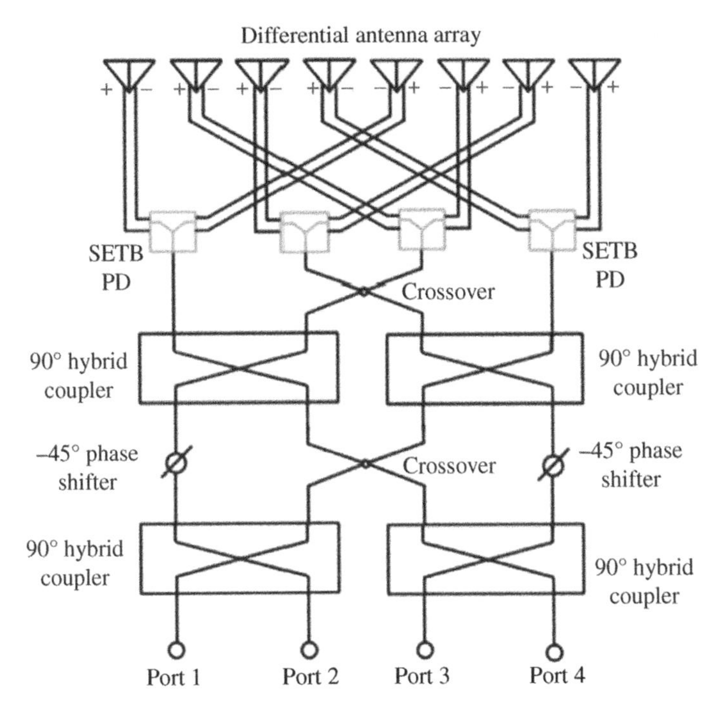

Figure 5.42 Configuration of a 4 × 8 Butler matrix feeding an eight-element LP differential antenna array. *Source*: Based on [23] / IEEE.

and 2 × 8 BMs can achieve high CM and mode-conversion suppression levels, attractive features facilitated by the developed SETB PDs.

The presented design topology provides a general design approach to realize 2^n-way differential BMs utilizing the SETB PDs. Furthermore, since the input ports of differential BMs are usually single-ended and the output ports are balanced, the extension from a 2^n-way BM to a 2^{n+1}-way version can be realized by utilizing 2^{n-1} more SETB PDs and 2^{n-1} more single-ended PDs. The associated methodology can also be extended to realize larger-scale BMs with more beams.

As an example, Figure 5.42 displays a 4 × 8 BM configuration that feeds an eight-element LP differential array. Similar to the 2 × 4 case, the 180° phase shifters are not necessary because the ports, port + and port −, can be connected in reverse order to the feed points of the fifth, sixth, seventh, and eighth elements as displayed in Figure 5.42. The simulated radiation patterns are given in Figure 5.43. Four beams are produced with maximum gains of 15–17 dBi. The X-pol level is less than −40 dB for any observation angle θ. Thus, the X-pol patterns are not presented.

5.5 Conclusion

A variety of differential devices and antenna systems were presented. They include differential-fed LP and CP antennas, differential BFNs facilitated by newly developed SETB DPs, and single and multi-beam differential-fed arrays excited with BM versions of DFNs. The advantageous CM suppression and very low DC-to-CM conversion associated with the differential systems were emphasized. Comparisons between single feed and differential-fed antennas and antenna arrays

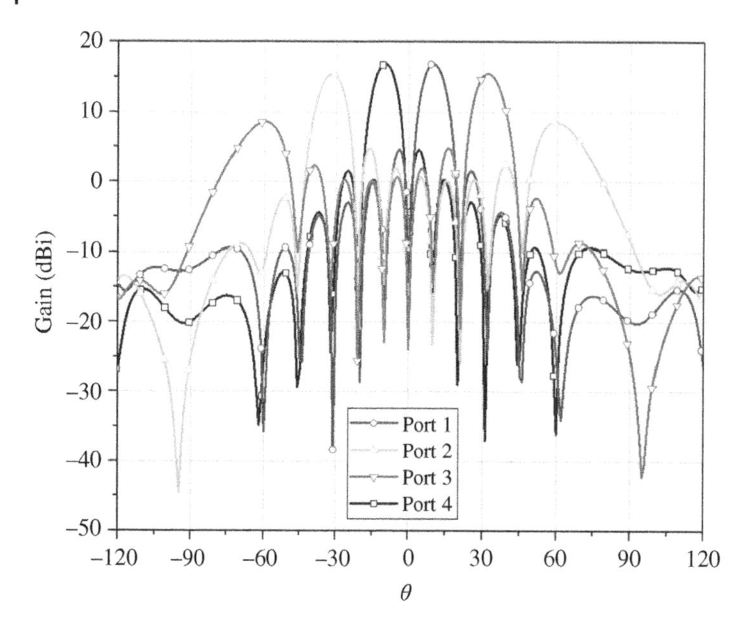

Figure 5.43 Simulated E-plane radiation patterns for an eight-element LP differential array fed by the extended 4 × 8 Butler matrix. *Source*: Based on [23] / IEEE.

illustrated the consequent advantages. Similarly, it was demonstrated that the differential BMs had superior performance characteristics in comparison to traditional BMs. These presented differential BFNs have simpler structures, more compact sizes, and better performance characteristics. It is expected that differential beamforming networks and differential antenna arrays will find wide applications in current 5G and evolving 6G and beyond wireless communication systems.

References

1. Li, B. and Leung, K.W. (2008). On the differential-fed rectangular dielectric resonator antenna. *IEEE Trans. Antennas Propag.* **56** (2): 353–359.
2. Bisharat, D.J., Liao, S., and Xue, Q. (2016). High gain and low cost differential-fed circularly polarized planar aperture antenna for broadband millimeter-wave applications. *IEEE Trans. Antennas Propag.* **64** (1): 33–42.
3. Stutzman, W.L. and Thiele, G.A. (2012). *Antenna Theory and Design*, 3e. New York: Wiley.
4. Garg, R., Bhartia, P., Bahl, I., and Ittipiboon, A. (2001). *Microstrip Antenna Design Handbook*. Boston, MA, USA: Artech House.
5. Wu, J., Yin, Y., Wang, Z., and Lian, R. (2015). Broadband circularly polarized patch antenna with parasitic strips. *IEEE Antennas Wirel. Propag. Lett.* **14**: 559–562.
6. Chen, X., Qin, P.-Y., Guo, Y.J., and Fu, G. (2017). Low-profile and wide-beamwidth dual-polarized distributed microstrip antenna. *IEEE Access* **5**: 2272–2280.
7. de Lera Acedo, E., Garcia, E.E., Gonzalez-Posadas, V. et al. (2010). Study and design of a differential-fed tapered slot antenna array. *IEEE Trans. Antennas Propag.* **58** (1): 68–78.
8. Hu, H.T., Chen, F.C., Qian, J.F., and Chu, Q.X. (2017). A differential filtering microstrip antenna array with intrinsic common-mode rejection. *IEEE Trans. Antennas Propag.* **65** (12): 7361–7365.

9. Jin, H., Che, W., Chin, K.S. et al. (2018). Millimeter-wave TE_{20}-mode SIW dual-slot-fed patch antenna array with a compact differential feeding network. *IEEE Trans. Antennas Propag.* **66** (1): 456–461.
10. Xue, Q., Liao, S.W., and Xu, J.H. (2013). A differentially-driven dual-polarized magneto-electric dipole antenna. *IEEE Trans. Antennas Propag.* **61** (1): 425–430.
11. Tang, H., Chen, J.-X., Yang, W.-W. et al. (2017). Differential dual-band dual-polarized dielectric resonator antenna. *IEEE Trans. Antennas Propag.* **65** (2): 855–860.
12. Tang, Z., Liu, J., Lian, R. et al. (2019). Wideband differential-fed dual-polarized planar antenna and its array with high common-mode suppression. *IEEE Trans. Antennas Propag.* **67** (1): 131–139.
13. Wen, L.-H., Gao, S., Luo, Q. et al. (2019). A low-cost differentially driven dual-polarized patch antenna by using open-loop resonators. *IEEE Trans. Antennas Propag.* **67** (4): 2745–2750.
14. Lin, W., Chen, S.L., Ziolkowski, R.W., and Guo, Y.J. (2018). Reconfigurable, wideband, low-profile, circularly polarized antenna and array enabled by artificial magnetic conductor ground. *IEEE Trans. Antennas Propag.* **66** (3): 1564–1569.
15. Ali, M.M. and Sebak, A. (2019). Printed RGW circularly polarized differential feeding antenna array for 5G communications. *IEEE Trans. Antennas Propag.* **67** (5): 3151–3160.
16. Yadav, A.N. and Bhattacharjee, R. (2007). Gysel type unbalanced-to-balanced equal power divider. *Proceedings of the 47th IEEE European Microwave Conference*, Nuremberg, Germany (10–12 October 2017).
17. Yu, Y. and Sun, L. (2015). A design of single-ended to differential-ended power divider for X band application. *Microw. Opt. Technol. Lett.* **57** (11): 2669–2673.
18. Muralidharan, S., Wu, K., and Hella, M. (2015). A compact low loss single-ended to two-way differential power divider/combiner. *IEEE Microw. Wireless Compon. Lett.* **25**: 103–105.
19. Zhang, W., Liu, Y., Wu, Y. et al. (2017). Novel planar compact coupled-line single-ended-to balanced power divider. *IEEE, Trans. Microwave Theory Techn.* **65** (8): 2953–2963.
20. Wu, Y., Zhuang, Z., Kong, M. et al. (2018). Wideband filtering unbalanced-to-balanced independent impedance-transforming power divider with arbitrary power ratio. *IEEE Trans. Microwave Theory Techn.* **66** (10): 4482–4496.
21. Feng, W., Zhao, Y., Che, W. et al. (2018). Single-ended-to-balanced filtering power dividers with wideband common-mode suppression. *IEEE Trans. Microwave Theory Techn.* **66** (12): 5531–5542.
22. Gupta, K.C., Garg, R., Bahl, I., and Bhartia, P. (1996). *Microstrip Lines and Slotlines*, 2e. Norwood, MA: Artech House.
23. Zhu, H., Sun, H.-H., Jones, B. et al. (2019). Wideband dual polarized multiple beam-forming antenna arrays. *IEEE Trans. Antennas Propag.* **67** (3): 1590–1604.
24. Zhu, H. and Guo, Y.J. (2020). Circularly-polarized differential antenna array fed by single-ended-to-balanced power dividers with high common-mode rejection. *Proceedings of 4th Australian Microwave Symposium (AMS)*, Sydney, Australia.
25. Berg, M., Chen, J., and Parssinen, A (2019). Radiation characteristics of differential-fed dual circularly polarized GNSS antenna. *IEEE 13th European Conference on Antennas and Propagation (EuCAP 2019)*, Krakow, Poland.

6

Conformal Transmitarrays

In response to the increasing demand for wireless connectivity, wireless communication technologies are moving toward the integration of terrestrial networks with airborne and spaceborne networks. This network fusion would enable vast communication coverages for people and vehicles in remote and rural areas, as well as in the air and on water. The creation of this Integrated Space and Terrestrial Network (ISTN) is of critical importance to industries associated with logistics, mining, agriculture, fishery, and defense. Consequently, ISTNs serve as ambitious targets in the development of sixth-generation (6G) mobile wireless communication networks [1, 2].

A typical ISTN is a combination of space, aero, and terrestrial networks. Aero networks consist of nodes and user terminals mounted on airborne platforms. As a result, conformal high-gain arrays with beam scanning capabilities are highly desired. Not only can they meet aerodynamic requirements at lower costs, but they also facilitate air-to-air and air-to-ground high data-rate links. Conformal transmitarray antennas are an appealing choice since they can be designed to follow the shapes of the various platforms on which they are mounted. Moreover, unlike reflectarray systems, their feeds are placed behind the array and, hence, within the platform.

In this chapter, a systematic study of conformal transmitarrays is presented. The challenges facing the design of conformal transmitarrays are discussed first in Section 6.1. A conformal transmitarray that employs thin triple-layer slot elements is characterized in Section 6.2. A new mechanical method to achieve beam steering is described in Section 6.3. An ultrathin dual-layer conformal transmitarray is then presented in Section 6.4, which achieved significantly improved efficiency when compared to the triple layer element one presented in Section 6.2. An elliptically conformal multi-beam transmitarray with wide spatial coverage and small gain variations is described in Section 6.5. The chapter concludes in Section 6.6 with some directions for future research.

6.1 Conformal Transmitarray Challenges

6.1.1 Ultrathin Element with High Transmission Efficiency

Transmitarrays typically consist of multiple layers of flat elements and an illuminating feed. The phases of these elements are individually designed in order to provide appropriate phase responses to transform the spherical phase front from the feed into a planar phase front. By taking advantages of both lens antennas and microstrip phased arrays, they can achieve high gains without using complex and lossy feed networks and provide beam-steering abilities [3–5]. To be conformal, a transmitarray must be designed to follow the shape of the platform on which it is mounted, e.g., aircrafts

Advanced Antenna Array Engineering for 6G and Beyond Wireless Communications, First Edition.
Y. Jay Guo and Richard W. Ziolkowski.

and unmanned aerial vehicles (UAVs). Considering current manufacturing technologies, one of the most feasible methods to implement a conformal transmitarray is to employ very thin array elements printed on a substrate whose thickness is about 0.5–1.0 mm. This choice facilitates various bending methods to achieve the desired conformal configuration without sacrificing the realization of sufficient phase changes to achieve the desired planar wave fronts. Most of the transmitarrays reported to date employ multilayer array elements, i.e., at least three metal layers printed on two dielectric substrates separated by air gaps or dielectric materials. Generally, the total thickness of those structures varies from 0.4 to 1.0 λ_0 (λ_0 being the signal wavelength of interest in the free space).

Recent research efforts have been largely devoted to reducing the thickness of the transmitarray elements and, therefore, the overall array structure to facilitate conformal applications. A three-layer bandpass frequency selective surface (FSS)-based element was presented in [6] that is 0.36 λ_0 thick. Another three-layer element with a thickness of 0.22 λ_0 has been demonstrated that consists of a split circular ring connected by a narrow strip in the middle layer and two polarizers in the upper and bottom layers [7]. A folded transmitarray was developed in [8] using a three-metal-layer polarizer with a thickness of 1.0 mm (0.1 λ_0 at 30 GHz). All of these elements were designed for a flat transmitting surface, making them suitable for planar transmitarray antennas. Elements whose thickness is greater than 0.1 λ_0 are applicable to the realization of conformal transmitarrays operating at or above 30 GHz ($\lambda_0 = 10$ mm) because they are easily bent into the desired shapes. However, transmitarrays working below 30 GHz are highly desired for many applications, e.g., aerial and satellite systems and 5G. Therefore, ultrathin elements which can be employed for conformal transmitarrays operating below 30 GHz are urgently needed.

It is important to note that any ultrathin element design should avoid the use of vias because they can lead to breakages when the array surface is bent. Although using high dielectric substrates can reduce the thickness of the array elements, the efficiency of the antenna could be significantly affected by their associated high losses and often they are brittle materials that easily break when bent. Moreover, the total cost and weight of the array will be increased. Another straightforward method to achieve thin elements is to simply reduce the total thickness of the multilayers. However, this brute force thinning may reduce the transmission efficiency significantly as there are always tradeoffs between the thickness of the array and its efficiency. As reported in [9], a three-layer metallic unit without air gaps has a low profile, 1.0 mm (0.033 λ_0) overall thickness at 10 GHz and was employed in a transmitarray design that achieved only a 36% aperture efficiency. Similarly, a thin transmitarray with a 37.9% aperture efficiency was developed in [10] using three-metal-layer antenna elements whose thickness was 1.6 mm (0.07 λ_0 at 13.5 GHz). The efficiency of almost all reported transmitarrays is only around 35% or even less when the thickness of their elements is smaller than 0.1 λ_0. Therefore, a key challenge for conformal designs is to develop very thin array elements with high transmission efficiencies.

Although three-layer metal structures have been successfully developed to reduce the thickness of transmitarrays, the precise alignment and complicated assembly of their multiple layers are challenging in practice and, hence, costly, especially at high frequencies. Therefore, dual-layer structures are much preferred. In fact, they have a high potential to further reduce the thickness of the array elements. Consequently, another key challenge is the realization of a dual-layer ultrathin conformal transmitarray with high transmission efficiency. To date, only a few dual-layer planar transmitarray elements have been reported. A 1.5-mm-thick (0.1 λ_0 at 20 GHz) transmitarray element was developed in [11]. It consists of two modified Malta crosses printed on the two sides of a dielectric substrate with four vertically plated through vias. However, as mentioned earlier, elements with vias are not preferred for conformal designs.

Huygens elements [12–14] are able to realize near-zero reflection and total transmission with sub-wavelength thicknesses. While they make good candidates to provide high-efficiency and low-profile transmitarrays, most of the currently reported Huygens elements are multilayer structures. The elements in [13] consist of three metal layers using two vias to connect the first and the third layers to create a current loop for the magnetic response. The second layer is for the electric response. The total thickness is 3 mm at 10 GHz (0.1 λ_0). The elements in [14] consist of three-layer patterned metallic surfaces that mimic one electric dipole and one magnetic dipole printed on two bonded substrates. The thickness of the elements is 0.4 mm (0.1 λ_0 at 77 GHz). There are a few reported two-layer Huygens elements [15, 16] with only two metal layers on which the requisite electric and magnetic currents are induced separately. The main limitation of this design is that discrete printed circuit board tiles have to be constructed for each array element; and the boards then have to be stacked into an array. Compared with fully planar structures [13, 14], this assembly aspect makes it more difficult to realize a large surface.

6.1.2 Beam Scanning and Multi-Beam Operation

For 5G and beyond networks, conformal transmitarrays are required to cover a predefined angular range. Thus, they must be able to scan their beams or, even better, to generate a number of concurrent, but independent, directive beams. One typical approach to achieve electronic beam scanning from flat transmitarray antennas is to tune the transmission phases of the antenna elements using either PIN [17–19] or varactor [20, 21] didoes. The former achieves discrete beam steering; the latter achieves continuous beam steering. These structures belong to the category of reconfigurable antennas. If the curvature of the conformal transmitarray antenna is large, however, it becomes very challenging to integrate active elements onto the surface. One solution is to steer the beam by mechanically rotating the excitation structure, e.g., a horn antenna. Compared to electronic beam scanning, this solution can avoid the extra losses and costs associated with using a very large number of active elements and complicated control circuits.

The requirements and challenges for the multi-beam operation of conformal transmitarrays are the same as those for their planar counterparts. They both must be able to achieve a wide-angle coverage with very small gain variation. Only a few multi-beam planar transmitarrays have been reported to date that have attempted to address this challenge [22–28]. For example, a metamaterial-based thin planar transmitarray was presented in [23] that achieved a spatial coverage of $\pm 27^\circ$ with a gain variation of 3.7 dB. A $\pm$ 30° spatial coverage was realized with a 3 dB gain variation in [25]. Roughly speaking, the current state-of-the-art multi-beam planar transmitarrays are only capable of achieving a $\pm$ 30° spatial coverage with less than a 3 dB scan loss. Unfortunately, there are few reports of multi-beam conformal transmitarrays.

6.2 Conformal Transmitarrays Employing Triple-Layer Elements

6.2.1 Element Designs

Consider the transmitarray element shown in Figure 6.1 that consists of three layers of identical square ring slots. This three-layer slot element was employed in [29] for a planar multiple-polarization transmitarray. The width and the length of each slot are w and L, respectively. The unit periodicity is P and the height of the whole unit is h. The dimensions w and P are chosen, respectively, to be 0.8 and 5.6 mm. The transmission phase of the element can be varied with L over

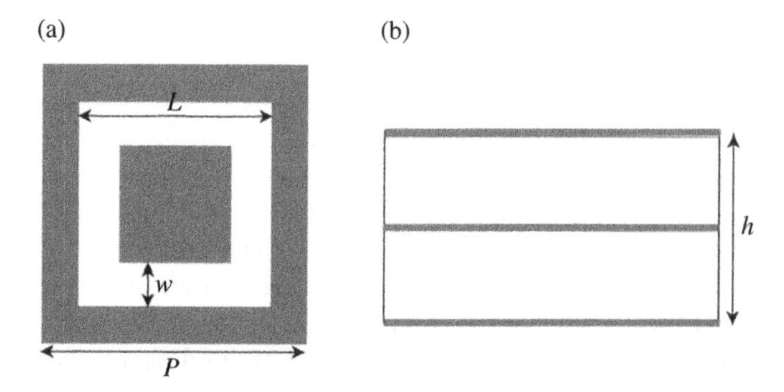

Figure 6.1 FSS multilayered element (Red represents metal and white represents substrate) reported in [30] (a) Top view. (b) Side view. *Source*: From [30] / with permission of IEEE.

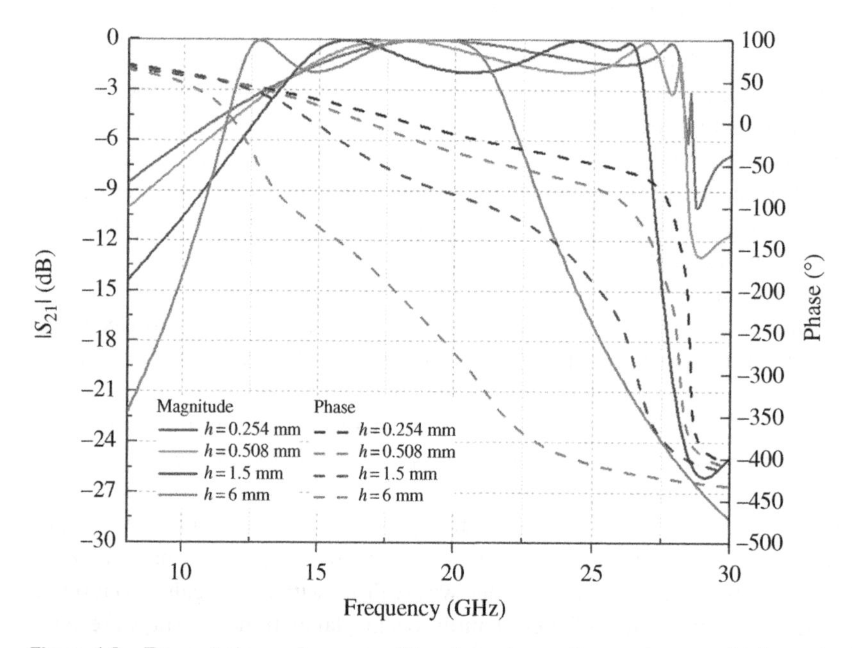

Figure 6.2 Transmission performance of the planar transmitarray element with L equal to 5.0 mm for different values of its height h. *Source*: From [30] / with permission of IEEE.

a range of 3.5–5.41 mm. However, the total thickness of the element is 30 mm, which is more than half of the wavelength at its operating frequency of 6 GHz.

In order to utilize this three-layer element in a conformal design, a parametric study of the effect of the height h on the transmission coefficient was conducted. The simulated results are given in Figure 6.2. The element performance was calculated with the 3D electromagnetic (EM) simulation software HFSS (high frequency structure simulator) using its Floquet method with master–slave boundaries. When h was varied, the other parameters were left unchanged. Figure 6.2 shows that the operating frequency increases and the phase range decreases when h is reduced. When h equals 6.0 mm, which is 0.5 λ_0 at 25 GHz, the phase range can cover more than 360°. However, the phase range is decreased significantly for a 0.254 mm thickness, which would otherwise be ideal for a conformal design at that frequency. Therefore, a compromise must be made between the thickness and the phase range. It was found that printed circuit boards (PCBs) with a thickness of more than 1 mm are easily broken when they are bent. Therefore, h was chosen to be 0.508 mm at 25 GHz, which is about 0.04 λ_0.

Since this planar triple-layer square ring slot element was designed for a conformal transmitarray, its transmission coefficient was then examined when it was bent with different curvatures. The results shown in Figure 6.3a are given for different values of the bend angle α, as defined in the figure. When α is equal to zero, the element recovers its planar behavior. The element was bent with α equal to 7.2° for the conformal transmitarray reported in [30]. Figure 6.3b depicts the amplitude and phase of the transmission coefficient versus the slot length L. When L varies from 3.5 to ~5.41 mm, a transmission phase range of 330° is achieved for the flat element. The transmission loss for most of the values of L is lower than 3 dB. The worst case is 3.6 dB when L is equal to 5.36 mm. The phase range and the loss remain almost the same as the flat one for small values of α. The transmission loss increases slightly, but only when α becomes very large, i.e., 60°. Nevertheless, it is still lower than 3.6 dB.

It is interesting to find that the phase range is nearly unchanged even for this large angle. This behavior occurs because the phase change is determined by the length of the slot; it is not affected by the curvature of the surface. Note that the periodic boundary conditions used in the simulations only mimic an infinitely large planar surface, not a finite one. Therefore, there may exist some discrepancy in the element performance between the infinite and the actual finite-sized surfaces. Nevertheless, as the bend angle used for the transmitarray design is only 7.2°, these simulation results were used as a reasonably accurate reference when the phase response of the elements was calculated.

The transmission phase of the element was also examined for different incident angles. These results are shown in Figure 6.3c. The oblique incidence was examined for the element bent with $\alpha = 7.2°$. It can be seen that the losses increase with the angle of incidence for L between 4.0 and 5.2 mm. When the incidence angle equals 35°, which was the largest one considered, the maximum loss increased to 4.0 dB when $L = 5.36$ mm. On the other hand, the phase range remained effectively unchanged for different incidence angles. Considering the ultralow profile of the structure, the 330° phase range and the maximum 4.0 dB loss at the maximum 35° incidence angle achieved by the element are quite acceptable performance characteristics for a conformal transmitarray antenna.

6.2.2 Conformal Transmitarray Design

In order to generate a beam in a particular direction from any transmitarray, the output phase of each array element should be designed to compensate for the spatial phase delays associated with the distance from the source to that element. This compensation should produce the phase distribution necessary to generate a beam in the specified direction [31].

A rectangular horn is taken to be the feed. This choice anticipates the realized prototype to be described later in this chapter. Figures 6.4a and 6.4b depict a three-dimensional (3D) view of the entire system and a side view of it, respectively. The feed horn is located at the prime focus of the cylindrical transmitarray surface and is pointed at the middle of the surface. The unbent transmission surface is displayed in Figure 6.5. The different colors represent its unit elements and their different, discrete transmission phases.

To accurately calculate the required compensation phase at different positions on the cylindrical transmitarray, the phases of the elements on the cylindrical arc in the *yz*-plane, i.e., the middle cut of

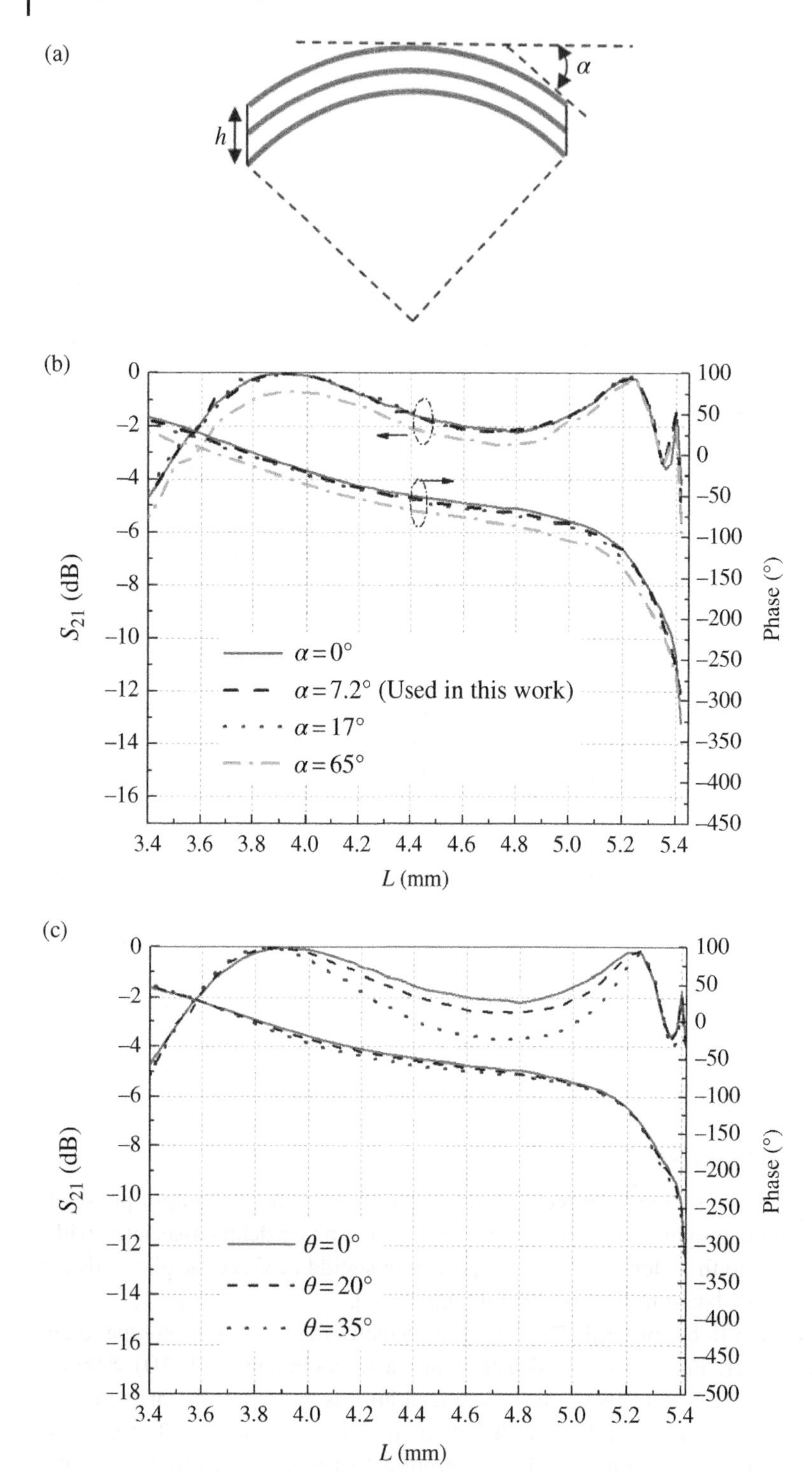

Figure 6.3 Bent transmitarray element slot length L parameter studies. (a) The element with a bend angle α. (b) Simulated transmission amplitude and phase. (c) Transmission performance of the curved element with α equal to 7.2° at 25 GHz for normal and oblique incidence. *Source*: From [30] / with permission of IEEE.

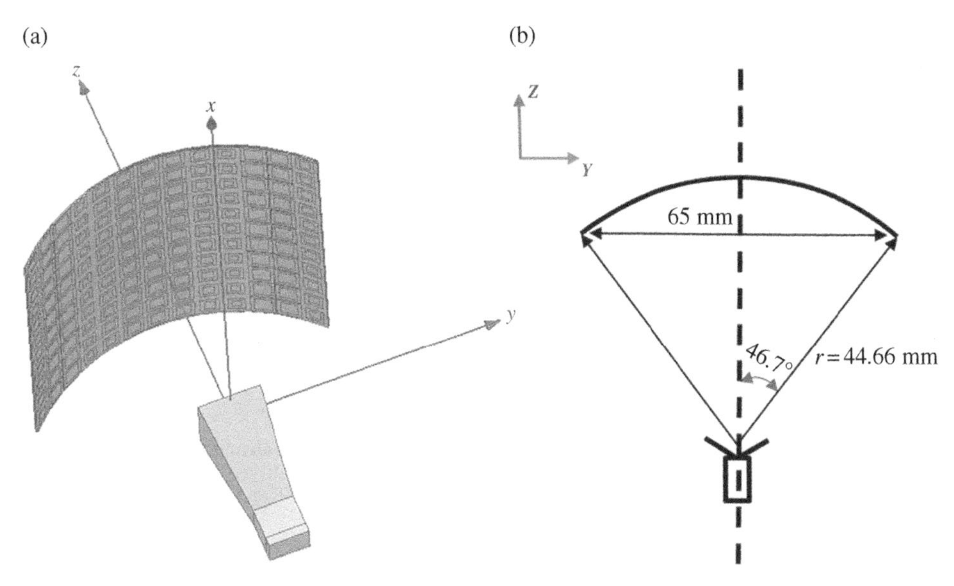

Figure 6.4 Conformal transmitarray configuration. (a) 3D perspective view. (b) Side view of the conformal transmitarray. *Source*: From [30] / with permission of IEEE.

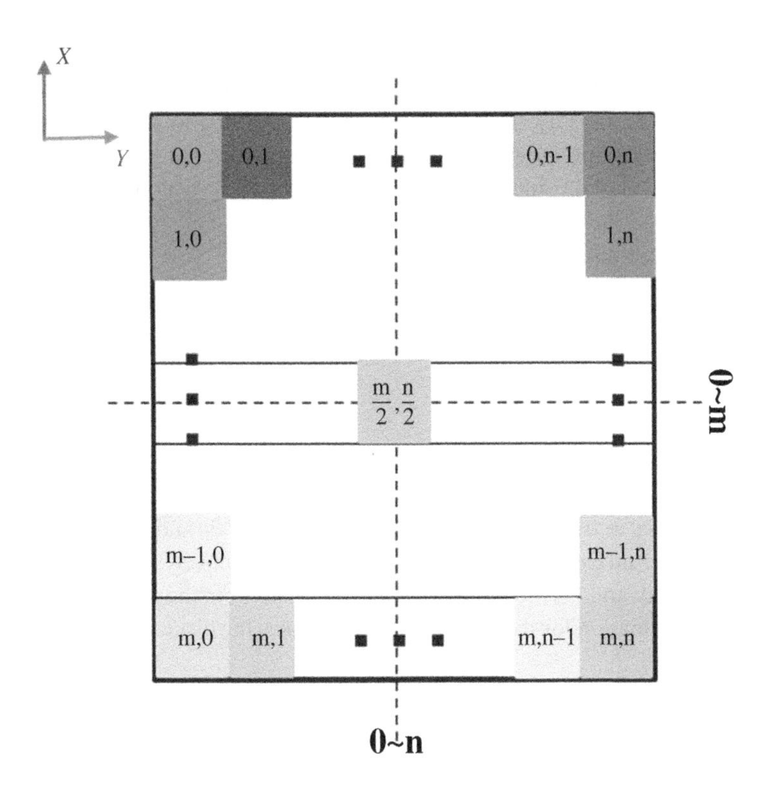

Figure 6.5 Phase distribution on the conformal transmitarray before bending. *Source*: From [30] / with permission of IEEE.

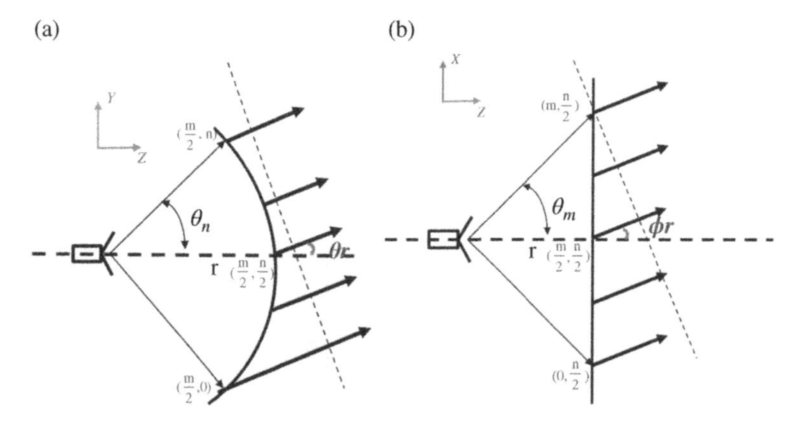

Figure 6.6 Graphical depiction of the phase of the elements on the conformal transmitarray: (a) *y0z* plane. (b) *x0z* plane.

the surface, are considered first. As seen from Figure 6.6a, the phases of the elements on this curved line can be calculated using Eq. (6.1) for a given beam direction θ_r:

$$\varphi_{\frac{m}{2},n} = \varphi_{\frac{m}{2},\frac{n}{2}} + r(\cos\theta_r - \cos(\theta_n - \theta_r)) * \frac{2\pi}{\lambda_0} \tag{6.1}$$

These elements are also used in each column of the transmitarray along the *x*-axis. The phase calculation for the elements in each column of Figure 6.5 is the same as the phase calculated according to Figure 6.6b, i.e., the phase value is that obtained along the straight line from the horn to the corresponding location on the flat transmitarray. This yields the phase required to achieve the desired phase distribution on the tangent plane shown in Figure 6.6a. For example, the phase along the middle column of Figure 6.5 is calculated using Eq. (6.2) for a given beam direction φ_r:

$$\varphi_{m,\frac{n}{2}} = \varphi_{\frac{m}{2},\frac{n}{2}} + r\left(\frac{1}{\cos\theta_m} - \tan\theta_m * \sin\varphi_r - 1\right) * \frac{2\pi}{\lambda_0} \tag{6.2}$$

Finally, Eq. (6.3) gives the phase φ_{mn} for any unit element in Figure 6.5:

$$\varphi_{mn} = \varphi_{m,\frac{n}{2}} + r(\cos\theta_r - \cos(\theta_n - \theta_r)) * \frac{2\pi}{\lambda_0} = \varphi_{\frac{m}{2},\frac{n}{2}}$$
$$+ \mathrm{r}\left(\frac{1}{\cos\theta_m} + \cos\theta_r - \cos(\theta_n - \theta_r) - \tan\theta_m * \sin\varphi_r - 1\right) * \frac{2\pi}{\lambda_0} \tag{6.3}$$

where θ_r and φ_r are the radiation angles in the *y0z* and *x0z* planes, respectively; θ_n and θ_m are the half subtended angles of the cylindrical surface in the *y0z* and *x0z* planes, respectively; $\varphi_{m/2,\ n/2}$ is the transmission phase of the center unit; and *r* is the distance from the feed point to the center unit. In order to achieve a −10 dB edge illumination for the transmitarray aperture, the distance *r* is chosen to be 44.66 mm.

A conformal transmitarray antenna prototype using the developed elements was designed, fabricated, assembled, and tested. It consisted of 13 × 11 = 143 elements. The cross section size of the transmission surface is 65.0 mm × 61.6 mm. A standard gain horn, the LB-28-10-C-KF horn from A-INFO, was used as the feed. It was placed at the focus and centered on the cylindrical surface with its peak gain direction pointed along the normal at the center of it as depicted in Figure 6.7. The aperture edge illumination level was around −10 dB and it had a half subtended angle of 46.7°.

Each layer of the unbent two-layered transmission surface was fabricated using standard PCB technology on low-cost Wangling F4B substrates whose relative dielectric constant is 2.2 and loss tangent is tan $\delta = 0.007$. The two surfaces were laminated together and were attached to a 3D-printed cylindrical frame. The entire bent transmission surface was mounted on a U-shaped metallic frame for measurement purposes. Photographs of the mounted prototype in the anechoic measurement chamber are shown in Figure 6.7.

Figure 6.8 shows the simulated and measured input reflection coefficients of the prototype as functions of the frequency. The input reflection coefficients are below −10 dB across the frequency

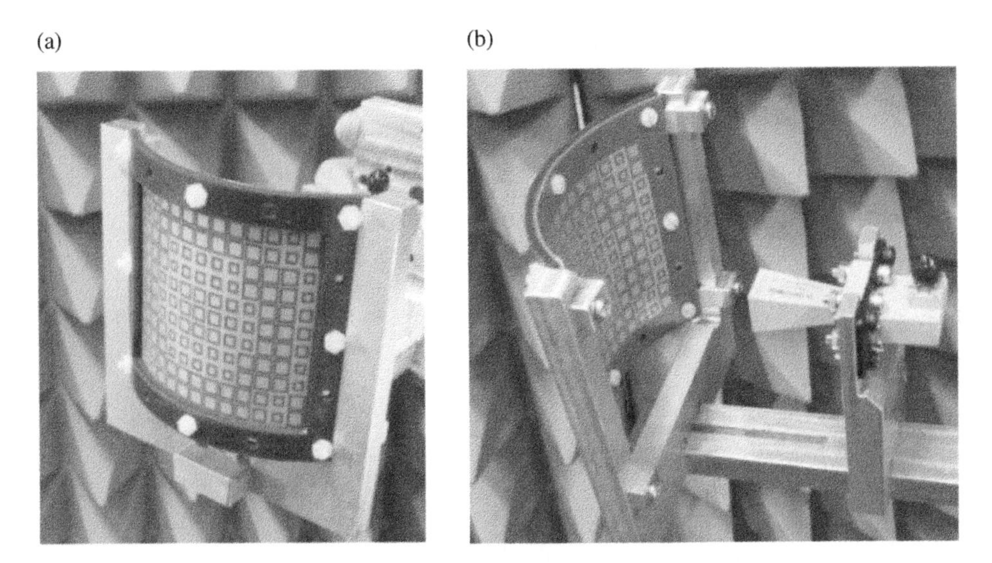

Figure 6.7 Photographs of the conformal transmitarray prototype. (a) Front view. (b) Back view.

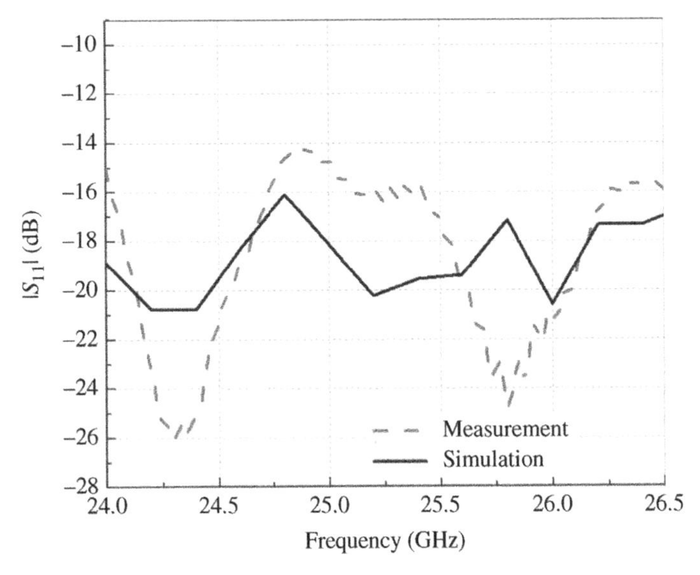

Figure 6.8 Simulated and measured input reflection coefficients of the conformal transmitarray as functions of the source frequency. *Source*: From [30] / with permission of IEEE.

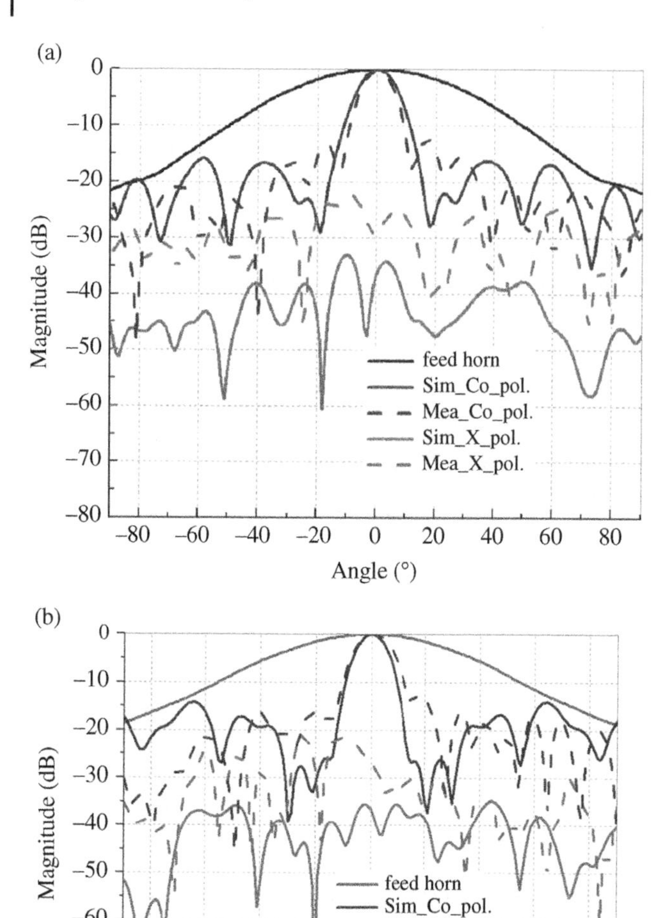

Figure 6.9 Simulated and measured radiation patterns at 25.5 GHz. (a) E-plane. (b) H-plane. *Source:* From [30] / with permission of IEEE.

band from 24 to 26.5 GHz. A reasonably good agreement between the simulated and measured results was achieved. The far-field radiation patterns were measured using a Microwave Vision Group (MVG) compact range antenna measurement system.

Figures 6.9a and 6.9b show, respectively, the simulated and measured E- and H-plane radiation patterns at 25.5 GHz. Compared to the feed horn antenna that has a gain of 10.6 dBi, the simulated peak gain of the conformal transmitarray is 20.5 dBi at 25.0 GHz, while the measured one is 19.6 dBi at 25.5 GHz. The measured 3-dB gain bandwidth is 6.7%. In contrast to the horn pattern, the 3-dB beamwidth of the prototype is reduced from 54° to 12° in both planes. The measured cross-polarization level is less than −20 dB.

The simulated and measured peak realized gain values versus the operating frequency are shown in Figure 6.10a. Taking into account that the gain was calculated with the size of the aperture cross

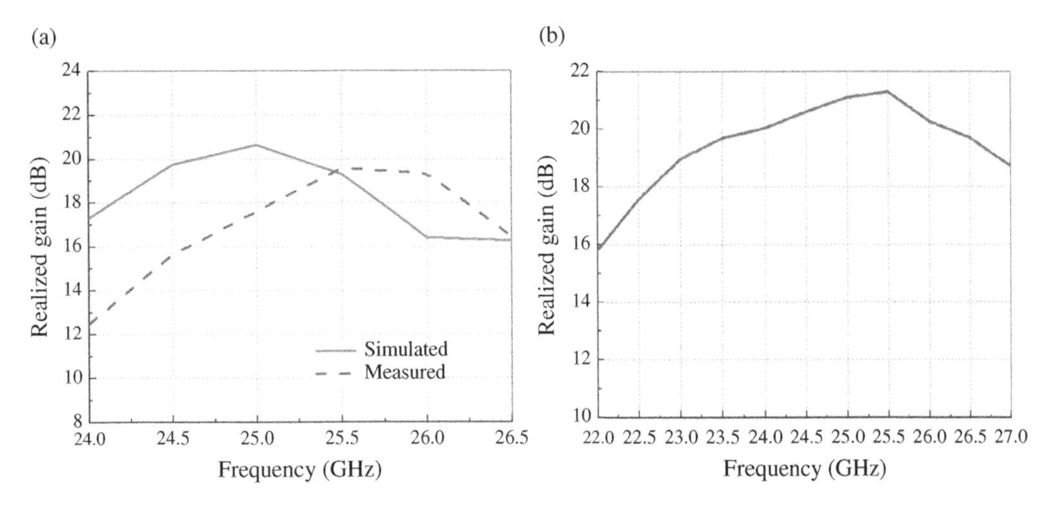

Figure 6.10 Simulated and measured realized gain versus frequency: (a) Conformal transmitarray with total thickness h = 0.508 mm. *Source*: From [30] / with permission of IEEE. (b) Conformal transmitarray with total thickness h = 2.5 mm.

section, the measured aperture efficiency was 25.1%. This low efficiency and the small gain bandwidth are a result of using the very thin unit elements in the conformal transmitarray. As can be seen from Figure 6.2, the phase curve of the array element is sharper when the thickness is smaller. This leads to large phase-correcting errors which reduce the realized gain of the transmitarray and limit its gain bandwidth. It was found that the efficiency and the gain bandwidth of the transmitarray can be increased to around 43 and 17%, respectively when the total thickness of the element is increased to 2.5 mm. The simulated realized gain versus frequency is shown in Figure 6.10b with the realized gain being 21.1 dB at 25 GHz. Recall that the choice of using the 0.508-mm-thick element for the prototype was purposely made for the ease of bending the transmitarray to conform to a cylinder. It was made simply to verify the design concept. In principle, a conformal transmitarray using 2.5-mm-thick elements could also be fabricated using 3D-printing technology which can accommodate more complicated platform shapes.

It is noticed that there were some discrepancies between the simulated and measured input reflection coefficients. They are mostly attributed to the following factors. First, there were some inaccuracies in the fabrication process of the elements and from alignment errors in the assembly of the layers. Second, the PCB board used to fabricate the antenna was a low-cost one. Consequently, there were expected variations from its data sheet in the actual dielectric constant and loss tangent values.

6.3 Beam Scanning Conformal Transmitarrays

Another important requirement for transmitarray antennas is the ability to scan their output beams. It is highly desired that the beam scanning be realized using simple feed networks, thereby reducing their overall costs and losses, especially at millimeter-wave or higher frequencies [32]. One typical approach to achieve electronic beam scanning with a flat transmitarray antenna is to change the transmission phase of its elements by using PIN didoes [17–19] or varactor didoes [20, 21]. The former (latter) achieves discrete (continuous) beam steering.

For conformal transmitarray antennas, however, it is very challenging to integrate active elements on the surface of a platform if its curvature is very large. One potential solution is to steer the beam by mechanically rotating the feed horn. Compared to electronic beam scanning, this solution avoids the extra losses and costs associated with the use of a very large number of active elements and their requisite, complicated control circuits. A method to achieve mechanical beam scanning of the conformal transmitarray prototype design is described in this section. It involves simply rotating the feed. This is a practical solution for most conformal applications because the transmission surface is the part of the communications platform and, hence, should not be moved.

6.3.1 Scanning Mechanism

Each array element on the transmission surface of a non-reconfigurable transmitarray is designed to radiate its output beam toward a specific direction, e.g., at the broadside angle 0°. In order to achieve beam scanning by rotating the feed, the transmitting surface reported in Section 6.2 is divided into two parts, one on either side of its centerline as illustrated in Figure 6.11a. If one part of the transmitting surface is designed to direct the beam into the angle φ_1 and the other part into φ_2, the combined beam is directed toward the angle $(\varphi_1 + \varphi_2)/2$. Consequently, in order not to produce a split beam, the two output beams cannot be separated by too much of an angle nor be too narrow themselves. It has been found that a step of about 10° serves as an optimum value based on a variety of simulated results.

The transmitting surface that was designed to enable the beam scanning feature is shown in Figure 6.11b and compared to the fixed one in Figure 6.11a. It consists of six sections labeled as $a1$, $a2$, $a3$, $b1$, $b2$, and $b3$. They are each designed to radiate into different beam directions with respect to the z-axis, as depicted in Figure 6.12.

The subtended angles of the parts $a1$, $a2$, $b1$, and $b2$ are 36°; and they are 18° for $a3$ and $b3$. Therefore, the total angle subtended by the transmitting surface is 180°, making it a half-circle. The feed, again the LB-28-15-C-KF from A-INFO horn, is placed at the focus and center of the assumed cylindrical transmission surface. It has a gain of 12.2 dBi at 25.0 GHz. For each operating state of the

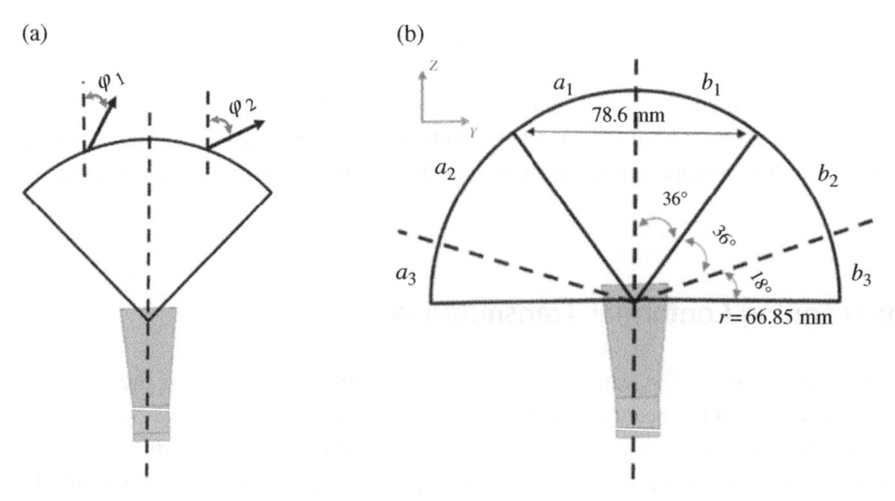

Figure 6.11 Side views of two different conformal transmitarrays. (a) Passive version from Section 6.2.2. (b) Reconfigurable version considered in this section. *Source*: From [30] / with permission of IEEE.

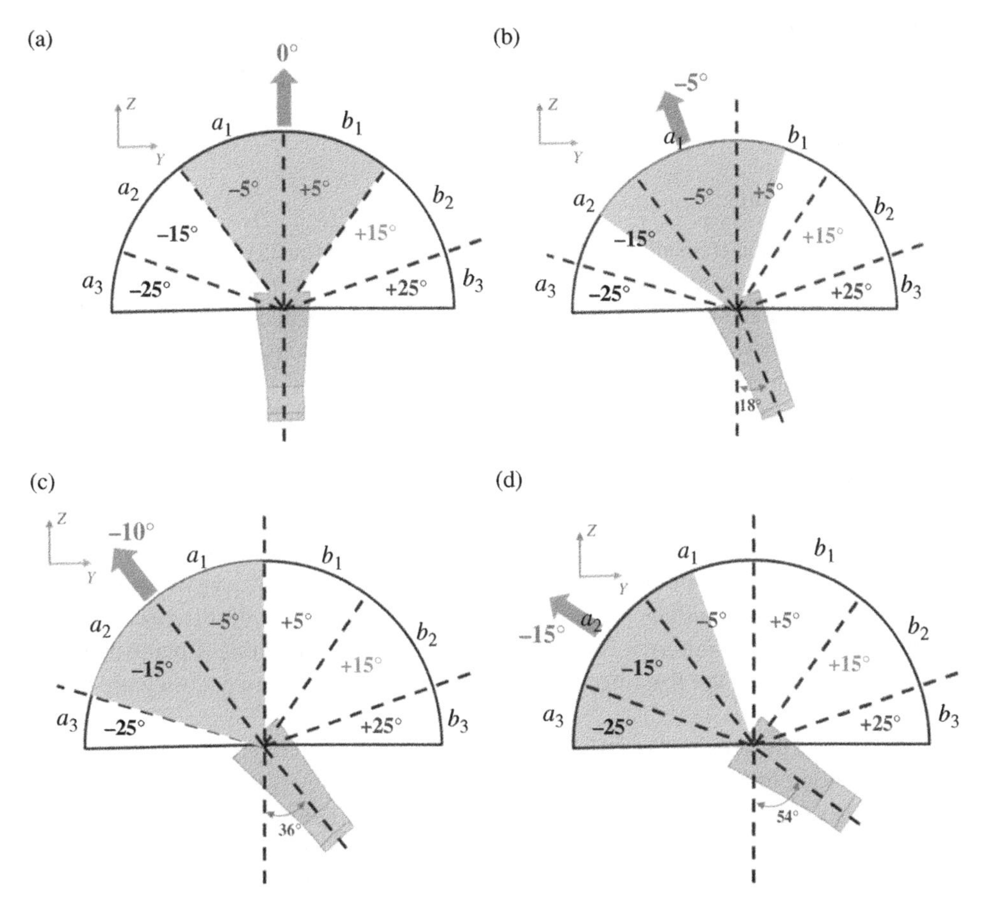

Figure 6.12 Different operating states of the beam scanning conformal transmitarray. Each feed horn points into different directions referred to the z-axis: (a) 0°, (b) −18°, (c) −36°, and (d) −54°. The resulting outgoing beam directions are (a) 0°, (b) −5°, (c) −10°, and (d) −15°. *Source*: From [30] / with permission of IEEE.

transmitarray, the horn is rotated to point at the center of an arc of its surface that has a 72° subtended angle. The level of the illumination at the edge of this arc is again −10 dB less than its center value. The adjacent, complementary parts of the transmitarray surface have only minimal effects on the output beams. The transmitarray element size is the same as that of the fixed case reported in Section 6.2. Each 72° transmitting segment of the surface consists of 15 × 13 elements. The cross section size is 78.6 mm × 72.8 mm.

When the feed horn is pointed at the center of the transmission surface, its two parts about this direction are illuminated as indicated by the green area shown in Figure 6.12a. These two parts direct their beams toward the angles −5° and +5°, respectively. The combined beam thus points toward 0°. If the horn is rotated anti-clockwise by 18°, the active surface changes as shown in Figure 6.12b. Three parts of the transmission surface are now illuminated, namely $a1$, half of $a2$, and half of $b1$. The resulting combined beams are then directed into the angle $(\varphi_{a1} + [\varphi_{a2}/2 + \varphi_{b1}/2])/2 = (-5° + [-15°/2 + 5°/2])/2 = -5°$. In the same manner, when the feed horn is rotated by another 18°, as shown in Figure 6.12c, the segments of the transmission surface $a1$ and $a2$ become active and the output beam is directed into the angle −10°. Finally, the output beam is directed into the direction −15° when the feed is rotated by yet another 18° as shown in

Figure 6.12d. When the feed horn is rotated clockwise, the output beams are then directed into the angles 5°, 10°, and 15°, respectively, since the transmission surface is symmetrical with respect to the z-axis.

6.3.2 Experimental Results

The reconfigurable conformal transmitarray design was verified with the 25 GHz prototype transmission surface described in Section 6.2.2. In this case, the two-layer surface was mounted on a platform that enabled a rotatable feed horn. An angle protractor was printed on the back surface of the platform so that the horn could be rotated to any required angle. Photographs of the prototype in the measurement setup are shown in Figure 6.13.

The input reflection coefficients and the radiation patterns of the reconfigurable transmitarray prototype were measured for seven different positions of the feed horn. It is seen from Figure 6.14 that the measured input reflection coefficients are below −10 dB for all these seven states from 24 to 26 GHz. The measured and simulated realized gain patterns in the H-plane of the prototype are compared at 25 GHz in Figure 6.15. The simulated realized gain value at boresight is 20.3 dBi and around 19.5 dBi for other scanning angles. Stable realized gain values of at least 18.7 dBi were achieved in all seven states. It shows that the scan loss of this mechanically reconfigurable conformal transmitarray is very small. Since the corresponding E-plane patterns did not change from those of the fixed-beam transmitarray given in Figure 6.8, they are omitted here.

The absolute values of the cross-polarization levels at 25 GHz are given in Figure 6.16. The maximum cross-polarization level is 5.0 dB for the 15° output beam state. Thus, the relative cross-polarization level is lower than −14 dB for all of the seven states. For each working state with its 72° transmission surface, the simulated aperture efficiency is 17.8%. The measured value is 14.8%. The efficiency of the scanned transmitarray is lower than that of the fixed one. This decrease is mainly due to the surface and consequent beam combinations used to generate the desired output beam angle. As a result, the realized gain of the scanned transmitarray is lower than that of the fixed one.

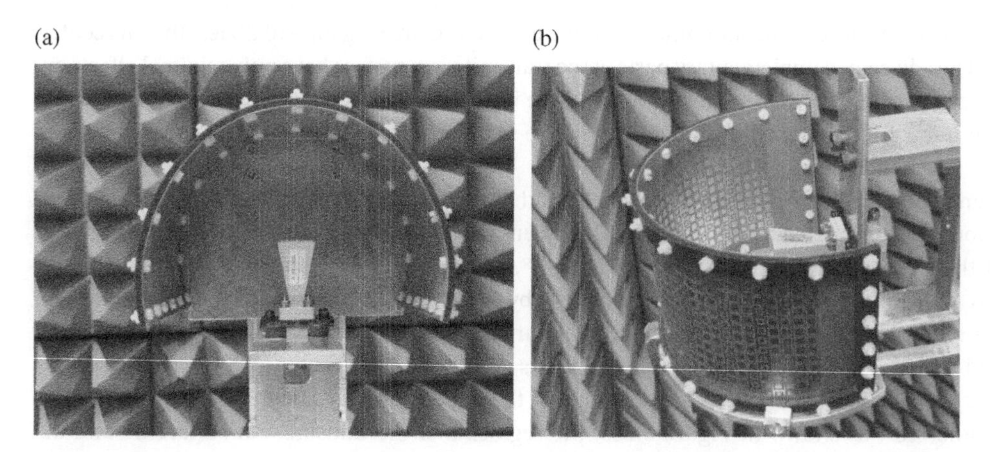

Figure 6.13 Photographs of the reconfigurable transmitarray prototype in the measurement setup. (a) Front view. (b) Back view.

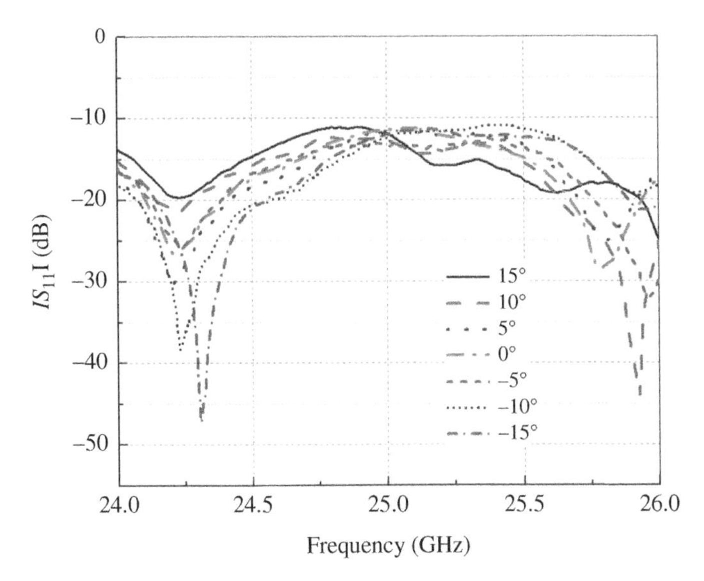

Figure 6.14 Measured input reflection coefficients of the reconfigurable conformal transmitarray for the different output beam states. *Source*: From [30] / with permission of IEEE.

6.3.3 Limits of the Beam Scanning Range

Each main beam of the prototype reconfigurable conformal transmitarray is a combination of the output beams of two or three parts of the transmission surface. Therefore, the beam scanning range is related to the output beam direction of each part θ_{rn} (with respect to the z-axis), the subtended angle of each part θ_h, and the number of parts, $2n$, where n covers each half of the transmission surface. The angle θ_h was chosen as 36° to deal with the −10 dB edge illumination associated with the feed horn pattern. As a result, the maximum value of n for the 180° subtended surface is 180°/36° = 5. The beam directions for the parts $a1$ and $b1$ are chosen to be −5° and +5°, respectively, resulting in a 10° step. If this step is increased, the combined output beam will split and its maximum gain will drop. If the step is decreased, the total scanning range will be reduced. Therefore, the 10° step serves as an optimum tradeoff value based on the simulation studies using the low-profile unit elements.

According to this angular step, the parts $a2$ and $a3$ radiate, respectively, in the directions of −15° and −25° with respect to the z-axis. As is known [33], the directivity drops significantly when a beam is scanned far away from the direction normal to the transmitting surface. The normal direction of part $a3$ is the $-y$-axis. If the whole segment $a3$ was employed, it would need to radiate in the direction 65° from its normal direction as shown in Figure 6.17. This choice is not effective. For this reason, only half of $a3$ is used in the array to achieve the −15° output beam direction shown in Figure 6.12d. Consequently, the subtended angles of $a3$ and $b3$ are only half of the other parts. Moreover, there is no need to include more parts for larger beam angles since the realized gain will be reduced even further. One possible method to increase the beam scanning range would be to reduce the subtended angle of each part of the transmission surface by including more surface parts.

Compared to the two-part fixed-beam conformal transmitarray, e.g., one which would consist of only the two parts $a1$ and $b1$, the total size of the reconfigurable one is about 2.5 times larger. On the

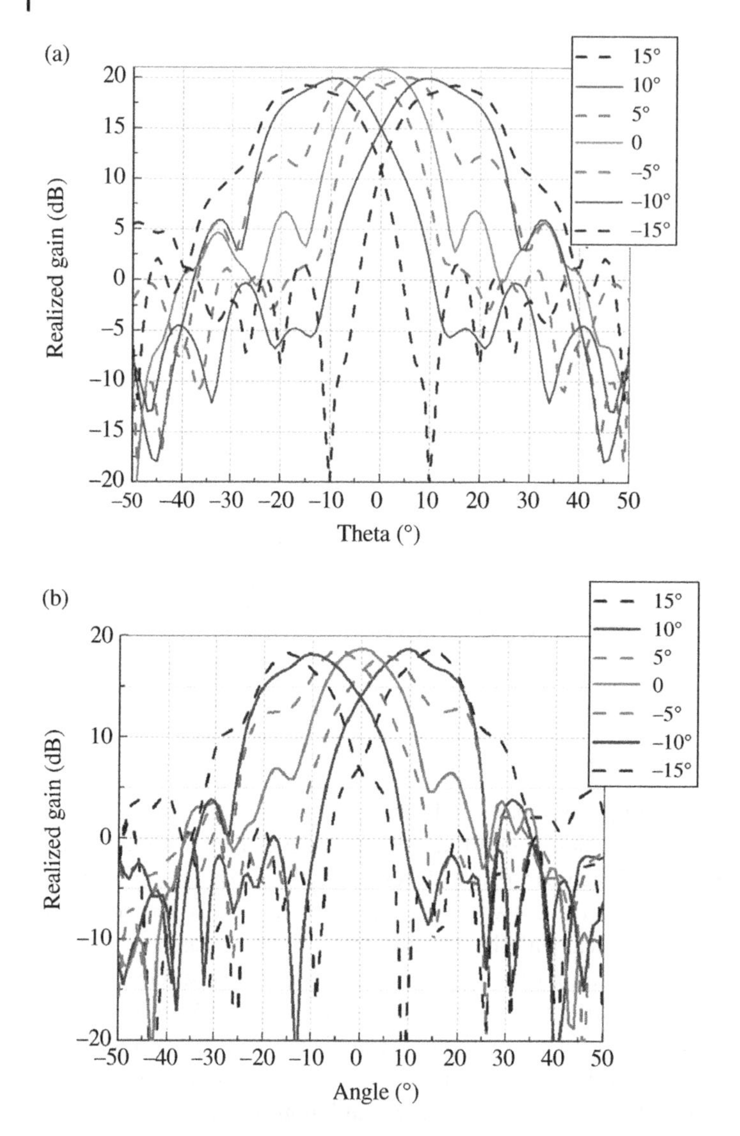

Figure 6.15 H-plane realized gain patterns at 25.0 GHz for the different output beam states of the reconfigurable conformal transmitarray. (a) Simulated results. (b) Measured results. *Source*: From [30] / with permission of IEEE.

other hand, seven beam directions were achieved. For some communication platforms, where the size of the surface is not a primary constraint, the reported beam scanning method has interesting advantages.

It should be noted that a straightforward way to achieve beam scanning over larger angles is to employ several identical transmitting surfaces and rotate the feed horn to the center of each surface. This concept is depicted in Figure 6.18. However, the number of beam directions is dependent on the number of transmitarray antenna sections employed. Moreover, the angular step is usually large. For example, if three identical transmitting surfaces were used, as shown in Figure 6.18, three output beams pointing into the angles 72°, 0°, and 72° would be achieved when the feed horn is rotated to point at the center of each section of the transmitarray.

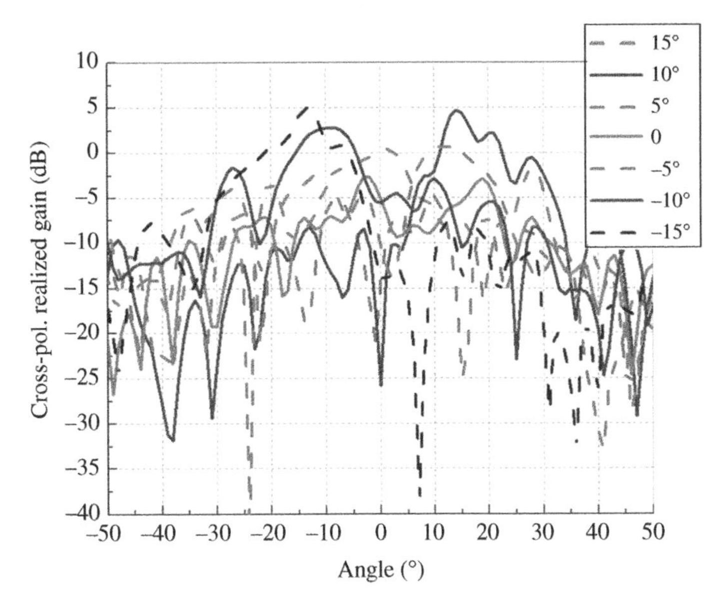

Figure 6.16 Measured cross-polarization realized gain levels in the H-plane of the reconfigurable conformal transmitarray at 25.0 GHz for its different output beam states. *Source*: From [30] / with permission of IEEE.

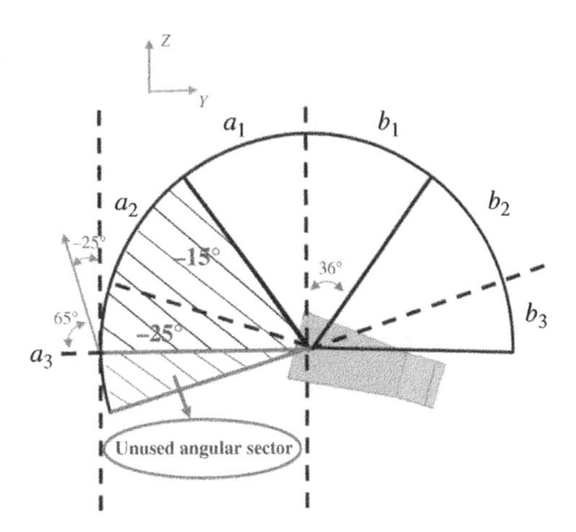

Figure 6.17 Beam steering range limit of the prototype conformal transmitarray antenna. *Source*: From [30] / with permission of IEEE.

6.4 Conformal Transmitarray Employing Ultrathin Dual-Layer Huygens Elements

For many three-layer Huygens elements, the metal traces on the first and third layers are used to generate a current loop that is equivalent to a magnetic dipole and the trace on the middle layer is designed to be equivalent to an electric dipole. To facilitate the conformal design, an ultrathin Huygens element with only two metal layers has been developed. It leads to a significantly thinner transmitarray surface. The element's thickness is 0.5 mm, $\lambda_0/60$ at 10 GHz. It consists of a pair of symmetrical "I"-shaped patches and adjacent capacitively loaded strips on the top and bottom layers [34]. These elements generate magnetic and electric dipoles, respectively. Eight elements

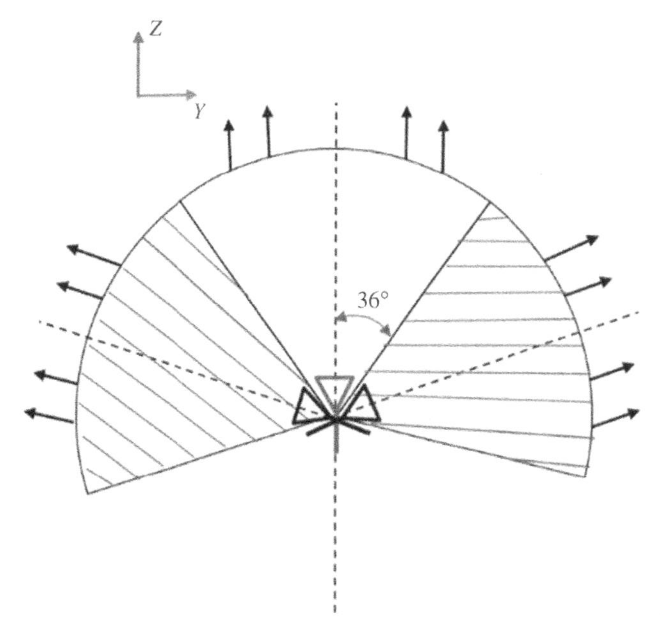

Figure 6.18 Three transmitarray antennas to achieve the three beam directions: −72°, 0°, and 72°. *Source*: From [30] / with permission of IEEE.

with different dimensions to cover a quantized 360° phase range have been developed. The highest element loss is 1.67 dB. A cylindrically conformal transmitarray is developed employing these Huygens elements. The measured aperture efficiency is found to be 47% [35], which is much higher than the conformal transmitarrays that use triple-layer frequency-selective surface (FSS) elements [30] and other transmitarrays whose thicknesses are about 0.1 λ_0.

6.4.1 Huygens Surface Theory

As shown in Figure 6.19a, both electric, J_s, and magnetic, K_s, currents are induced on a Huygens surface when an EM wave impinges on it. Once these steady-state E- and M-currents are excited on the surface, each can be treated as an independent source as illustrated in Figure 6.19b. Assume the surface locally resides on the zx-plane. The source is assumed to be in the region $z < 0$. The combinations of the fields generated by the E- and M-sources on it, in turn, represent the reflected and transmitted fields on opposite sides of the surface in the half-spaces $z < 0$ and $z > 0$, respectively [35–38].

To establish the fields in both half-spaces, consider the surface to be locally planar. For an E-current surface, the scattered fields generated by $\boldsymbol{J}_s$ on both sides of it must satisfy the EM boundary conditions:

$$\boldsymbol{z} \times [\boldsymbol{E}_{J2}(z = 0_+) - \boldsymbol{E}_{J1}(z = 0_-)] = 0 \tag{6.4}$$

$$\boldsymbol{z} \times [\boldsymbol{H}_{J2}(z = 0_+) - \boldsymbol{H}_{J1}(z = 0_-)] = \boldsymbol{J}_s \tag{6.5}$$

where $\boldsymbol{E}_{J1}$ and $\boldsymbol{H}_{J1}$ are the electric and magnetic fields generated by the E-source $\boldsymbol{J}_s$ in the region $z < 0$, and $\boldsymbol{E}_{J2}$ and $\boldsymbol{H}_{J2}$ are the corresponding ones in the region $z > 0$. The coordinates $z = 0_\pm$ represent the limit of z going to zero from the region $z > 0$ and $z < 0$, respectively.

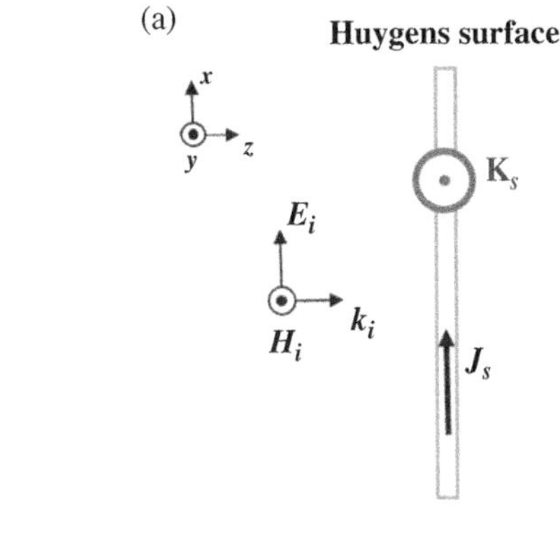

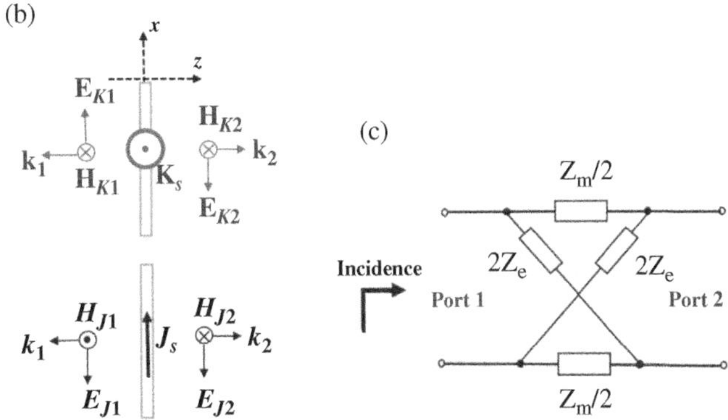

Figure 6.19 Huygens surface. (a) Sketch of the EM field and the induced currents associated with the excitation process. (b) Fields separately generated by the induced E- and M-currents. (c) Equivalent circuit model. *Source*: From [35] / with permission of IEEE.

Equation (6.4) means the tangential components of the electric fields in both regions are continuous across the surface:

$$(\boldsymbol{E}_{J1})_t(z = 0_+) = (\boldsymbol{E}_{J1})_t\,(z = 0_-) \tag{6.6}$$

Since it is an impedance surface, the electric current on it is induced by the total electric field driving it, i.e., taking into account Eq. (6.6),

$$Z_e \boldsymbol{J_s} = (\boldsymbol{E}_i + \boldsymbol{E}_{J1})_t(z = 0) = (\boldsymbol{E}_i + \boldsymbol{E}_{J2})_t(z = 0) \tag{6.7}$$

where $\boldsymbol{E}_i$, $\boldsymbol{H}_i$ are the source-generated EM fields incident on the surface. Assuming that the local fields are plane-wave-like and taking into account the vector orientation choices in Figure 6.19, the scattered electric and magnetic fields on both sides of the surface, which propagate in opposite directions, are related by

$$[\boldsymbol{H}_{J2}(z = 0_+)]_t = -\left[\frac{\boldsymbol{E}_{J2}(z = 0_+)}{\eta}\right]_t \tag{6.8}$$

$$[\boldsymbol{H}_{J1}(z = 0_-)]_t = +\left[\frac{\boldsymbol{E}_{J1}(z = 0_-)}{\eta}\right]_t \tag{6.9}$$

where η is the free space wave impedance. Consequently, Eqs. (6.5)–(6.7) yield

$$(\boldsymbol{E}_i + \boldsymbol{E}_{J1})_t(z=0) = Z_e \boldsymbol{J}_s = \frac{Z_e}{\eta}[-\boldsymbol{E}_{J2}(z=0_+) - \boldsymbol{E}_{J1}(z=0_-)]_t$$
$$= -2\frac{Z_e}{\eta}[\boldsymbol{E}_{J1}(z=0)]_t \tag{6.10}$$

Then, again taking into account the vector orientation choices in Figure 6.19, the ratio

$$\frac{\boldsymbol{E}_{J1}\cdot\hat{\boldsymbol{x}}}{\boldsymbol{E}_i\cdot\hat{\boldsymbol{x}}} = \frac{-1}{1+2\frac{Z_e}{\eta}} = \frac{-\eta}{\eta+2Z_e} \tag{6.11}$$

Similarly, if the surface has an M-current induced on it,

$$-\boldsymbol{z}\times[\boldsymbol{E}_{K2}(z=0_+) - \boldsymbol{E}_{K1}(z=0_-)] = \boldsymbol{K}_s \tag{6.12}$$
$$\boldsymbol{z}\times[\boldsymbol{H}_{K2}(z=0_+) - \boldsymbol{H}_{K1}(z=0_-)] = \boldsymbol{0} \tag{6.13}$$

where $\boldsymbol{E}_{K1}$and $\boldsymbol{H}_{K1}$ are the electric and magnetic fields generated by the M-source $\boldsymbol{K}_s$ in the region $z < 0$, and $\boldsymbol{E}_{K2}$ and $\boldsymbol{H}_{K2}$ are the corresponding ones in the region $z > 0$. The tangential components of these magnetic fields are now continuous across the surface:

$$(\boldsymbol{H}_{K1})_t(z=0_+) = (\boldsymbol{H}_{K1})_t(z=0_-) \tag{6.14}$$

The surface now is considered to have an impedance Z_m, and the magnetic current on it is induced by the total magnetic field driving it. This means

$$\boldsymbol{K}_s = Z_m(\boldsymbol{H}_i + \boldsymbol{H}_{K1})_t(z=0) = Z_m(\boldsymbol{H}_i + \boldsymbol{H}_{K2})_t(z=0) \tag{6.15}$$

Again, assuming that the local fields are plane-wave-like, the scattered electric and magnetic fields on both sides of the surface, which are propagating in opposite directions, are related as

$$[\boldsymbol{E}_{K2}(z=0_+)]_t = +\eta[\boldsymbol{H}_{K2}(z=0_+)]_t \tag{6.16}$$
$$[\boldsymbol{E}_{K1}(z=0_+)]_t = -\eta[\boldsymbol{H}_{K1}(z=0_+)]_t \tag{6.17}$$

Therefore, taking into account the vector orientation choices in Figure 6.19, Eqs. (6.12), (6.15)–(6.17) yield

$$Z_m\left(\frac{\boldsymbol{E}_i}{\eta} - \frac{\boldsymbol{E}_{K1}}{\eta}\right)_t(z=0) = \boldsymbol{K}_s = [\boldsymbol{E}_{K2}(z=0_+) + \boldsymbol{E}_{K1}(z=0_-)]_t$$
$$= +2[\boldsymbol{E}_{K1}(z=0)]_t \tag{6.18}$$

Consequently, the corresponding ratio for the magnetic current generated fields is given by

$$\frac{\boldsymbol{E}_{K1}\cdot\hat{\boldsymbol{x}}}{\boldsymbol{E}_i\cdot\hat{\boldsymbol{x}}} = \frac{1}{1+\frac{2\eta}{Z_m}} = \frac{Z_m}{Z_m+2\eta} \tag{6.19}$$

Finally, these results are combined to obtain the amplitude reflection and transmission coefficients, R and T, for the complete Huygens surface with both E and M currents and the field orientations shown in Figure 6.19a. They are the sum of those associated with each of them:

$$R = R_J + R_M = \frac{\boldsymbol{E}_{J1}\bullet\hat{\boldsymbol{x}}}{\boldsymbol{E}_i\bullet\hat{\boldsymbol{x}}} + \frac{\boldsymbol{E}_{K1}\bullet\hat{\boldsymbol{x}}}{\boldsymbol{E}_i\bullet\hat{\boldsymbol{x}}} = \frac{-\eta}{2Z_e+\eta} + \frac{Z_m}{Z_m+2\eta} \tag{6.20}$$

$$T = T_J + T_M = \frac{\boldsymbol{E}_{J2}\bullet\hat{\boldsymbol{x}}}{\boldsymbol{E}_i\bullet\hat{\boldsymbol{x}}} + \frac{\boldsymbol{E}_{K2}\bullet\hat{\boldsymbol{x}}}{\boldsymbol{E}_i\bullet\hat{\boldsymbol{x}}} = \frac{\boldsymbol{E}_{J1}\bullet\hat{\boldsymbol{x}}}{\boldsymbol{E}_i\bullet\hat{\boldsymbol{x}}} + \left(1 - \frac{\boldsymbol{E}_{K1}\bullet\hat{\boldsymbol{x}}}{\boldsymbol{E}_i\bullet\hat{\boldsymbol{x}}}\right)$$

$$= \frac{-\eta}{2Z_e + \eta} + \left[1 - \left(\frac{Z_m}{Z_m + 2\eta}\right)\right] = \frac{2Z_e}{2Z_e + \eta} - \frac{Z_m}{Z_m + 2\eta} \tag{6.21}$$

Manipulating Eqs. (6.20) and (6.21) to obtain the surface impedances in terms of R and T, one has

$$Z_e = \frac{\eta}{2} \times \left[\frac{1 + R + T}{1 - R - T}\right] \tag{6.22}$$

$$Z_m = 2\eta \times \left[\frac{1 + R - T}{1 - R + T}\right] \tag{6.23}$$

The Huygens surface is capable of realizing zero reflection and unity (full) transmission, i.e., it can realize $R = 0$ and $T = e^{j\varphi_t}$, where φ_t is the transmission phase φ_t whose value can be varied by the specific design of the surface elements. Consequently, the corresponding electric and magnetic surface impedances can be rewritten as

$$Z_e = \frac{j\eta}{2\tan(\varphi_t/2)} \tag{6.24}$$

$$Z_m = -j2\eta \tan\left(\frac{\varphi_t}{2}\right) \tag{6.25}$$

It is then understood that the transmission phase strongly depends on the realization of particular surface impedances.

As developed in [35, 36], a Huygens surface is locally equivalent to the circuit model shown in Figure 6.19c. Thus, the surface impedances can be defined by the Z matrix from microwave network theory as:

$$Z_e = \frac{Z_{11} + Z_{21}}{2} \tag{6.26}$$

$$Z_m = 2 * (Z_{11} - Z_{21}) \tag{6.27}$$

6.4.2 Ultrathin Dual-Layer Huygens Elements

A two-layer Huygens transmission surface was developed. Its ultrathin square unit cell is shown in Figure 6.20. It consists of two metallization layers. A pair of elements is printed on the top and bottom surfaces of a 0.5 mm-thick substrate whose relative dielectric constant is 3.55 and loss tangent $\tan\delta = 0.0027$. The "I"-shaped patches on the top and bottom layers have exactly the same dimensions and are centered in the unit cell. A capacitively loaded strip (CLS) [35] is also printed on each surface. Each is centered between the I-patch and the edge of the unit cell. However, they lie on opposite sides of the I-patch on the two surfaces. The period of the lattice is the same as the length of the unit cell, $P = 8.5$ mm.

Assume an x-polarized wave impinges on the element, i.e., along the length of both the I-shaped patch and the CLS. Because the I-shaped patches on the top and bottom surfaces lie directly over each other, they produce currents along the x-axis but with opposite orientation. Consequently, the pair mimics a current loop, which is equivalent to a magnetic dipole. On the other hand, currents are induced on the separated CLSs along the x-axis with the same direction. They are equivalent to electric dipoles. These CLS elements have no effect on the magnetic response of the unit cell, while the I-shaped patches do contribute to its electric response. Therefore, the I-shaped patches were designed first. The CLSs were then designed to achieve the electric response that yields the desired Huygens behavior.

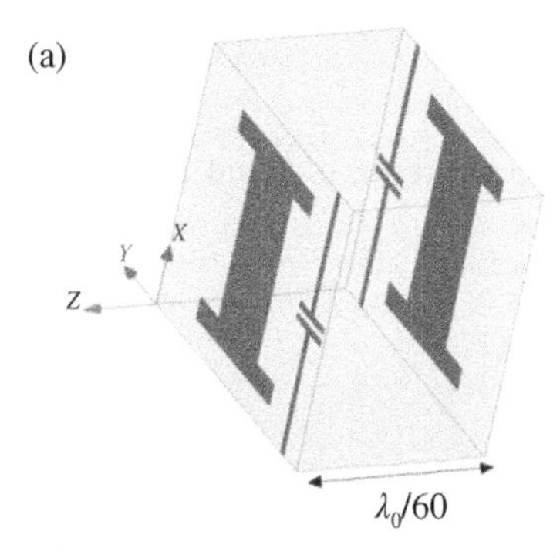

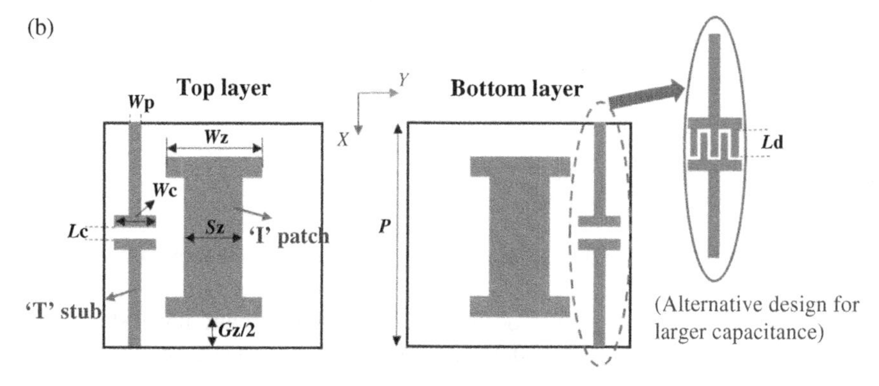

Figure 6.20 Developed Huygens element unit cell. (a) 3D view. (b) Top and bottom layers as viewed from the $-z$-axis. *Source*: From [35] / with permission of IEEE.

The magnetic response of the unit cell and, hence, the magnetic surface impedance, Z_m, can be adjusted by changing the capacitive and inductive properties of the I-shaped patches. They are related to the patch dimensions: Wz, Gz, and Sz, shown in Figure 6.20b. Similarly, the major electric response of the unit cell, and, hence, the electric surface impedance Z_e, is manipulated by varying the dimensions of the CLSs: Lc, Wc, and Wp, also shown in Figure 6.20b. For some designs, a large capacitance must be attained from the CLSs. This is achieved by introducing interdigitated strips of length Ld into their gaps, as shown in the inset of Figure 6.20b.

Since the dimensions of the I-shaped patches and the CLSs determine the surface impedances, they also control the transmission phase of the Huygens element. However, φ_t cannot be continuously changed when only one or two dimensions of the elements are varied. An optimization of every dimension of the unit cell and its elements is necessary and this generally makes the transmission surface design very complicated. Consequently, it is more practical to employ a quantized phase distribution instead. Eight elements were designed to achieve a three-bit quantized phase distribution. The detailed phase values and the corresponding Z_m and Z_e values calculated from Eqs. (6.24) and (6.25) are listed in Table 6.1.

The dimensions of the unit cell and its components were optimized with the 3D EM simulation software HFSS. The simulations employed Floquet ports at the input and output faces of the simulation model. Master–slave boundaries were specified on its remaining four sides.

The magnetic response of the unit cell was simulated first with only the I-shaped patches being present in it. The two CLSs were then added to the unit cell model to determine its electric response. After many parameter studies, it was decided for simplicity to vary the Wz values to tune the

Table 6.1 Theoretical specifications of the Huygens surface's impedance elements to achieve a quantized distribution of the transmission phase covering 360°.

Element no.	Phase (°)	$\text{Im}(Z_e)/\text{k}\Omega$	$\text{Im}(Z_m)/\text{k}\Omega$
1	−15	−1.43	0.1
2	−60	−0.33	0.44
3	−105	−0.14	0.98
4	−150	−0.05	2.81
5	−195	0.02	−5.72
6	−240	0.11	−1.31
7	−285	0.25	−0.58
8	−330	0.7	−0.2

Source: From [35] / with permission of IEEE.

magnetic response while fixing $Sz = 2.3$ mm and $Gz = 1.7$ mm. Similarly, Lc and Wc were varied to tune the electric response with $Wp = 0.2$ mm. Figure 6.21 shows the results of these parameter studies. All other dimensions were left unchanged as one parameter sweep was performed. As shown in Figure 6.21a, the Z_m curve moves to a lower band as Wz increases. This behavior occurs because it causes a higher capacitance between the I-shaped patches in the unit cell. As Figures 6.21b and 6.21c indicate, the Z_e curve also shifts to lower frequencies as Wc increases. This trend is the same as when Lc is decreased. Thus, Z_e moves to a lower frequency band when the capacitance of the CLSs increases. Furthermore, Figures 6.22a and 6.22b indicate that Wc and Lc have almost no effect on the magnetic response of the unit cell. This feature of the unit cell is the reason why its magnetic and electric responses can be designed independently.

An iterative method based on these parametric studies was employed to obtain the dimensions of the eight Huygen elements to achieve the desired eight transmission phase states in Table 6.1 at 10 GHz. First, the Z parameters of the initial element were simulated with periodic boundary conditions. The values of Z_m and Z_e were obtained based on Eqs. (6.26) and (6.27). The dimensions of the unit cell components were further adjusted to make Z_m and Z_e be close to their values listed in Table 6.1. The surface impedances, the design parameter values, and the phases and magnitudes of S_{21} of the eight optimized elements are given in Table 6.2. Note that the two CLS elements are not needed in Element 2 since the I-shaped patch provides a sufficient electric response. While an ideal Huygens element realizes unity transmission without any loss, the developed two-metal layer unit cells are lossy. The losses arise from the copper traces and the lossy substrate. In fact, the losses were found to be comparable to those of the three-layer Huygens elements [11]. Moreover, once some of the specific phase values were attained, it was found that the transmission loss of those elements was greater than 2 dB. Small variations of the phase values were then made to lower the losses. For example, for Element 5, the required ideal phase is −195°. On the other hand, the synthesized phase is −187°. Consequently, the unit cells were finalized with a compromise between the phase error and the amplitude loss.

After obtaining each of the optimized values of the element's design parameters for each of the eight phase values, the transmission coefficients for these elements were determined. For example, the $|S_{21}|$ results for Element 1 are given in Figure 6.23. They show that the element's loss is only 0.16 dB at 10 GHz and the corresponding phase value is −14°, which is close to the desired

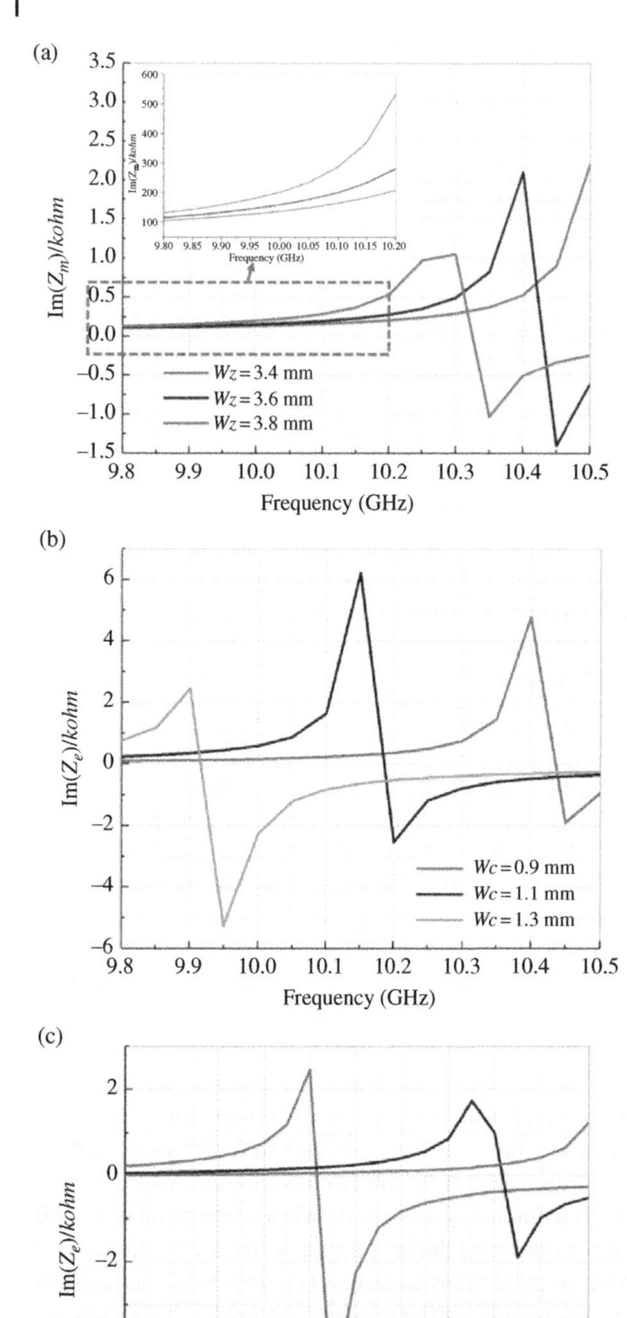

Figure 6.21 Simulated surface impedance values as functions of the source frequency for different design parameters. (a) Z_m for different *Wz*. (b) Z_e for different *Wc*. (c) Z_e for different *Lc*. *Source*: From [35] / with permission of IEEE.

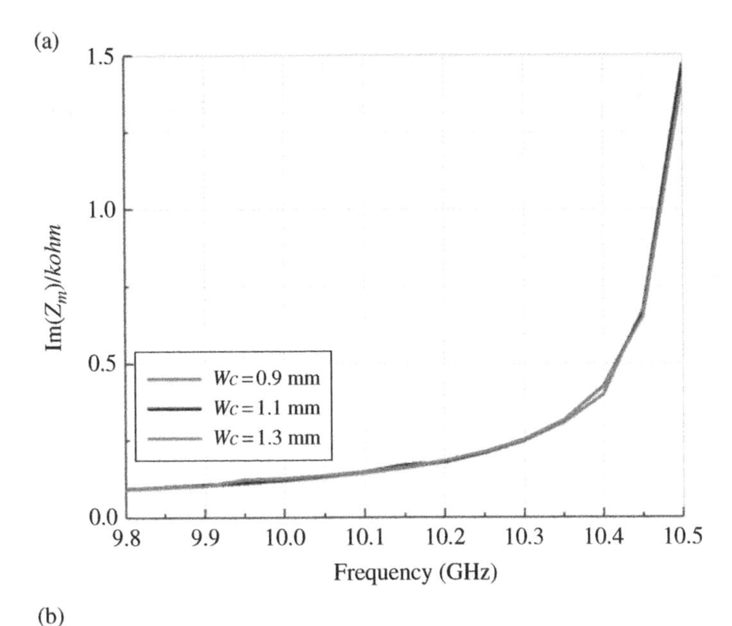

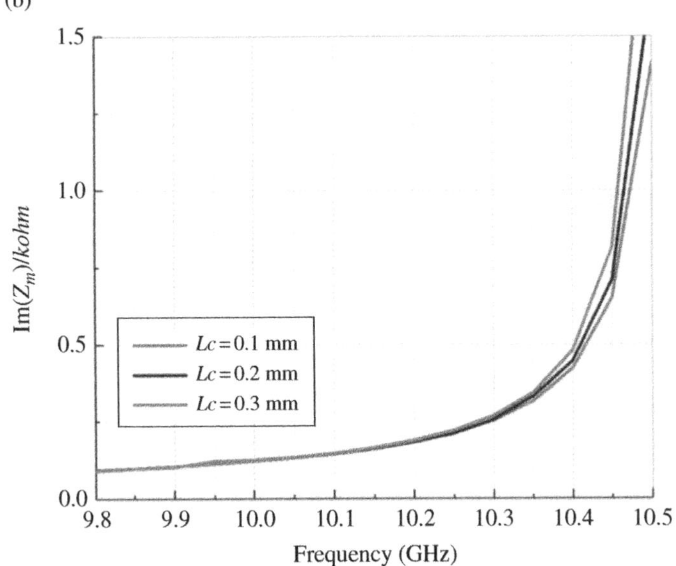

Figure 6.22 Simulated Z_m values as functions of the source frequency for different design parameter values. (a) Different *Wc*. (b) Different *Lc*.

value, −15°, listed in Table 6.1. Moreover, the current distributions on the traces in Element 1 at 10 GHz are shown in Figure 6.24 at different times in one source period T. It can be seen clearly at t = T/4 and 3 T/4 that the currents on the two I-shaped patches form a loop mode. Hence, they produce the expected magnetic dipole response. The phase interval between the maximum electric and magnetic magnitude current responses is 90°, which is a necessary condition to attain the balanced electric and magnetic dipole responses required to achieve a nonreflective Huygens element [12, 38].

Table 6.2 Properties of the optimized unit cells.

Element no.	Im(Z_e)/kΩ	Im(Z_m)/kΩ	$\|S_{21}\|$ (dB)	$\angle S_{21}$(°)	*Wz* (mm)	*Wc* (mm)	*Lc* (mm)
1	−2.3	0.13	−0.16	−14	3.6	1.3	0.1
2	−0.35	0.17	−0.42	−41	3.8	1.6	0.1
3	−0.12	0.64	−1.00	−100	4.1	/	/
4	−0.04	1.60	−1.67	−153	4.19	1.2	0.6
5	−0.02	−0.44	−1.66	−187	4.2	1.2	0.37
6	0.10	−1.10	−1.36	−241	4.25	1.5	0.4
7	0.22	−0.52	−0.86	−284	4.3	1.62	0.4
8	0.94	−0.25	−0.50	−330	4.4	1.62	0.35

Source: From [35] / with permission of IEEE.

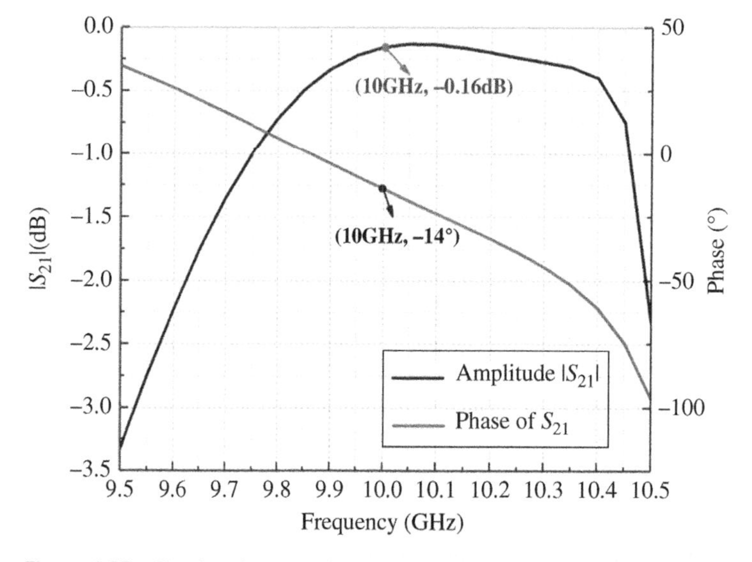

Figure 6.23 Simulated amplitude and phase of $|S_{21}|$ for Element 1 as functions of the source frequency.

6.4.3 Conformal Transmitarray Design

A cylindrically conformal transmitarray prototype of the optimized two-layer Huygens surface was constructed and tested to verify the simulation results. It is illustrated in Figure 6.25. It has $16 \times 17 = 272$ elements. The radius of the transmission surface and the subtended half-angle of its consequent aperture are illustrated in Figure 6.25b. The cross section size of the transmission surface is 121.3 mm × 144.5 mm. The source is the same standard horn, the LB-75-10-C-SF from A-INFO, used in the experiments described earlier. It was placed at the focal point of the surface with its main beam pointed along the normal to it at its middle. A -10 dB illumination is again attained at the edges of the surface. The equivalent planar structure used to design the source-distance compensated output response is shown in Figure 6.25c.

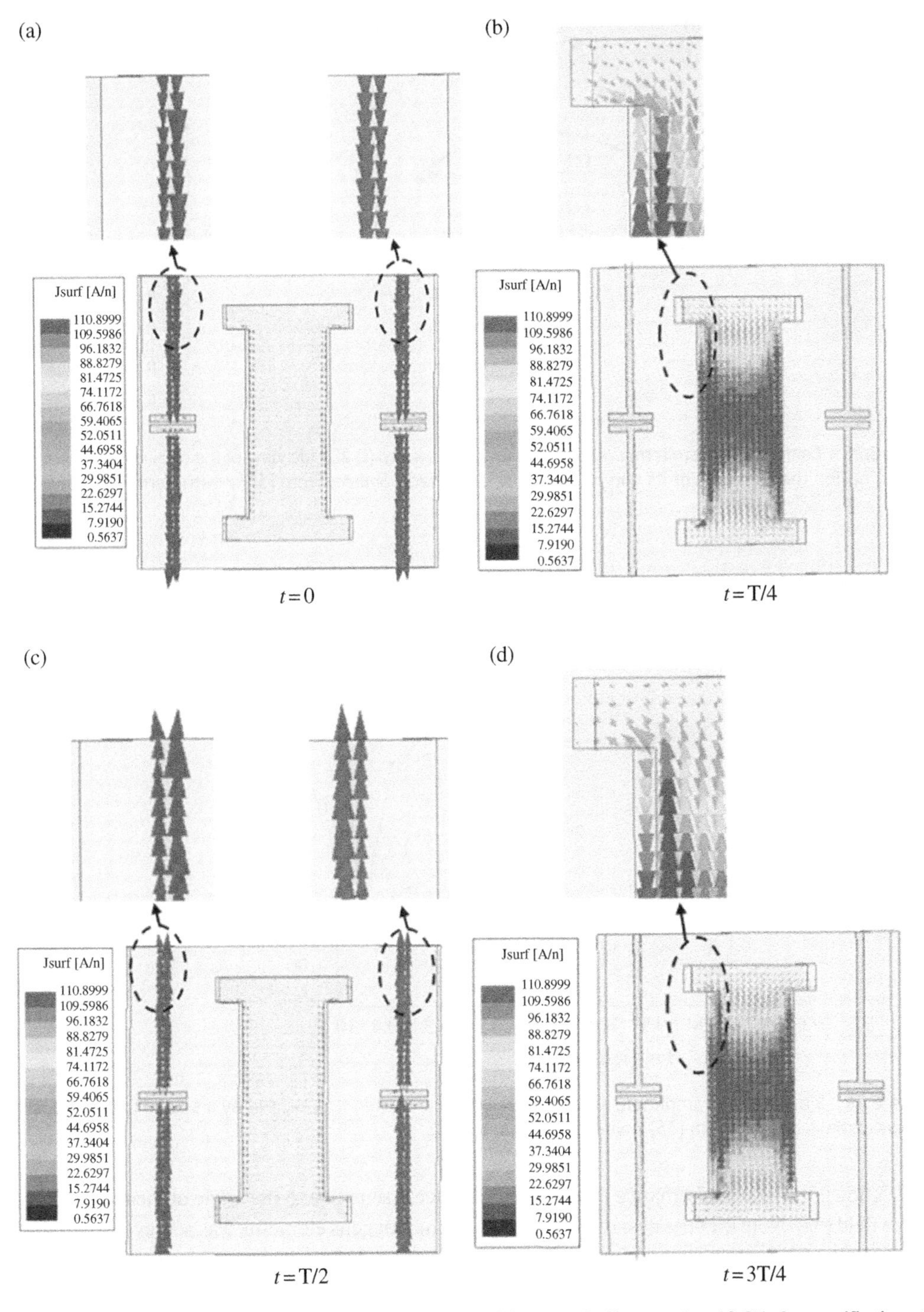

Figure 6.24 Simulated current distributions on the metallic traces in Element 1 at 10 GHz for specific times in one source period T. *Source*: From [35] / with permission of IEEE.

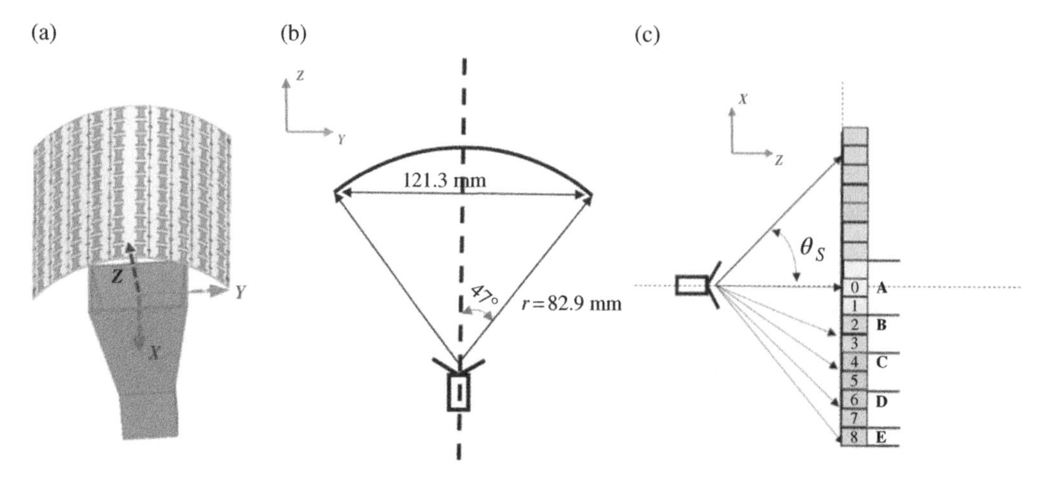

Figure 6.25 Conformal Transmitarray. (a) 3D view. (b) 2D side view. (c) 2D side view of the unbent planar version used to design the elements of its curved transmission surface. *Source*: From [35] / with permission of IEEE.

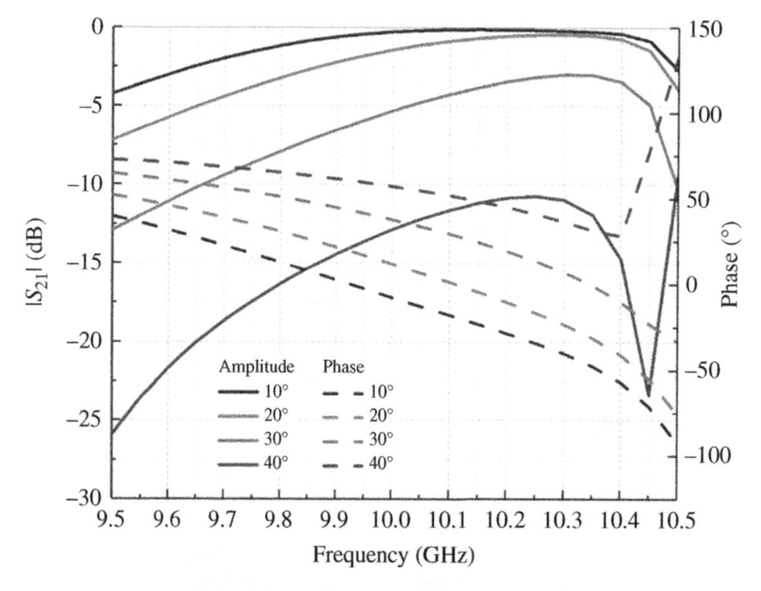

Figure 6.26 Simulated S_{21} amplitude and phase of Element 1 when it is excited by a source field obliquely incident upon it. *Source*: From [35] / with permission of IEEE.

It is known that the performance of a Huygens surface is sensitive to the angle of incidence of the source field [32]. Note that the design of each of the eight Huygens elements was achieved assuming the excitation field was normally incident upon it. However, before construction of the prototype, the response of each element was considered for different angles of incidence. As shown in Figure 6.26, the simulated amplitudes and phases of S_{21} when Element 1 is excited by a source field having different incidence angles change substantially.

As shown in Figure 6.25c, the equivalent planar design in the *zx*-plane can be broken into five straight segments, denoted as A, B, C, D, and E. Each element listed in Table 6.2 was then re-simulated under four different oblique angles of incidence, i.e., 14° (Zone B), 25° (Zone C), 33° (Zone D), and 39° (Zone E). The unbent transmission surface was fabricated using standard

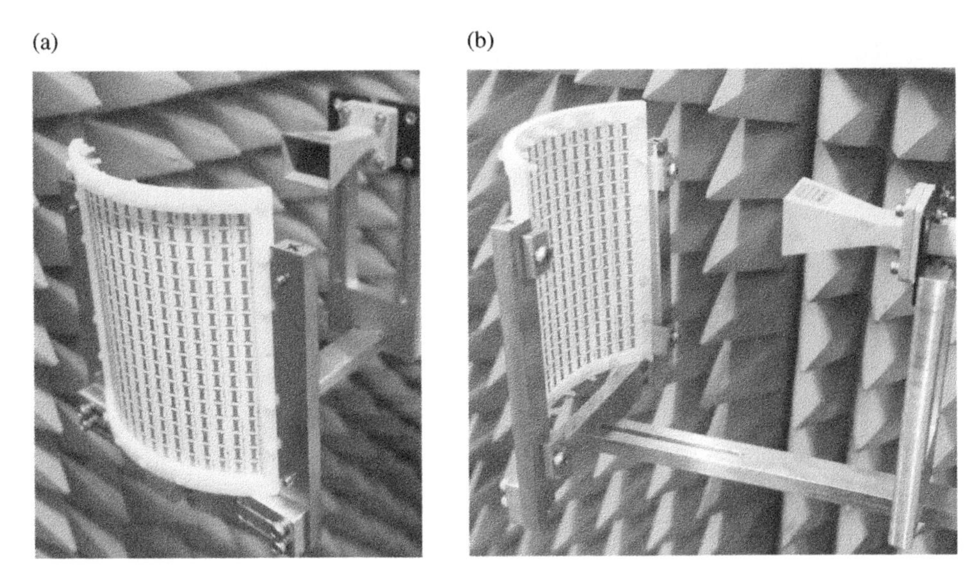

Figure 6.27 Photographs of the conformal transmitarray prototype in the measurement chamber: (a) Front view. (b) Back view.

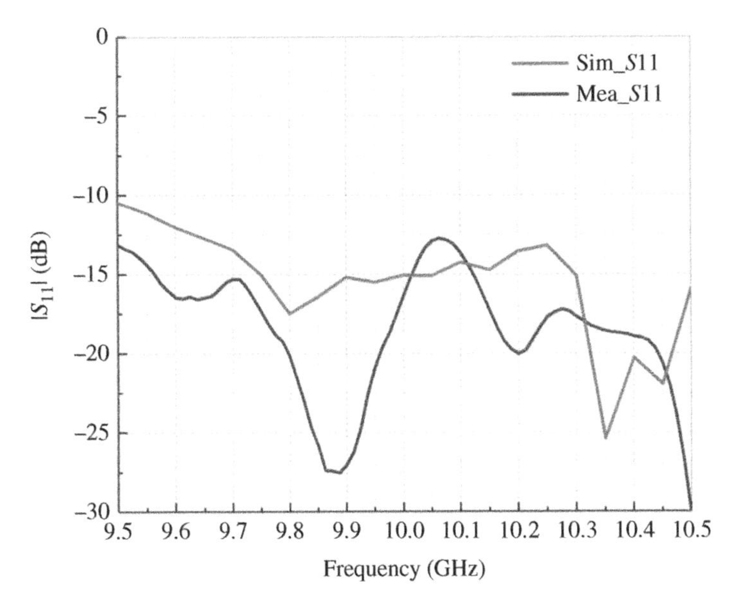

Figure 6.28 Simulated and measured values of the input reflection coefficients of the prototype Huygens transmitarray as functions of the source frequency.

PCB technology on low-cost Wangling F4B substrates whose relative dielectric constant is 3.55 and loss tangent tan $\delta = 0.0027$. Due to its ultrathin profile, 0.5 mm, which is $\lambda_0/60$ at 10 GHz, the fabricated prototype was easily bent to fit into a 3D-printed cylindrical frame. Photographs of the transmitarray prototype in the measurement chambers are shown in Figure 6.27.

The input reflection coefficients of the transmitarray were measured with a vector network analyzer. They are compared with their simulated values in Figure 6.28. They are below −10 dB from 9.5 to 10.5 GHz.

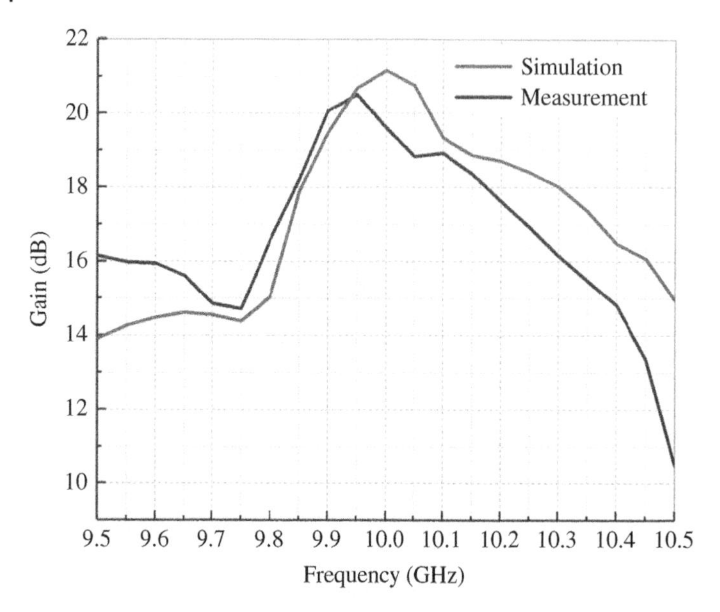

Figure 6.29 Simulated and measured values of the boresight-realized gain of the prototype Huygens transmitarray as functions of the source frequency.

The realized far-field gain patterns were measured using the same Global Big Data Technogies Centre's (GBDTC's) MVG compact range antenna measurement system located at the University of Technology Sydney, Ultimo, Australia. The simulated and measured realized gain values as functions of the source frequency are compared in Figure 6.29. The simulated peak realized gain, 21.2 dBi, appears at 10 GHz. Thus, the transmitarray has a 54% antenna efficiency. The measured results exhibit their maximum realized gain, 20.6 dBi, at 9.95 GHz, with the corresponding antenna efficiency being 47%. The simulated and measured E- and H-plane-realized gain patterns at 9.95 GHz are compared in Figures 6.30a and 6.30b, respectively. Good agreement was achieved except for a slight beam tilt of about 1° in the measured results. The measured cross-polarization levels for these two principal planes are lower than −15 dB. The simulated cross-polarization levels are very low and, as a consequence, are not shown.

The slight beam tilt and the discrepancy of the peak realized gain values can be mostly attributed to fabrication inaccuracies, particularly in the 3D-printed cylindrical frame. Errors in the frame affected the curvature of the transmitting aperture. Moreover, there were some small alignment errors and the low-cost PCB board may not have had a constant dielectric constant everywhere. Furthermore, the dielectric constant may be slightly different from its datasheet value.

6.5 Elliptically Conformal Multi-Beam Transmitarray with Wide-Angle Scanning Ability

A multi-beam conformal antenna that can generate a number of concurrent, but independent, directive beams with high gain values is presented below. As was discussed in Section 6.1, the main challenge for multi-beam conformal transmitarrays is to achieve large beam scanning coverage

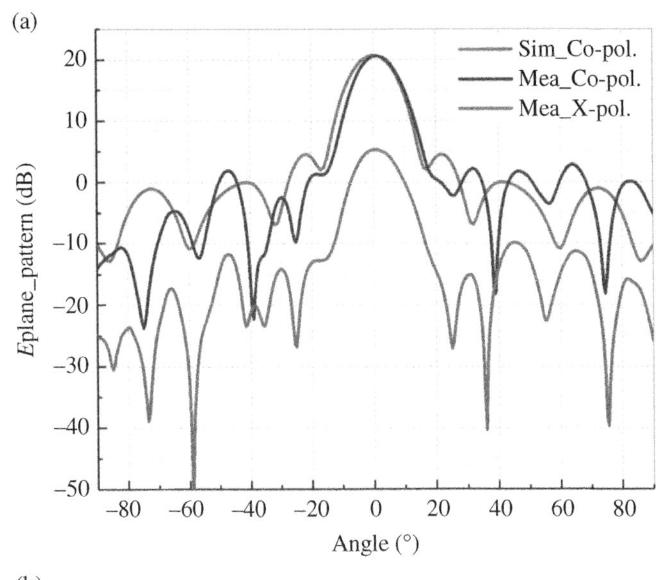

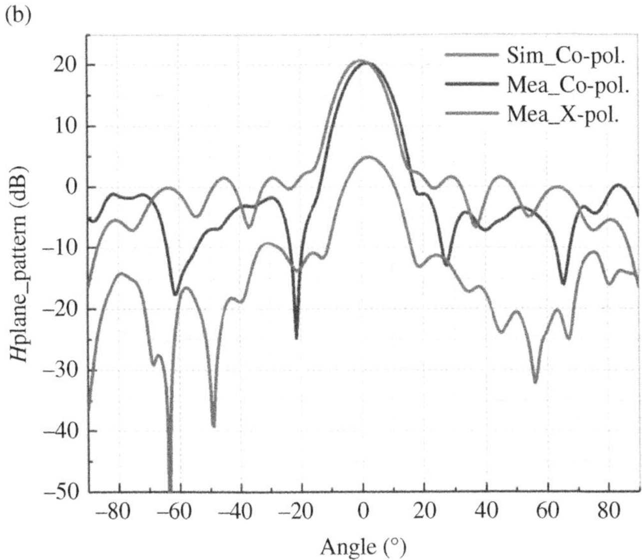

Figure 6.30 Simulated and measured realized gain patterns of the prototype Huygens transmitarray. (a) E-plane. (b) H-plane. *Source*: From [35] / with permission of IEEE.

with only a small gain variation. Based on the 2-D Ruze lens theory [39], we have introduced a new method for realizing multiple beams with greater beam coverage. An elliptic cylinder transmitarray was developed that employs a new method to locate the feeds and a novel phase compensation method to minimize the aberrations [40]. An antenna based on an optimized design was fabricated and tested. A beam scanning coverage of ±45° with a small gain variation has been verified experimentally.

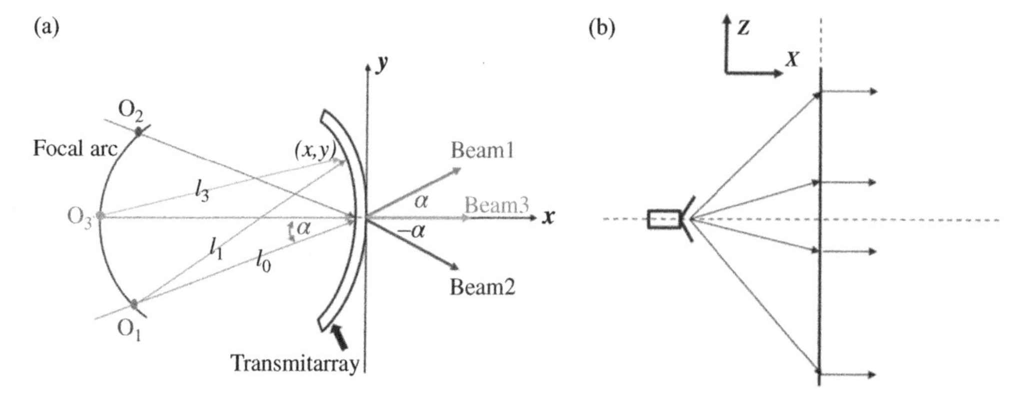

Figure 6.31 Transmitarray contour design. (a) Elliptical section. (b) Corresponding straight section.

6.5.1 Multi-Beam Transmitarray Design

a) Transmitarray Contour and Phase Calculation

The phase compensation of the transmitarray aperture of the design was calculated based on the predefined largest beam angles. As illustrated in Figure 6.31a, the points O_1 and O_2 are the focal points of two symmetrical radiated beams in the $x0y$ plane. These beams are denoted as *beam*1 and *beam*2; they are directed at the angles $\pm\alpha$, which are equal to the feed offset angles. The element phase compensation for *beam*1 must satisfy the relation:

$$k_0 l_1 - \varphi_{t1}(x, y) = k_0(l_0 + x * \cos\alpha + y * \sin\alpha) - \varphi_{t1}(0, 0) \tag{6.28}$$

where l_1 represents the distance between the focal point and a position (x, y) on the aperture. The focal length of O_1 is l_0, and k_0 represents the propagation constant in free space. The element phase compensation value at (x, y) is $\varphi_{t1}(x, y)$, and $\varphi_{t1}(0, 0)$ is the element phase compensation value at $(0, 0)$.

The distance l_1 is calculated with the expression:

$$l_1 = \sqrt{(x + l_0 * \cos\alpha)^2 + (y + l_0 * \sin\alpha)^2} \tag{6.29}$$

Defining u as:

$$u = \frac{\Delta\varphi_{t1}}{k_0} = \frac{\varphi_{t1}(x, y) - \varphi_{t1}(0, 0)}{k_0} \tag{6.30}$$

and substituting (6.29) in (6.28), one obtains:

$$(x + l_0 * \cos\alpha)^2 + (y + l_0 * \sin\alpha)^2 = (l_0 + x * \cos\alpha + y * \sin\alpha + u)^2 \tag{6.31}$$

After some manipulations, one obtains:

$$x^2 * \sin^2\alpha + y^2 * \cos^2\alpha - 2xy * \sin\alpha * \cos\alpha = u^2 + 2ul_0 + 2uy * \sin\alpha + 2ux * \cos\alpha \tag{6.32}$$

Similarly, considering the focal point at O_2 for *beam*2 with the angle $-\alpha$, one also has:

$$x^2 * \sin^2\alpha + y^2 * \cos^2\alpha + 2xy * \sin\alpha * \cos\alpha = u^2 + 2ul_0 - 2uy * \sin\alpha + 2ux * \cos\alpha \tag{6.33}$$

Combining Eqs. (6.32) and (6.33), one obtains the phase compensation equation in the $x0y$ plane along with the ideal transmitarray contour. They are given by the expressions:

$$\Delta\varphi_{t1} = k_0 u = -k_0 x \times \cos\alpha \tag{6.34}$$

$$\left(\frac{x}{l_0 \times \cos\alpha} + 1\right)^2 + \left(\frac{y}{l_0}\right)^2 = 1 \tag{6.35}$$

As can be seen from Eq. (6.35), the ideal transmitarray contour is elliptical in $x0z$ plane. As with the previous designs, the phase compensation values to achieve boresight radiation are calculated along the straight lines shown in Figure 6.31b.

b) Refocusing Design

The multi-beam transmitarray with multiple feeds was designed initially assuming that the feeds would be placed along the focal arc with radius l_0 and with the pivot at (0, 0) as shown in Figure 6.31a. Recall that the phase compensation values along the transmitarray were calculated relative to those for the maximum oblique angles $\pm\alpha$ instead of at 0° (along x-axis). Therefore, there is a phase error for the feed located at the center of the focal arc, the feed point at O_3, which points to the center of the transmission surface and generates the boresight beam pointing toward 0°. The phase error for this beam, labeled *beam*3 in Figure 6.31a, is described below to establish how a correction for it can be made.

The ideal phase compensation value for *beam*3 is obtained from the derived elliptical contour of the transmitarray aperture and the relevant form of (6.28), i.e.,

$$k_0 l_3 - \varphi_{t3}(x, y) = k_0(l_0 + x) - \varphi_{t3}(0, 0) \tag{6.36}$$

where l_3 denotes the distance from the focal point O_3 to any point (x, y) on the aperture, and l_0 is the focal length specifically at O_3. The term $\varphi_{t3}(x, y)$ is the compensating phase value of the element at (x, y), the distance l_3 is calculated with the relation:

$$l_3 = \sqrt{(x + l_0)^2 + (y)^2} \tag{6.37}$$

Therefore, the relative phase compensation value relative to O_3 is:

$$\Delta\varphi_{t3} = \varphi_{t3}(x, y) - \varphi_{t3}(0, 0) = k_0 \times \left(\sqrt{(x + l_0)^2 + y^2} - x - l_0\right) \tag{6.38}$$

The phase error at point O_3 is then:

$$\delta = \Delta\varphi_{t3} - \Delta\varphi_{t1} = k_0 \times \left(\sqrt{(x + l_0)^2 + y^2} - x(1 - \cos\alpha) - l_0\right) \tag{6.39}$$

Now consider the relationship between x and y from (6.35). Employing a Taylor series expansion for the value y in (6.39) at (0, 0), one finds that the first- and third-order derivatives of δ with respect to y are both zero. Thus, (6.39) can be approximated with as second-order term as:

$$\delta \approx \delta_1 = k_0 \frac{\sin^2\alpha}{2l_0} \times y^2 \tag{6.40}$$

The corresponding plots of $\frac{\delta}{k_0 l_0}$ and $\frac{\delta_1}{k_0 l_0}$ versus $\frac{y}{l_0}$ are given in Figure 6.32 for the two cases in which the maximum beam angles α are 60° and 45°, respectively. There is a quite good agreement between δ and δ_1 for both; and, therefore, the approximation (6.40) is acceptable.

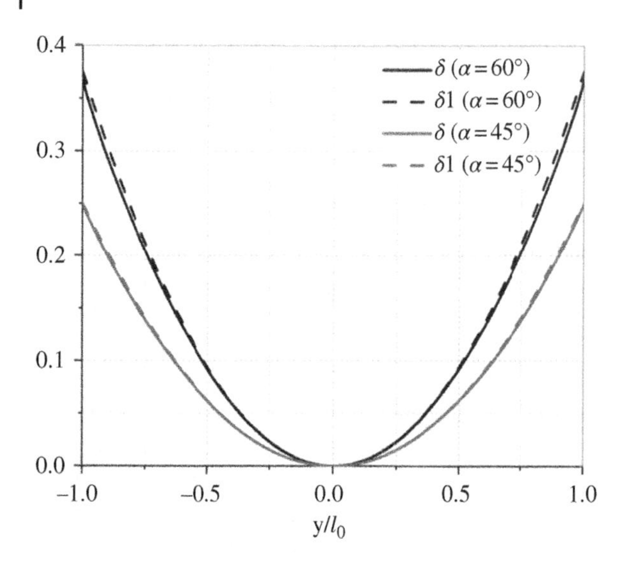

Figure 6.32 Comparison of the exact and approximated phase error.

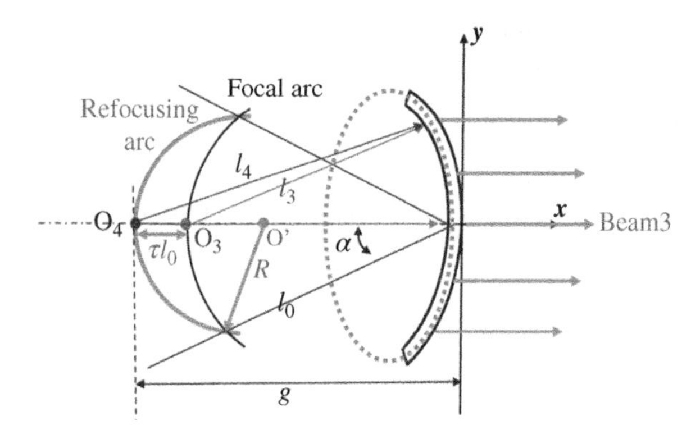

Figure 6.33 Refocusing schematic.

To compensate for the phase error at O_3, the feed point O_3 is moved away from the transmitarray aperture by τl_0 to point O_4, as illustrated in Figure 6.33. The relative phase compensation value at O_4 is:

$$\Delta\varphi_{t4} = \varphi_{t4}(x, y) - \varphi_{t4}(0, 0) = k_0 \times \left[\sqrt{(x + (1+\tau)l_0)^2 + y^2} - x - (1+\tau)l_0\right] \tag{6.41}$$

The phase correction resulting from this refocusing to O_4 is given by:

$$\sigma = \Delta\varphi_{t4} - \Delta\varphi_{t3} = k_0 \times \left(\sqrt{(x + (1+\tau)l_0)^2 + y^2} - \sqrt{(x + l_0)^2 + y^2} - \tau l_0\right) \tag{6.42}$$

With the Taylor series expansion of (6.42), one obtains:

$$\sigma \approx \sigma 1 = -k_0 \frac{1}{2l_0} \times \frac{\tau}{1+\tau} \times y^2 \tag{6.43}$$

Therefore, the phase aberration for the boresight *beam*3 is ideally fixed by refocusing the feed source to O_4, as long as $\sigma 1 = -\delta 1$. This means:

$$\tau = \tan^2 \alpha \tag{6.44}$$

A new focal arc, labeled as the refocusing arc in Figure 6.33, is thus produced. Consequently, the pivot point is moved from (0, 0) to point O′, i.e., to the point $(R - g, 0)$, where $g = (1 + \tau)\, l_0$ is the new focal length at O_4. The radius R is then calculated with the law of cosines as:

$$R = \frac{2gl_0 \cos \alpha - l_0^2 - g^2}{2l_0 \cos \alpha - 2g} \tag{6.45}$$

Finally, a continuously beam scan can be realized in the range of $-\alpha$ to $+\alpha$ as the feed is rotated along the refocusing arc.

c) Phase Compensation Along z-Axis

Since the conformal transmitarray is designed for a cylindrical elliptic contour, the phase compensation calculation is divided into two parts, i.e., for points along the elliptical arc in the *x*0*y* plane and those along the *z*-axis. Points lying in the *x*0*y* and *x*0*z* planes shown in Figure 6.34 are labeled with *i* and *j*, respectively. Multiple feed horns, numbered as $-N$ to $+N$, are placed in the $z = 0$ plane. The distances between each horn and the transmitarray element at (0, 0) are labeled as d_{00_-N} to d_{00_+N}, i.e., where $i = 0$. The distance between each horn and an element in the *x*0*z* plane is labeled as d_{0j_-N} to d_{0j_+N}. The phase distribution along the elliptical aperture in the *x*0*y* plane is specified for different positions of the feed in order that the output beams point at different angles. However, only boresight beams are radiated along the *x*0*z* plane.

The phase distribution along this *x*0*z* plane is analyzed as follows. The *zero* column (center column along the *x*0*z* plane) corresponds to the requisite phase compensation along the *z*-axis. It is illustrated in Figure 6.34c. This phase compensation value:

$$\Delta\varphi_s = k_0 \times (d_{0j} - d_{00}) \tag{6.46}$$

is thus related to the focal length d_{00}. As shown in Figure 6.34b, the focal lengths for horn 0, d_{00_0}, and for horns $-N$ and $+N$, d_{00_-N} and d_{00_+N}, are different after refocusing. As a result, the phase

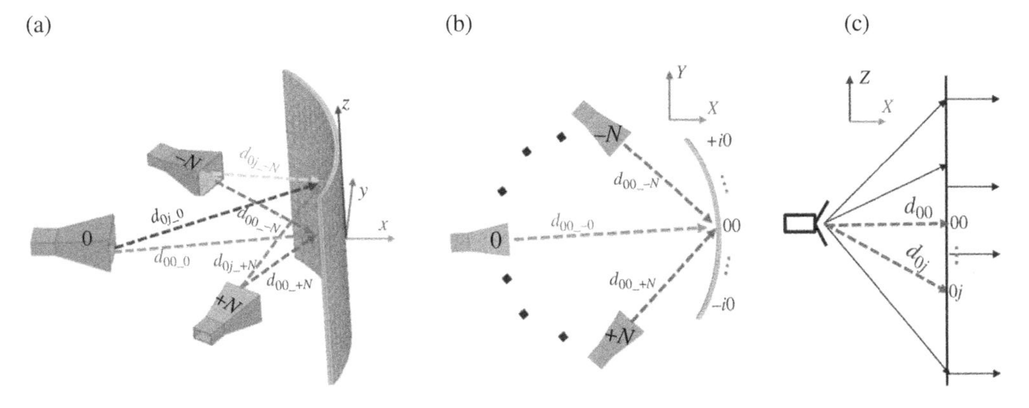

Figure 6.34 Transmitarray configuration having feed system with *2 N + 1* horns. (a) 3D illustration. (b) Top view. (c) Side view.

distribution along the z-axis for each feed horn would be different. Therefore, a virtual value of d_{00} must be determined so that the transmitarray will meet the system requirements. For example, if the goal is to maximize the gain of the boresight beam, then d_{00} must be close to d_{00_0}. On the other hand, if the goal is to minimize the gain difference between the boresight beam and the beam with the maximum-steered angle, then d_{00} should be the average of d_{00_0} and d_{00_+N}.

6.5.2 Concept Verification Through Simulation

For other phase element columns along the z-axis, the values of d_{i0} may be different from those associated with the central column d_{00}. Ideally, d_{i0} should be calculated for each column according to the system requirements as noted above. For simplicity, however, the value of d_{00} for the central column is used as d_{i0} for all of the other columns in the design. Three conformal transmitarray prototypes with different d_{00} have been developed. The simulated realized gains for their boresight and steered beams are presented. They illustrate the effects of d_{00} on the multi-beam performance.

a) Unit Cell

The unit cell employed for the multi-beam transmitarray design is a triple-layer structure that consists of three of the square slotted rings shown in Figure 6.1. The rings are identical and printed on the surfaces of two identical substrates. Each substrate has a relative dielectric constant $\varepsilon_r = 2.2$ and loss tangent of $\tan\delta = 0.0009$.

As analyzed in [27], a trade-off always exists between the element thickness h and the resulting phase range. To realize a 360° phase compensation range with a maximal 3 dB insertion loss, h is chosen to be 3.0 mm, i.e., 0.21 λ_0 at 21 GHz. Two of the optimized design parameters are: $P = 7.2$ mm and $w = 1.4$ mm. The HFSS simulations of the magnitude and phase performance versus the slot length L were obtained and the results are presented in Figure 6.35. A 340° phase variation was realized with a maximum 3 dB transmission loss when L was varied from 4.4 to 6.96 mm.

b) Transmitarray Contour and Refocusing for Boresight Radiation

An elliptic cylinder transmitarray has been designed with its maximum beam angles at ±45° and with its focal length l_0 chosen to be 150 mm at the corresponding horn locations O_1 and O_2. The ellipse in the $x0y$ plane is defined by Eq. (6.35):

$$\left(\frac{x}{75\sqrt{2}}+1\right)^2+\left(\frac{y}{150}\right)^2=1 \tag{6.47}$$

The phase compensation along this contour is given by the relation:

$$\Delta\varphi_e=-\frac{2\pi}{\lambda_0}\times\frac{\sqrt{2}}{2}x \tag{6.48}$$

To precisely arrange the elements in the transmission surface and to calculate the phase compensation value for each element, the elliptical arclength is taken to be integer multiples of the element periodicity P. The element positions along the arc are then straightforward to specify. If the coordinates x and y in (6.47) are expressed as:

$$\begin{cases} x = 75\sqrt{2}\,(\cos\varphi - 1) \\ y = 150\sin\varphi \end{cases} \tag{6.49}$$

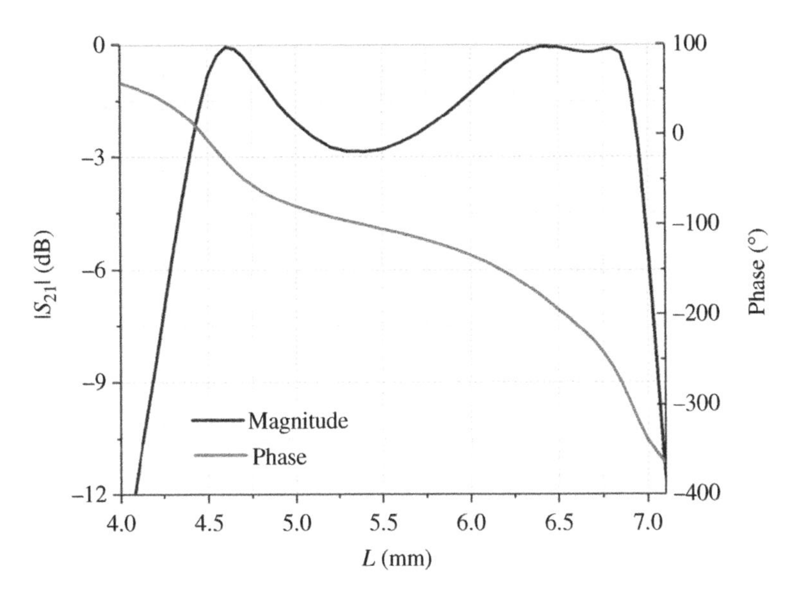

Figure 6.35 Simulated magnitude and values of $|S_{21}|$ as functions of the slot length L.

then the associated arclength is:

$$l = \int_0^{\varphi} \sqrt{\left(75\sqrt{2}\right)^2 \times \sin^2\varphi + (150)^2 \times \cos^2\varphi}\ d\varphi \tag{6.50}$$

Thus, the element positions can be specified in terms of this arclength. For example, the arclength of the position of the *i-th* element in the *x*0*y* plane is $l = iP$. The related phase compensation value is obtained from Eqs. (6.48)–(6.50).

To eliminate the phase error along the boresight direction, the refocusing method illustrated in Figure 6.33 is applied. The feed position O_4 is obtained from Eq. (6.44) with $\tau = 1.0$. One obtains $g = 2\,l_0 = 300$ mm. Several feed horns need to be arranged along the refocusing arc as shown in Figure 6.34 to realize the desired multiple beams.

c) Phase Distribution on the Conformal Transmitarray

The calculation of the phase distributions for the entire structure is divided into two parts. Eq. (6.48) is used to calculate the phase along the elliptical arc in the *x*0*y* plane. The phase values along the *z*-axis, as discussed above, require a specified virtual focal length d_{00} to attain the specified performance characteristics.

An elliptic cylinder transmitarray model with $29 \times 25 = 725$ cells was constructed in HFSS using the stacked three-layer element model described in Section 6.5.2.A. The same standard gain horn, LB-51-10-C-SF from A-INFO, acts as the source and is placed at each focal point on the elliptical contour. The gain of this source horn, as specified by its data sheet, is 13.52 dBi at 21 GHz. The size of the aperture cross section is 199.2 mm × 180.0 mm. Assuming the central column to be along the *z*-axis, these dimensions were chosen to have the aperture edge illumination be about −10 dB when the horn is located at points O_1 and O_2.

The feed horn positions for the boresight and −45° beams are $(x, y) = (-300\text{ mm}, 0\text{ mm})$ and $(-106.1\text{ mm}, -106.1\text{ mm})$, respectively. The focal length is 300 mm for the 0° beam, and the −45° beam configuration then has a 150 mm focal length. Initially, the virtual value $d_{00} = 205$ mm

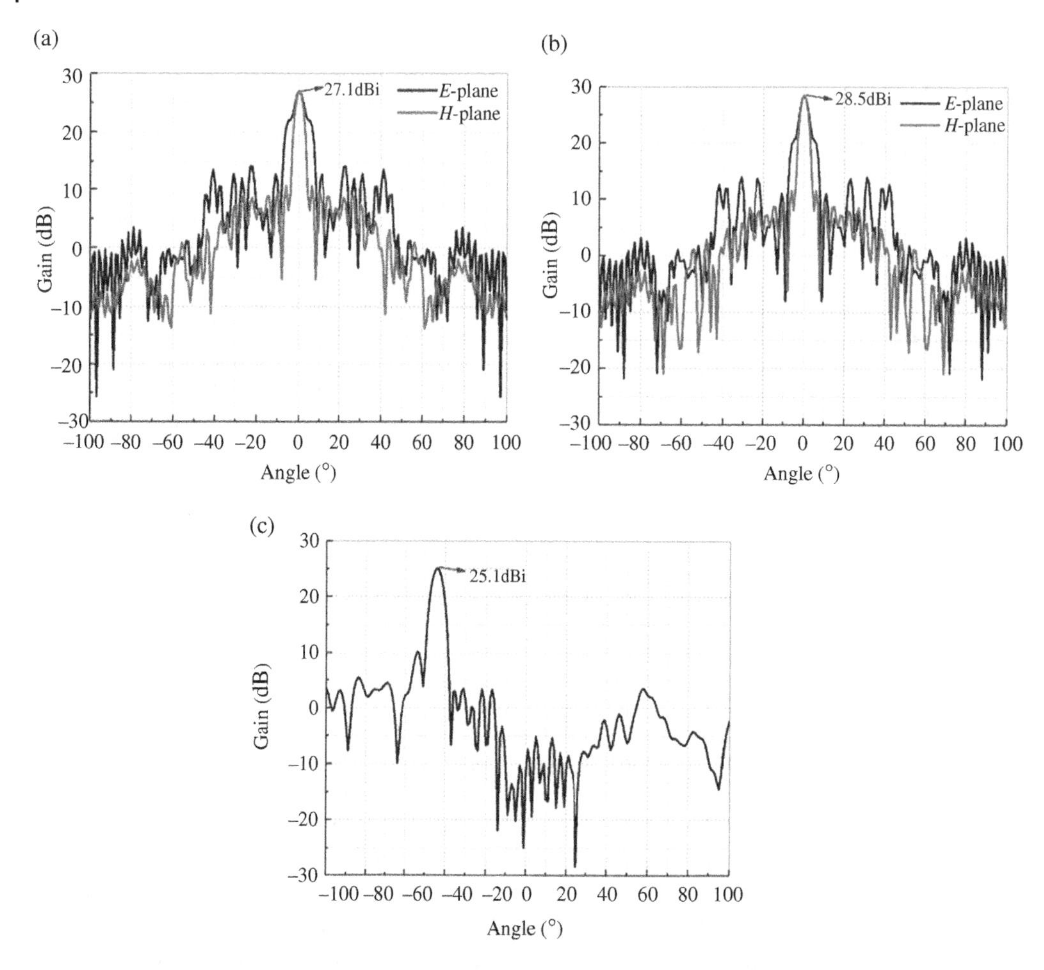

Figure 6.36 Simulated E- and H-plane realized gain patterns with d_{OO} = 205 mm. (a) Boresight beam with g = 300 mm. (b) Boresight beam and (c) −45° beam with g = 275 mm.

was chosen to be approximately the average focal length value. The phase compensation along the z-axis was obtained with Eq. (6.46). After combining this result with Eq. (6.48), the entire transmitarray configuration was derived.

The simulated boresight radiation patterns along the $x0y$ (H-plane) and $x0z$ (E-plane) planes are given in Figure 6.36a. Note that the −10 dB E-plane beamwidth is wide because the chosen d_{OO} value deviates substantially from the actual focal length for 0° radiation, i.e., $d_{OO\_0} = 300$ mm. Therefore, the refocusing position (−g, 0) for horn *zero*, i.e., the gain horn in the boresight position, was tuned to balance the E-plane and H-plane radiation performance. Moving the horn *zero* position toward the aperture by changing g to 275 mm, the realized gain patterns along the E-plane were improved as demonstrated in Figure 6.36b. The peak gain was increased from 27.1 to 28.5 dBi. The corresponding radiation pattern for the −45° beam is shown in Figure 6.36c.

Following the same design procedure, another two d_{OO} values were chosen to design two different conformal transmitarrays. They also were simulated with HFSS to study the effects of d_{OO} on the overall radiation performance. The realized gain results for all three prototypes are compared in Table 6.3. It can be seen that when the d_{OO} phase compensation value along the z-axis is chosen

Table 6.3 Simulation results of the transmitarray designs with different choices of d_{OO}.

d_{OO} (mm)	g (mm)	Peak gain at 0° (dBi)	Peak gain at −45° (dBi)	Gain difference (dBi)
235	280	29.1	24.5	4.6
205	275	28.5	25.1	3.4
180	240	27.6	25.6	2

to be larger, the peak gain at 0° increases, while it decreases for the −45° beam. This occurs because the larger d_{OO} is, the closer it is to the real focal length for 0° (d_{OO_0}). On the other hand, it is further away from d_{OO_+N} for the −45° beam. Also note that the peak gain difference between the 0° and −45° reduces as d_{OO} decreases.

d) Feed System Arrangement for Multi-Beam Realization

Finally, since the design target was to realize a stable gain for different beam angles, the value of d_{OO} was selected to be 180 mm for the final conformal transmitarray design. In this case, the final refocusing position for horn *zero* was optimized to be $(-g, 0) = (-240\text{ mm}, 0)$. The refocusing arc was optimized for the three positions of O_1, O_2, and O_4. The feed position for each desired beam was then determined.

Figure 6.37 illustrates that for beam radiation at a general angle β, its horn location O_n has the same offset angle β. The focal length is denoted as L_β. With $g = 240$ mm, $l_0 = 150$ mm, and designed maximal radiation angle $\alpha = 45°$, the radius of the refocusing arc was calculated from Eq. (6.45) to be $R = 109$ mm. Then the focal length L_β was derived from the triangle in Figure 6.37 specified by R, g-R, and β. With it expressed as $(-L_\beta \cos\beta, -L_\beta \sin\beta)$, all of the calculated feed positions are listed in Table 6.4.

By exciting a feed horn at the feed location specified by Table 6.4, the multiple beams shown in Figure 6.38a were achieved at 21 GHz. The corresponding peak gain values are also listed in Table 6.4. Note that the beam can be scanned to ±45° with only a 2.6 dB drop from the maximum realized gain. The side lobe level is lower than −13 dB in all cases. The peak realized gain of the 0° beam as a function of the source frequency is presented in Figure 6.38b. A 3-dB bandwidth of 16% was achieved.

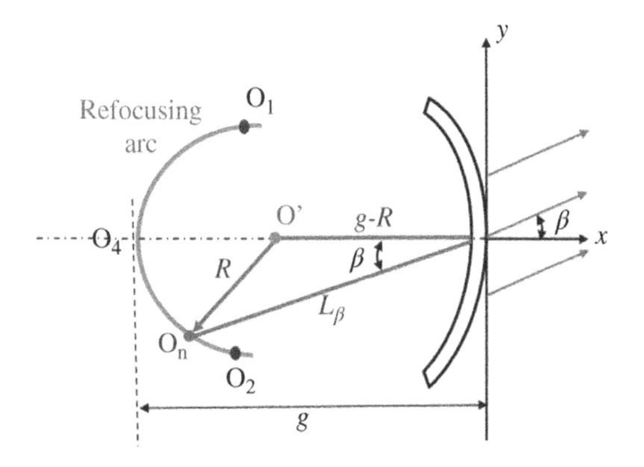

Figure 6.37 Geometry used to calculate the horn feed position.

Table 6.4 Feed positions for different beam directions.

Beam angle (β, °)	L (mm)	O_n (x_n, y_n) (mm, mm)
0	240.0	(−240, 0)
10	235.6	(−232, −41)
20	222.5	(−209.1, −76.1)
30	200.6	(−173.7, −100.3)
40	169.6	(−130, −109)
50	150.0	(−106.1, −106.1)

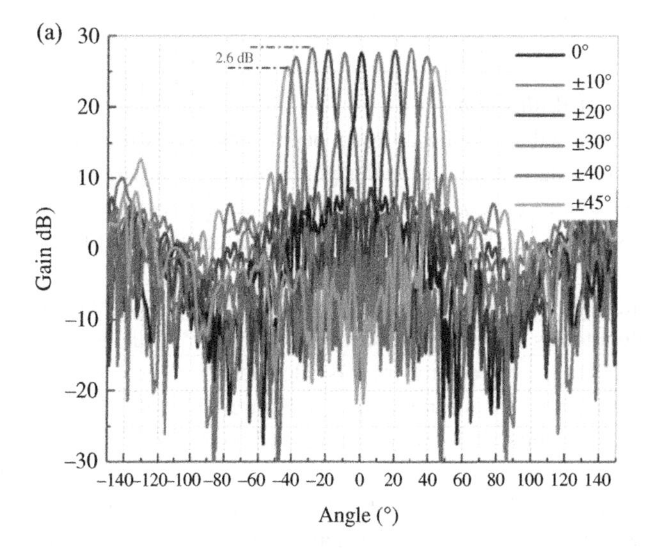

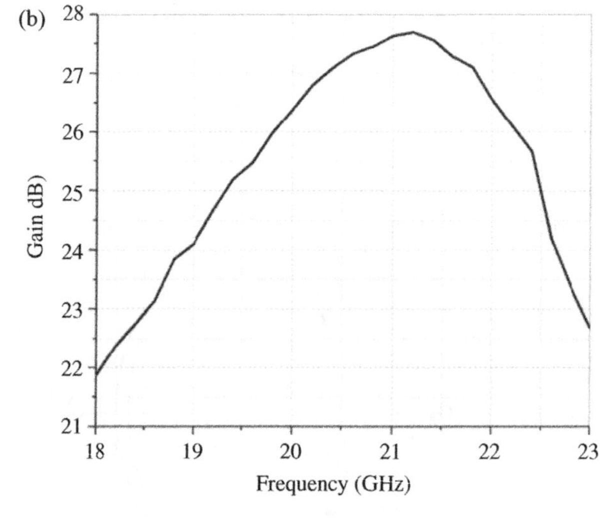

Figure 6.38 Simulated gain of the multi-beam transmit array at 21 GHz. (a) Multi-beam realized gain patterns for different output beam angles. (b) Gain in the 0° direction as a function of the source frequency.

e) Design Procedure

According to the theoretical analyses and prototype simulations discussed in this chapter, the multi-beam transmitarray design procedure can be summarized as follows:

Step 1: Choose a proper focal length l_0 and the required maximum beam angles $\pm\alpha$.

Step 2: The formula of elliptical contour for the transmitarray is derived from Eqs. (6.34) and (6.35). Then obtain the specific phase compensation values along it.

Step 3: Adopt the refocusing design based on Eq. (6.44) to find the feed position of the gain horn for boresight radiation.

Step 4: Choose the value of d_{00} and calculate the reference focal length for phase compensation along the z-axis, aiming to balance the beam performance at 0° and ±45°.

Step 5: Combine phase compensation along the elliptical arc and along the z-axis to determine the initial conformal transmitarray configuration.

Step 6: Optimize the refocusing position for boresight radiation to obtain the best 0° radiation pattern.

Step 7: Draw the refocusing arc with three known feed positions and locate the various feed horns along this arc to realize multiple output beams pointed between $-\alpha°$ and $\alpha°$.

6.6 Conclusions

Owing to their aerodynamic performance, it is expected that conformal transmitarrays will find a wide range of applications to airborne, spaceborne, and satellite networks, such as UAVs, high-altitude platforms (HAPs), and aircraft, especially in the context of 6G networks. In this chapter, we have discussed the technical challenges in designing single- and multi-beam conformal transmitarrays and offered a number of practical solutions. In particular, three antenna configurations were presented to achieve wide beam scanning angles; a thin and high transmission efficiency Huygens' surface; and wide-angle multi-beams. To date, research in conformal transmitarrays is still in its infancy. Consequently, the ideas presented are meant to inspire more innovations in the field. New thin, high transmission efficiency and ultra-wideband elements are required to improve the performance of transmitarrays [41]. Another important topic in individually steerable multi-beam conformal transmit arrays is the realization of their low-cost, highly integrated smart feeds. We expect to see some progress to be made in this area in the near future.

References

1. Khan, F. (2016). Multi-comm-core architecture for terabit-per-second wireless. *IEEE Commun. Mag.* **54** (4): 124–129.
2. Liu, J., Shi, Y., Fadlullah, Z.M., and Kato, N. (2018). Space-air-ground integrated network: A survey. *IEEE Commun. Surveys Tuts.* **20** (4): 2714–2741, 4th Quart.
3. Abdelrahman, A.H., Yang, F., Elsherbeni, A.Z. et al. (2017). *Analysis and Design of Transmitarray Antennas*. San Francisco, CA, USA: Morgan & Claypool.
4. Dussopt, L. (2018). *Transmitarray antennas*. In: *Aperture Antennas for Millimeter and Sub-Millimeter Wave Applicatons* (eds. A. Boriskin and R. Sauleau), 191–220. Springer.
5. Gaber, S., Zainud-Deen, S.H., and Malhat, H.A.E. (2014). *Analysis and design of reflectarrays/transmitarrays antennas*. LAP LAMBERT Academic Publ.

6. Luo, Q., Gao, S., Sobhy, M., and Yang, X. (2018). Wideband transmitarray with reduced profile. *IEEE Antennas Wireless Propag. Lett.* **17** (3): 450–453.
7. Feng, P.-Y., Qu, S.-W., and Yang, S. (2018). Octave bandwidth transmitarrays with a flat gain. *IEEE Trans. Antennas Propag.* **66** (10): 5231–5238.
8. Ge, Y., Lin, C., and Liu, Y. (2018). Broadband folded transmitarray antenna basedon an ultrathin transmission polarizer. *IEEE Trans. Antennas Propag.* **66** (11): 5974–5981.
9. Tian, C., Jiao, Y.-C., and Zhao, G. (2017). Circularly polarized transmitarray antenna using low-profile dual-linearly polarized elements. *IEEE Antennas Wireless Propag. Lett.* **16**: 465–468.
10. Luo, Q., Gao, S., Sobhy, M. et al. (2019). A hybrid design method for thin panel transmitarray antennas. *IEEE Trans. Antennas Propag.,* Early access **67** (10): 6473–6483.
11. An, W., Xu, S., Yang, F., and Li, M. (2016). A double-layer transmitarray antenna using malta crosses with vias. *IEEE Trans. Antennas Propag.* **64** (3): 1120–1125.
12. Glybovski, S.B., Tretyakov, S.A., Belov, P.A. et al. (2016). Metasurfaces: From microwaves to visible. *Phys. Rep.* **634**: 1–72.
13. Wong, J.P.S., Selvanayagam, M., and Eleftheriades, G.V. (2014). Design of unit cells and demonstration of methods for synthesizing Huygens metasurfaces. *Photonics Nanostruct. Fundam. Appl.* **12** (4): 360–375.
14. Pfeiffer, C. and Grbic, A. (2013). Millimeter-wave transmitarrays for wavefront and polarization control. *IEEE Trans. Microw. Theory Techn.* **61** (12): 4407–4417.
15. Pfeiffer, C. and Grbic, A. (2013). Metamaterial Huygens' surfaces: Tailoring wave fronts with reflectionless sheets. *Phys. Rev. Lett.* **110** (19): 197401.
16. Chen, K., Feng, Y., Monticone, F. et al. (2017). A reconfigurable active Huygens' metalens. *Adv. Mater.* **29** (17): 1606422.
17. Palma, L., Clemente, A., Dussopt, L. et al. (2017). Circularly-polarized reconfigurable transmitarray in Ka-band with beam scanning and polarization switching capabilities. *IEEE Trans. Antennas Propag.* **65** (2): 529–540.
18. Huang, C., Pan, W., and Luo, X. (2016). Low-loss circularly polarized transmitarray for beam steering application. *IEEE Trans. Antennas Propag.* **64** (16): 4471–4476.
19. Wang, M., Xu, S., and Yang, F. (2017). Design of a Ku-band 1-bit reconfigurable transmitarray with 16 × 16 Slot coupled elements. *2017 IEEE AP-S*, San Diego, CA, USA.
20. Lau, J.Y. and Hum, S.V. (2012). A wideband reconfigurable transmitarray element. *IEEE Trans. Antennas Propag.* **60** (3): 1303–1311.
21. Nicholls, J.G. and Hum, S.V. (2016). Full-space electronic beam-steering transmitarray with integrated leak-wave feed. *IEEE Trans. Antennas Propag.* **64** (8): 3410–3422.
22. Xi, B., Xue, Q., Bi, L. et al. (2018). Design of multi-beam transmitarray antenna using alternating projection method. *IEEE International Conference on Computational Electromagnetics (ICCEM)*, Chengdu, China, pp. 1–3, March 2018.
23. Jiang, M., Chen, Z.N., Zhang, Y. et al. (2017). Metamaterial based thin planar lens antenna for spatial beamforming and multibeam massive MIMO. *IEEE Trans. Antennas Propag.* **65** (2): 464–472.
24. Abdelrahman, A.H., Nayeri, P., Elsherbeni, A.Z., and Yang, F. (2016). Single-feed quad-beam transmitarray antenna design. *IEEE Trans. Antennas Propag.* **64** (3): 953–959.
25. Liu, G., Kodnoeih, M.R.D., Pham, T.K. et al. (2019). A millimeter-wave multibeam transparent transmitarray antenna at Ka-band. *IEEE Antennas Wireless Propag. Lett.* **18** (4): 631–635.
26. Dussopt, L. et al. (2017). A V-band switched-beam linearly polarized transmit-array antenna for wireless backhaul applications. *IEEE Trans. Antennas Propag.* **65** (12): 6788–6793.
27. Beccaria, M., Massaccesi, A., Pirinoli, P. et al. 2018. Multibeam transmitarrays for 5G antenna systems. *2018 IEEE Seventh International Conference on Communications and Electronics (ICCE)*. IEEE, 2018, pp. 217–221.

28. Hou, Y., Chang, L., Li, Y. et al. (2018). Linear multibeam transmitarray based on the sliding aperture technique. *IEEE Trans. Antennas Propag.* **66** (8): 3948–3958.

29. Zhong, X., Chen, L., Shi, Y., and Shi, X. (2016). Design of multiple-polarization transmitarray antenna using rectangle ring slot element. *IEEE Antennas Wireless Propag. Lett.* **15**: 1803–1806.

30. Qin, P.-Y., Song, L.-Z., and Guo, Y.J. (2019). Beam steering conformal transmitarray employing ultra-thin triple-layer slot elements. *IEEE Trans. Antennas Propag.* **67** (8): 5390–5398.

31. Huang, J. and Encinar, J.A. (2008). *Reflectarray Antennas, by Institute of Electrical and Electronics Engineers.* Wiley.

32. Hum, S.V. and Perruisseau-Carrier, J. (2014). Reconfigurable reflectarrays and array lens for dynamic antenna beam control: A review. *IEEE Trans. Antennas Propag.* **62** (1): 183–198.

33. Al-Joumayly, M.A. and Behdad, N. (2011). Wideband planar microwave lenses using sub-wavelength spatial phase shifters. *IEEE Trans. Antennas Propag.* **59** (12): 4542–4552.

34. Cheng, C.Y. and Ziolkowski, R.W. (2003). Tailoring double-negative metamaterial responses to achieve anomalous propagation effects along microstrip transmission lines. *IEEE Trans. Microwave Theory Tech.* **51** (12): 2306–2314.

35. Song, L., Qin, P., and Jay Guo, Y. (2021). A high-efficiency conformal transmitarray antenna employing dual-layer ultra-thin Huygens element. *IEEE Trans. Antennas Propag.* **69** (2): 848–858. https://doi.org/10.1109/TAP.2020.3016157.

36. Epstein, A. and Eleftheriades, G.V. (2016). Huygens' metasurfaces via the equivalence principle: Design and applications. *J. Opt. Soc. Am. B* **33** (2): A31–A50.

37. Selvanayagam, M. and Eleftheriades, G.V. (2013). Circuit modelling of Huygens surfaces. *IEEE Antennas Wireless Propag. Lett.* **12**: 1642–1645.

38. Lin, W., Ziolkowski, R.W., and Huang, J. (2019). Electrically small, low-profile, highly efficient, Huygens dipole rectennas for wirelessly powering internet-of-things devices. *IEEE Trans. Antennas Propag.* **67** (6): 3670–3679.

39. Ruze, J. (1950). Wide-angle metal-plate optics. *Proc. IRE* **38** (1): 53–69.

40. Song, L., Qin, P., Chen, S., and Jay Guo, Y. An elliptical cylindrical shaped transmitarray for wide-angle multibeam applications. *IEEE Trans Antennas Propag.* https://doi:10.1109/TAP.2021.3083828.

41. Song, L., Qin, P., Maci, S., and Guo, Y.J. Ultrawideband conformal transmitarray employing connected slot-bowtie elements. *IEEE Trans. Antennas Propag.* **69** (6): 3273–3283. https://doi.org/10.1109/TAP.2020.3037785.

7

Frequency-Independent Beam Scanning Leaky-Wave Antennas

Compared with conventional antenna arrays, leaky-wave antennas (LWAs) have a number of attractive features. These include simple feeding structures, low profiles, and inherent scanning abilities [1–13]. Beam scanning is conventionally achieved with an LWA by sweeping the source frequency. While this approach may be well suited for certain remote sensing applications, it is not for wireless communication. This drawback is due to the fact that most wireless communications systems have predefined bandwidths for the operations.

In order to realize beam scanning at fixed frequencies, one can reconfigure the boundaries or inner structures of the LWA to modify the propagation constant of the waves in its guiding structure [14, 15]. Based on this concept, there has been rapid progress in research on reconfigurable LWAs to achieve single beam scanning and, most recently, multi-beam scanning [16–26]. Because reconfigurable LWAs are compact structures that feature low power consumption and costs, it is expected that with effective fixed-frequency scanning and multi-beam features, they will find wide applications in both wireless and satellite communications systems, particularly for moving platforms where there is a limited energy supply.

In this chapter, we introduce four types of advanced reconfigurable LWA structures for fixed-frequency beam scanning, i.e., a reconfigurable Fabry–Pérot (FP) LWA [20], a period-reconfigurable substrate-integrated-waveguide (SIW) LWA [25], a reconfigurable composite right-/left-handed (CRLH) LWA [26], and a uniplanar two-dimensional multi-beam LWA suited for millimeter-wave operations. These antenna configurations and their operating mechanisms and performance are described. Some future directions of research in reconfigurable LWAs are discussed to conclude the chapter.

7.1 Reconfigurable Fabry–Pérot (FP) LWA

The earliest reconfigurable LWA with fixed-frequency scanning was based on a waveguide structure. It was composed of a waveguide with its top metal surface replaced with a tunable partially reflective surface (PRS) and its bottom metal surface replaced with a tunable high impedance surface (HIS). This configuration is shown in Figure 7.1 [20, 27]. The operating principles of these surfaces enable this reconfigurable FP LWA design.

In particular, the HIS is made of electronically tunable patches. As shown in Figure 7.1b, their reconfigurability is facilitated by loading them with varactor diodes. The Fabry–Pérot (FP) cavity resonance is thus changed by tuning these HIS elements. As a consequence, the pointing angle of

Advanced Antenna Array Engineering for 6G and Beyond Wireless Communications, First Edition.
Y. Jay Guo and Richard W. Ziolkowski.

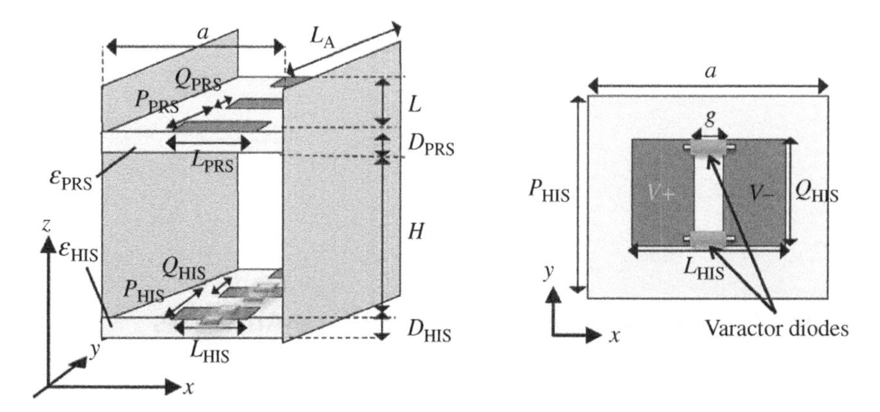

Figure 7.1 1-D reconfigurable FP LWA. (a) 3D view of a short portion of the LWA. (b) Unit cell of the HIS. *Source*: From [20] / with permission of IEEE.

the output beam is also changed by tuning their resonant length. Similarly, the PRS is also made of electronically tunable patches. The leakage rate of the LWA is controlled by adjusting their lengths.

The transverse resonance method (TRM) is a simple, yet accurate, method to obtain the complex propagation constant of the LWA. The TRM is used for the analysis and design of the reconfigurable LWA. The dispersion curves obtained using this method not only give a clear understanding of the LWA's operating principles, but also provide a theoretical foundation for beam steering in LWAs.

7.1.1 Analysis of 1-D Fabry–Pérot LWA

A transverse equivalent network (TEN) can be used to analyze a one-dimensional (1-D) FP LWA [20]. The cross section of the antenna and the TEN for the reconfigurable FP LWA are shown in Figure 7.2. The most important parts of the TEN model are the equivalent admittance of the PRS, Y_{PRS}, and the equivalent admittance of the HIS, Y_{HIS}. Since the HIS involves varactor diodes whose capacitance can be varied with different applied voltages, the admittance of the HIS is reconfigurable since it is a function of the varactor junction capacitance C_j.

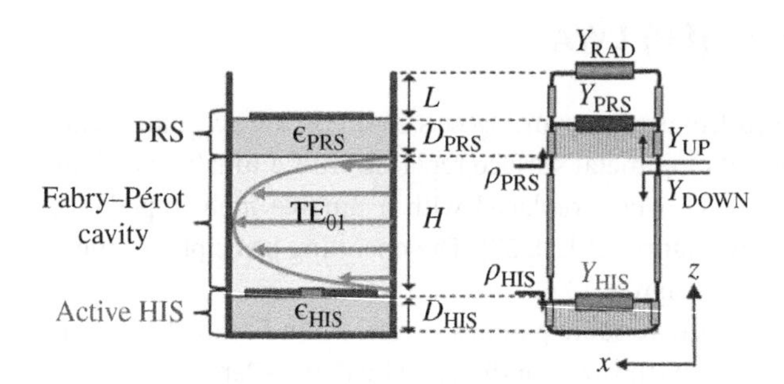

Figure 7.2 Reconfigurable 1-D FP LWA cross section (left) and the transverse equivalent network, TEN (right). *Source*: From [20] / with permission of IEEE.

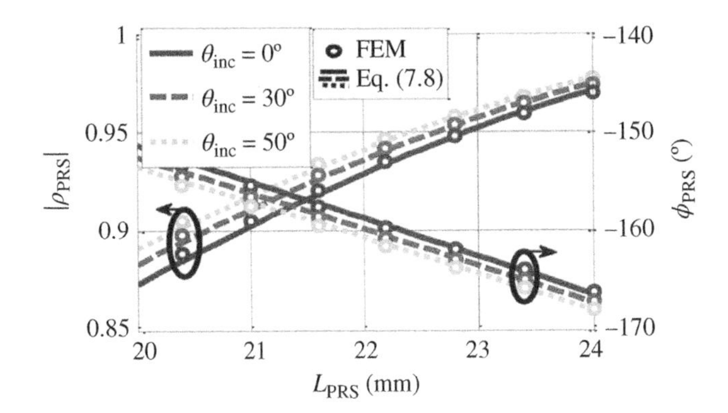

Figure 7.3 Phase and magnitude of the reflection coefficient associated with a plane wave incident on the PRS at different angles as a function of the length L_{PRS}. *Source*: From [20] / with permission of IEEE.

When the PRS is fixed, the length L_{PRS} of the conducting patches can be efficiently determined using the following analytical pole-zero expression for the PRS admittance (Y_{PRS}):

$$Y_{PRS}(k_y, L_{PRS}) = j\frac{L_{PRS}\left[L_{PRS} - L_{PRS_{z1}}(k_y)\right]\ldots\left[L_{PRS} - L_{PRS_{zn}}(k_y)\right]}{\left[L_{PRS} - L_{PRS_{p1}}(k_y)\right]\ldots\left[L_{PRS} - L_{PRS_{pn}}(k_y)\right]} \tag{7.1}$$

where the longitudinal wave number k_y determines the location of the poles and zeros. The wave number k_y is related to the leaky-mode angle (incident or radiating) by:

$$\sin\theta_{inc} = \frac{\text{real}(k_y)}{k_0} \tag{7.2}$$

where k_0 is the wave number in free space. The value of L_{PRS} determines the reflectivity of the PRS and, hence, it impacts the radiation efficiency.

The phase and magnitude of the reflection coefficient at 5.6 GHz when a plane wave is incident on the PRS at three different angles as L_{PRS} changes are shown in Figure 7.3. The values of the other design parameters of the PRS were fixed in this particular parameter study as follows: patch width $Q_{PRS} = 18$ mm, patch distance $P_{PRS} = 20$ mm, permittivity of the substrate $\varepsilon_{rPRS} = 4.4$, and thickness of the substrate $D_{PRS} = 0.8$ mm. It is seen from Figure 7.3 that the results obtained from the analytical pole-zero model of the PRS admittance agree well with those obtained from the full-wave finite element method (FEM)-based simulator ANSYS HFSS (high frequency structure simulator). The value of L_{PRS} was selected to be 22 mm to obtain a reflectivity greater than 0.9 for all scanning angles.

To enable electronic control, the patches on the HIS were divided into two parts along the x-direction as shown in Figure 7.1, leaving a gap (g) of 1.0 mm between them. Two varactors were placed across the edges of two neighboring patches. The value of the patch length L_{HIS} can be varied effectively by electrically controlling the varactors' junction capacitance C_j. The values of the remaining HIS design parameters were fixed: patch width $Q_{HIS} = 14$ mm, patch distance $P_{HIS} = 30$ mm, permittivity of the substrate $\varepsilon_{rHIS} = 3.0$, and thickness of the substrate $D_{HIS} = 1.524$ mm. The pole-zero expression for the electronically tunable HIS is dependent on the junction capacitance (C_j) of the varactor, and can be expressed as [28]:

$$Y_{HIS}(k_y, C_j) = j\frac{\left[C_j - C_{j_{z1}}(k_y)\right]\ldots\left[C_j - C_{j_{zn}}(k_y)\right]}{\left[C_j - C_{j_{p1}}(k_y)\right]\ldots\left[C_j - C_{j_{pn}}(k_y)\right]} \tag{7.3}$$

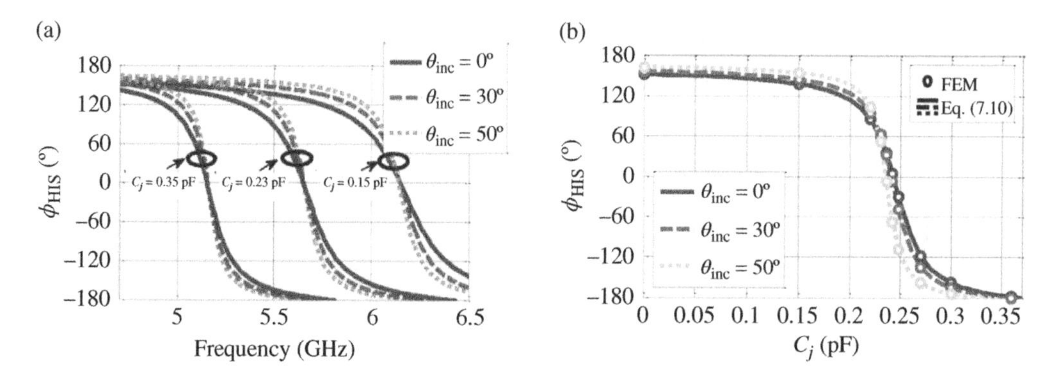

Figure 7.4 Phase of the PRS reflection coefficient for plane waves incident on it from different angles. (a) Variable C_j with frequency. (b) Frequency fixed at 5.6 GHz and as a function of C_j. *Source*: From [20] / with permission of IEEE.

By varying the physical resonance length of its patches, the HIS-scattering properties can be changed. This serves as a direct mechanism to control the TE_{01} leaky-mode cutoff frequency and, hence, leads to beam scanning at a fixed-frequency.

Figure 7.4a shows the phase of the reflection coefficient of the PRS arising from plane waves with different incidence angles scattering from it. The given results are obtained using the analytical pole-zero expression given in Eq. (7.3) for different incidence angles (θ_{inc}) as functions of the frequency. It is seen that the PMC condition ($\varphi_{HIS} = 0$) is attained with the tunable HIS at 5.15, 5.65, and 6.15 GHz when the value of C_j is 0.35, 0.25, and 0.15 pF, respectively.

At the fixed frequency of 5.6 GHz, the response of Y_{HIS} (k_y, C_j) can be modeled analytically as a function of the junction capacitor C_j. The reflection phase as a function of C_j for different θ_{inc} is shown in Figure 7.4b. It is observed that the HIS can be tuned to behave either as a grounded dielectric slab, a PMC sheet, or a PEC sheet by controlling the junction capacitance and the incidence angle. For example, when $C_j = 0$ pF and $\varphi_{HIS} = 150°$, the HIS acts as a dielectric slab. It acts as a PMC sheet with $C_j = 0.23$ pF and $\varphi_{HIS} = 0°$, and it behaves as a PEC sheet for $C_j = 0.35$ pF and $\varphi_{HIS} = 180°$. A similar response is obtained with a larger value of C_j and a larger value of L_{HIS}. This indicates a direct relationship between the effective resonant length of the HIS patches and the junction capacitance C_j. It is now clear that the effective resonant length of the HIS can be controlled by controlling C_j and, hence, the beam direction of the LWA at a fixed frequency.

7.1.2 Effect of C_j on the Leaky-Mode Dispersion Curves

Obtaining the leaky-mode dispersion curves as a function of C_j is essential for the electronic beam steering application. The unknown leaky-mode complex wave number can be obtained from the TEN model shown in Figure 7.2 by solving the corresponding transverse resonance equation (TRE) [28]

$$Y_{UP}(f, k_y, L_{PRS}) + Y_{DOWN}(f, k_y, C_j) = 0 \tag{7.4}$$

Figure 7.5 highlights the frequency dispersion property of the normalized phase and leakage constants as C_j is varied from 0.1 to 0.23 pF. As expected, the TE_{01} leaky-mode cutoff frequency decreases as C_j increases. This behavior results in a continuous rise of the leaky-mode phase constant at a fixed frequency as C_j is increased; it is illustrated in Figure 7.5c for 5.6 GHz. Again, very good agreement is obtained between the TEN and the full-wave FEM results, further validating the TEN model. As Figure 7.5c indicates, the normalized propagation constant along the length of the

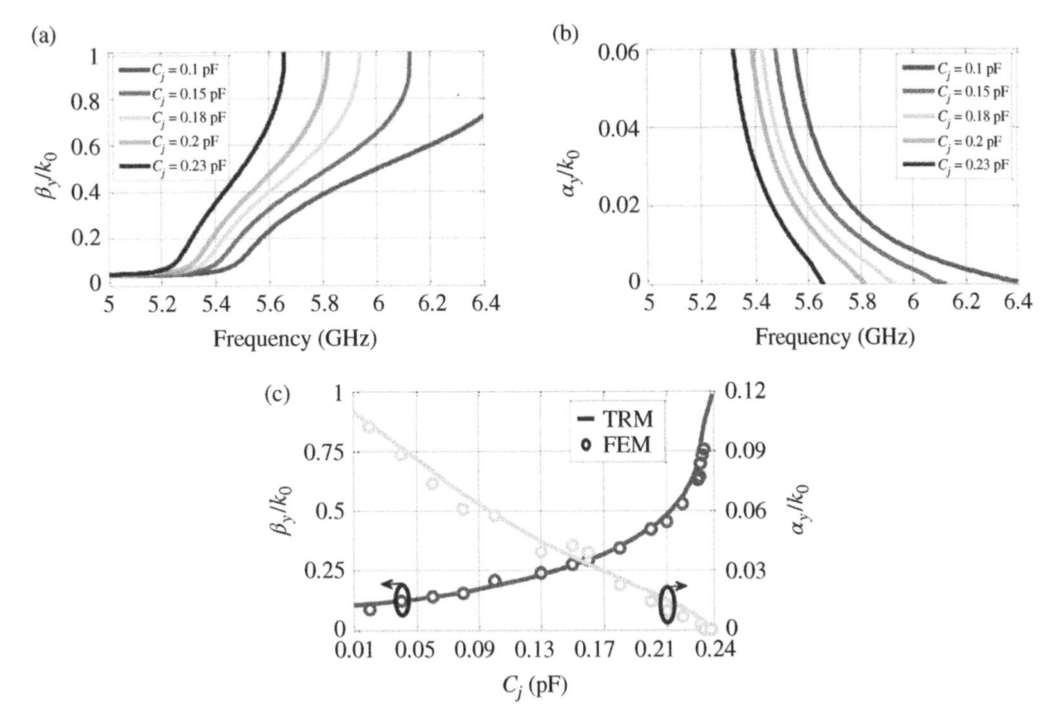

Figure 7.5 Leaky-mode dispersion curves (L_{PRS} = 22 mm). (a) and (b) Frequency dispersion as a function of C_j. (c) Wave number and attenuation constant behavior at 5.6 GHz as functions of C_j. *Source*: From [20] / with permission of IEEE.

array, β_y/k_0, is varied from values close to zero when C_j = 0.01 pF, to values close to one when C_j = 0.23 pF. Since the leaky-mode pointing angle is given by sin $\theta_{RAD} \approx \beta_y/k_0$, one finds that fixed-frequency beam scanning from broadside to endfire is realized by controlling C_j in the aforementioned range [0.01, 0.23] pF.

Figure 7.6 illustrates the TE_{01} leaky-mode electric field distribution in the cross section of the FP cavity at the design frequency of 5.6 GHz for different values of C_j. This parameter study gives more physical insights into the operating principles of the reconfigurable FP LWA. The HIS behaves as a grounded slab for low values of C_j as shown in Figure 7.4b. The leaky-mode shown in Figure 7.6 resonating in the metallic cavity of height H provides a maximum horizontal field at $Z = H/2$ for C_j = 0.01 pF. However, as C_j is increased to 0.23 pF, the HIS tends to behave as a PMC sheet which induces the maximum electric field. The leaky mode in the resonant FP cavity changes significantly as C_j is varied. In particular, its transverse wavelength λ_z is modified from $\lambda_z = 2H$ when C_j = 0.01 pF to $\lambda_z = 4H$ when C_j = 0.23 pF. This enlargement in the transverse wavelength λ_z implies an associated reduction in the longitudinal wavelength λ_y. Therefore, a rise in the leaky-mode longitudinal phase constant ($\beta_y = 2\pi/\lambda_y$) occurs as C_j is increased. This behavior is illustrated in Figure 7.5c.

Once the leaky-mode complex propagation constant has been obtained as a function of C_j, the associated H-plane pattern can be directly obtained. The simulated patterns for an LWA of length $L_A = 5\lambda_0$ at 5.6 GHz shown in Figure 7.7 for different values of C_j clearly demonstrate beam scanning. In particular, Figure 7.7 shows the results for C_j = 0.01 pF ($\beta_y/k_0 \approx 0.09$, $\theta_{RAD} = 5°$), C_j = 0.2 pF ($\beta_y/k_0 \approx 0.42$, $\theta_{RAD} = 25°$), and C_j = 0.23 pF ($\beta_y/k_0 \approx 0.65$, $\theta_{RAD} = 40°$). The pointing angle θ_{RAD} is swept from nearly broadside toward the endfire direction as C_j is increased, as

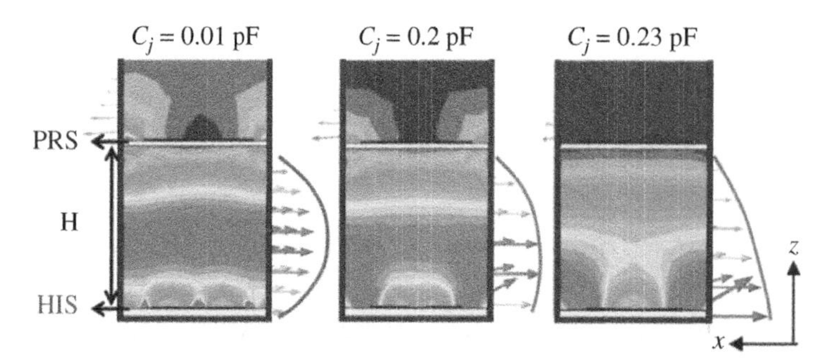

Figure 7.6 HFSS-simulated leaky-mode electric-field pattern inside the FP PRS-tunable HIS cavity at 5.6 GHz for different values of C_j. *Source*: From [20] / with permission of IEEE.

anticipated from the β_y/k_0 curve plotted in Figure 7.5c. The beam direction and the beamwidth predicted from the TRM model are in very good agreement with the HFSS simulations of the whole reconfigurable LP LWA. Note that a strong reflected lobe appears for $C_j = 0.23$ pF, due to the very low leakage-rate associated with high values of C_j (see Figure 7.5c). Thus, the radiation efficiency is poor and a large amount of energy is reflected at the termination end of the LWA. This fact limits the maximum scanning angle of this reconfigurable FP LWA to 40°.

7.1.3 Optimization of the FP Cavity Height

The physical height, H, of the FP cavity is a key design parameter which determines the scanning range and the sensitivity of the reconfigurable FP LWA. All of the results presented in the previous subsections were computed using the optimum value $H = H_{opt} = 25.2$ mm. Figure 7.8 shows that a wider scanning range is obtained in this case. This cavity height makes the leaky-mode cutoff condition ($\theta_{RAD} = 0°$) coincident with the minimum value of C_j. Consequently, the dynamic range of the varactors is fully utilized and the minimum scanning angle is close to the broadside direction. On the other hand, as discussed in [29], the maximum achievable pointing angle in this type of 1-D LWA is produced when the HIS realizes a PMC resonance, which in our design corresponds to the value $C_j = 0.23$ pF.

A sudden fall of the leakage rate and a rapid divergence of the phase constant occur above the HIS PMC resonance. This behavior is shown in Figure 7.5c. It limits the scan range to higher angles. Figure 7.8 shows that for $H = H_{opt}$, the scan angle can be swept from $\theta_{RAD} = 5°$ for $C_j = 0.01$

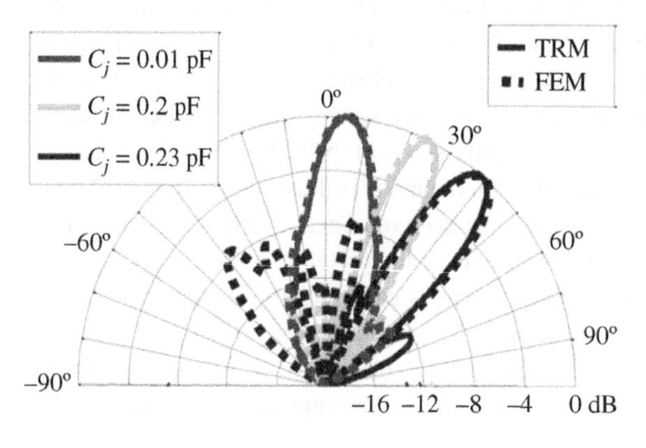

Figure 7.7 Simulated normalized H-plane patterns for the reconfigurable FP LWA at 5.6 GHz for different C_j (antenna length $L_A = 5\lambda_0$). *Source*: From [20] / with permission of IEEE.

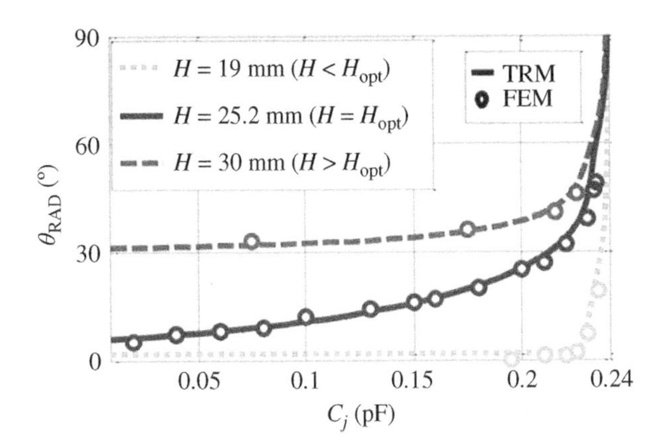

Figure 7.8 Leaky-mode pointing angle of the reconfigurable FP LWA as a function of C_j for different values of the cavity height, H, at 5.6 GHz. *Source*: From [20] / with permission of IEEE.

pF to $\theta_{RAD} = 40°$ for $C_j = 0.23$ pF, as was previously described. However, if $H > H_{opt}$, the minimum value of $C_j = 0$ does not correspond to the leaky-mode cutoff point ($\theta_{RAD} \approx 0°$). Rather, it corresponds to a larger pointing angle. As a result, the minimum scan angle is located further from broadside, e.g., $\theta_{RAD} = 30°$ at $C_j = 0.01$ pF in Figure 7.8 for $H = 30$ mm. Since the maximum pointing angle given by the HIS PMC resonance ($C_j \approx 0.23$ pF) is $\theta_{RAD} = 40°$, the scan range is thus reduced to (30°, 40°) instead of (5°, 40°). Moreover, it is observed that the sensitivity of θ_{RAD} with C_j is lower for this nonoptimal case and is due to the nonlinear response of the varactor.

7.1.4 Antenna Prototype and Measured Results

Figure 7.9a shows the antenna prototype, including separate images of its parallel plates and PRS, that was fabricated and tested to confirm its simulated performance characteristics at 5.6 GHz. The length $L_A = 5\lambda_0$. A horizontal coaxial probe was used to excite the TE_{01} leaky mode of the FP cavity. The phase-agile cell used in the HIS and a detailed photograph of the biasing network are shown in Figure 7.9b. MGV125-08 varactor diodes were used for the tunable HIS. Their junction capacitance varies between 0.055 and 0.6 pF with an applied DC reverse bias voltage between 20 and 2 V_{DC} and they had a tolerance of ±0.05 pF according to the datasheet provided by the manufacturer.

The measured reflection coefficients of the prototype are shown in Figure 7.10 from 5 to 6 GHz for different reverse bias voltages, V_R. The probe dimensions were optimized for best input impedance matching for $C_j = 0.15$ pF, which is the operating range midpoint. It is observed that as V_R decreases, i.e., C_j increases, the matched band shifts to lower frequencies. Shifting of the matched band to lower frequencies with C_j increasing yields a behavior similar in nature to the dispersion curve results presented in Figure 7.5. The simulated and measured S-parameters of the antenna as functions of V_R at 5.6 GHz are given in Figure 7.11. Both the reflection ($|S_{11}|$) and transmission ($|S_{21}|$) coefficients vary as V_R changes. Since the probe dimensions were optimized for $V_R = 7.17$ V_{DC}, which correspond to $C_j = 0.15$ pF, poorer impedance matching is observed at other operating points.

The experimental setup for the measurements of the performance characteristics of the reconfigurable 1-D FP LWA is shown in Figure 7.12. Figure 7.13 gives the measured normalized patterns at 5.6 GHz for different values of V_R. The main beam scans from 9.2° to 34.2° when V_R varies from 18.2 V_{DC} (corresponds to $C_j = 0.06$ pF) to 4.5 V_{DC} (corresponds to $C_j = 0.245$ pF). The measured gain, directivity, and total efficiency values are shown in Figure 7.14a together with their simulated

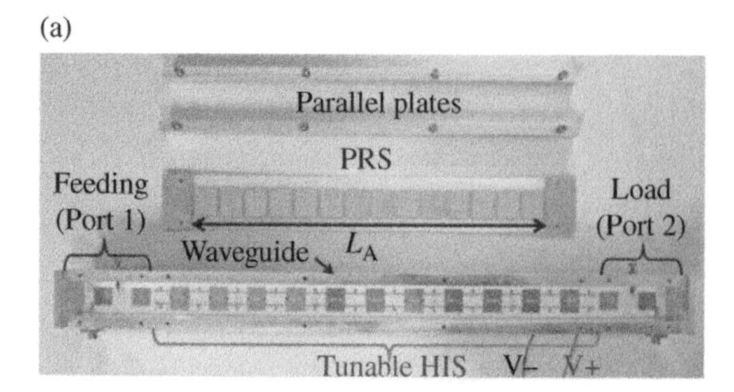

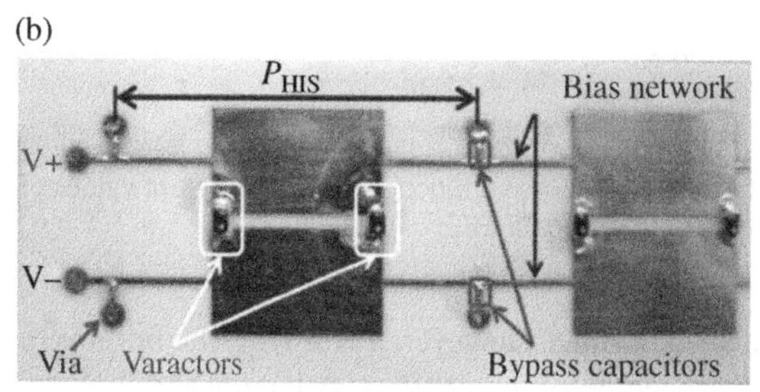

Figure 7.9 Reconfigurable FP LWA prototype. (a) Top views of the prototype and its parallel plates and PRS. (b) Top view of the tunable HIS cell and its bias network. *Source*: From [20] / with permission of IEEE.

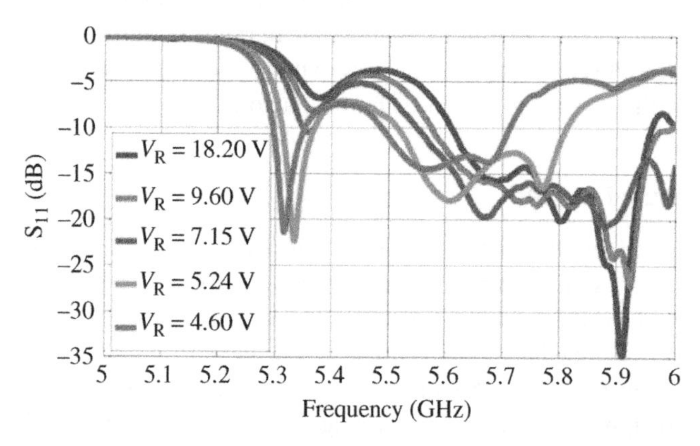

Figure 7.10 Measured reflection coefficient values as functions of the source frequency for different reverse bias voltages applied to the varactors. *Source*: From [20] / with permission of IEEE.

values. The maximum measured gain is 12.95 dBi when $V_R = 10.6\ V_{DC}$; this corresponds to $C_j = 0.1$ pF and $\theta_{RAD} = 12°$. A significant drop in the gain curve is observed when V_R is lower than the optimum point. For example, the measured gain for $V_R = 4.2\ V_{DC}$ is −3.55 dBi ($C_j = 0.26$ pF and $\theta_{RAD} = 35°$). This phenomenon prevents large scan angles. The abrupt degradation of the gain is due to a higher mismatch loss (see Figure 7.11) and a fall of the leakage rate (Figure 7.5b) associated with the behavior of the PMC resonance of the HIS at high θ_{RAD}. The gain is stable and greater than 10 dBi for beam angles below 25°, i.e., for $V_R > 6\ V_{DC}$.

The total antenna efficiency is comprised of the mismatch (η_{MIS}), leaky-mode radiation (η_{RAD}), and ohmic (η_Ω) efficiencies and is expressed as their product:

$$\eta_{TOT} = \eta_{MIS} \cdot \eta_{RAD} \cdot \eta_\Omega \tag{7.5}$$

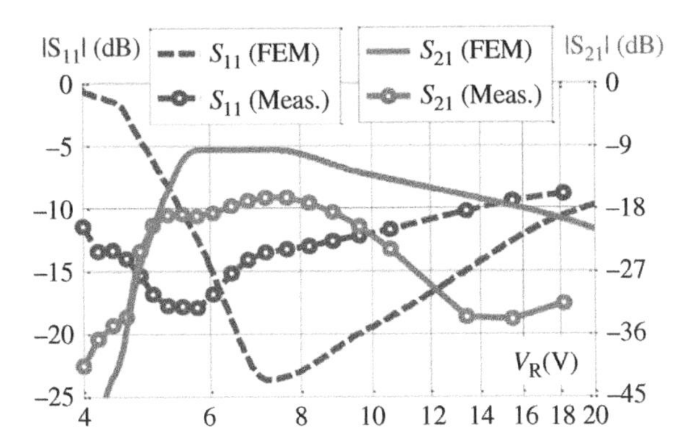

Figure 7.11 Simulated and measured *S*-parameters as functions of the reverse bias voltage V_R at 5.6 GHz. *Source*: From [20] / with permission of IEEE.

Figure 7.12 Photograph of the experimental setup of the pattern measurements of the reconfigurable FP LWA. *Source*: From [20] / with permission of IEEE.

Figure 7.13 Measured normalized patterns at 5.6 GHz for different reverse bias voltages applied to the varactor diode. *Source*: From [20] / with permission of IEEE.

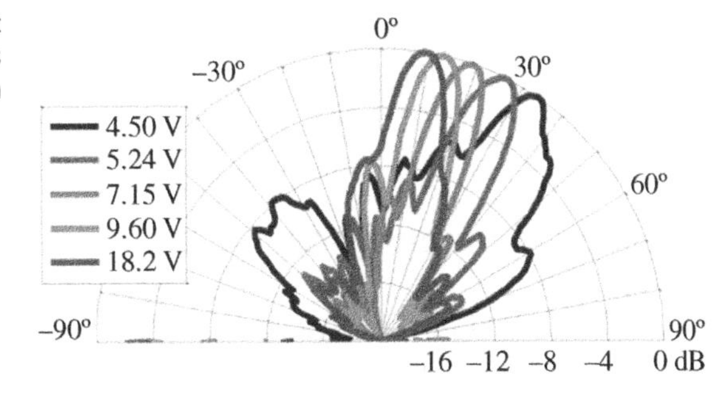

where η_{MIS} is computed from the measured reflection coefficient as: $\eta_{MIS} = 1 - |S_{11}|^2$, and η_{RAD} is estimated using the leakage rate: $\eta_{RAD} = 1 - e^{-2\alpha_y L_A}$. The ohmic efficiency η_Ω can be further decomposed into the dielectric losses (η_{DIE}) and the power dissipation in the varactor's series resistance, η_{VAR}. Therefore, Eq. (7.5) can be expressed as:

$$\eta_{TOT} = \eta_{MIS} \cdot \eta_{RAD} \cdot \eta_{DIE} \cdot \eta_{VAR} \tag{7.6}$$

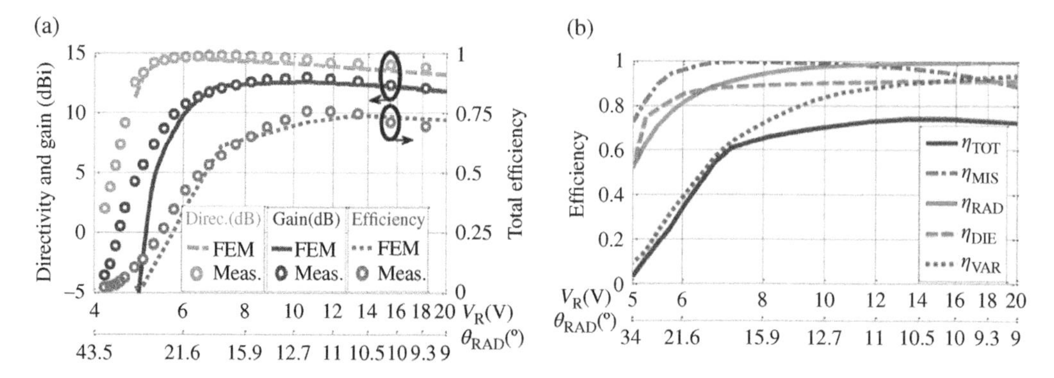

Figure 7.14 Measured and simulated performance characteristics of the reconfigurable FP LWA prototype at 5.6 GHz. (a) Directivity, gain, and total efficiency values. (b) Estimated efficiencies as functions of V_R and θ_{rad}. *Source*: From [20] / with permission of IEEE.

The total efficiency of the reconfigurable FP LWA is greater than 50% and reaches a maximum of 75% when V_R varies between 10 and 14 V_{DC}. For $V_R = 5\ V_{DC}$, i.e., when $\theta_{RAD} > 30°$, the total efficiency is around 3%. The reason behind the gain drop for large scan angles is explained by Eq. (7.6). The estimated component efficiencies are shown in Figure 7.14b as functions of V_R and θ_{RAD}. It is observed that the mismatch efficiency (η_{MIS}) is greater than 85% throughout the dynamic range of the varactor. On the other hand, η_{RAD} is 100% when $\theta_{RAD} = 9°$ ($V_R = 20\ V_{DC}$) and decreases to 52% when the beam points at the larger angle $\theta_{RAD} = 30°$ ($V_R = 5\ V_{DC}$). As the beam scans from 9° to 30°, the dielectric loss efficiency η_{DIE} falls from 90% to 50% because the electric field is concentrated in the substrate of the HIS in the PMC regime. Moreover, the density of the current flowing through the diodes increases, which leads to higher losses associated with the diode's series resistance. This behavior, in turn, significantly decreases η_{VAR} as θ_{RAD} increases, as depicted in Figure 7.14b. When the beam scans toward endfire, i.e., as θ_{RAD} increases, the HIS absorption increases and the $|\rho_{HIS}|$ drops abruptly. The value of $|\rho_{HIS}|$ is 0.3 for $\theta_{RAD} = 34°$. This outcome is related to a large change in the HIS's phase when the antenna operates very close to the PMC resonance. Because of the degradation of leaky-mode radiation efficiency and the increase of the varactor resistance losses, the gain drops at large scan angles. This feature, of course, restricts the beam scanning to angles far away from the endfire direction.

7.2 Period-Reconfigurable SIW-Based LWA

The beam radiated by the reconfigurable 1-D FP LWA can only scan in the forward direction because the propagation constant is positive. In order to provide beam scanning in both the forward and backward directions, the propagation constant needs to have both positive and negative values when the physical structure is reconfigured. A period-reconfigurable LWA provides such a solution.

An infinitely long periodic LWA with period P would have aperture fields that would be spatially periodic. The phase constant for the n-th harmonic can be expressed as:

$$\beta_n = \beta_0 + \frac{2n\pi}{P} \tag{7.7}$$

where β_0 and β_n are the phase constants of the basic harmonic mode and n-th order harmonic mode, respectively. Hence, the main beam angle for each space harmonic can be approximately calculated as:

$$\sin \theta_n = \frac{\beta_n}{k_0} \tag{7.8}$$

where θ_n is the beam direction of the n-th harmonic. One finds from Eq. (7.7) that if a slow wave harmonic, $n = -1$ for instance, was selected and the period P was varied, then the propagation constant would change from negative to positive when P is increased. Thus, the beam direction θ_n would change from the backward to the forward direction. Furthermore, if one could suppress all the modes except the desired one, a single beam LWA capable of backward to forward scanning would be attained.

7.2.1 Antenna Configuration and Element Design

A period-reconfigurable LWA based on a substrate integrated waveguide (SIW) configuration is depicted in Figure 7.15a [25]. The substrates 1 and 2 are Rogers RT/Duroid™ 6006 ($\varepsilon_r = 6.15$) with thicknesses of 2.54 and 1.27 mm, respectively. The traveling-wave feed structure is based on an SIW that consists of metal layers 1 and 2, and substrate 1 as depicted in Figure 7.15b. The thickness of the copper (35 μm) is also considered in the simulations of this model. Two arrays of patch elements are

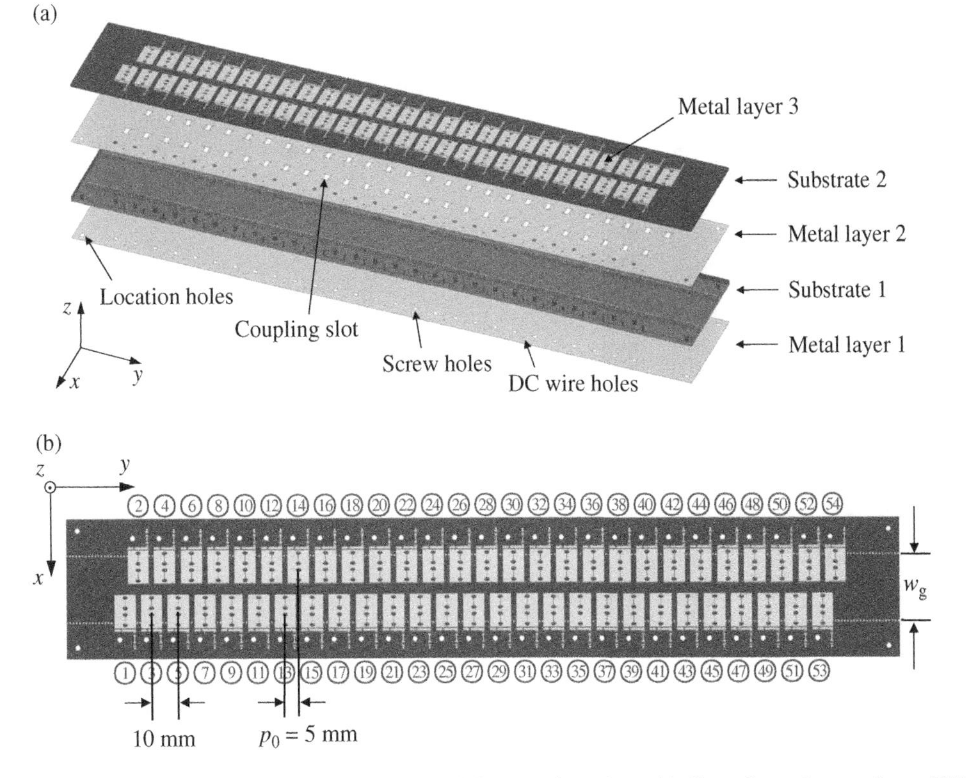

Figure 7.15 Period-reconfigurable LWA. (a) Perspective view. (b) Top view. *Source*: From [25] / with permission of IEEE.

located on metal layer 3; they are the radiating elements and are excited by the fields leaking out from the SIW through the coupling slots on metal layer 2. There are 54 patch elements in total. The period for one row of elements is 10 mm. By introducing an offset of 5 mm for the other row, an equivalent period $p_0 = 5$ mm in the y direction is achieved. This length is half of the original period, 10 mm. It is a very favorable choice because the smaller the p_0 is, the more flexibility one has to achieve a different period P by reconfiguring the unit cell length, thus providing greater beam steering flexibility.

The structure of each patch element is shown in Figure 7.16a. There are two symmetrical dumbbell-shaped slots etched on the patch, which is excited by an H-shaped coupling slot etched on the top surface of the SIW (metal layer 2). This particular structure has the advantage of being smaller in size when compared to conventional patches. Its design follows from the concepts associated with the electrically small metamaterial-inspired antennas reported in [30, 31]. The patch efficiently realizes the miniaturization through the presence of the different shaped slots etched on it [32]. An equivalent capacitance is introduced by the narrow slot of the dumbbell, and an equivalent inductance is produced by the metal portion beside the ellipses of the dumbbell-shaped slot. A parasitic strip is located beside the patch. The shorted post at one end of the strip increases the equivalent resonant length of the strip by generating a strong coupling to the patch and thus lowers its resonance frequency. The second role of the shorted post is to act as a common electrode connected in parallel for the biasing of the PIN diodes (BAR50-02L) on the parasitic strip, i.e., it also works as a binary switch.

The equivalent circuit model shown in Figure 7.16b was developed, and its resonance frequency was readily obtained. When the diode is forward-biased (using a resistance of 4 Ω in the

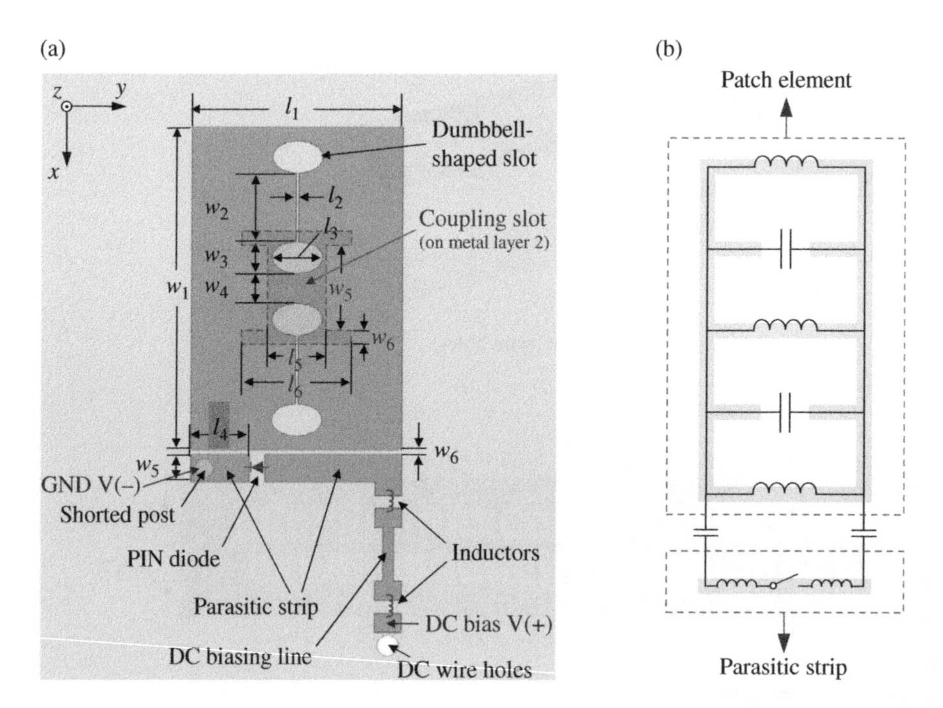

Figure 7.16 Unit cell of the period-reconfigurable LWA. (a) Configuration. (b) Its equivalent circuit model. The design parameters in millimeters are: $w_1 = 12.0$, $w_2 = 2.55$, $w_3 = 1.2$, $w_4 = 1.1$, $w_5 = 3.0$, $w_6 = 0.2$, $l_1 = 8.0$, $l_2 = 0.1$, $l_3 = 1.92$, $l_4 = 2.2$, $l_5 = 2.1$, $l_6 = 4.0$. *Source*: From [25] / with permission of IEEE.

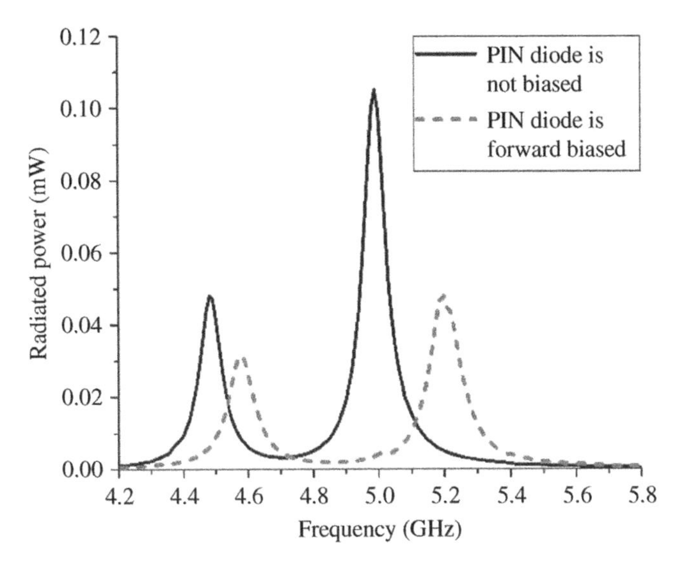

Figure 7.17 Radiation efficiency of one unit cell when the diode is forward biased or is not biased. *Source*: From [25] / with permission of IEEE.

simulation), the switch in Figure 7.16b is on. A strong coupling occurs between the patch and the parasitic strip to yield a shift of the patch's operating frequency. On the other hand, if the diode is not biased, it is represented by a parallel combination of a capacitance of 0.08 pF and a resistance of 3.5 kΩ. When the switch in Figure 7.16b is off, the currents induced on the parasitic strip are very weak; they have little effect on the patch's operating frequency. The DC biasing line is laid in the x direction and is broken into sections with two inductors (0402HP-8N7X) in order to choke the RF currents on it and to reduce the cross-polarization.

The power radiated from one element excited with an input power of 1 mW is depicted in Figure 7.17 as a function of the source frequency with the PIN diode ON and OFF. If the PIN diode is OFF, i.e., not biased, the operating frequency of the element is 5 GHz and the peak value of the radiated power is obtained (solid black curve). If the PIN diode is forward biased (dashed red curve), the operating frequency shifts to 5.2 GHz and the power radiated at 5 GHz is very low. Consequently, it can be regarded approximately as a non-radiative state at 5 GHz. Therefore, the element can be switched between its binary states, i.e., the activated state "1" (the diode is not biased) and the non-activated state "0" (the diode is forward biased), respectively. The advantage of this element design is that when it is activated, i.e., the diode is in its OFF state. Therefore, the ohmic loss at 5 GHz is very low because the current through the diode is very small. On the other hand, when the element is not activated, i.e., the diode is forward biased, the loss caused by the diode is large. However, this shifts the peak frequency to 5.2 GHz where the peak value is much lower. As a consequence, the total loss at the operating frequency, 5 GHz, remains small and, in fact, can be limited to a negligible level.

Figure 7.17 also indicates that there is another peak at 4.48 GHz when the diode is not biased and at 4.58 GHz when it is forward biased, respectively. These peaks are caused by another resonant mode of the patch. Figure 7.18 shows the current distribution on the patch at 5.0 and 4.48 GHz, respectively, when the diode is not biased. Figure 7.18a shows the main mode is resonant in the y direction at 5 GHz. Figure 7.18b shows that the second resonance is in the x direction. Its resonance frequency is mainly determined by the length w_1.

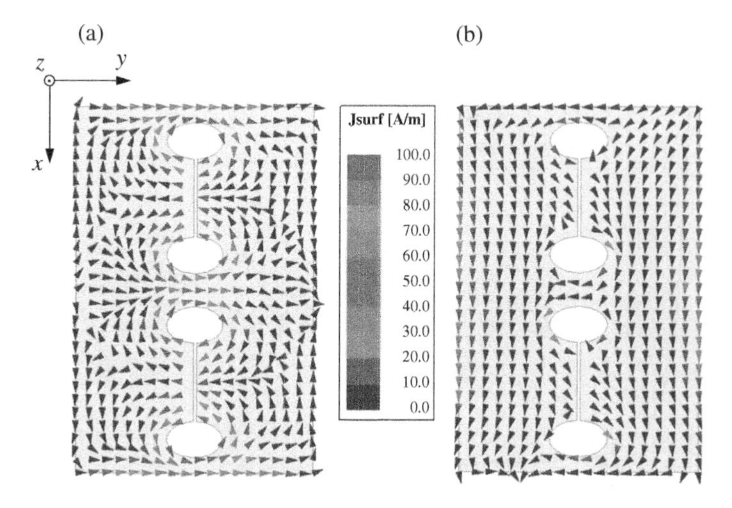

Figure 7.18 Current distributions on the patch when the diode is not biased. (a) 5 GHz. (b) 4.48 GHz. *Source:* From [25] / with permission of IEEE.

7.2.2 Suppression of Higher-Order Harmonics

The transmission behavior of the SIW in Figure 7.15 is similar to that of a rectangular waveguide with an equivalent width of 23.6 mm. The TE_{10} mode is the only transmission mode based on the dimensions of the SIW. It has the phase constant $\beta_0 = 223.2$ rad/m at 5.0 GHz. With the two constraints

$$\begin{cases} -k_0 < \beta_{-1} < k_0 \\ \beta_{-2} < -k_0 \end{cases}, \tag{7.9}$$

where $\beta_n = \beta_0 + n(2\pi/P)$ is the propagation constant of the n-th mode, together with Eq. (7.7), yield the condition on the period P to obtain mono-harmonic radiation:

$$\frac{2\pi}{\beta_0 + k_0} < P < \min\left\{\frac{2\pi}{\beta_0 - k_0}, \frac{4\pi}{\beta_0 + k_0}\right\} \tag{7.10}$$

Substituting the values of β_0 and k_0 at 5.0 GHz into Eq. (7.10), the mono-harmonic condition becomes 19.2 mm $< P <$ 38.3 mm.

Figure 7.19 shows the relations between the scan angle θ_n and P obtained with Eq. (7.8) for the four harmonics: $n = -1, -2, -3, -4$. The gray box indicates the mono-harmonic region. The corresponding scanning range is from −90 to 34°. Once P is greater than 38.3 mm, i.e. $\theta_{-1} >$ 34°, the $n = -2$ harmonic will appear and the mono-harmonic condition is violated. Thus, the maximum scan angle of this mono-harmonic system cannot exceed 34°.

The normalized beam steering patterns within the mono-harmonic region are depicted in Figure 7.20. The period P must be an integer multiple of the element period p_0. The values of P were selected to be $4p_0$ (20 mm), $5p_0$ (25 mm), $6p_0$ (30 mm), $7p_0$ (35 mm), $8p_0$ (40 mm), and $9p_0$ (45 mm), respectively, to illustrate the beam scanning. It is noticeable that the beams ⑤ and ⑥ have obvious $n = -2$ harmonics included with those peaks appearing at −60° and −32°, respectively, since $P = 8p_0, 9p_0$ falls outside of the mono-harmonic region. Therefore, only the four beams ① to ④ can be generated before the $n = -1$ harmonic region is exceeded and the $n = -2$ harmonic occurs. Thus, scan angle range is limited to the range: −60° to −25°. In order to overcome the

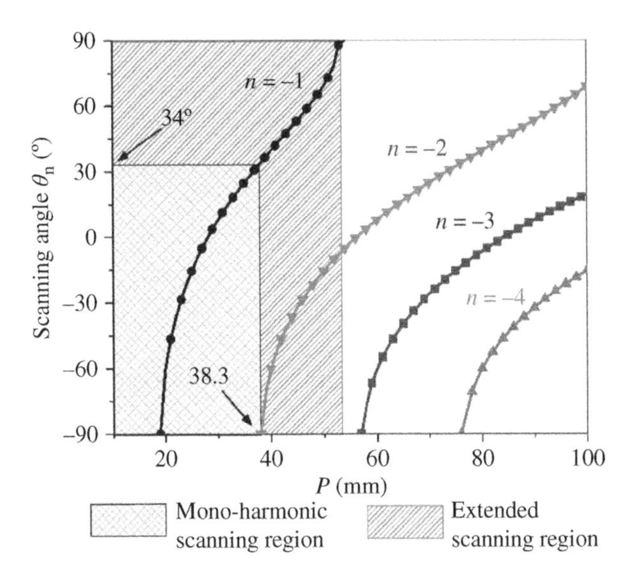

Figure 7.19 Variation of the scan angle θ_n as a function of the period P for four harmonics. *Source*: From [25] / with permission of IEEE.

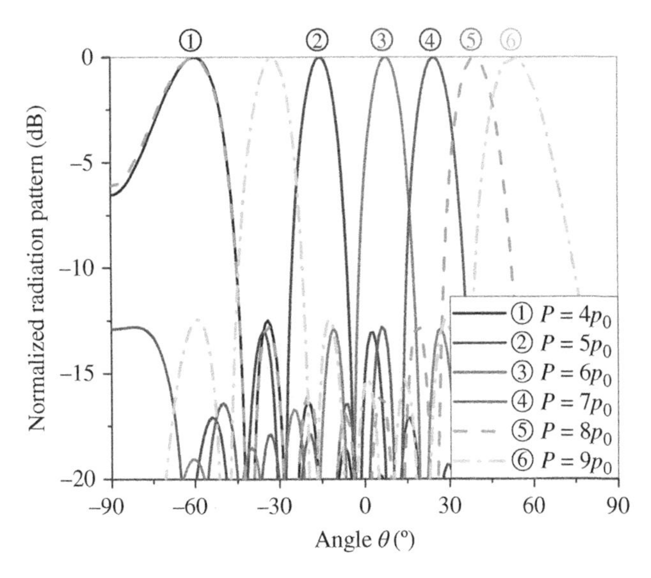

Figure 7.20 Normalized patterns of the period-reconfigurable LWA when its period P is changed. *Source*: From [25] / with permission of IEEE.

limitation of mono-harmonic radiation, a method for suppressing the higher-order space harmonics has been developed. The aim being to extend the mono-harmonic scanning range to the red region in Figure 7.19. Note that all of the patterns under discussion here in Section 7.2 are based on the assumption of each element of the array being an isotropic point source with omnidirectional pattern. If the radiation pattern of each element is taken into account, the scanning range will be reduced.

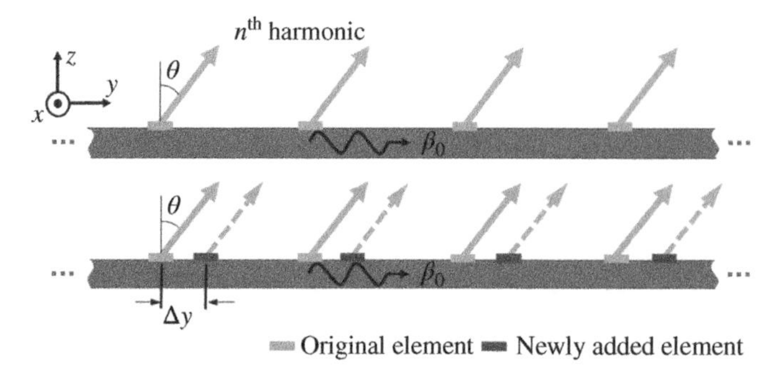

Figure 7.21 Illustration of introducing a second array shifted from the original to suppress the higher-order space harmonics. *Source*: From [25] / with permission of IEEE.

The principle of the method developed to extend the scan angle is illustrated in Figure 7.21. A new array of elements (green boxes with dashed arrows). The location of each of its elements is shifted by Δy from the original array elements (in red). Assume that the n-th harmonic is to be eliminated. The phase relationship between the radiated fields of the original array and the newly added array is obtained in the beam direction of the n-th harmonic as:

$$k_0 \Delta y \sin\theta_n - \beta_0 \Delta y = \pm (2m+1)\pi, \quad m = 0, 1, 2, 3, \ldots \tag{7.11}$$

where the phase difference is an odd multiple of π. Substituting Eqs. (7.7) and (7.8) into Eq. (7.11), the shift distance becomes

$$\Delta y = \pm \frac{(2m+1)P}{2n_{\text{sup}}}, \qquad n_{\text{sup}} = \pm 1, \pm 2, \pm 3, \ldots \tag{7.12}$$

where n_{sup} is the order of the harmonic that is to be suppressed. The values of Δy for several negative harmonics are summarized in Table 7.1. Consequently, the far-field radiation of the n_{sup} harmonic can be eliminated by introducing the indicated location shift of the additional element from the original array.

For example, if $n_{\text{sup}} = -1$ mode is to be suppressed, then the new array of elements should be introduced with a location shift of $\Delta y = \mp \frac{P}{2}$. Similarly, the suppression of the $n_{\text{sup}} = -2$ harmonic is realized by choosing $\Delta y = \mp \frac{P}{4}$ or $\mp \frac{3P}{4}$. This choice is very useful for common periodic LWAs since it will extend their scanning ranges beyond the mono-harmonic region. Furthermore, if

Table 7.1 Location shift of the newly added array.

n_{sup}	Δy
−1	$\mp\frac{P}{2}$
−2	$\mp\frac{P}{4}, \mp\frac{3P}{4}$
−3	$\mp\frac{P}{6}, \mp\frac{P}{2}, \mp\frac{5P}{6}$
−4	$\mp\frac{P}{8}, \mp\frac{3P}{8}, \mp\frac{5P}{8}, \mp\frac{7P}{8},$
−5	$\mp\frac{P}{10}, \mp\frac{3P}{10}, \mp\frac{P}{2}, \mp\frac{7P}{10}, \mp\frac{9P}{10}$

Source: From [25] / with permission of IEEE.

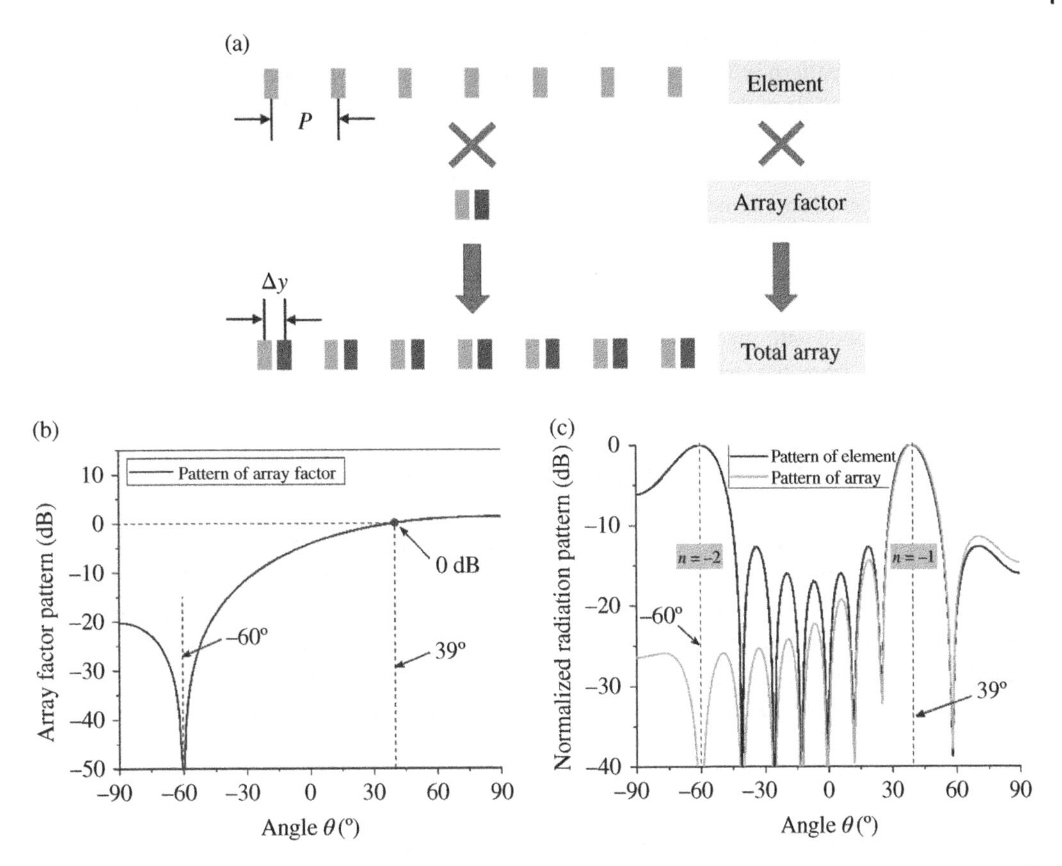

Figure 7.22 Array factor (AF) explanation on the harmonic suppression method. The suppression of the $n = -2$ harmonic is illustrated. (a) Sketch of the "sub-array" as a large element principle. (b) Pattern of the array factor when $P = 8p_0 = 40$ mm. (c) Element and total array patterns. *Source*: From [25] / with permission of IEEE.

the $n = -2$ and -3 harmonics exist simultaneously, the suppression of the $n_{\text{sup}} = -3$ harmonic will help add more mono-harmonic beams to those generated by the $n = -2$ harmonic.

It is also noted that all of the odd-order harmonics can be suppressed using $\Delta y = \mp \frac{P}{2}$. Theoretically, full-space continuous scanning from −90° to 90° can be achieved using this method if isotropic sources were employed and an arbitrary period P was realizable. However, these two assumptions are difficult to realize in practice. Therefore, beam scanning can be obtained only in a limited range, and the beams can only occur at discrete angles with period-reconfigurable LWAs.

The harmonic suppression method can also be explained from the perspective of an array factor (AF). This point of view is illustrated in Figure 7.22. Taking the $P = 8p_0$ case as an example, seven radiating elements are selected among the 54 elements of the LWA and are depicted in Figure 7.15. This set is approximately equivalent to the seven-element array shown in Figure 7.22a (red blocks) with element spacing $P = 8p_0$ and a series-fed phase constant $\beta_0 = 223.2$ rad/m. A new array (additional green boxes) is then introduced with its location shifted by Δy from the original array. If these two arrays are then regarded as two large "elements," a new 1×2 array has been achieved with a spacing $\Delta y = \frac{P}{4}$ between them. Therefore, the radiation pattern of the total 14 element array is obtained as the product of the large "element" pattern and the AF.

Figure 7.22b shows the AF pattern, while Figure 7.22c depicts the patterns of the "element" and total array. The AF pattern in Figure 7.22b has a distinct null in the direction $\theta = -60°$ that corresponds to the beam angle of the $n = -2$ harmonic shown in Figure 7.22c. Therefore, the product of the element and AF patterns will result in a null in the total array pattern in the $\theta = -60°$ direction. This result (red curve) is confirmed in Figure 7.22c. Also note that in Figure 7.22c, the beam generated at $\theta = 39°$ from the $n = -1$ harmonic is not influenced by the AF because, as indicated in Figure 7.22b, it is 0 dB at $\theta = 39°$. Thus, having suppressed the $n = -2$ harmonic beam, the $n = -1$ harmonic has become the only radiating mode.

To illustrate the efficacy of the suppression technique further, the removal of the $n = -1$ harmonic is explained with the AF approach in Figure 7.23 A new array is introduced with a location shift $\Delta y = \frac{P}{2}$ from the original array. The composite array has the AF pattern shown in Figure 7.23a. A null is generated in the direction $\theta = 39°$, which corresponds to the beam angle of the $n = -1$ harmonic shown in Figure 7.23b. Thus, the $n = -1$ harmonic has been suppressed as is also illustrated in Figure 7.23b. On the other hand, the beam from the $n = -2$ harmonic at $\theta = 60°$ remains and is increased as a result of the AF shown in Figure 7.23a. Therefore, the $n = -2$ harmonic becomes the only radiating mode, and mono-harmonic radiation is again realized.

In summary, the introduction of an additional array has been demonstrated to be an efficient method to overcome the limitation of mono-harmonic radiation. The mono-harmonic beam scanning region can be expanded with this approach by suppressing higher-order harmonics and, as a consequence, can significantly increase the beam scanning range of periodic LWAs. While it has been illustrated here with an ideal model, this method can be employed to expand the scanning range of practical periodic LWAs. Moreover, it also is suitable for the suppression of grating lobes in other types of antenna arrays.

7.2.3 Element Activation States and Scanning Properties

In order to suppress a given harmonic in practice, the radiating elements of the structure must be configured to have well-defined states and the means to activate and deactivate them must be present. Two examples based on a $P = 8p_0$ and a $P = 15p_0$ array are given to illustrate the activation principles. Assume that "1" and "0" indicate whether the element is activated or not.

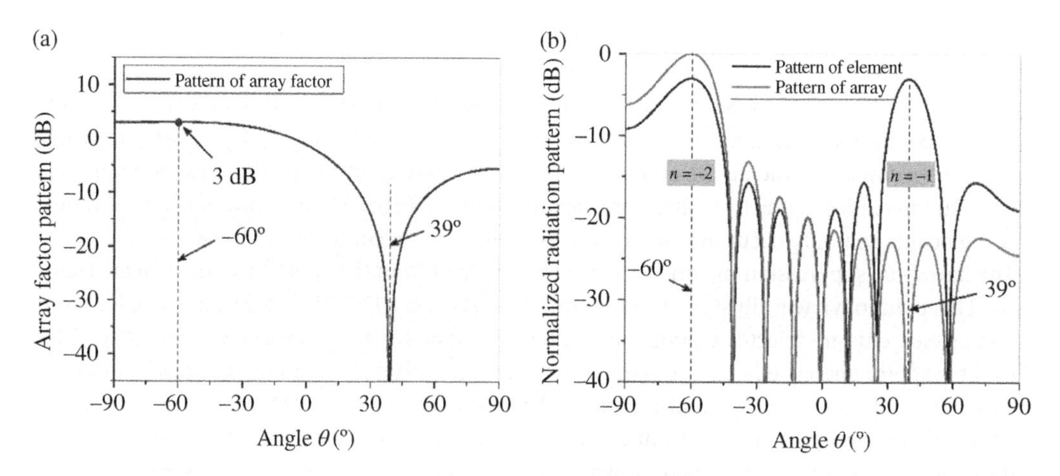

Figure 7.23 The harmonic suppression method is used to remove the $n = -1$ harmonic from the period-reconfigurable LWA with $P = 8p_0 = 40$ mm and $\Delta y = P/2$. (a) AF pattern. (b) Element and total array patterns. *Source*: From [25] / with permission of IEEE.

Example 7.1 *Suppressing the $n = -2$ harmonic when $P = 8p_0$*

Assume that there are two states of the elements with the same period $P = 8p_0$:

State 1: 1 0 *1* 0 0 0 0 0 1 0 *1* 0 0 0 0 0...
State 2: 1 1 *1 1* 0 0 0 0 1 1 *1 1* 0 0 0 0...

This means that one unit is activated in each period of eight units in State 1 and two in State 2. The newly activated elements (italic "1") for both states occur at the distance $\Delta y = \frac{P}{4} = 2p_0$ from the original array. Therefore, both states can generate the same normalized radiation pattern with the $n = -2$ harmonic being suppressed. However, their radiation strengths are different because the two states have a different total number of activated elements.

With the $n = -2$ harmonic suppressed, more beams can be achieved in both states and the mono-harmonic radiation region is extended as was explained in connection to Figure 7.19.

Example 7.2 *Suppressing $n = -3$ harmonic when $P = 15p_0$*

While the scanning range has been extended in Example 7.1, there is still more potential to achieve yet more beams to cover the full space if the mono-harmonic radiation is realized with the $n = -2$ harmonic. The LWA with $P = 15p_0$ is taken as an example to illustrate how the $n = -3$ harmonic can be suppressed to realize a mono-harmonic radiation region based on the $n = -2$ harmonic. The original state for $P = 15p_0$ is selected to be:

100000000000000100000000000000...

The normalized radiation pattern in this state is shown in Figure 7.24 (black curve). It has three obvious beams pointing at $\theta = 32, -16, -90°$. They are generated by the $n = -2, -3, -4$ harmonics, respectively. Although the $n = -4$ harmonic was a non-radiative mode in Figure 7.19, it generates a strong backward lobe as a sidelobe at $\theta = -90°$ for this larger $P = 15p_0$-based array.

In order to realize the mono-harmonic radiation range based on the $n = -2$ harmonic, the $n = -3$ harmonic needs to be suppressed. Moreover, the performance would be improved if the lobe generated by $n = -4$ harmonic can be reduced at the same time. Consequently, the following four states are proposed for this purpose.

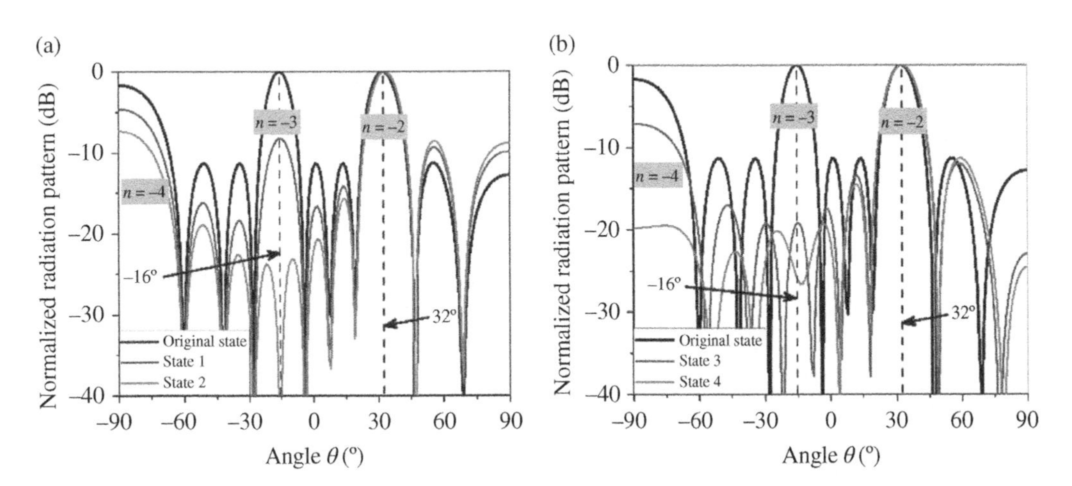

Figure 7.24 Normalized patterns of the period-reconfigurable when the $n = -3$ harmonic is suppressed with $P = 15p_0$. (a) $\Delta y = P/6$ is employed in States 1 and 2. (b) $\Delta y = P/2$ is employed in States 3 and 4. *Source*: From [25] / with permission of IEEE.

State 1: 10*11*00000000000 10*11*00000000000...
State 2: 11*111*0000000000 11*111*0000000000...
State 3: 11100000*11*00000 11100000*11*00000...
State 4: 11110000*111*0000 11110000*111*0000...

A shift $\Delta y = P/6 = 2.5p_0$ is required in States 1 and 2 to suppress the $n = -3$ harmonic. However, "$2.5p_0$" is not an integer multiple of the element spacing p_0. Consequently, two new elements have to be activated in State 1 within one period of 15 elements (as indicated) to produce an equivalent "center" with a distance of $2.5p_0$ from the original activated element. Similarly, in State 2, there are three newly activated elements to generate an equivalent "center" with a distance of $2.5p_0$ from the "center" of the first two elements.

The normalized patterns generated by States 1 and 2 are illustrated in Figure 7.24a. It is observed that the beam generated by the $n = -3$ harmonic at $\theta = -16°$ is suppressed in both of these two states. Nevertheless, their radiation fields cannot be completely canceled by each other in the beam direction of the $n = -3$ harmonic because the number of newly activated elements is different from that of the original elements. Note that the reduction of the $n = -3$ harmonic is only -8.2 dB in State 1 and is much better in State 2. This improvement occurs because the number of newly activated elements in State 2 is relatively closer to that of the original number, i.e., the number ratio is 3/2 rather than 2/1 as it is State 1.

The location shift $\Delta y = P/2 = 7.5p_0$ of the newly activated elements in States 3 and 4 is employed to suppress the $n = -3$ harmonic. As with States 1 and 2, only the equivalent "centers" attain this new spacing arrangement. The normalized patterns generated by State 3 and 4 are illustrated in Figure 7.24b. Comparing the pattern for State 3 to the one of State 2 in Figure 7.24a, one finds similar effects on the suppression of the beam at $\theta = -16°$. Therefore, it is verified that both $\Delta y = P/6$ and $\Delta y = P/2$ can be used to suppress the $n = -3$ harmonic emissions.

Note that not only the $n = -3$ harmonic but also the $n = -4$ harmonic is dramatically suppressed in State 4. This effect is explained by dividing its first four elements into two groups (elements 1 and 2 and elements 3 and 4) whose "centers" have a spacing of $2p_0$. This shift is close to the requirement of the suppression on the $n = -4$ harmonic, i.e., $\Delta y = P/8 = 1.875p_0$. However, because of this difference between the practical Δy ($2p_0$) and the ideal Δy ($1.875p_0$), the lobe of the $n = -4$ harmonic cannot be completely eliminated. Finally, State 4 is selected for the suppression of $n = -3$ harmonic when $P = 15p_0$.

These examples yield the final activation scheme given in Table 7.2. For each period P, more mono-harmonic beams can be achieved by suppressing specific higher-order space harmonics. A total of 15 beams were selected to cover the scanning range. Each beam is marked with a sequence number to make it easily distinguished from the other ones. Figure 7.25 shows the normalized patterns of the mono-harmonic steered beams. The scan angles for all of these beams are summarized in Figure 7.26. It shows that scanning can occur in the range from $-60°$ to $78°$, which is much wider than the case shown in Figure 7.22. Nevertheless, note again that this scan range would be reduced if the actual element patterns were taken into account.

Finally, it can be concluded that the presented method for suppressing higher-order harmonics can significantly increase the beam scanning range for periodic LWAs. Note that, in principle, more beams could be generated if other non-integer periods could be employed. Moreover, if smaller elements with a smaller period p_0 could be generated, they would bring more flexibility to achieve more beams matched to the desired directions. While this method was verified only with the ideal model of an array of point (isotropic) radiators, it has been successfully employed in a practical periodic LWA. It is further noted that this suppression method is also suitable for eliminating grating lobes in array patterns.

Table 7.2 Activation states of elements for different beams.

Beam sequence number	Period length	Activation states
①	$P = 4p_0$	11001100110011001100110011001100110011001100110011
②	$P = 4.25p_0$	10001000110011000100010001100110001000100011001100 0100
③	$P = 9p_0\,(n_{sup} = -1)$	10001100010001100010001100010001100010001100010001 1000
④	$P = 4.75p_0$	10001100011000100001000110001100010000100011000110 0010
⑤	$P = 5p_0$	11000110001100011000110001100011000110001100011000 1100
⑥	$P = 11p_0$ ($n_{sup} = -3$)	11100011000111000110001110001100011100011000111000 1100
⑦	$P = 6p_0$	11000011000011000011000011000011000011000011000011 0000
⑧	$P = 13p_0$ ($n_{sup} = -3$)	11000011100001100001110000110000111000011000011100 0011
⑨	$P = 7p_0$	11100000111000001110000011100000111000001110000011 100
⑩	$P = 15p_0$ ($n_{sup} = -3$)	00111100000111000001111000001110000011110000011100 001111000
⑪	$P = 8p_0\,(n_{sup} = -2)$	01111000011110000111100001111000011110000111100001 1110
	$P = 9p_0\,(n_{sup} = -2)$	00111100000011110000001111000000111100000011110000 001111000
⑬	$P = 9.5p_0$ ($n_{sup} = -2$)	11110000000111000000011110000000111000000011110000 000111000
⑭	$P = 10p_0$ ($n_{sup} = -2$)	11010000000110100000001101000000011010000000110100 00001101
⑮	$P = 10.5p_0$ ($n_{sup} = -2$)	01111100000001111100000001111100000001111100000001 1111000000

Source: From [25] / with permission of IEEE.
Note: (1) "1" represents an element that is activated (the diode is not biased).
"0" represents an element that is not activated (the diode is forward-biased).
(2) The sequence of activation states coincides with the sequence number of the elements.

7.2.4 Results and Discussion

The radiation performance of the actual elements developed for a prototype of the period-reconfigurable LWA is analyzed and the experimental results obtained with the entire prototype are described in detail. The conductor and dielectric losses have been considered in the simulations to be presented. An FPGA module was developed to control the operating states of the array. The measured results verify its design principles and simulated performance characteristics.

7.2.4.1 Element Pattern and Antenna Prototype

Returning to the period-reconfigurable LWA unit cell shown in Figure 7.16, the electric field from the guiding structure is concentrated mainly across the straight section of the dumbbell-shaped slot with length w_2 due to the capacitive effects associated with it. Therefore, the field radiated by the patch is more alike to that radiated by the equivalent magnetic currents in the slot. The normalized pattern of the unbiased unit cell at 5 GHz is shown in Figure 7.27 and is compared with the patterns radiated by an isotropic point source (omni-directional pattern, black semi-circle curve) and an infinitesimal magnetic current element source (blue closed curve) near to the ground plane. The

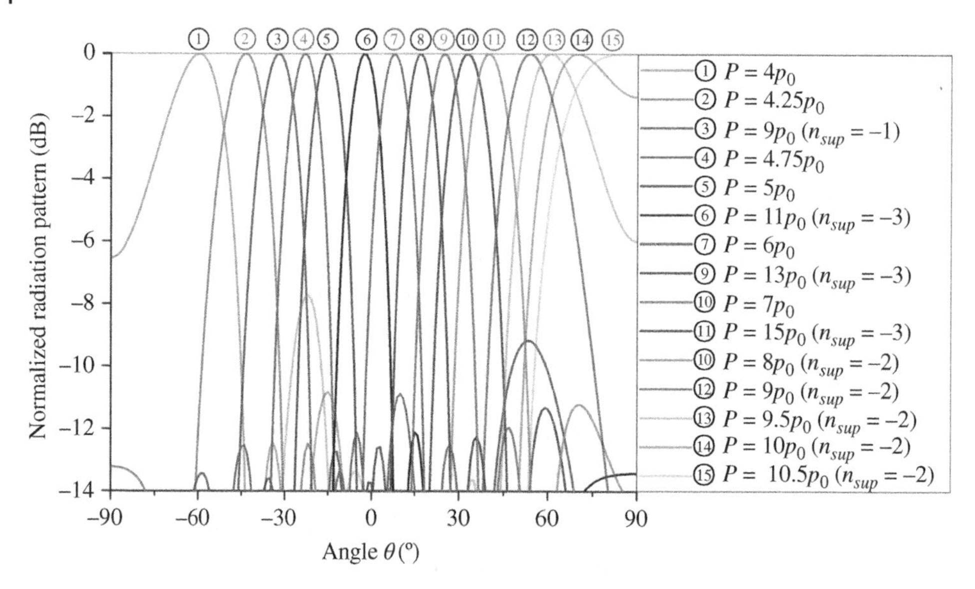

Figure 7.25 Beam scanning normalized radiation patterns of the isotropic-point-source period-reconfigurable array when the harmonic suppression method is employed (n_{sup} is the order of the harmonic which is suppressed). *Source*: From [25] / with permission of IEEE.

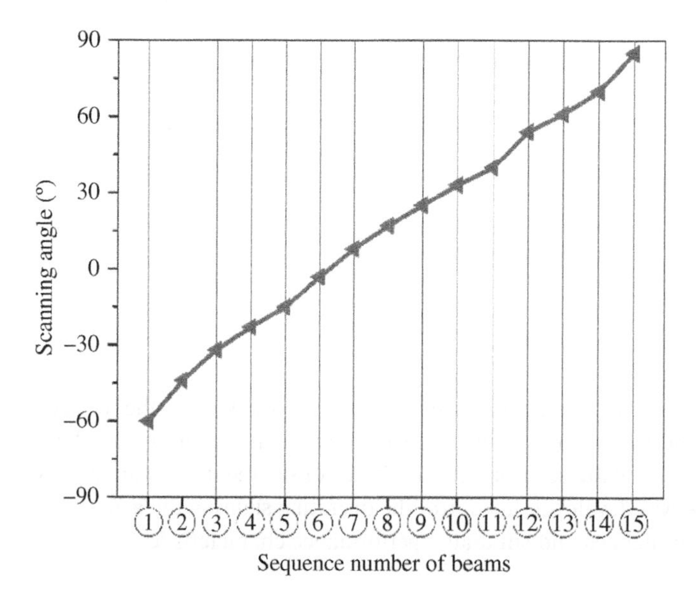

Figure 7.26 Beam scanning angle for each beam generated by the period-reconfigurable array. *Source*: From [25] / with permission of IEEE.

pattern of the patch is actually nearer to the isotropic source pattern; it has a very wide 3 dB beamwidth, 148°. This feature is important to achieving the desired wide-angle beam scanning.

The optimized period-reconfigurable beam scanning LWA introduced in Figure 7.15 was fabricated with PCB processing. The photos of this prototype are presented in Figure 7.28. Substrates 1 and 2 are shown in Figure 7.28a, together with the FPGA control platform. Figure 7.28b shows the

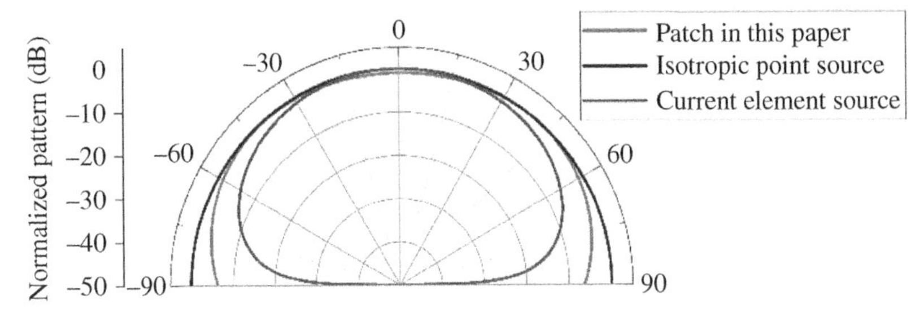

Figure 7.27 Comparison of the normalized patterns of the electric field radiated by the patch of the unit cell, an isotropic-point source, and an elemental magnetic current element source. *Source*: From [25] / with permission of IEEE.

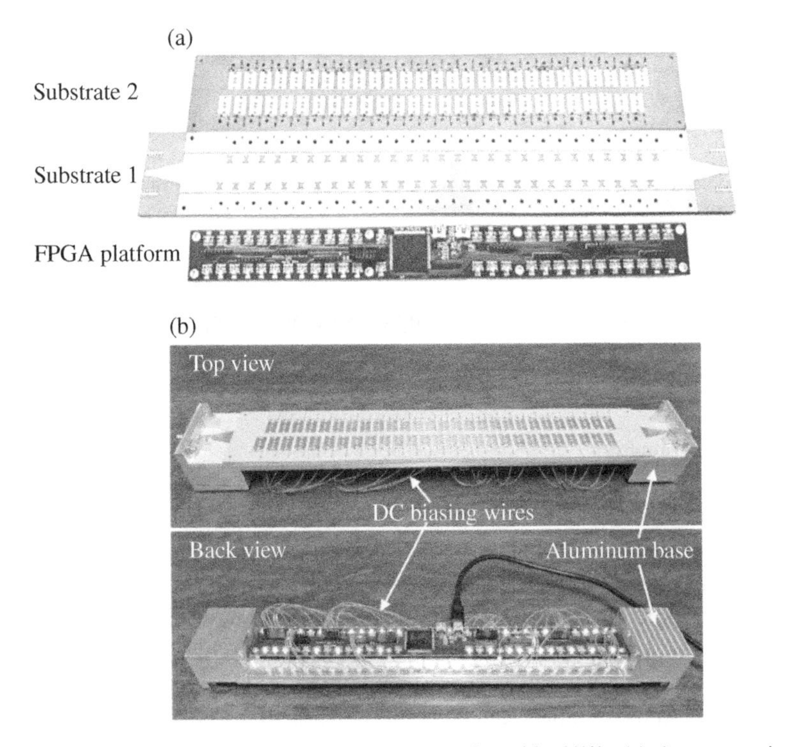

Figure 7.28 Prototype of the period-reconfigurable LWA. (a) Antenna substrates and FPGA platform. (b) Perspective views of the top and bottom of the antenna. *Source*: From [25] / with permission of IEEE.

top and back views of the assembled antenna. The two substrates are mounted on an aluminum base for easy measurement. Two rows of nylon screws are aligned along the edges to reinforce its mechanical stability.

The FPGA platform was specially developed to control the states of the antenna. As shown in Figure 7.28b, it is mounted on the bottom of an aluminum base for compact integration. The FPGA device (Xilinx XC3S50AN-4TQG144C) was selected. Its 54-way voltage outputs were generated to control the 54 patch elements. Since all of the diodes in the unit cells are connected in parallel, they all share the same negative pole of the biasing circuit through the shorted post in each element. The driving voltages are applied to the positive poles of diodes. 54 LED lamps were used for ease of

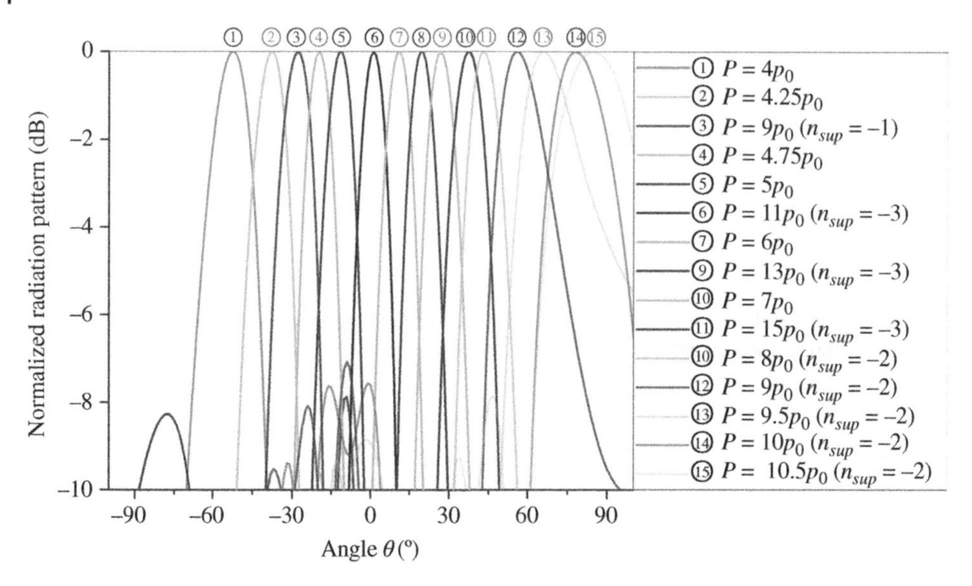

Figure 7.29 Simulated normalized patterns of the scanned beams radiated by the period-reconfigurable LWA at 5 GHz. *Source*: From [25] / with permission of IEEE.

monitoring the activation states of the diodes. Current-limiting resistors were employed to ensure the stability of LED lamps.

7.2.4.2 Radiation Patterns and *S*-Parameters

The beams radiated by the antenna were scanned by controlling the states of the PIN diodes. It realized 15 beams over a 137° scanning range, from −52° back toward the source to 85° in the forward direction. The simulated and measured normalized patterns are shown, respectively, in Figures 7.29 and 7.30. Beam scanning through broadside was achieved. The simulated beam ⑥ (black curve) points in the broadside direction. The operating frequency corresponding to the measured peak gain was found to be shifted to 4.87 GHz for all of the activation states. This outcome was attributed to the inaccurate PCB processing. Moreover, the chokes may have not provided the ideal isolation of RF currents, which would have impacted the operating frequency. The actual chokes (inductors) operate as parallel LC-resonators and generate very large impedances, while the adjacent parts on the board such as the solder joints primarily influence the capacitance. Consequently, the measured normalized patterns in Figure 7.30 are given at 4.87 GHz. It is observed that the measured scanning range was from −60° to 73°, which is smaller than simulated one. This outcome was attributed to the large shift in the operating frequency.

Note that only 15 beams are shown to clearly exhibit the beam scanning performance of this period-reconfigurable LWA. In fact, more beams can be produced to yield finer coverage of a given angular range by choosing the period P appropriately and using the developed method for suppressing any undesirable harmonics. The results would be more steered beams at much denser angular intervals.

The simulated gain values in the main beam directions for all of the patterns at 5 GHz shown in Figure 7.30 are given in Figure 7.31. The maximum gain is 12.5 dBi at $\theta = 11°$ corresponding to beam ⑦. The gain in the broadside direction, beam ⑥, is 12.0 dBi. The slight degradation of the maximum value at broadside is caused by the well-known open-stop band effects. Nevertheless, the

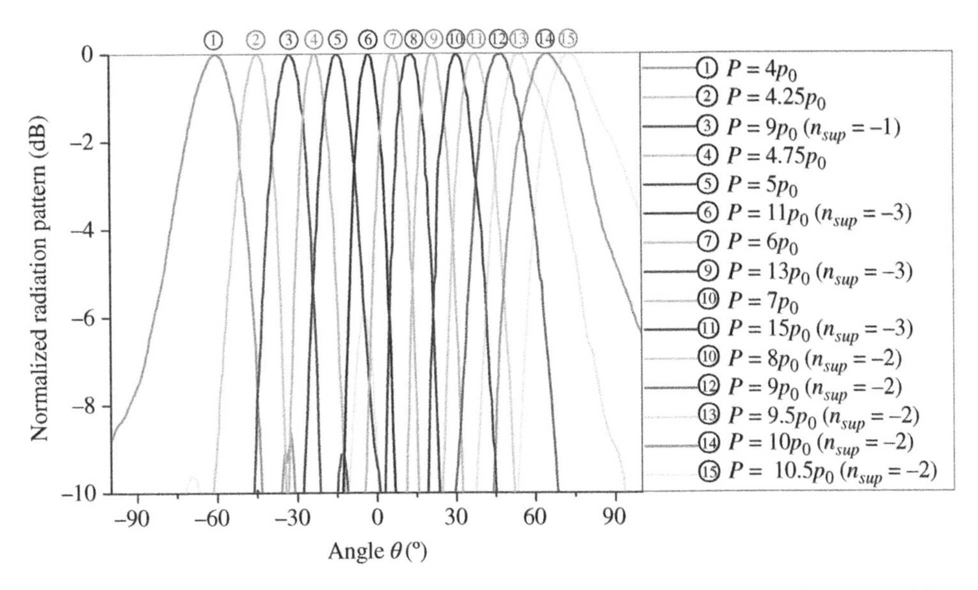

Figure 7.30 Measured results of the beam scanning radiation patterns normalized patterns of the scanned beams radiated by the period-reconfigurable LWA at 4.87 GHz. *Source*: From [25] / with permission of IEEE.

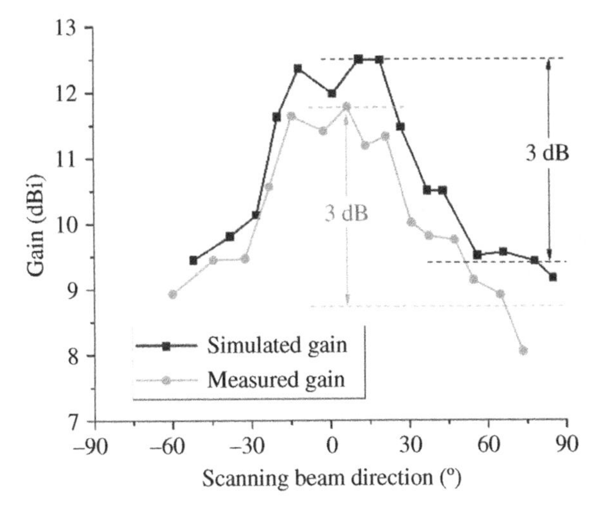

Figure 7.31 The simulated (at 5 GHz) and measured (at 4.87 GHz) gain values at all of the beam angles in the scanning range. *Source*: From [25] / with permission of IEEE.

degradation of the maximum gain values for the indicated 15 steered beams does not exceed 3-dB except for the last beam ⑮. The full 3 dB scanning range is 130°, from −52 to 78°.

The corresponding measured maximum gain values at 4.87 GHz are also given in Figure 7.31. The measured gain is lower than the simulated gain, but the degradation is smaller than 1 dB for most of the beam angles. The measured gain for the last beam did show a degradation that exceeds 3 dB. The measured scanning range is 125°, from −60° to 65°. The simulated and measured scan angles for the 14 beams whose maximum gain value did not exceed the 3-dB variation are compared in

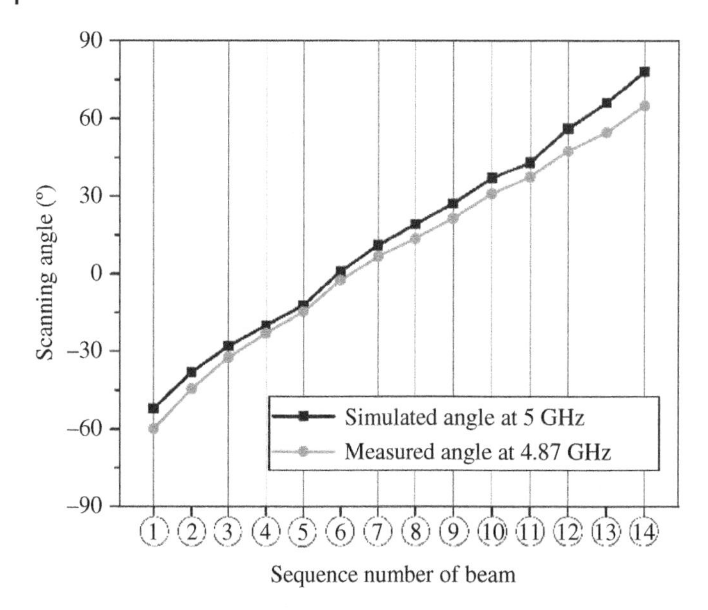

Figure 7.32 The simulated and measured beam angles of the 14 scanned beams within the 3-dB maximum gain variation. *Source*: From [25] / with permission of IEEE.

Figure 7.32. Note that the measured beam directions are shifted slightly to smaller angles when compared with the simulated results. This is due to the fact that the measured patterns were obtained at the lower frequency.

The simulated and measured maximum sidelobe levels (SLLs) of the patterns associated with each of the scanned beams are given in Figure 7.33. The sidelobes of the patterns were not included in Figures 7.29 and 7.30 to avoid the visual difficulty of distinguishing the many overlapping patterns. The SLLs for the beams ④ to ⑨ (beam angles from −20° to 27°) are lower than −10 dB, while they are only lower than −7 dB for all of the other beams.

A drawback of many LWA structures is their inherent dispersive nature which causes the directions of their beams to change as the frequency in their operating band changes. The direction of the beams radiated by the period-reconfigurable LWA shifts about 6° near boresight and about 9° for beams at the widest scan angles when the frequency changes from 4.9 to 5.1 GHz. As a consequence, this type of LWA is more suitable for operations within a narrower sub-band, which are not uncommon in modern communications systems. Further research is needed to make this class of antennas suitable for broadband operation.

Figure 7.34 shows the simulated reflection coefficients of the period-reconfigurable LWA for five different period P values. Only representative five states are shown in Figure 7.34a to demonstrate their behavior again to avoid a messy overlap of curves. Their reflection coefficients at the operating frequency 5 GHz are all below −10 dB. The measured reflection coefficients at the operating frequency 4.87 GHz are shown in Figure 7.34b. They too are all under −10 dB. The fact that $|S_{11}|$ for all 15 states is below −10 dB at the operating frequency for both the simulated and measured results is illustrated again in Figure 7.34c. On the other hand, the simulated $|S_{21}|$ values fluctuate between −7 and −10 dB; the measured ones are only slightly lower than their simulated values.

The open-stopband effect is not obvious from Figure 7.34. The reason is believed to be the two-layer-substrate nature of the structure. Substrate 1 has a relatively large thickness and yields a smaller reflection at each slot in comparison to that of a conventional single-layer LWA. Moreover, the impedance analysis in [33] tells one that because the H-shaped coupling slot on the top of

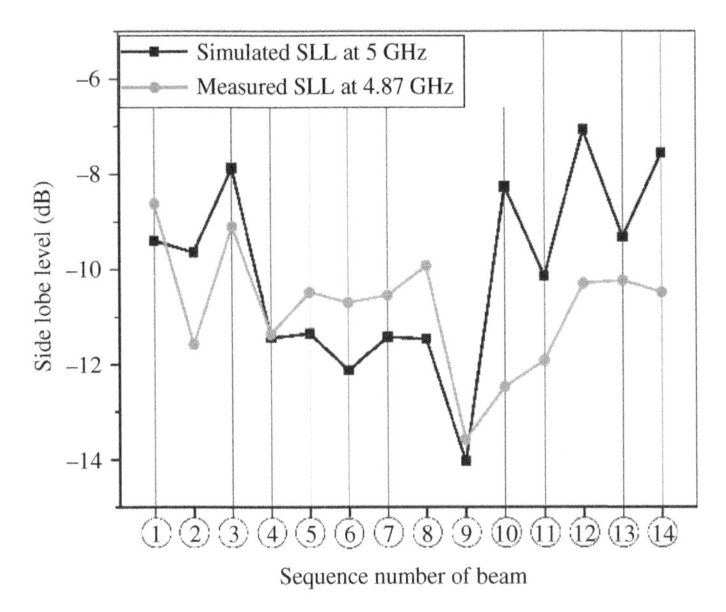

Figure 7.33 The simulated and measured sidelobe levels of each of the 14 scanned beams within the 3-dB maximum gain variation. *Source*: From [25] / with permission of IEEE.

the SIW consists of both transverse and longitudinal slots, the impedance-matching performance is better than when only one transverse slot is present and might be improved with further optimization. However, the degradation of the gain at broadside is clearly observed in Figure 7.31 even though the reflection coefficient requirement of −10 dB has been satisfied.

Other methods might be appropriate to improve the broadside performance of this design. For instance, a metallic post could be introduced inside substrate 1 near the coupling slot as a tuning element (shunt inductance). It could compensate the capacitance of the coupling slot for better impedance matching [34]. It is expected that using such methods, the open-stopband effect can be suppressed to improve the broadside gain.

The attenuation constant α corresponding to each beam is calculated using the values of S_{21}, i.e., $\alpha \approx -\ln |S_{21}|/L$. The simulated and measured normalized result (α/k_0) for all 14 beams is depicted in Figure 7.35. It can be observed that the simulated values of α/k_0 fluctuate around 0.03, which is a relatively small value and is beneficial to achieving the high directivity. The measured curve shows a similar trend, but has slightly higher values than the simulated results.

The power loss of the period-reconfigurable LWA has been simulated; the results are shown in Figure 7.36. The power balance for the antenna can be divided into four parts: (i) the power reflected at the input port, which can be calculated from $|S_{11}|$; (ii) the power radiated into free space, which can be calculated from an integration of the fields on the radiation boundary of the HFSS simulation region; (iii) the power absorbed at the terminal load, which can be calculated from $|S_{21}|$; and (iv) the power lost on the structure, which includes the diodes, conductors, and dielectrics. The power lost is calculated by subtracting the sum of (i), (ii), and (iii) from the total input power. Figure 7.36 shows the power percentages of parts (iii) and (iv). It is observed that the power absorbed at the terminal load, part (iii), is somewhat high, varying from 16 to 21% for the different beam directions. The power lost on the diodes, conductor, and dielectric, part (iv), is high, around 40–46%. The reason for these high values is that the resistance of the diodes generates high ohmic losses. Additional obvious losses are caused by the resonant fields on the copper patch and from the dielectric substrates, which have a relatively high loss tangent, 0.0027. It is generally quite difficult to

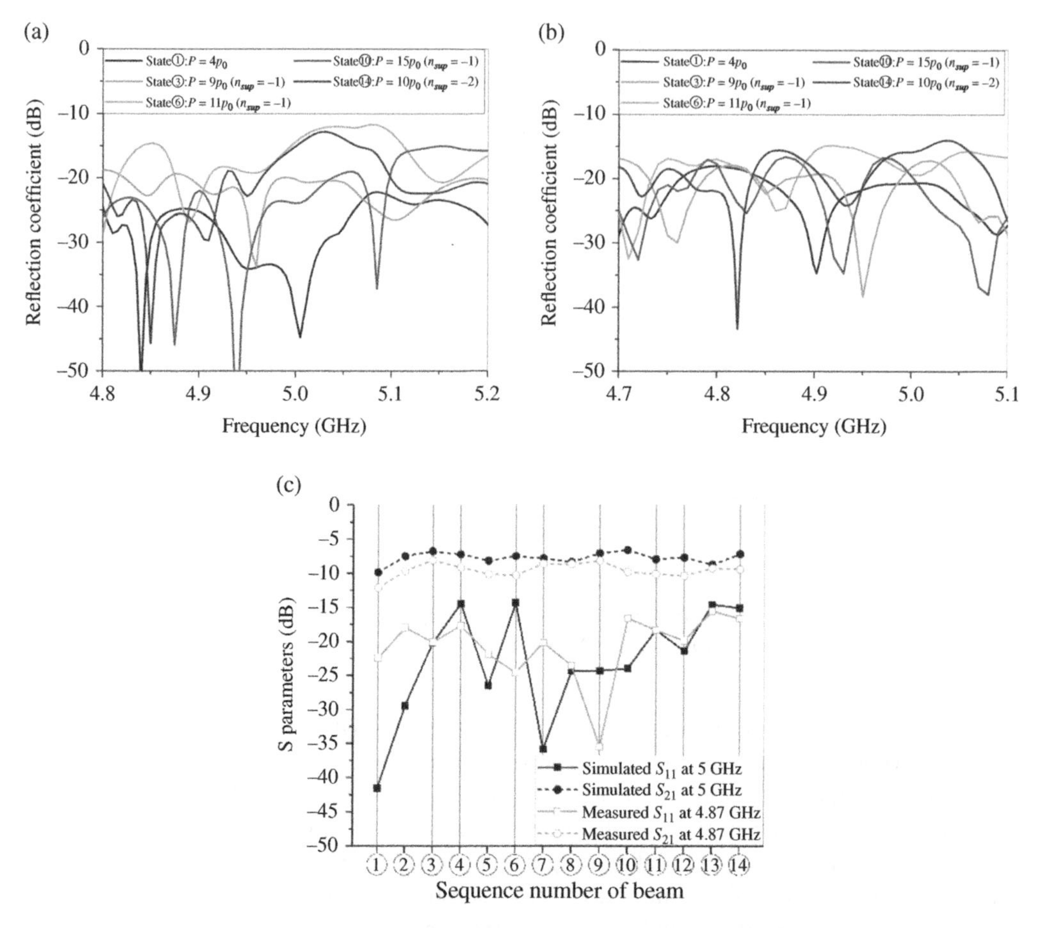

Figure 7.34 *S*-parameters of the period-reconfigurable LWA. (a) Simulated and (b) measured reflection coefficient values for different periods *P*. (c) Simulated and measured *S*-parameters at the indicated frequencies for each scanned beam angle of the 14 beams within the 3-dB maximum gain variation. *Source*: From [25] / with permission of IEEE.

reduce the power loss associated with part (iv) in a practical antenna. Note that it is difficult to calculate the individual losses associated with the diodes, conductors, and dielectrics separately.

The antenna's overall efficiency is also shown in Figure 7.36. It varies from 36% to 43%. There are some methods to increase the antenna efficiency. First, a certain amount of the input power remains at the end of the antenna (16–21%). The magnitude of S_{21} can be made much lower to improve the efficiency if the antenna length is increased and more patch elements are added. Second, the antenna efficiency can be improved if low loss dielectric and better diodes were used. Third, the percentage of the radiated power could be improved by further optimizing the shapes of the coupling slot and the patch.

7.3 Reconfigurable Composite Right/Left-Handed LWA

Another way to produce a full-range scanning LWA is through the introduction of metamaterial constructs, i.e., artificial materials with exotic properties that can be tailored to a given application [35]. A classical T-type equivalent circuit that represents the well-known one-dimensional (1-D)

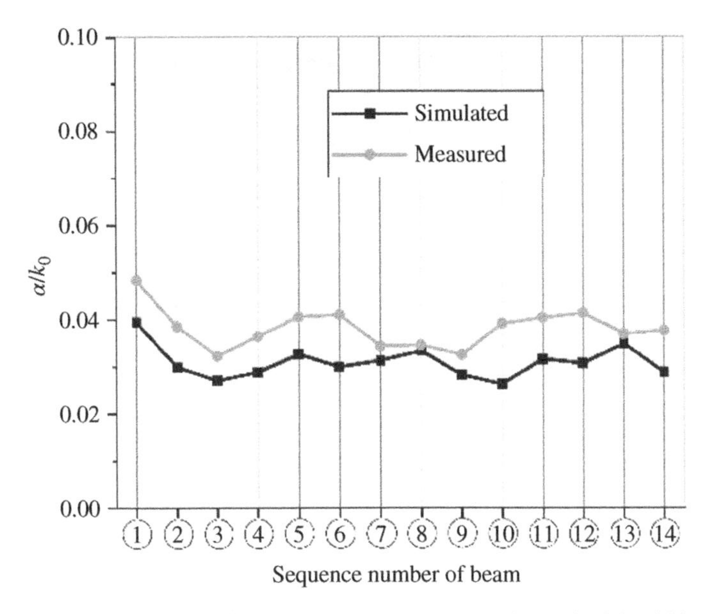

Figure 7.35 Normalized attenuation constant α/k_0 for each of the 14 beams within the 3-dB maximum gain variation. *Source*: From [25] / with permission of IEEE.

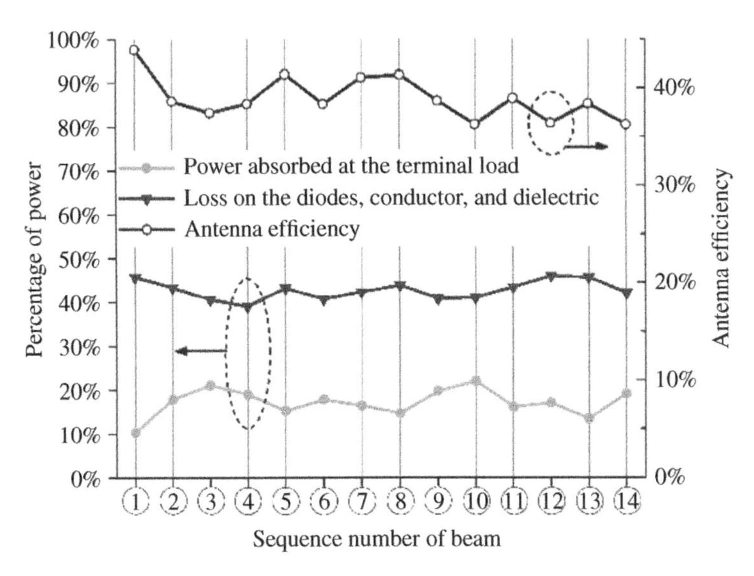

Figure 7.36 Power percentages and antenna efficiency for each of the 14 beams within the 3-dB maximum gain variation. *Source*: From [25] / with permission of IEEE.

CRLH unit structure is depicted in Figure 7.37 [36–38]. The inductance L_R and capacitance C_R are associated with the properties of a conventional right-handed (RH) transmission line, and the placement of the inductance L_L and capacitance C_L lead to left-handed (LH) transmission line properties. The resistor R_s denotes radiation losses from the unit cell.

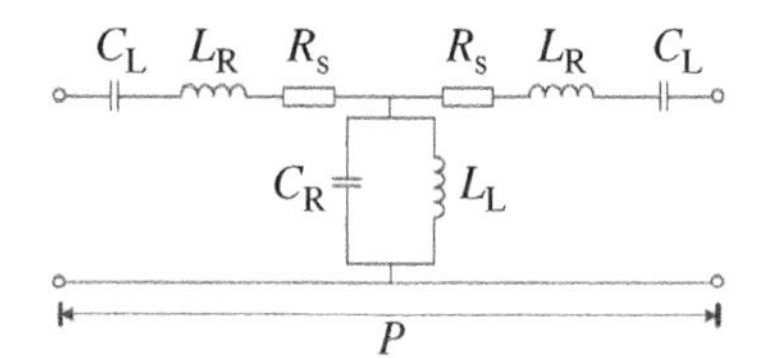

Figure 7.37 T-type equivalent circuit of a 1-D conventional CRLH unit cell. *Source*: From [26] / with permission of IEEE.

The phase constant β of a lossless CRLH unit cell in an infinite periodic transmission line:

$$\beta = \frac{1}{P}\cos^{-1}\left(1-\omega^2 L_R C_R - \frac{1}{\omega^2 L_L C_L} + \frac{C_R}{C_L} + \frac{L_R}{L_L}\right) \tag{7.13}$$

is employed for the theoretical analysis of low loss versions [26]. The balanced condition when there is no gap between the RH and LH parts of the dispersion curve is achieved when $L_R/C_R = L_L/C_L$ [36–38].

Consider an LWA constructed as a finite series of CRLH unit cells. Eq. (7.13) actually indicates that the main beam can be scanned at a fixed frequency f_0 and period P by changing the values of $\widetilde{L}_R$, $\widetilde{C}_R$, $\widetilde{L}_L$, $\widetilde{C}_L$ in a consistent manner. This becomes more apparent if one expresses the phase constant $\widetilde{\beta}$ at a specific frequency f_0 as:

$$\widetilde{\beta}(f_0) = \frac{1}{P}\cos^{-1}\left(1-\omega_0^2 \widetilde{L}_R \widetilde{C}_R - \frac{1}{\omega_0^2 \widetilde{L}_L \widetilde{C}_L} + \frac{\widetilde{C}_R}{\widetilde{C}_L} + \frac{\widetilde{L}_R}{\widetilde{L}_L}\right) \tag{7.14}$$

where $\omega_0 = 2\pi f_0$. Then the main beam angle $\widetilde{\theta}$ can be expressed as:

$$\widetilde{\theta} = \sin^{-1}\left[\frac{\widetilde{\beta}(f_0)}{k_0(f_0)}\right] \tag{7.15}$$

where $k_0(f_0)$ is the free space wave number at f_0.

7.3.1 Parametric Analysis

It is observed from Eqs. (7.14) and (7.15) that the propagation constant $\widetilde{\beta}$ and, therefore, the scan angle of the LWA can be changed by varying either the operating frequency f_0 or the CRLH circuit parameters $\widetilde{L}_R$, $\widetilde{C}_R$, $\widetilde{L}_L$, and $\widetilde{C}_L$. The latter can be realized by introducing a set of switching devices to vary their values such as a set of PIN diodes or varactor diodes and controlling them electronically. To illustrate the concept, the values of the circuit parameters are fixed to be $\widetilde{L}_R = 1.6$ nH, $\widetilde{C}_R = 2.0$ pF, $\widetilde{L}_L = 0.508$ nH, $\widetilde{C}_L = 0.635$ pF, and the period (unit cell length) is chosen as $P = 14$ mm. Calculating the propagation constant as a function of the frequency, a dispersion curve is obtained. The propagation constant $\widetilde{\beta}$ attains both positive and negative values for the CRLH configuration. There is a bandgap in the frequencies when $\widetilde{\beta} = 0$ unless the balanced condition is met. As the values of the unit cell's circuit parameters are changed, one obtains a different dispersion curve for each new set of circuit parameters.

Figure 7.38 shows the dispersion diagrams for different values of $\widetilde{L}_L$. Note that only positive values of the propagation constant are shown. All of the points above the air (light) line correspond to

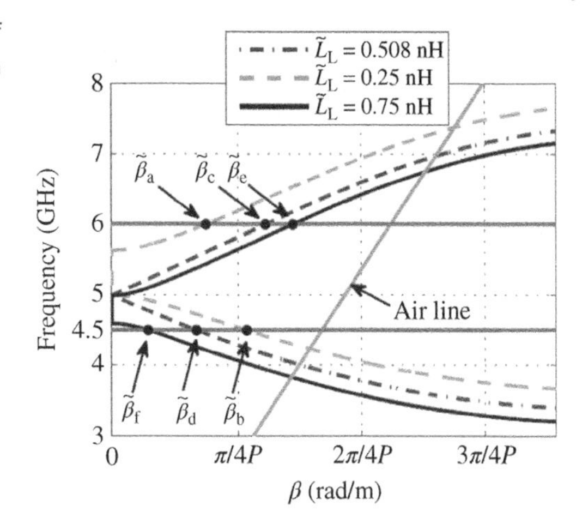

Figure 7.38 Dispersion curves for different values of $\tilde{L}_L$ in the CRLH unit cell. *Source*: From [26] / with permission of IEEE.

waves that will propagate in air away from the array. The left-handed (LH) (right-handed [RH]) portions are the curves for which the frequency is decreasing (increasing) as β is increasing. One observes that when $\tilde{L}_L = 0.508$ nH, a continuous transition at 5.0 GHz is obtained without any gap between the LH and RH portions. This is the balanced point for this case. Thus, the equivalent CRLH LWA would be able to radiate a broadside beam as well as beams in both the forward and backward directions. When $\tilde{L}_L$ is reduced from 0.508 to 0.25 nH, both the RH and LH portions of the dispersion curve shift up in frequency. On the other hand, when $\tilde{L}_L$ increases from 0.508 to 0.75 nH, both shift down. It is noted that a "nonoptimal" inductance and, hence, an unbalanced unit cell would lead to a gap in the dispersion curve and, hence, a "stop gap" region around the broadside direction occurs, i.e., no radiation is emitted normal to the LWA. Furthermore, note that the two green horizontal lines representing two different fixed frequencies cross several RH or LH portions. Consequently, one can achieve fixed-frequency forward *or* backward scanning at two *different* frequencies.

Figure 7.39 shows the dispersion curves obtained when both $\tilde{L}_L$ and $\tilde{C}_L$ are varied. Fix the operating frequency of the antenna at 5.0 GHz. When $(\tilde{L}_L, \tilde{C}_L) = (0.508$ nH, 0.635 pF), one obtains $\tilde{\beta}_1 = 0$ at 5 GHz and broadside radiation would be realized. When $(\tilde{L}_L, \tilde{C}_L)$ are decreased respectively to (0.25 nH, 0.55 pF), both the LH and RH portions of the dispersion curves shift upward and a band gap appears. Thus, $\tilde{\beta}_2$ at 5.0 GHz gives a backward main beam direction. On the other hand, when $(\tilde{L}_L, \tilde{C}_L)$ are increased to (0.75 nH, 0.8 pF), both the LH and RH portions shift downward indicating $\tilde{\beta}_3$ at 5.0 GHz would realize a forward main beam direction. These examples demonstrate that one can achieve continuous beam scanning from the backward to the forward directions through broadside at a fixed frequency by simultaneously changing the values of $\tilde{L}_L$ and $\tilde{C}_L$.

Finally, the effects on the dispersion curves have also been investigated when three circuit parameters are changed simultaneously as shown in Figure 7.40. When $(\tilde{L}_L, \tilde{C}_L, \tilde{L}_R) = (0.508$ nH, 0.635 pF, 1.6 nH), the beam points in the broadside direction at 5.0 GHz with the phase constant $\tilde{\beta}_1 = 0$. When all of their values are decreased to (0.35 nH, 0.55 pF, 1.3 nH), both the LH and RH dispersion curves shift upward yielding $\tilde{\beta}_2$ and a backward beam direction at 5.0 GHz. On the other hand, when $\tilde{L}_L$, $\tilde{C}_L$, and $\tilde{L}_R$ are increased to (1.3 nH, 1.1 pF, and 1.1 nH), $\tilde{\beta}_3$ corresponds to a forward beam direction at 5.0 GHz. Thus, we conclude that more flexibility in

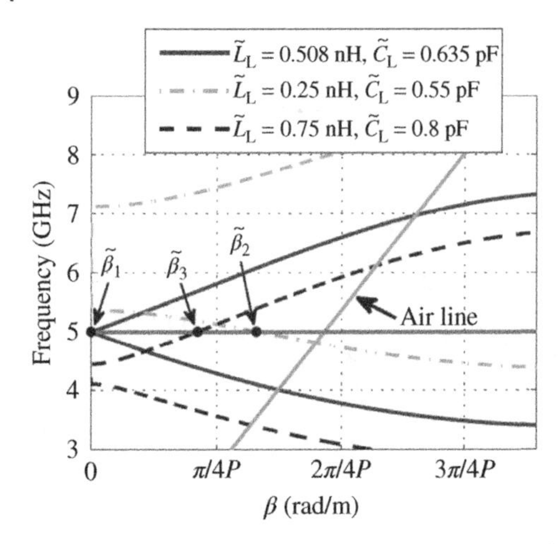

Figure 7.39 Dispersion curves for different values of $\widetilde{L}_L$ and $\widetilde{C}_L$ in the unit cell. *Source*: From [26] / with permission of IEEE.

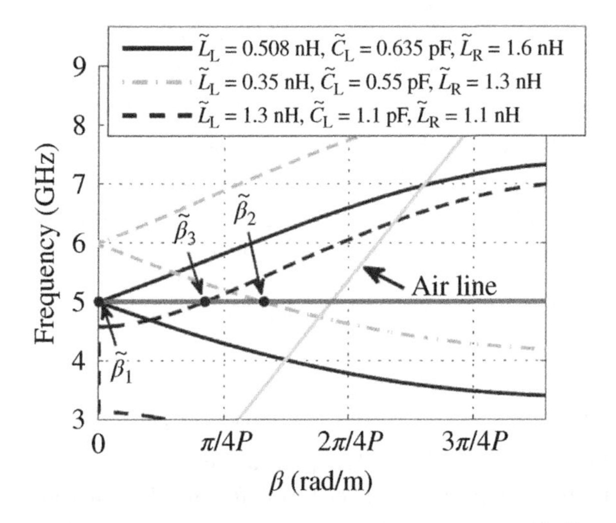

Figure 7.40 Dispersion curves for different values of $\widetilde{L}_L$, $\widetilde{C}_L$, and $\widetilde{L}_R$ in the unit cell. *Source*: From [26] / with permission of IEEE.

controlling the beam direction is realized by varying three circuit parameters simultaneously. However, it is also recognized that this flexibility comes at the cost of an increased number of switching elements and the associated complexity of their biasing network.

These parameter studies allow us to draw the following guidelines for reconfiguring a CRLH circuit to achieve fixed-frequency beam scanning:

a) If only one circuit parameter value is changed while the other three parameters remain fixed, beam scanning at a fixed operating frequency is limited to either the forward or backward direction.

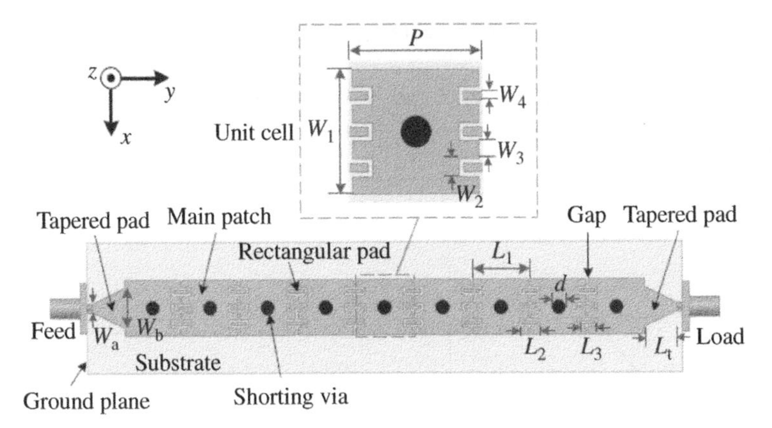

Figure 7.41 Single-layer frequency-based continuous beam scanning CRLH LWA designed and tested to confirm its operating principles and predicted performance characteristics. The top inset represents a unit cell of the antenna. *Source*: From [26] / with permission of IEEE.

b) When combinations of two to four parameter values are changed simultaneously, the main beam can be continuously scanned at a specific frequency from the backward to the forward direction through broadside.
c) There is a trade-off between flexibility and complexity when the number of tunable elements increases. A decision based on fabrication and assembly difficulties and costs will have to be made.

7.3.2 Initial Frequency-Scanning CRLH LWA

A simple single-layer frequency-based continuous beam scanning CRLH LWA was designed, fabricated, assembled, and tested to verify the presented concepts and simulation results. The optimized design is shown in Figure 7.41; it corresponds to the T-type equivalent circuit in Figure 7.37. A two-layer copper-clad substrate is utilized for the LWA. The substrate has the relative permittivity $\varepsilon_r = 2.2$ and loss tangent $\tan\delta = 0.001$. The antenna height, width, and length are 1.575, 33.0, and 154.5 mm, respectively. The main patches, rectangular pads, and tapered pads are printed on the top layer. The copper-clad bottom surface acts as a solid ground plane.

The antenna consists of nine unit cells and the configuration of each unit cell is shown in the top inset in Figure 7.41. Its main patch of each unit cell is connected to the solid ground plane with a centrally located shorting via. Meander gaps are etched at both ends of this main patch. Rectangular pads are inserted into the meander gaps. This unit cell evolved naturally from a conventional RH microstrip transmission line. Two widely used, yet simple and effective, methods are employed to achieve the LH properties, i.e., the center shorting vias produce the LH inductance and the meander gaps yield the LH capacitance.

To obtain the values of the capacitances and inductances to achieve the balanced condition, preliminary dimensions of the antenna's unit cell were selected and calculated using the microstrip equations for the capacitance and inductance given in [29, 39]. The unit cells and matching elements in Figure 7.41 were analyzed, simulated, and optimized with HFSS parameter studies. The optimized values of the antenna parameters are given in Table 7.3.

The length and width of each patch are L_1 and W_1, respectively. The diameter of the shorting vias is d. The dimensions of the meander gap and the rectangular pads are represented by L_2, L_3, W_2, W_3,

Table 7.3 Optimized design parameter values (in mm) of the frequency-scanning CRLH LWA.

Parameters	L_1	L_2	L_3	W_1	W_2	W_3	W_4	P	d	W_a	W_b	L_t
Values (mm)	14.5	5.1	4.1	14	2	2	1	14.7	3.5	2.5	10	8

Source: From [26] / with permission of IEEE.

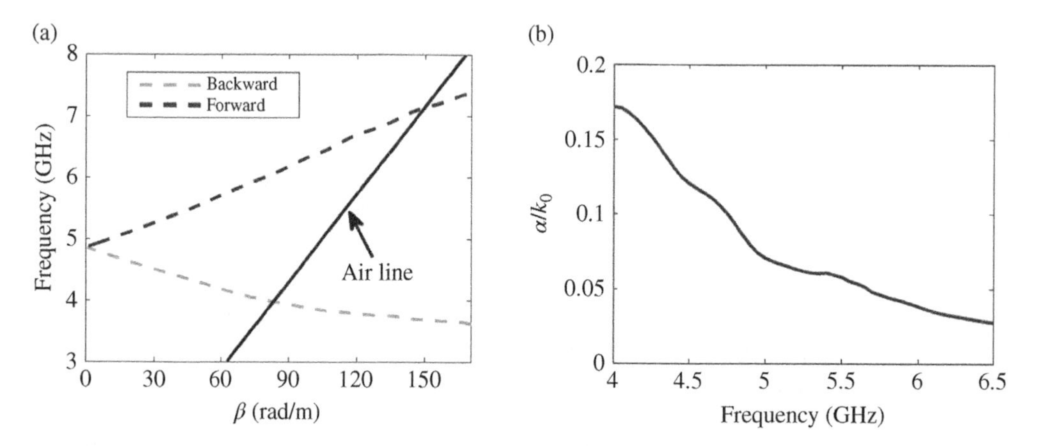

Figure 7.42 The properties of the unit cell of the frequency-based continuous beam scanning CRLH LWA. (a) Dispersion curves. (b) Normalized attenuation constant. *Source*: From [26] / with permission of IEEE.

and W_4. The length of the unit cell is P. The antenna is fed by a 50-Ω source from the left and terminated in a 50-Ω load on the right. Tapered pads are symmetrically printed at both ends of the antenna. They provide impedance matching of the feed and load SMAs to the central set of unit cells over a wide frequency bandwidth. The dimensions of the tapered pad are W_a, W_b, and L_t.

The phase and the attenuation constant properties of its unit cell are shown in Figures 7.42a and 7.42b, respectively. Figure 7.42a shows that a continuous transition from the backward to the forward dispersion curves occurs at 4.86 GHz, i.e., the balanced condition was obtained and there is no bandgap. Furthermore, Figure 7.42b shows the attenuation constant steadily decreasing as the source frequency increases, and the values around the transition frequency are small but nonzero. This behavior demonstrates that there is no open stop band (OSB) for this antenna and, hence, it can continuously scan the main beam from the backward to the forward direction through broadside [40–42].

Figure 7.43a shows the corresponding simulated S-parameters as functions of the operating frequency. The simulated $|S_{11}| \leq -10$ dB bandwidth was determined to be 2.01 GHz, from 4.02 to 6.03 GHz. The values actually remain below −8.8 dB down to 4.0 GHz and below −5.8 dB up to 6.5 GHz. The values of $|S_{21}|$ increase stably with the operating frequency. Their peak value, −5.7 dB, occurs at 6.35 GHz.

The fields radiated by the meander gaps between the adjacent unit cells are the main contributors to the antenna's far-field radiation. Thus, the co-polarized component of the electrical field in the YZ plane is E_θ, and its cross-polarized component is E_φ. Figure 7.43b shows the simulated realized gain and beam angle values as functions of the operating frequency. The realized gain exhibits a stable performance with a variation only between 8.2 dBi (at 4 GHz) and 10.8 dBi (at 5.95 GHz).

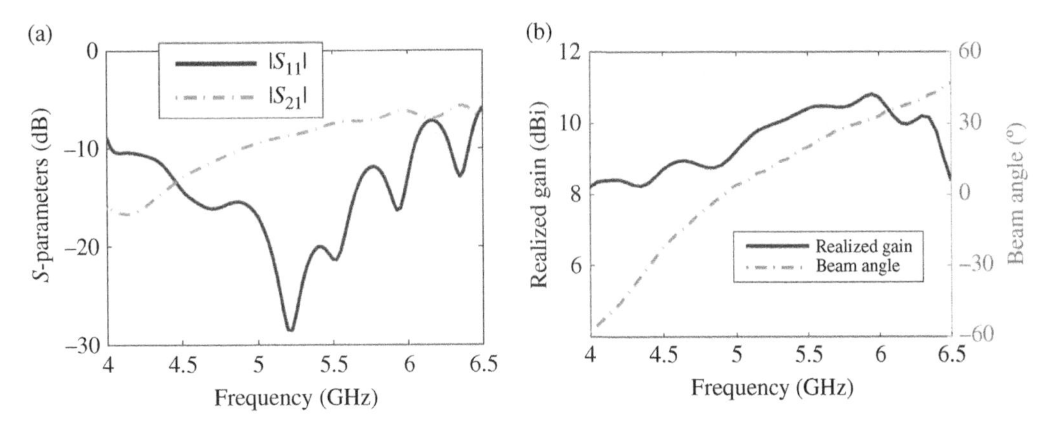

Figure 7.43 The frequency-based continuous beam scanning CRLH LWA. (a) *S*-parameters. (b) Realized gain and beam angles. *Source*: From [26] / with permission of IEEE.

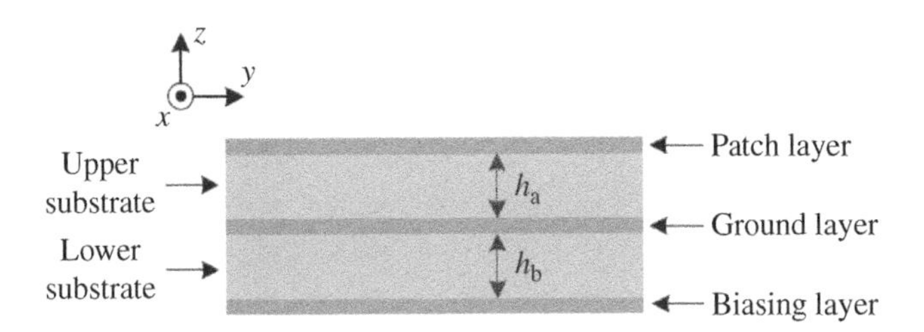

Figure 7.44 Side view (not to scale) of the developed fixed-frequency beam scanning CRLH LWA. *Source*: From [26] / with permission of IEEE.

The main beam angle varies from −58° to +47° when the frequency sweeps from 4 to 6.5 GHz. The broadside radiation occurs at 4.9 GHz with a realized gain of 8.8 dBi.

7.3.3 Reconfigurable Fixed-Frequency Scanning CRLH LWA

While the CRLH LWA presented above scans as a function of frequency, it is non-reconfigurable. Nevertheless, it provides the basis for realizing fixed-frequency beam scanning via reconfigurability. In particular, varactor diode switches are introduced into the antenna to electronically tune the CRLH parameters and, hence, to change the dispersion properties of its unit cells and to scan the beam at a fixed frequency. A practical layout strategy for the DC biasing lines associated with these switches is necessary in any reconfigurable system and is presented below.

7.3.3.1 Antenna Configuration

The two circuit parameters, i.e., $\widetilde{L}_L$ and $\widetilde{C}_L$, were selected for reconfigurability which was enabled with two groups of varactor diodes. The resulting multilayered antenna configuration is illustrated in Figure 7.44 [26]. The two substrates employed have the same material properties: relative permittivity $\varepsilon_r = 2.2$, and $\tan\delta = 0.001$. They have the physical sizes: width = 33.0 mm and length = 154.5 mm to accommodate 10 unit cells plus the necessary feed and termination ends.

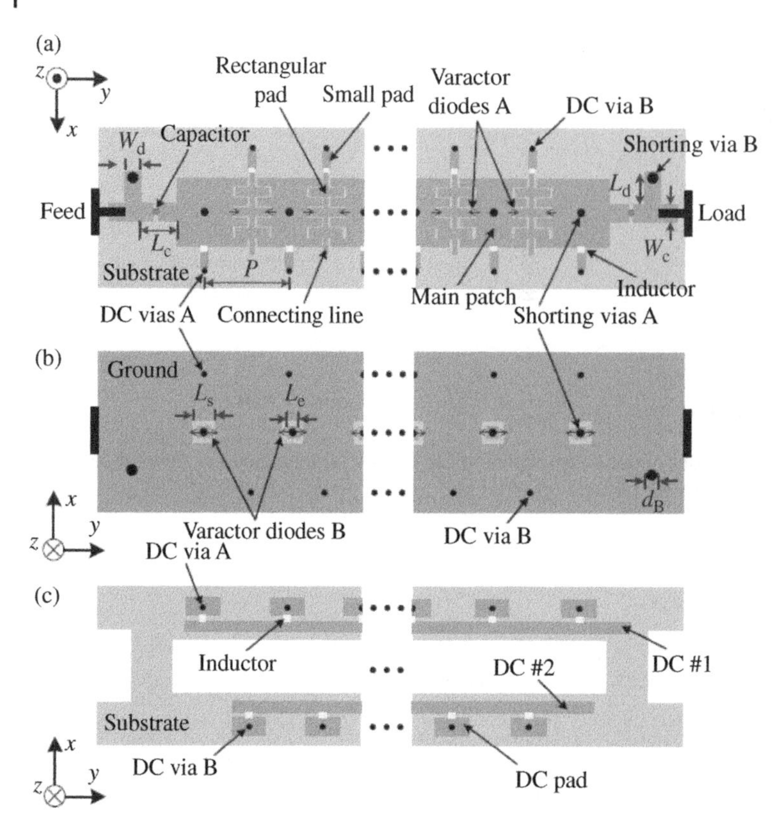

Figure 7.45 The fixed-frequency beam scanning CRLH LWA configuration. (a) Top view of the patch layer. (b) Bottom view of the ground layer. (c) Bottom view of the biasing layer. *Source*: From [26] / with permission of IEEE.

The heights of the upper substrate and the lower substrate are 1.575 and 0.508 mm, respectively. It has three metal layers including the patch layer, the ground layer, and the biasing layer.

Figures 7.45a–c show the details of the patch, ground, and biasing layers, respectively. The patch layer consists of the rectangular pads in the meander gaps of the unit cells, traces that connect the rectangular pads between each unit cell, and the source and termination feedlines with their shorted stubs. One group of the varactor diodes, A, is placed between the central rectangular pads and the main patches. Because these meander gaps introduce the left-handed capacitances $\widetilde{C}_L$, placing varactor diodes here makes it possible to reconfigure it. The varactor diodes in group A are biased through the trace connecting the three rectangular pads in each meander gap. Note that the directions of each pair of the varactor diodes in group A, which is placed at the two ends of each central rectangular pad, is reversed. This way, they can be biased simultaneously with a single DC source.

As shown in Figure 7.45b, square-ring-shaped slots are etched into the ground layer, i.e., the middle metal layer. Smaller square pads are centered in these slots and, consequently, are isolated from the ground plane. The dimensions of the square-ring-shaped slots and the square pads are represented by L_s and L_e. Two varactor diodes B are then reversely soldered between each square pad and the ground plane. Their orientations are opposite to those on the upper layer. Shorting vias A are centrally located with respect to both the main patches on the upper layer and the square pads

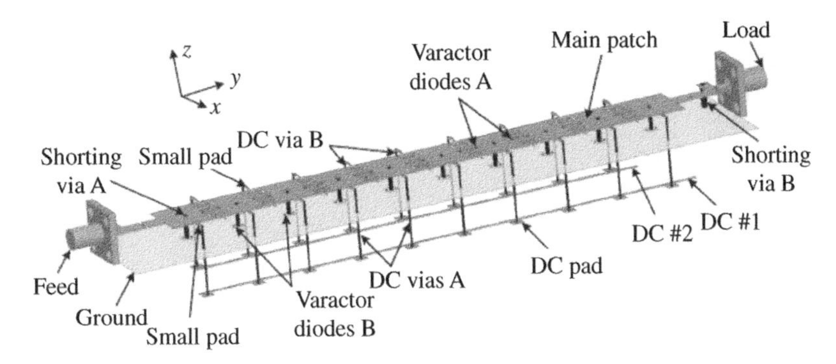

Figure 7.46 A 3D view (not to scale) of the fixed-frequency beam scanning CRLH LWA without the substrates being present. *Source*: From [26] / with permission of IEEE.

on the bottom layer. Consequently, the connections between the shorting vias A and the ground plane are controlled by group B varactor diodes. Thus, the left-handed inductance $\widetilde{L}_{\mathrm{L}}$ is introduced when the shorting vias A are turned on. Furthermore, this inductance can be tuned by varying the capacitance of the varactor diodes in group B. Since the prototype antenna has nine unit cells, there are 18 varactor diodes in group A and 20 varactor diodes in group B to realize fixed-frequency beam scanning.

As shown in Figure 7.45c, two groups of DC pads and two long DC biasing lines, DC #1 and DC #2, are printed on the biasing layer. The varactor diodes in group B are all soldered to the ground layer. Because this soldering causes significant bumps, a continuous rectangular slot was cut out of the bottom layer between the DC bias lines in order to have it lie flat on the ground layer. Similar rectangular cuts were made at the ends of this layer to facilitate the placement of the feed and load SMA connectors.

This reconfigurable antenna was fed by a 50-Ω source from its left side and terminated in a 50-Ω load on its right side. However, in contrast to the initial design, tapered termination pads are not employed for matching purposes. The microstrip structures illustrated in Figure 7.45a were used instead. A shorted stub with a shorting via B was introduced into these terminating lines to facilitate the required impedance matching between these feedlines and the periodic portion of the antenna. The dimensions (in millimeters) of the shorted stub and the shorting via B are denoted as L_{c}, W_{c}, L_{d}, W_{d}, and d_{B}.

7.3.3.2 DC Biasing Strategy

For a clear understanding of the DC biasing network, a three-dimensional (3D) view of the entire system is depicted in Figure 7.46 with the substrates removed. Two capacitors are located between the feedlines and the first and last main patches to isolate the DC signals while maintaining the continuity of the RF signals. Inductors are employed to choke the RF signal whilst maintaining the continuity of the DC signal.

The positive pole of each of the varactor diodes in group A on the patch layer is connected to a main patch. The negative pole is connected to the rectangular pads in the meander gaps. The positive pole of each of the varactor diodes in group B on the ground layer is connected to a small square pad, and then connected to a main patch through a shorting via A. The negative pole is connected directly to the ground plane. Note that the varactor diodes in group A and varactor diodes in group B share the same positive poles. A set of small rectangular strips are distributed on the patch

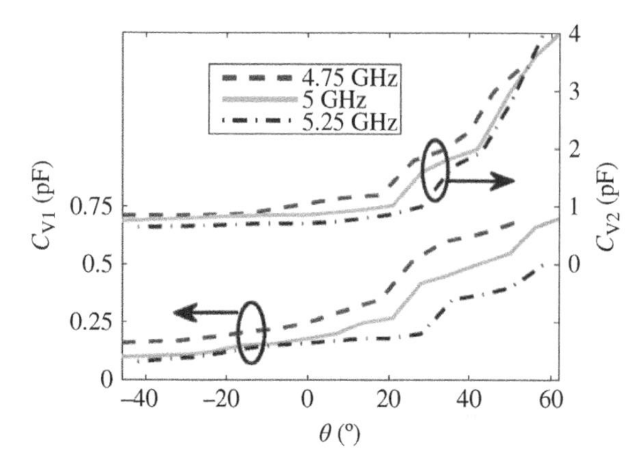

Figure 7.47 Capacitance values of C_{V1} and C_{V2} as functions of the simulated main beam direction at 4.75, 5, and 5.25 GHz. *Source*: From [26] / with permission of IEEE.

layer. They are placed close to the outside (horizontal) edges of the main patches. Each one is connected to the main patch of a unit cell through an inductor. Furthermore, these small pads are also connected to the DC pads on the biasing layer through a DC via A. Each of the latter is connected to the bias line DC #1 through an inductor. Thus, the positive pole of each varactor diode A and also each varactor diode B are connected to DC #1. Similarly, the connecting lines and, hence, the rectangular pads in the meander slots are all connected to the bias line DC #2 through the small pads on the upper surface, the DC vias B, and the DC pads on the bottom layer. Thus, the negative poles of each varactor diode A are connected to DC #2. Consequently, when a DC voltage is applied between the bias lines DC #1 and DC #2, all 18 varactor diodes A can be biased simultaneously with a single DC source. When a DC voltage is applied between the bias lines DC #1 and the ground plane, all 20 varactor diodes B are biased simultaneously. This DC biasing strategy proves to be simple, yet effective.

7.3.3.3 Simulation Results

The performance of the fixed-frequency beam scanning CRLH LWA was optimized with a series of HFSS simulations. The simulation results indicate that the optimized LWA would operate over a 10% frequency bandwidth, from 4.75 to 5.25 GHz, and for each frequency point within the bandwidth, the main beam can be continuously scanned from the backward to forward directions through broadside.

The simulated performance of the optimized LWA working at 4.75, 5, and 5.25 GHz is shown in Figures 7.47–7.49. Figure 7.47 shows the values of the capacitances C_{V1} and C_{V2} as functions of the simulated main beam direction. One notes that with increasing values of (C_{V1}, C_{V2}), the main beam can be continuously scanned from −46° to +54°, −46° to +62°, and −42° to +58°, at 4.75, 5, and 5.25 GHz, respectively. It is further noted that the slopes of these curves are relatively flat for the backward main beam angles and are steep for the forward angles. This behavior indicates that the main beam angles in the backward direction are less sensitive to variations in the values of C_{V1} and C_{V2} as compared to the ones in the forward direction.

Figure 7.48 shows the simulated S-parameters as functions of the simulated main beam directions. The values of $|S_{11}|$ are close to or below −10 dB and the values of $|S_{21}|$ increase as the main beam scans from the backward to forward directions for most of the main beam angles. Simulations

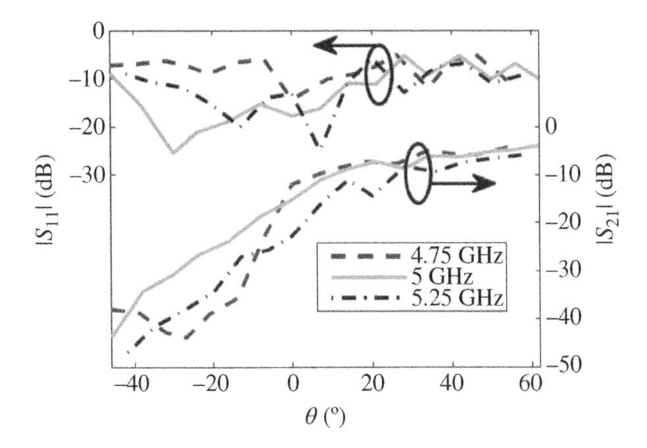

Figure 7.48 *S*-parameters as functions of the simulated main beam direction at 4.75, 5, and 5.25 GHz. *Source*: From [26] / with permission of IEEE.

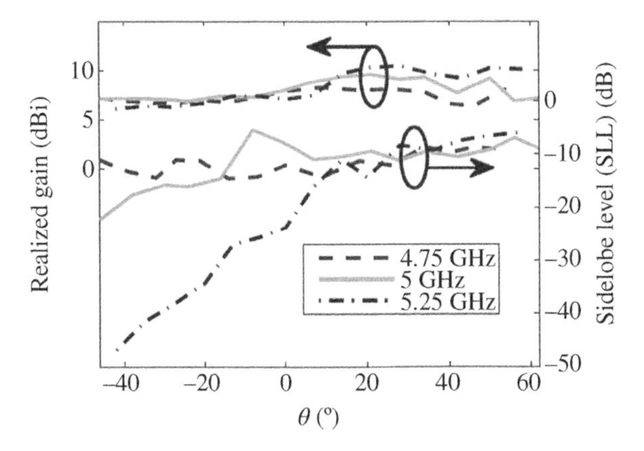

Figure 7.49 Sidelobe level (SLL) and realized gain as functions of the simulated main beam direction at 4.75, 5, and 5.25 GHz. *Source*: From [26] / with permission of IEEE.

also indicate that the radiation efficiency decreases with an increase of the main beam angle. The radiation efficiency is over 90% for the backward main beam angles, and is more than 60% within the whole scanning range. The sidelobe level (SLL) and realized gain as functions of the simulated main beam direction at 4.75, 5.0, and 5.25 GHz are shown in Figure 7.49.

The varactor diodes A and B were assumed to be lossless for the above simulations. It was anticipated that while the lossless models would predict gain values higher than the actual ones, they would not cause a large difference between the predicted and measured beam angles and radiation patterns. This hypothesis was confirmed with the testing of the prototype system.

The simulated radiation patterns for six operating states of the antenna, i.e., States 1–6, at 5 GHz are shown in Figure 7.50. It is noted that the half-power beamwidths (HPBWs) of the back directed beams are wider than those of the forward-directed ones. For example, the HPBW is 47° for State 1 (C_{V1}, C_{V2}) = (0.1 pF, 0.75 pF) and is 24° for State 5 (C_{V1}, C_{V2}) = (0.45 pF, 1.8 pF). This is due to the fact that the radiation in the first few states (backward pointing) is primarily attributed to only the first few unit cells of the LWA whereas most of the unit cells contribute to the forward beams.

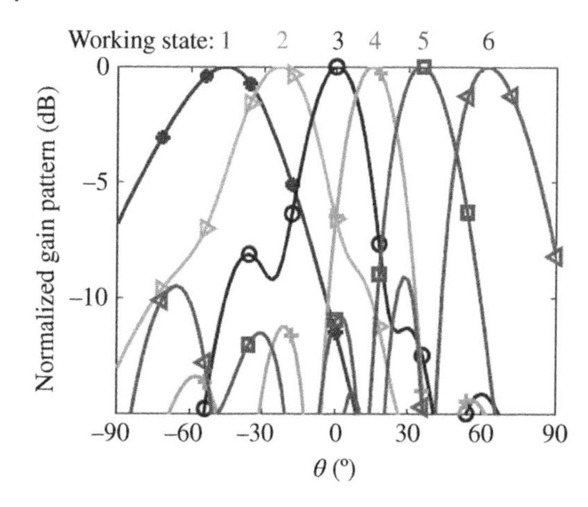

Figure 7.50 Simulated radiation patterns for the antenna's six operating states at 5 GHz. *Source*: From [26] / with permission of IEEE.

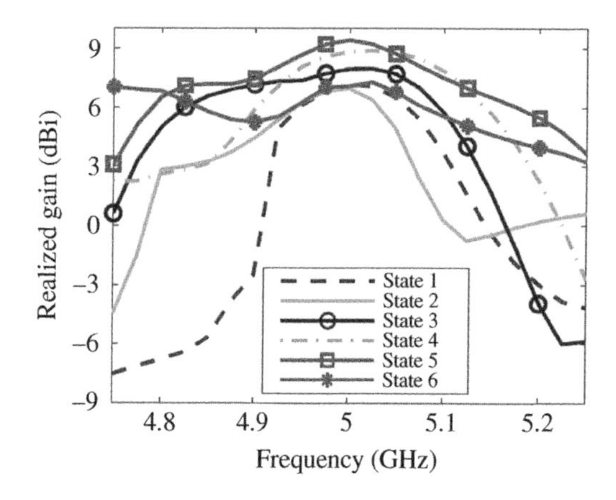

Figure 7.51 Realized gain as a function of the frequency for six operating states of the antenna at 5 GHz. *Source*: From [26] / with permission of IEEE.

Figure 7.51 shows the realized gain as a function of the frequency for the above six operating states at 5.0 GHz. One observes that the antenna's 3-dB operating bandwidth increases as the main beam direction increases.

7.3.3.4 Measured Results

The optimized design was fabricated and tested. Figures 7.52a and 7.52b show photographs of this prototype before and after assembly, respectively. Two types of commercial varactors were utilized to realize the capacitance tuning. A flip-chip varactor diode MA46H120 produced by MACOM company was used for the varactor diodes A. Its datasheet indicates that when the reverse biasing voltage is varied from 0 to 18 V, the capacitance value is reduced from 1.3 to 0.15 pF. The associated series resistance is approximately 2 Ω at 0.5 GHz. On the other hand, a silicon abrupt junction varactor diode SMV1405 from Skyworks company was used for the varactor diodes B. Its datasheet indicates that when the reverse biasing voltage is varied from 0 to 30 V, the capacitance value is reduced from 2.67 to 0.63 pF. The associated series resistance is smaller, approximately 0.8 Ω at 0.5 GHz. Unfortunately, these were the lowest loss diodes that we could purchase. Their capacitance ranges fall short of the simulated values and only four (5.25 GHz) and five (5.0 GHz) of the six simulated states could be measured effectively. Every inductor was the surface-mount inductor 0402HP-8N7X (8.7 nH) from Coilcraft. According to its datasheet and confirmed by

Figure 7.52 Photographs of the fabricated fixed-frequency beam scanning CRLH LWA prototype. (a) Before assembly. (b) After assembly. *Source*: From [26] / with permission of IEEE.

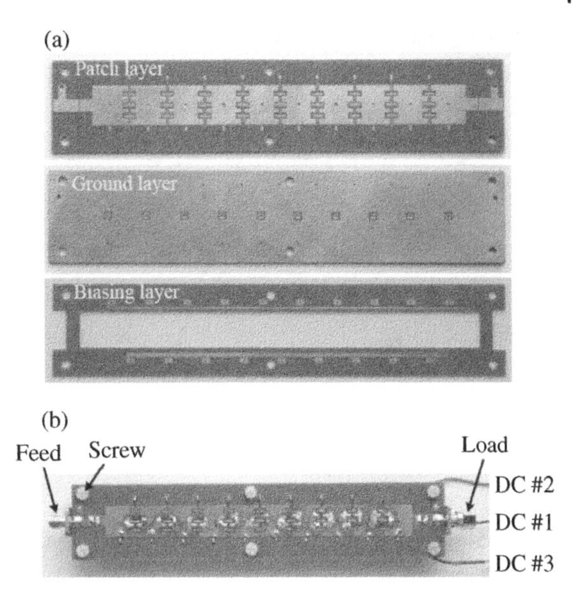

Table 7.4 Measured performance characteristics of the fixed-frequency beam scanning CRLH LWA operating at 5 GHz.

	Working state				
Antenna performance	**1**	**2**	**3**	**4**	**5**
Voltage V_1 (V)	18	18	10.5	3.1	2.4
Voltage V_2 (V)	16	13	11	2	0.1
$\|S_{11}\|$	−16.8	−12.9	−10.7	−9.6	−10.0
$\|S_{21}\|$	−39.9	−34.7	−35.3	−18.3	−20.7
Main beam angle (°)	−37	−19	0	+24	+32
Realized gain (dBi)	5.7	4.9	5.4	5.7	4.6

Source: From [26] / with permission of IEEE.

communications with Coilcraft's technical support, it has a self-resonant frequency at 5.0 GHz and sufficient bandwidth to choke the RF signals within the frequency band of interest while maintaining the DC continuity. Every DC signal-blocking capacitor was selected to be the 600L-0402 capacitor from American Technical Ceramics (ATC) company with a value of 8.2 pF.

To verify the fixed-frequency beam scanning performance, the far-field radiation characteristics of the prototype were measured using the spherical near-field (SNF) antenna measurement system NSI-700S-50. Due to the limitations of this facility, measurements were only conducted at 5.0 and 5.25 GHz. Table 7.4 presents the measured performance of the fixed-frequency beam scanning CRLH LWA's five working states at 5.0 GHz. All of the measured $|S_{11}|$ values are below −9.6 dB, and those of $|S_{21}|$ are below −18.3 dB. The measured radiation patterns for those five working states are depicted in Figure 7.53. The measured beam scanning range is from −37° to +32° when (V_1, V_2) is varied from (18 V, 16 V) to (2.4 V, 0.1 V). When (V_1, V_2) = (10.5 V, 11 V), the system is in State 3 and the measured main beam points in the broadside direction with the peak realized gain,

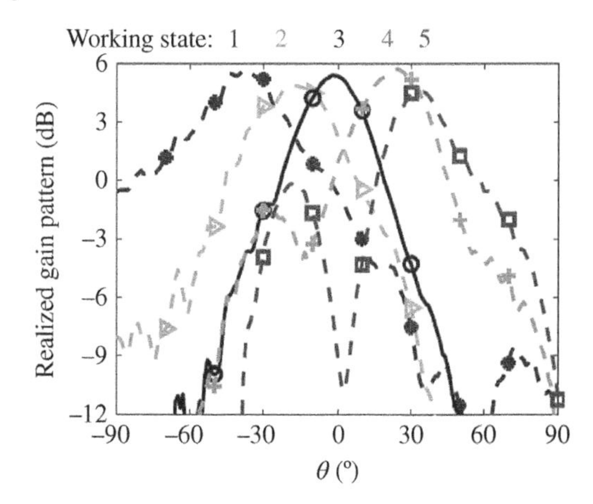

Figure 7.53 Measured realized gain patterns for five of the antenna's operating states at 5.0 GHz. *Source*: From [26] / with permission of IEEE.

Table 7.5 Measured performance characteristics of the fixed-frequency beam scanning CRLH LWA operating at 5.25 GHz.

	Working state			
Antenna performance	**1**	**2**	**3**	**4**
Voltage V_1 (V)	18	18	18	3.1
Voltage V_2 (V)	20	16	13	2
$\|S_{11}\|$	−27.5	−9.6	−15.9	−16.5
$\|S_{21}\|$	−43.7	−37.1	−45.7	−19.0
Main beam angle (°)	−15	0	+13	+34
Realized gain (dBi)	5.0	6.4	4.5	6.1

Source: From [26] / with permission of IEEE.

5.4 dBi. Note that the maximum forward beam angle (+32°) of the measured results for State 5 is smaller than the simulated State 6 value, +62°. As discussed, the missing measured state six arises from the limited capacitance tuning values C_{V2} associated with the SMV1405 varactors, i.e., the required maximum value of C_{V2} is 4.0 pF, whereas the practical maximum value is only 2.67 pF.

Table 7.5 gives the measured performance of the fixed-frequency beam scanning CRLH LWA in its four working states at 5.25 GHz. The measured $|S_{11}|$ values are below −9.6 dB, and those of $|S_{21}|$ are below −19.0 dB. The measured patterns in those four working states are shown in Figure 7.54. The measured beam scans from −15° to +34° when (V_1, V_2) is varied from (18 V, 20 V) to (3.1 V, 2 V). It points in the broadside in State 2 with the peak realized gain value, 6.4 dBi. The reduced measured scanning range is again due to the limitations of the practical tuning capacitance values of both C_{V1} and C_{V2}.

7.3.3.5 Discussions

One clearly sees that the measured realized gains are lower than their simulated values. The maximum gain difference is around 4.0 dB. It is also noticed that the measured $|S_{21}|$ values are lower than the simulated ones, especially for the forward main beam angles. As anticipated,

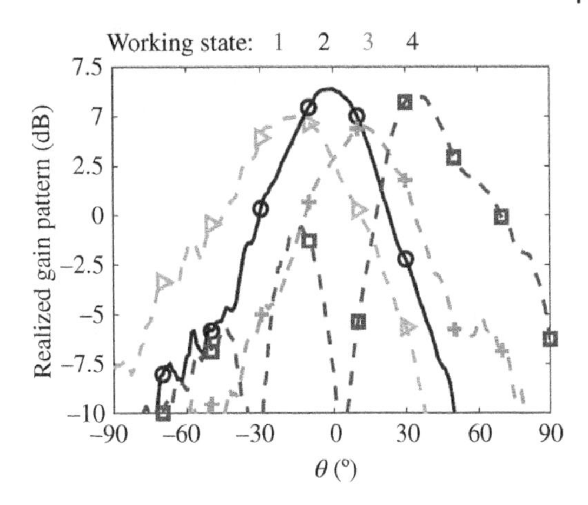

Figure 7.54 Measured realized gain patterns for four of the antenna's operating states at 5.25 GHz. *Source*: From [26] / with permission of IEEE.

Table 7.6 Realized gain values for the six operating states with different values of R_{V1} and R_{V2}.

		Working states					
Resistor values (R_{V1}, R_{V2})	**Antenna performance**	**1**	**2**	**3**	**4**	**5**	**6**
(0 Ω, 0 Ω)	Realized gain (dBi)	7.2	6.9	8.3	9.7	9.3	7.2
	Beam angle (°)	−46	−24	0	+14	+34	+62
(0.5 Ω, 0.2 Ω)	Realized gain (dBi)	6.1	5.8	7.8	8.3	8.0	5.8
	Beam angle (°)	−42	−22	+4	+16	+36	+62
(1 Ω, 0.4 Ω)	Realized gain (dBi)	5.7	5.4	7.7	8.1	7.5	5.2
	Beam angle (°)	−42	−22	+4	+16	+36	+62
(2 Ω, 0.8 Ω)	Realized gain (dBi)	4.7	4.6	7.1	7.4	6.4	4.5
	Beam angle (°)	−40	−22	+4	+16	+36	+62

Source: From [26] / with permission of IEEE.

these outcomes occur because lossless capacitances C_{V1} and C_{V2} were used for the simulations. The resistance associated with each varactor was not included. To improve the diode model predictions, the simulated realized gain and beam angles were obtained for different loss values of the varactor diodes A and B. The loss was included in the HFSS simulations by incorporating a resistor in series with each lumped element capacitor representing the varactor diodes A and B. In particular, diode A (B) is represented by a resistor R_{V1} (R_{V2}) in series with the capacitance C_{V1} (C_{V2}).

Table 7.6 gives the realized gain values for the six operating states at 5.0 GHz for different combinations of R_{V1} and R_{V2}. The realized gains obtained when (R_{V1}, R_{V2}) = (0, 0) denote the lossless case. Table 7.6 indicates that when (R_{V1}, R_{V2}) is varied from (0, 0) to (2.0, 0.8), the beam angle shift is less than 6° for all working states and the largest gain variation is around 2.9 dB. Thus, the variations of the resistors confirm both the anticipated nontrivial decreases in the realized gains and the small differences in the main beam angles. For States 1 and 2, the measured realized gains, 5.7 and 4.9 dBi, respectively, suggest that the equivalent (R_{V1}, R_{V2}) values are most likely (0.5, 0.2). On the other hand, for States 3–5, the measured gains are 5.4, 5.7, and 4.6 dBi, respectively. The equivalent

(R_{V1}, R_{V2}) values are most likely (2, 0.8). These outcomes confirm the fact that the resistor values of the commercial varactor diodes change with their biasing voltages. Obviously, this feature further complicates any precise modeling effort. Nonetheless, the lossless simulation results proved to be an effective guide to realize the prototype and to anticipate qualitatively its measured performance characteristics.

7.4 Two-Dimensional Multi-Beam LWA

Most research on LWAs to date has focused on 1-D single beam scanning. For most practical applications, however, one needs to have a two-dimensional (2-D) antenna that can generate multi-beams which cover a large portion of the hemisphere above it. One method to obtain such an antenna is to combine a number of reconfigurable 1-D LWAs using an orthogonal 1-D beamforming network [43].

Figure 7.55 shows the configuration of such an antenna in which a horizontal plane (H-plane) beam forming network (BFN) is combined with a set of 1-D vertical beam scanning LWAs. The BFN provides multi-beams in the *x0z* plane, and the pattern-reconfigurable antenna array is employed to steer the beam directions in the *y0z* plane. The planar topology allows the BFN and an array of pattern-reconfigurable antennas to be connected with each other conveniently. Furthermore, it successfully separates the vertical plane (V-plane) and H-plane beamforming functions into two different parts, namely the 1-D BFN and the pattern-reconfigurable antenna. In this manner, the beamforming function in the V- and H-planes can be controlled independently. This feature makes the design more flexible when choosing the desired beam number and array size. Moreover, the number of interconnections is reduced from $2N^2$ to N in comparison to employing traditional 2-D BFNs. This feature greatly simplifies the structure complexity and avoids bulky sizes and additional losses.

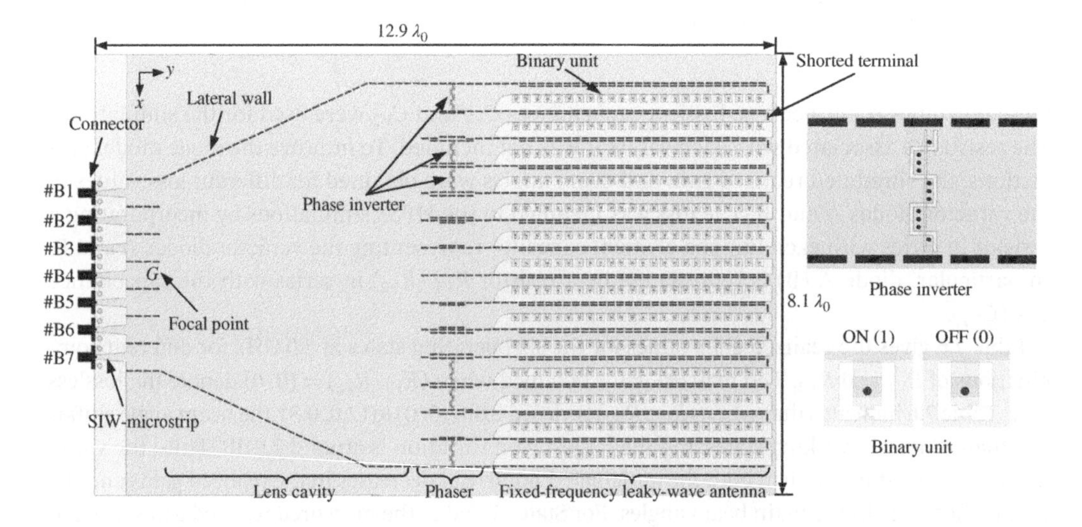

Figure 7.55 Simulation model of the designed 2-D scanning LWA. *Source:* From [43] / with permission of IEEE.

7.4.1 Antenna Design

The designed 2-D LWA is composed of two parts, namely a horn BFN with seven input ports and 14 output ports, and a series of fixed-frequency LWAs whose patterns can be steered individually by changing the state of its binary units. By switching the different input ports of the horn BFN, seven different beam orientations in the elevation plane are produced. On the other hand, each fixed-frequency LWA achieves five different scanned beam states (S1–S5) in the vertical plane. With different combinations of the orientations in the x- and y-directions, this 2-D scanning multi-beam antenna can generate $7 \times 5 = 35$ different beams in total with different elevation and azimuth angles.

The horn BFN design was developed from the concept proposed in [44]. Compared with other similar structures, such as the Rotman lens, the horn BFN has a much simpler structure and is easier to integrate with a linear array. However, the horn BFN proposed in [44] has shortcomings of relatively large-scanning gain drop and insufficient phase shifting capability. To alleviate these issues, a new phase-compensation method was developed. The difference between the new and the traditional methods is the introduction of a multi-port excitation, instead of single-port one. Using this new phase-compensation method, the gain drop is decreased from 3.9 to 2.2 dBi. The fixed-frequency LWA employed in this 2-D array is developed from the one reported in [18]. Since that antenna was designed with half-mode SIW (HMSIW) (or half-width microstrip line [45]) technologies, it cannot be connected to the SIW horn BFN directly. Therefore, transitions from the SIW to the HMSIW had to be introduced to excite the LWA array properly.

The resulting high-gain fixed-frequency 2-D multi-beam scanning LWA shown in Figure 7.55 is integrated in a single substrate whose relative permittivity $\varepsilon_r = 2.55$ and loss tangent $\tan\delta = 0.001$ at 10 GHz. The thicknesses of the dielectric and conductor layers are 1.0 mm and 35.0 μm, respectively. The operating frequency is 10 GHz. The overall size of this antenna is approximately 12.9 $\lambda_0 \times 8.1\ \lambda_0$, where $\lambda_0 = 30.0$ mm is the corresponding wavelength in free space. The design process is conducted using the commercial full-wave electromagnetic simulator HFSS.

7.4.1.1 Horn BFN

The simulated model of the horn BFN is shown in Figure 7.56. Compared with other lens-type BFNs, the horn BFN is a simple and symmetric structure. When the input port #B4 is placed at the focal point and excited, a cylindrical wave is generated within the parallel-plate waveguide. Some phase errors will be produced in the array plane due to the difference between propagation paths. To compensate for the phase errors, additional phase shifters are added after the BFN cavity in order to make all of the output fields in phase. Similar to the operating principles of a reflector lens [46–50], the beam orientation can be manipulated when the source is moved away from the middle because the outputs are no longer in phase.

The BFN cavity in Figure 7.56 is shaped mainly by three parameters: the width of the source plane d_s, the width of the array plane d_a, and the "focal distance" f_d. The width of the source plane is taken to be

$$d_s = n_b \times d_p \tag{7.16}$$

where n_b is the number of desired beams (hence, input ports) and d_p is the distance either between adjacent input ports or output ports. Similarly, the width of the array plane is calculated as:

$$d_a = n_a \times d_p \tag{7.17}$$

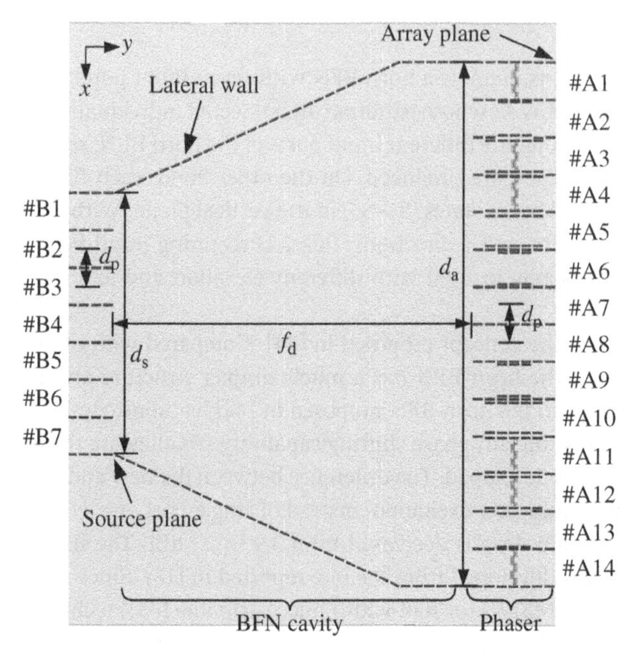

Figure 7.56 Simulation model of the horn BFN. Design parameters: d_p = 15.00, d_s = 104.20, d_a = 209.20, f = 150.00 (Units: mm). *Source*: From [43] / with permission of IEEE.

where n_a is the number of the output ports. A suitable n_a will balance the directivity and the pattern crossover level.

Assuming a uniformly excited linear array, the relationship between the directivity D and the half-power beamwidth, HPBW, is expressed as [51]:

$$D = 102/\text{HPBW} \tag{7.18}$$

Naturally, a higher directivity corresponds to a narrower beam. For a given angle difference, a narrower beamwidth means a decreased crossover level between adjacent beams. After determining the width of the array plane, an appropriate focal length is chosen to avoid any blind spots within the half-power coverage. The beam-switching angle can be estimated as [51]:

$$\theta(d_{\text{off}}) = \arcsin\left[-\frac{2d_{\text{off}}/f_{\text{d}}}{\sqrt{4 + (d_{\text{off}}/f_{\text{d}})^2}}\right] \tag{7.19}$$

where d_{off} represents the distance between the offset input and the focal point. The final parameters of this horn BFN are presented in Figure 7.56.

7.4.1.2 Phase-Compensation Method

To reduce drops in the gain at different beam angles, phase-compensation methods can be employed. Following the operating principles of a lens, phase compensation is naturally calculated in reference to its focal point. However, as it happens with a non-perfect lens, aberrations may occur when the source is offset, which may lead to gain drops. The following phase-compensation method has been successfully used to reduce them.

Let the orientation angle for the n-th input port be denoted as ψ_n $(1 \le n \le 7)$; the corresponding phase of the mth output port relative to the n-th input port be denoted as φ_{nm} $(1 \le m \le 14)$; and the required compensating phase shifts be denoted as δ_{nm} $(1 \le n \le 7,\ 1 \le m \le 14)$. For a given n, the values of the orientation angles ψ_n and the required phase correction shifts δ_{nm} are:

$$\psi_n = \sin^{-1}\left(-\frac{\Delta\varphi'_n}{kd}\right) \tag{7.20a}$$

$$\delta_{nm} = \varphi'_{nm} - \varphi_{nm} \tag{7.20b}$$

where φ'_{nm} is the actual phase shift for the n-th input port and m-th output port, $\Delta\varphi_n{}'$ is the required phase difference between the adjacent output ports related to the n-th input port, k represents the propagation constant in the guided wave structure, and d is the distance between adjacent antenna elements. With the given number of elements along each LWA, the following condition should be satisfied to maintain the symmetry of the output fields relative to the center of the antenna:

$$\delta'_{nm} = \delta'_{n(15-m)}, \quad 1 \le m \le 7 \tag{7.21}$$

The final phase shifts are then calculated with the following average:

$$\delta'_m = \delta'_{15-m} = \frac{1}{2}\left(\frac{1}{7}\sum_{n=1}^{7}\delta_{nm} + \frac{1}{7}\sum_{n=1}^{7}\delta_{n(15-m)}\right) \tag{7.22}$$

Compared to the ideal phase distribution, the proposed method decreases the phase errors for ports #1–#3 at the expense of increasing it for port #4.

This phase compensation method reduces the gain drop. The simulation data for the seven input-port 2-D LWA shown in Figure 7.56, including the beam directions and gains in those directions, are summarized in Table 7.7. The gain drop has been decreased from 3.9 to 2.2 dBi.

7.4.1.3 Phase Shifter Based on Phase Inverter

When designing the SIW-based phase shifters, the width of each path can be manipulated to obtain the desired phase shift. The relationship between the propagation constant β and the equivalent width w of the waveguide can be expressed as:

$$\beta(w) = \frac{2\pi f}{10^3 c} \times \sqrt{\varepsilon_r - \left(\frac{10^3 c}{2wf}\right)^2} \tag{7.23}$$

Table 7.7 Comparison between two phase compensation methods.

	New method			Traditional method		
Port	δ	ψ (°)	**Gain (dBi)**	Δ	ψ (°)	**Gain (dBi)**
#B1	58	26	14.1	68	25	12.9
#B2	39	16	15.4	52	17	14.2
#B3	28	8	15.3	33	9	14.8
#B4	25	0	16.3	3	0	16.8

Source: From [43] / with permission of IEEE.

where f is the frequency in GHz, c is the speed of light, and ε_r is the relative permittivity of the substrate in the waveguide. To make the horn BFN suitable for other types of antennas, a new phase shifter would need to be introduced within the horn BFN without introducing any changes to the radiation portion. One possible solution is to combine equal-length-and-unequal-width phase shifters and 180° phase inverters. It effectively avoids the problem of the limited phase shifting range of the SIW-based one. The design presented here is based on the compact and simple SIW phase inverter proposed in [52].

The simulation model of the phase inverter is displayed in Figure 7.57a. A meandered slot and a complete transverse slot are etched out of the top and bottom copper surfaces, respectively. A series of metallic via holes with diameter of 0.4 mm and spacing of 0.8 mm are introduced on both sides of these slots. These slots prevent the charged-based currents from propagating unrestricted along the top and bottom surfaces and, hence, from transmitting. The via holes conduct the current from one surface to the other one. With an optimized combination of these slots and via holes, these phase inverters can invert the phase while transmitting the TE_{10}-mode, i.e., they act like band-pass filters for the TE_{10}-mode. The phase shifting range of the phase inverter is less than 180° since the phase delay occurs when the surface current flows between the top and bottom surfaces [53]. To compensate for this delay, the SIW width d_4 can be adjusted slightly.

The performance of the 180° phase inverter is shown in Figure 7.57a. From 9 to 11 GHz, the return loss is above 18 dB and the insertion loss is around 0.4 dB. The phase shifting value is 180° at 10 GHz with a phase error of $\pm 11°$ within the 9–11 GHz range. The phase inversion is verified by the simulated E-field distribution shown in Figure 7.57b and compared with a section of normal SIW, i.e., the E-field direction is inverted after passing through the phase inverter. Based on the phase-compensation method and these phase shifters, a phase-corrected horn BFN with reduced gain drop was designed and optimized. It is shown in Figure 7.57c along with its dimensions. Owing to its symmetry, only the output guides #A1–#A7 are exhibited.

7.4.1.4 Fixed-Frequency Beam Scanning Leaky-Wave Antenna

An array of reconfigurable HMSIW-based LWAs is employed to provide the radiation part of the antenna. Since the horn BFN is designed in SIW technology, an SIW to HMSIW transition is required. The simulation model of this SIW-to-HMSIW-to-SIW transition and its predicted performance is shown in Figure 7.58. The reflection coefficient is lower than −23 dB at 10 GHz. The insertion loss of the SIW-to-HMSIW transition is estimated to be about 0.3 dB at 10 GHz.

The LWA designed for prototyping was made up of a section of HMSIW loaded with 24 binary unit cells. Their binary states facilitate changing the beam direction [18]. Overall, there are five different states of this LWA, denoted as S1–S5. The perspective view of the fixed-frequency LWA is displayed in Figure 7.59a with details of the binary units. For each binary unit cell, a patch is printed on the top conductor layer. The HMSI has a narrow 0.1 mm gap, which theoretically generates a gap capacitor. The patch is connected to a biasing pad printed on the bottom conductor layer using a via. The biasing pad is isolated from both the top and bottom conductor layers in order that direct current can be applied to each unit independently. Two switch pads are introduced for placement of the switching device.

Top and bottom views of the LWA are shown in Figure 7.59b. Its design parameters are indicated. The binary unit cells have a period of 6 mm. The width of the LWA is optimized to achieve the best radiation performance. The phase constant $\beta(w)$ and leakage rate $\alpha(w)$ of the HMSIW LWA are obtained from the relations:

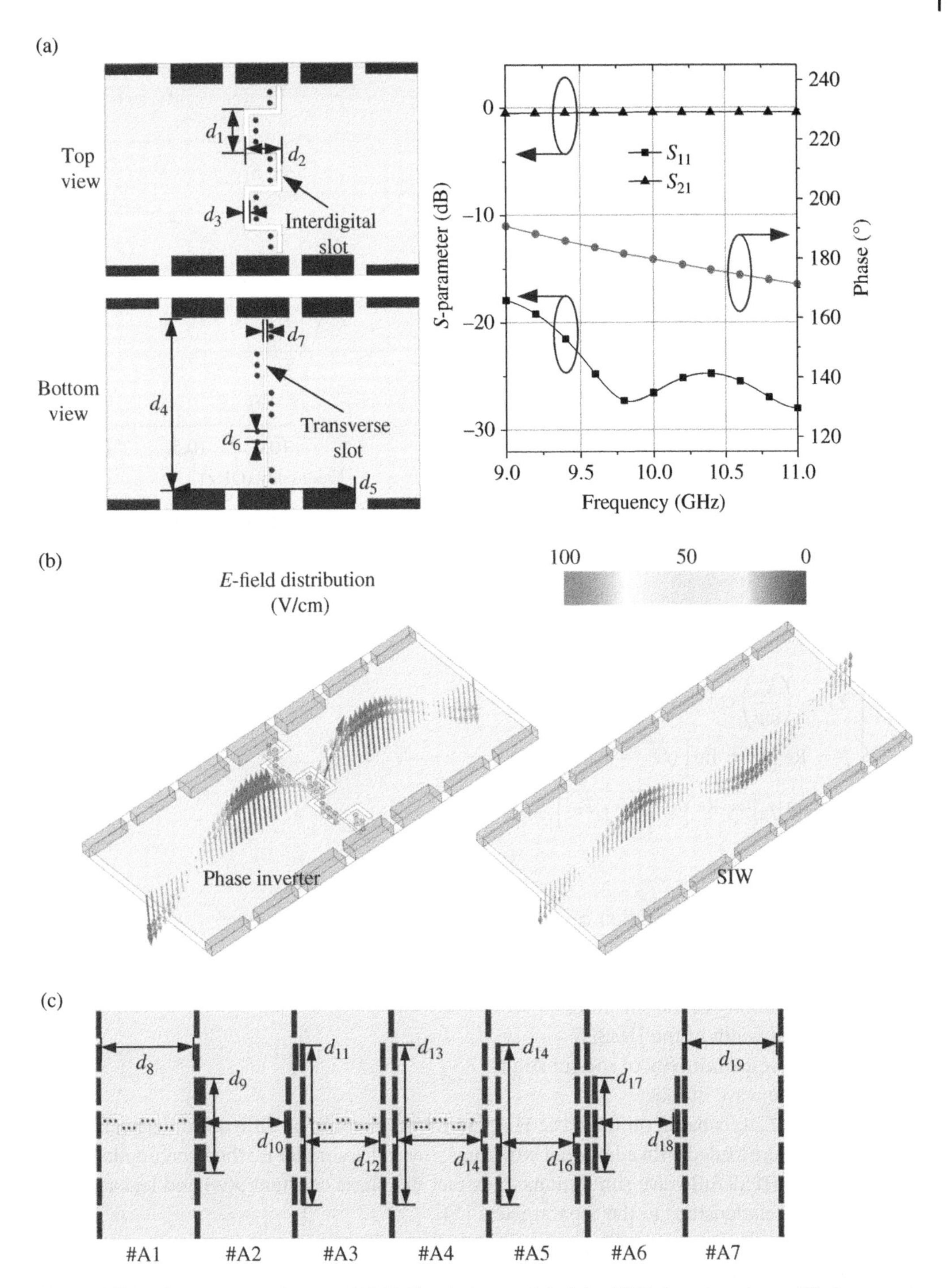

Figure 7.57 Phase inverter designs [43]. (a) Simulation model of the 180° phase inverter and its S-parameter results. (b) E-field distribution. (c) Simulated model of the phase inverters of the horn BFN. Design parameters: d_1 = 3.28, d_2 = 2.88, d_3 = 0.40, d_4 = 12.80, d_5 = 14.00, d_6 = 0.80, d_7 = 0.20, d_8 = 14.30, d_9 = 14.00, d_{10} = 12.40, d_{11} = 24.00, d_{12} = 11.60, d_{13} = 24.00, d_{14} = 13.10, d_{15} = 24.00, d_{16} = 11.20, d_{17} = 14.00, d_{18} = 11.90, d_{19} = 14.20 (Units: mm). *Source*: From [43] / with permission of IEEE.

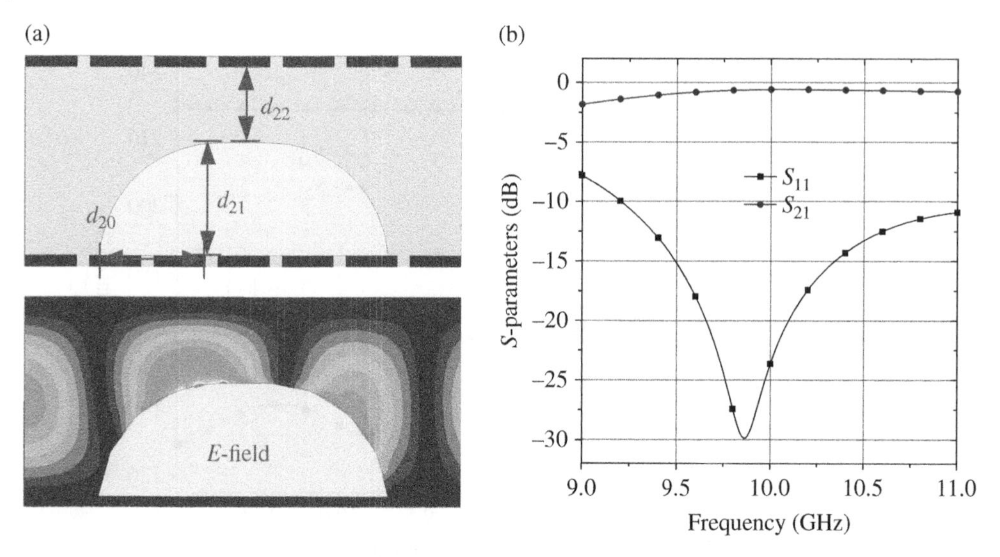

Figure 7.58 Simulated SIW-to-HMSIW transition model, *E*-field distribution and corresponding *S*-parameter results. Design parameters: d_{20} = 9.30, d_{21} = 8.50, d_{22} = 5.70 (Units: mm). *Source*: From [43] / with permission of IEEE.

$$\begin{cases} Y_t = j\left(\dfrac{k_x}{\omega\mu}\right)\cot(k_x w) \\ \beta = \mathrm{Re}\left[\gamma_z\right] = \mathrm{Im}\left[\sqrt{k_x^2 - \varepsilon_r k_0^2}\right] \\ \alpha = \mathrm{Im}\left[\gamma_z\right] = \mathrm{Re}\left[\sqrt{k_x^2 - \varepsilon_r k_0^2}\right] \end{cases} \tag{7.24}$$

where

Y_t: Edge admittance
γ_z: The longitudinal propagation constant
k_x: Transverse wave number
ω: Angular frequency
μ: The permeability of the substrate
w: Equivalent width of the HMSIW
ε_r: The relative permittivity of the substrate
k_0: Free-space wave number

Note that Eq. (7.24) is based on the basic HMSIW LWA. It would be difficult to determine the values of Y_t if it were loaded with additional structures. Instead, a simpler method was applied with the assistance of HFSS full-wave simulations to extract the phase constant $\beta(w)$ and leakage rate $\alpha(w)$ using their relationships to the *S*-parameters [54]:

$$\begin{cases} \beta = \mathrm{Re}\left[\gamma_z\right] = \mathrm{Re}\left[\dfrac{1}{p}\arccos\left(\dfrac{1 - S_{11}S_{22} + S_{12}S_{21}}{2S_{21}}\right)\right] \\ \alpha = \mathrm{Im}\left[\gamma_z\right] = \mathrm{Im}\left[\dfrac{1}{p}\arccos\left(\dfrac{1 - S_{11}S_{22} + S_{12}S_{21}}{2S_{21}}\right)\right] \end{cases} \tag{7.25}$$

The desired radiation angle θ_{rad} is then obtained with an appropriate choice of the width w:

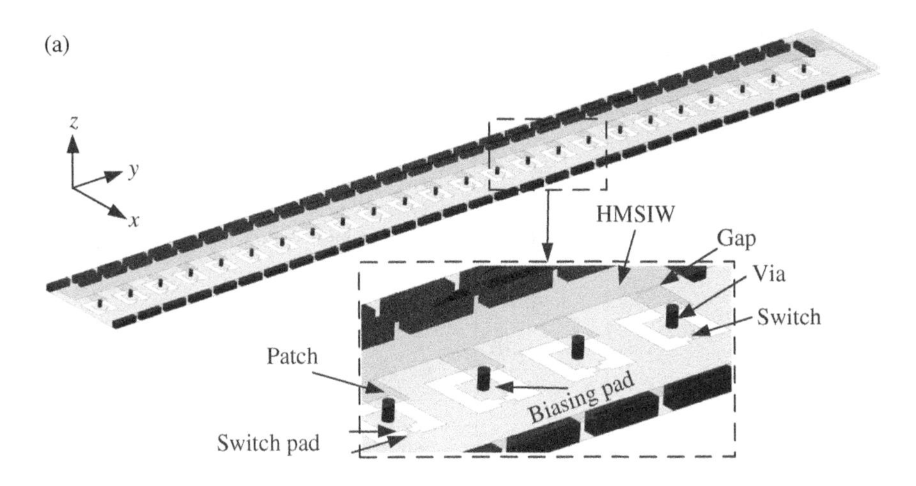

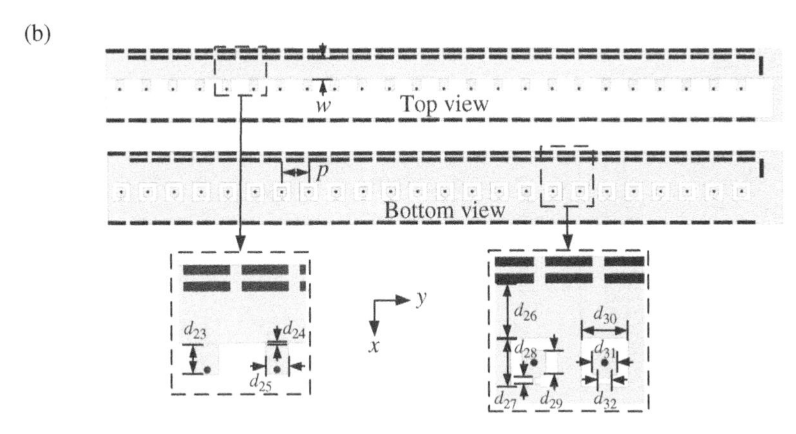

Figure 7.59 Simulation model of the HMSIW-based LWA [43]. (a) Perspective view. (b) Top and bottom views. Design parameters: d_{23} = 2.50, d_{24} = 0.10, d_{25} = 2.00, d_{26} = 4.50, d_{27} = 4.00, d_{28} = 0.48, d_{29} = 2.00, d_{30} = 4.00, d_{31} = 2.00, d_{32} = 1.00 (Units: mm). *Source*: From [43] / with permission of IEEE.

$$\theta_{\text{rad}} = \sin^{-1}\left[\frac{\beta(w)}{k_0}\right] \tag{7.26}$$

To analyze the impact of w on the performance of the LWA, the radiation patterns and S-parameters for different values of w are plotted in Figure 7.60. Only the results of state S1 and S5 are provided because the results for states S2–S4 are similar. In Figure 7.60a, the beam angle and gain would be increased with a larger w. The beam angle for state S1 increases faster than for state S5. This means a decreased beam scanning range between S1 and S5 will occur with a larger w. Consequently, the antenna gain and beam scanning range should be both taken into account when choosing the value of w.

According to Figure 7.60b, the reflection and transmission coefficients show different tendencies as w changes. To ensure that both $|S_{11}|$ and $|S_{21}|$ are lower than -10 dB, the value of w is found to be either 4.3 or 4.4 mm. In this design, the width w = 4.3 mm was chosen to obtain a lower $|S_{21}|$ to reduce the remaining energy received by the matched loads.

The terminal end of the LWA is connected to matched 50-Ω loads to absorb the remaining energy, i.e., the energy is not radiated. To simplify the design and also to reduce the cost of the antenna system,

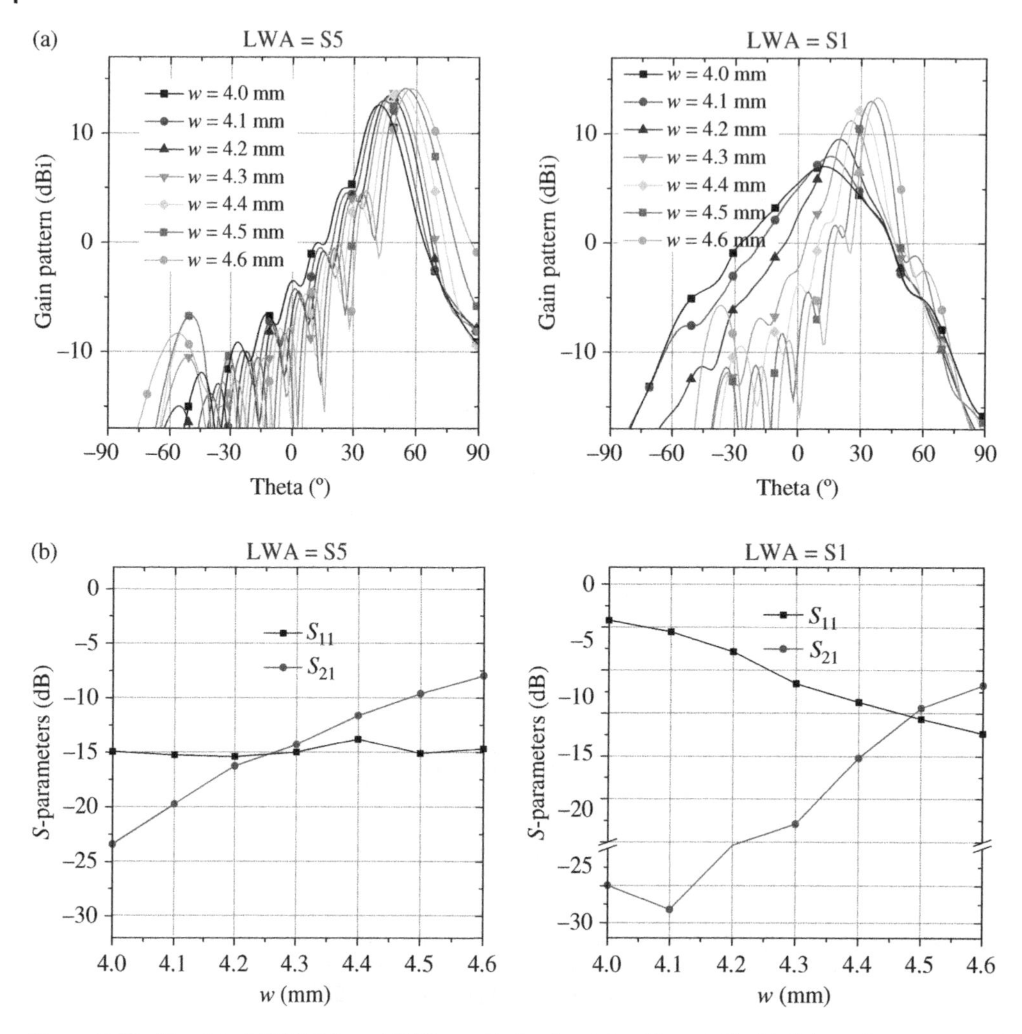

Figure 7.60 Simulated HMSIW-based LWA results for its states S1 and S5. (a) Gain patterns as *w* varies. (b) *S*-parameters. *Source*: From [43] / with permission of IEEE.

the terminal ends of the LWAs were shorted in this design. With this choice, it was expected that the remaining energy would be reflected and, hence, would generate a mirrored beam through the radiation aperture as it propagates back toward the source end. A suitable width *w* must be obtained to reduce this remaining energy as much as possible. The radiation patterns using 50-Ω matched loads and shorted terminals are plotted in Figure 7.61 for comparison. It can be seen that the patterns in the two cases agree well with each other and the mirrored beams are suppressed to be lower than −13 dB. The final results of the fixed-frequency HMSIW-based LWA are summarized in Table 7.8.

7.4.2 Performance and Discussion

The fabricated HMSIW-based LWA prototype is shown in Figure 7.62. The radiation patterns were measured using an NSI700S-50 spherical near-field system. The simulated and measured reflection and isolation coefficients at 10 GHz are plotted in Figure 7.63. Since there are many ports and states

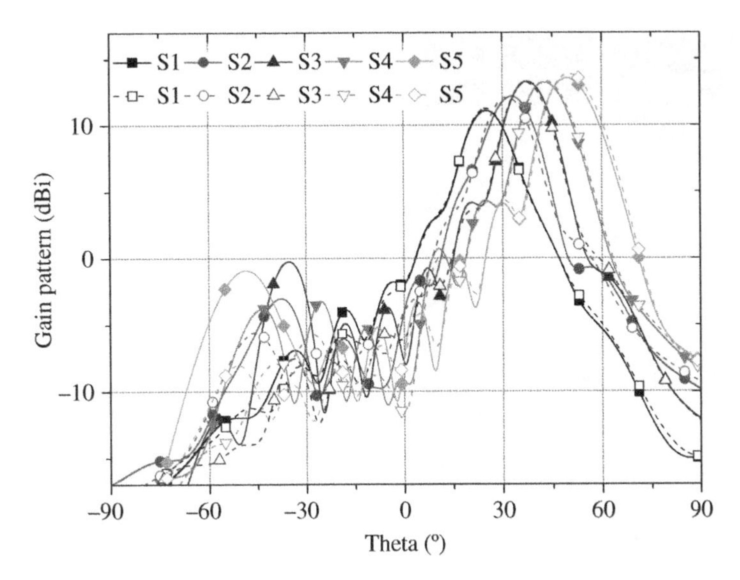

Figure 7.61 Simulated gain patterns of the HMSIW-based LWA. Solid line: with shorted terminal. Dashed line: without shorted terminal. *Source*: From [43] / with permission of IEEE.

Table 7.8 Simulated results of the states of the HMSIW-based LWA.

State	Binary pattern	Direction (°)	Gain (dBi)
S1	111111111111111111111111	25	11.1
S2	101011111010111110101111	33	12.2
S3	101010101010101010101010	37	13.3
S4	000010100000101000001010	43	13.2
S5	000000000000000000000000	49	13.5

Source: From [43] / with permission of IEEE.

in this design, only the isolation coefficients for port #1 and three states of the LWA (S1, S3, and S5) are provided. The simulated and measured *S*-parameters agree well with each other, although there are some slight differences between them. For all the ports, the reflection coefficients were lower than −14 dB and the isolation coefficients were less than −17 dB in both the simulated and measured results.

The BFN in the prototype could only provide seven different states, i.e., by changing the ports from #1 to #7. On the other hand, the LWAs have five states denoted as S1–S5. Thus, there were 35 different states. Their 3-dB contours are plotted in Figure 7.64. It is estimated that the 3-dB contours cover elevation angles from 15° to 75° and azimuth angles from 30° to 150°.

To demonstrate the 2-D scanning capability of the prototype, the pitch angle of the peak gain was first obtained. The azimuth-plane was then cut to check the elevation angles. The simulated and measured patterns corresponding to the plane with maximum gain are plotted in Figure 7.65. For brevity, only the results for ports #4–#7 related to the three states of the LWAs (S1, S3, and S5) are provided.

Figure 7.62 Fabricated prototype and measurement setup. *Source*: From [43] / with permission of IEEE.

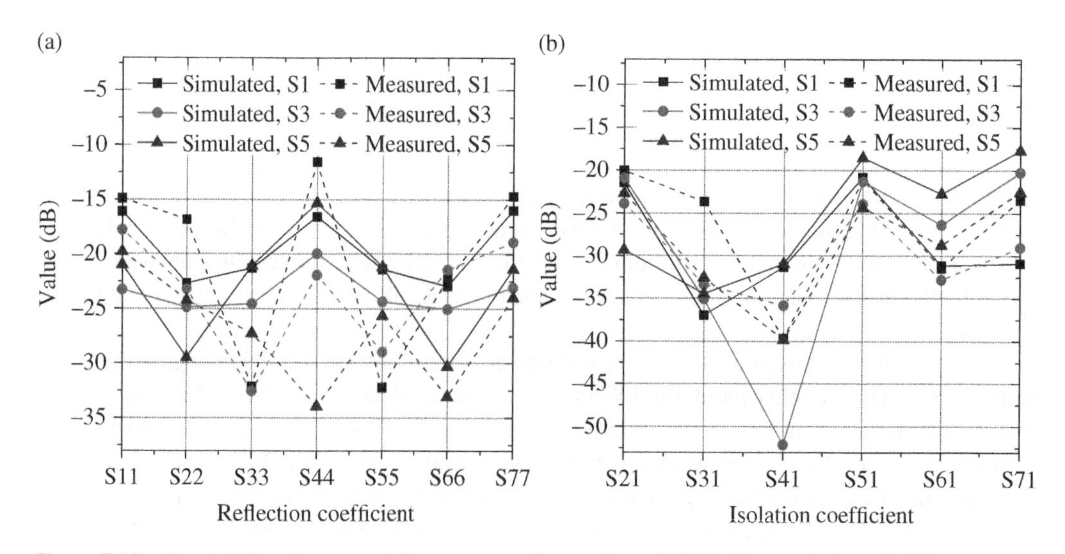

Figure 7.63 Simulated and measured *S*-parameters. *Source*: From [43] / with permission of IEEE.

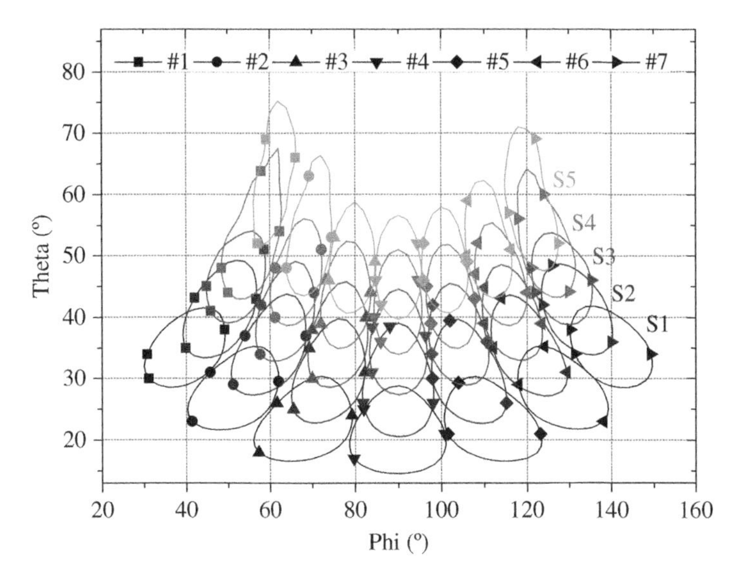

Figure 7.64 Simulated 3 dB contours showing the total coverage area of the HMSIW-based LWA. *Source*: From [43] / with permission of IEEE.

The remaining states have similar performance characteristics. The simulated and measured directivity and realized gain results are plotted in Figure 7.66. The differences between the simulated and measured directivities are less than 2 dBi. Due to limitations of the measurement system, the measured realized gain values were not acquired and only the simulated realized gain values are provided here. The simulated realized gain varies from 18.62 to 22.14 dBi. The beam angles and the corresponding realized gain values are summarized in Table 7.9.

The main purpose of proposing the new 2-D scanning topology is to overcome the design difficulties associated with traditional 2-D scanning topologies and to obtain high gain beams with a uniplanar configuration. The fixed-frequency HMSIW-based LWA functions as one type of pattern-reconfigurable antenna. The simulated and measured results of its prototype demonstrate the feasibility of this novel topology. Note that this fixed-frequency LWA is just but one design example. Other types of pattern-reconfigurable antennas may also be suitable for this topology.

This 2-D topology helps separate the 2-D beamforming function into the BFN and the radiation portions in contrast to the traditional ones [55]. These two parts can be advantageously designed independently. For example, if higher gain is required, the output ports of the horn BFN can be conveniently extended to more than 14 in number or the LWAs can be replaced with other types of antennas with higher gain. This topology is also more convenient for shaping the beam, e.g., to reduce the sidelobe levels.

7.5 Conclusions

Reconfigurable LWAs with fixed-frequency scanning and multi-beam ability are of both great academic interest and practical importance. These antennas typically have a low profile, simple feeding structures, and can be low-loss. By employing appropriate LW structures, such as SIW, these antennas can serve as valuable candidates for 5G and beyond systems particularly since the bandwidth requirements at millimeter-wave frequencies would be moderate and no wideband operation is

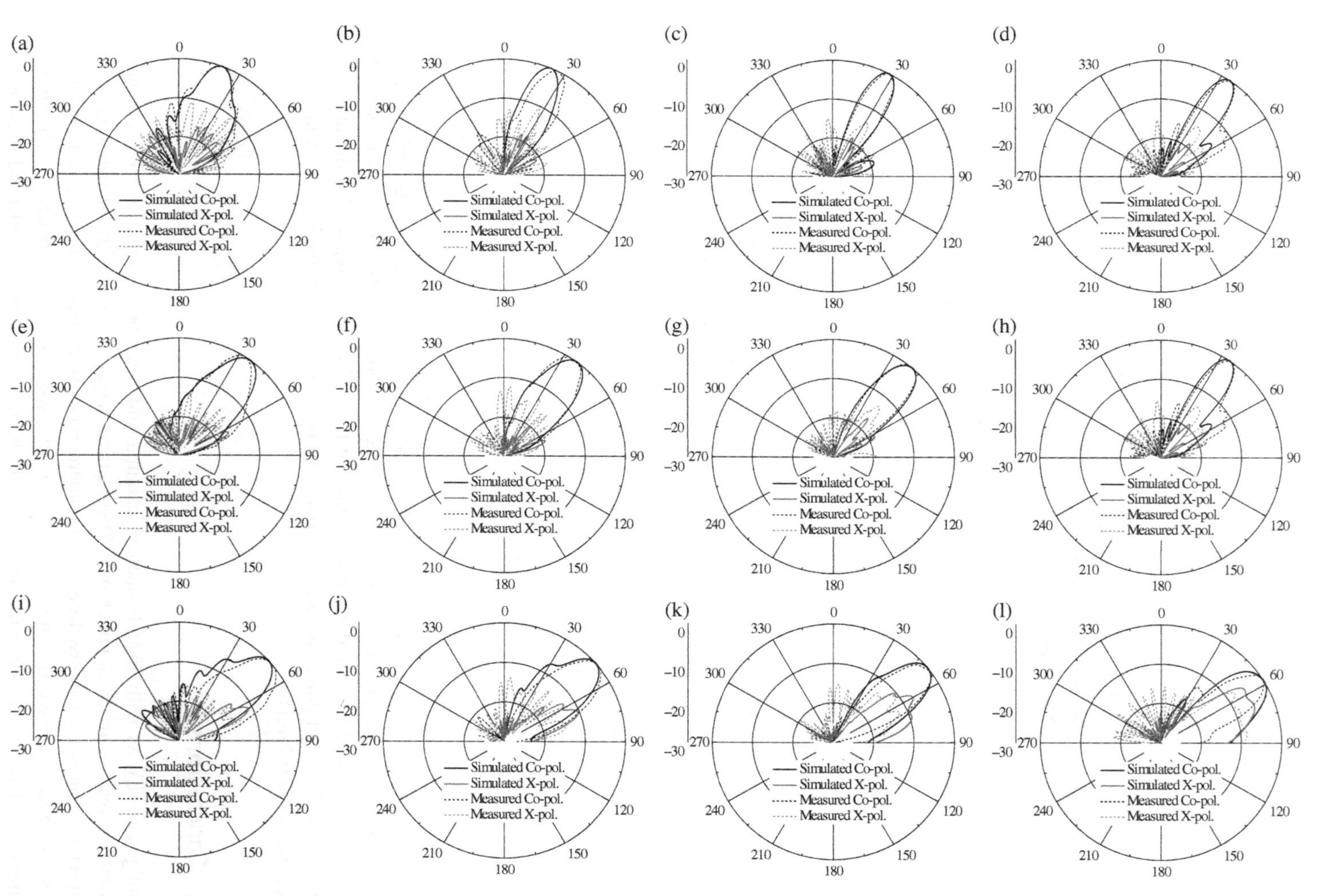

Figure 7.65 Simulated and measured normalized co- and cross-polarization patterns for several input ports and radiation states of the HMSIW-based LWA. (a) #4-S1. (b) #5-S1. (c) #6-S1. (d) #7-S1. (e) #4-S3; (f) #5-S3. (g) #6-S3. (h) #7-S3. (i) #4-S5. (j) #5-S5. (k) #6-S5. (l) #7-S5. *Source*: From [43] / with permission of IEEE.

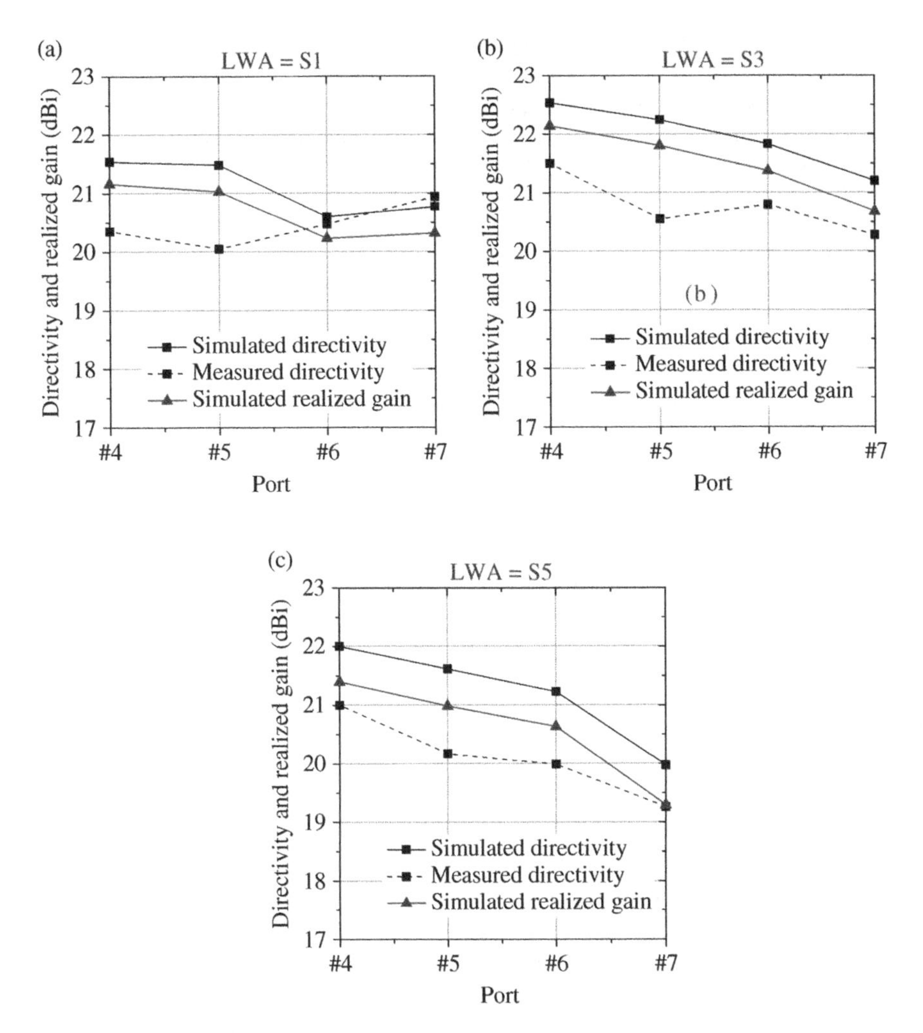

Figure 7.66 Directivity and realized gain results of the HMSIW-based LWA for ports #4–#7. (a) State 1. (b) State 2. (c) State 3. *Source*: From [43] / with permission of IEEE.

needed. In this chapter, four such antennas have been presented. They include a reconfigurable one-dimensional Fabry–Pérot antenna, a period-reconfigurable beam scanning antenna, a CRLH-based beam scanning antenna, and a two-dimensional multi-beam antenna. The reconfigurable one-dimensional Fabry–Pérot antenna can only radiate into forward directions. The reconfigurable CRLH LWA can continuously scan the main beam in a wide angular range. However, the gain losses on the switching devices, e.g., varactor diodes, are high and, hence, the antenna gains are lower compared to those of the period-reconfigurable LWAs. The latter can easily use the "off" state of the diodes as the radiation state. The two-dimensional multi-beam antenna presented divides the problem of generating 2-D multi-beams into two independent, one-dimensional beamforming pieces. This topology makes it possible to have a flat uniplanar structure. Since each linear reconfigurable LWA can effectively scan its beams in the vertical direction, one can then have multiple individually scanned beams. Naturally, these reconfigurable LWAs could be replaced by other radiating elements. It would be desirable for them to have multi-beam abilities as well. Another interesting research topic is the development of

Table 7.9 Beam direction (phi, theta) and corresponding realized gain (dBi) value.

LWA	#1	#2	#3	#4	#5	#6	#7
S1	(41°, 35°); 20.1	(52°, 28°); 20.0	(69°, 23°); 20.9	(90°, 21°); 21.2	(111°, 23°); 21.0	(128°, 28°); 20.2	(140°, 35°); 20.3
S2	(49°, 41°); 19.8	(61°, 35°); 20.6	(75°, 31°); 21.2	(90°, 30°); 21.4	(106°, 31°); 21.4	(120°, 35°); 20.7	(131°, 41°); 20.3
S3	(53°, 47°); 20.3	(64°, 41°); 21.1	(77°, 37°); 21.6	(90°, 36°); 22.1	(104°,37°); 21.8	(116°, 41°); 21.4	(128°, 46°); 20.7
S4	(56°, 51°); 19.4	(67°, 47°); 20.5	(78°, 44°); 21.5	(90°, 43°); 22.0	(102°, 44°); 21.8	(114°, 47°); 20.9	(124°, 52°); 20.2
S5	(61°, 60°); 18.6	(70°, 54°); 20.3	(80°, 50°); 20.7	(90°, 48°); 21.4	(101°, 49°); 21.0	(111°, 51°); 20.6	(119°, 62°); 19.3

Source: From [43] / with permission of IEEE.

new reconfiguration mechanisms which include, for instance, the use of different optically pumped switches [56], piezoelectric materials [57, 58], magnetic fields [59], and liquid metals [60, 61]. These are just some of the future research directions worth pursuing.

References

1. Jackson, D.R. and Oliner, A.A. (2008, ch. 7). Leaky-wave antennas. In: *Modern Antenna Handbook* (ed. C.A. Balanis), 325–368. New York: Wiley.
2. Lai, Q., Fumeaux, C., and Hong, W. (2012). Periodic leaky-wave antennas fed by a modified half-mode substrate integrated waveguide. *IET Microw. Antennas Propag.* **6** (5): 594–601.
3. Liu, J., Jackson, D.R., and Long, Y. (2012). Substrate integrated waveguide (SIW) leaky-wave antenna with transverse slots. *IEEE Trans. Antennas Propag.* **60** (1): 20–29.
4. Dong, Y. and Itoh, T. (2012). Substrate integrated composite right-/left-handed leaky-wave structure for polarization-flexible antenna application. *IEEE Trans. Antennas Propag.* **60** (2): 760–771.
5. Nasimuddin, N., Chen, Z.N., and Qing, X. (2013). Substrate integrated metamaterial-based leaky-wave antenna with improved boresight radiation bandwidth. *IEEE Trans. Antennas Propag.* **61** (7): 3451–3457.
6. Li, Z., Wang, J., Zhang, Z., and Chen, M. (2013). Far field computation of the traveling wave structures and a new approach for suppressing the sidelobe levels. *IEEE Trans. Antennas Propag.* **61** (4): 2308–2312.
7. Li, Z., Wang, J.H., Duan, J. et al. (2016). Analysis on the radiation property of the bounded modes of periodic leaky-wave structure with finite-length using a hybrid method. *Sci. Rep.* **6**: 22917.
8. Li, Z., Zhang, S., Wang, J. et al. (2019). A method of generating radiation null for periodic leaky-wave antennas. *IEEE Trans. Antennas Propag.* **67** (6): 4241–4246.
9. Karmokar, D.K., Guo, Y.J., Qin, P.-Y. et al. (2018). Substrate integrated waveguide-based periodic backward-to-forward scanning leaky-wave antenna with low cross-polarization. *IEEE Trans. Antennas Propag.* **66** (8): 3846–3856.
10. Karmokar, D.K., Guo, Y.J., Chen, S.-L., and Bird, T.S. (2020). Composite right/left-handed leaky-wave antennas for wide-angle beam scanning with flexibly chosen frequency sweep. *IEEE Trans. Antennas Propag.* **68** (1): 100–110.

11. Karmokar, D.K., Chen, S.-L., Bird, T.S., and Guo, Y.J. (2019). Single-layer multi-via loaded CRLH leaky-wave antennas for wide-angle beam scanning with consistent gain. *IEEE Antennas Wireless Propag. Lett.* **18** (2): 313–317.
12. Chen, S.-L., Karmokar, D.K., Li, Z. et al. (2019). Circular-polarized substrate-integrated-waveguide leaky-wave antenna with wide-angle and consistent-gain continuous beam scanning. *IEEE Trans. Antennas Propag.* **67** (7): 4418–4428.
13. Chen, S.-L., Karmokar, D.K., Qin, P.-Y. et al. (2020). Polarization-reconfigurable leaky-wave antenna with frequency-based continuous beam scanning through broadside. *IEEE Trans. Antennas Propag.* **68** (1): 121–133.
14. Lim, S., Caloz, C., and Itoh, T. (2004). Metamaterial-based electronically controlled transmission-line structure as a novel leaky-wave antenna with tunable radiation angle and beamwidth. *IEEE Trans. Microwave Theory Tech.* **52** (12): 2678–2690.
15. Guzman-Quiros, R., Gomez-Tornero, J.L., Weily, A.R., and Guo, Y.J. (2012). Electronic full-space scanning with 1-D Fabry–Pérot LWA using electromagnetic band-gap. *IEEE Antennas Wireless Propag. Lett.* **11**: 1426–1429.
16. Lim, S., Caloz, C., and Itoh, T. (2005). Metamaterial-based electronically controlled transmission-line structure as a novel leaky-wave antenna with tunable radiation angle and beamwidth. *IEEE Trans. Microw. Theory Tech.* **53** (1): 161–173.
17. Suntives, A. and Hum, S.V. (2012). A fixed-frequency beam-steerable half-mode substrate integrated waveguide leaky-wave antenna. *IEEE Trans. Antennas Propag.* **60** (5): 2540–2544.
18. Karmokar, D.K., Esselle, K.P., and Hay, S.G. (2016). Fixed-frequency beam steering of microstrip leaky-wave antennas using binary switches. *IEEE Trans. Antennas Propag.* **64** (6): 2146–2154.
19. Ouedraogo, R.O., Rothwell, E.J., and Greetis, B.J. (2011). A reconfigurable microstrip leaky-wave antenna with a broadly steerable beam. *IEEE Trans. Antennas Propag.* **59** (8): 3080–3083.
20. Guzman-Quiros, R., Gomez-Tornero, J.L., Weily, A.R., and Guo, Y.J. (2012). Electronically steerable 1-D Fabry–Pérot leaky-wave antenna employing a tunable high impedance surface. *IEEE Trans. Antennas Propag.* **60** (11): 5046–5055.
21. Fu, J.H., Li, A., Chen, W. et al. (2017). An electrically controlled CRLH-inspired circularly polarized leaky-wave antenna. *IEEE Antennas Wireless Propag. Lett.* **16**: 760–763.
22. Li, J., He, M., Wu, C., and Zhang, C. (2017). Radiation-pattern-reconfigurable graphene leaky-wave antenna at terahertz band based on dielectric grating structure. *IEEE Antennas Wireless Propag. Lett.* **16**: 1771–1775.
23. Wang, M., Ma, H.F., Zhang, H.C. et al. (2018). Frequency-fixed beam-scanning leaky-wave antenna using electronically controllable corrugated microstrip line. *IEEE Trans. Antennas Propag.* **66** (9): 4449–4457.
24. Chang, L., Li, Y., Zhang, Z., and Feng, Z. (2018). Reconfigurable 2-bit fixed-frequency beam steering array based on microstrip line. *IEEE Trans. Antennas Propag.* **66** (2): 683–691.
25. Li, Z., Guo, Y.J., Chen, S.-L., and Wang, J. (2019). A period-reconfigurable leaky-wave antenna with fixed-frequency and wide-angle beam scanning. *IEEE Trans. Antennas Propag.* **67** (6): 3720–3732.
26. Chen, S.-L., Karmokar, D.K., Li, Z. et al. (2019). Continuous beam scanning at a fixed frequency with a composite right/left-handed leaky-wave antenna operating over a wide frequency band. *IEEE Trans. Antennas Propag.* **67** (12): 7272–7284.
27. Garcia-Vigueras, M., Gomez-Tornero, J.L., Goussetis, G. et al. (2011). 1D-leaky wave antenna employing parallel-plate waveguide loaded with PRS and HIS. *IEEE Trans. Antennas Propag.* **59** (10): 3687–3694.

28. García-Vigueras, M., Gómez-Tornero, J.L., Goussetis, G. et al. (2010). A modified pole-zero technique for the synthesis of waveguide leaky-wave antennas loaded with dipole-based FSS. *IEEE Trans. Antennas Propagat.* **58** (6): 1971–1979.

29. Johnson, H.W. and Graham, M. (1993). *High-Speed Digital Design: A Handbook of Black Magic.* Englewood Cliffs, NJ: Prentice Hall.

30. Erentok, A. and Ziolkowski, R.W. (2008). Metamaterial-inspired efficient electrically small antennas. *IEEE Trans. Antennas Propag.* **56** (3): 691–707.

31. Lin, W. and Ziolkowski, R.W. (2018). Electrically small, low-profile, Huygens circularly polarized antenna. *IEEE Trans. Antennas Propag.* **66** (2): 636–643.

32. Lee, M.W.K., Leung, K.W., and Chow, Y.L. (2015). Dual polarization slotted miniature wideband patch antenna. *IEEE Trans. Antennas Propag.* **63** (1): 353–357.

33. Lyu, Y.L., Liu, X.X., Wang, P.Y. et al. (2016). Leaky-wave antennas based on noncutoff substrate integrated waveguide supporting beam scanning from backward to forward. *IEEE Trans. Antennas Propag.* **64** (6): 2155–2164.

34. Mikulasek, T., Lacik, J., Puskely, J., and Raida, Z. (2016). Design of aperture-coupled microstrip patch antenna array fed by SIW for 60 GHz band. *IET Microw. Antennas Propag.* **10** (3): 288–292.

35. Engheta, N. and Ziolkowski, R.W. (2006). *Metamaterials: Physics and Engineering Explorations.* Wiley.

36. Lai, A., Itoh, T., and Caloz, C. (2004). Composite right/left-handed transmission line metamaterials. *IEEE Microwave Mag.* **5** (3): 34–50.

37. Caloz, C. and Itoh, T. (2006). *Electromagnetic Metamaterials: Transmission Line Theory and Microwave Applications.* Wiley.

38. Eleftheriades, G.V. and Balmain, K.G. (2005). *Negative-Refraction Metamaterials: Fundamental Principles and Applications.* Wiley.

39. Mosallaei, H. and Sarabandi, K. (2007). Design and modeling of patch antenna printed on magneto-dielectric embedded-circuit metasubstrate. *IEEE Trans. Antennas Propag.* **55** (1): 45–52.

40. Comite, D., Podilchak, S.K., Baccarelli, P. et al. (2017). A dual-layer planar leaky-wave antenna designed for linear scanning through broadside. *IEEE Antennas Wireless Propag. Lett.* **16**: 1106–1110.

41. Liu, J., Zhou, W., and Long, Y. (2018). A simple technique for open-stopband suppression in periodic leaky-wave antennas using two nonidentical elements per unit cell. *IEEE Trans. Antennas Propag.* **66** (6): 2741–2751.

42. Lyu, Y.L., Meng, F.Y., Yang, G.H. et al. (2018). Leaky-wave antenna with alternately loaded complementary radiation elements. *IEEE Antennas Wireless Propag. Lett.* **17** (4): 679–683.

43. Lian, J.W., Ban, Y.L., Zhu, H., and Guo, Y.J. (2020). Uniplanar high-gain 2-D scanning leaky-wave multibeam array antenna at fixed frequency. *IEEE Trans. Antennas Propag.* **68** (7): 5257–5268. https://doi.org/10.1109/TAP.2020.2975285.

44. Lian, J.-W., Ban, Y.-L., Zhu, J.-Q. et al. (2018). SIW multibeam antenna based on modified horn beam-forming network. *IEEE Antennas Wireless Propag. Lett.* **17**: 1866–1870.

45. Nguyen-Trong, N. and Fumeaux, C. (2018). Half-mode substrate-integrated waveguides and their applications for antenna technology: a review of the possibilities for antenna design. *IEEE Antennas Propag. Mag.* **60** (6): 20–31.

46. Lian, J.-W., Ban, Y.-L., Chen, Z. et al. (2018). SIW folded Cassegrain lens for millimeter-wave multibeam application. *IEEE Antennas Wireless Propag. Lett.* **17** (4): 583–586.

47. Ettorre, M., Neto, M.A., Gerini, G., and Maci, S. (2008). Leaky-wave slot array antenna fed by a dual reflector system. *IEEE Trans. Antennas Propag.* **56** (10): 3143–3149.

48. Cheng, Y.J., Hong, W., Wu, K., and Fan, Y. (2011). Millimeter-wave substrate integrated waveguide long slot leaky-wave antennas and two-dimensional multibeam applications. *IEEE Trans. Antennas Propag.* **59** (1): 40–47.

49. Ettorre, M., Sauleau, R., and Coq, L.L. (2011). Multi-beam multi-layer leakywave SIW pillbox antenna for millimeter-wave applications. *IEEE Trans. Antennas Propag.* **59** (4): 1093–1100.
50. Ma, Z.L. and Chan, C.H. (2017). A novel surface-wave-based high-impedance surface multibeam antenna with full azimuth coverage. *IEEE Trans. Antennas Propag.* **65** (4): 1579–1588.
51. Mailloux, R.J. (2005). *Phased Array Antenna Handbook*, seconde. Norwood, MA: Artech House.
52. Zou, X., Geng, F.Z., Li, Y., and Leng, Y. (2017). Phase inverters based on substrate integrated waveguide. *IEEE Microw. Wireless Compon. Lett.* **27** (3): 227–229.
53. Lima, E.B., Matos, S.A., Costa, J.R. et al. (2015). Circular polarization wide-angle beam steering at Ka-band by in-plane translation of a plate lens antenna. *IEEE Trans. Antennas Propag.* **63** (12): 5443–5455.
54. Zhou, W., Liu, J., and Long, Y. (2018). Investigation of shorting vias for suppressing the open stopband in an SIW periodic leaky-wave structure. *IEEE Trans. Microw. Theory Techn.* **66** (6): 2936–2945.
55. Lian, J.-W., Ban, Y.-L., Zhu, J.-Q. et al. (2019). Planar 2-D scanning SIW multibeam array with low sidelobe level for millimeter-wave applications. *IEEE Trans. Antennas Propag.* **67** (7): 4570–4578.
56. Tawk, Y., Costantine, J., Hemmady, S. et al. (2012). Demonstration of a cognitive radio front-end using an optically pumped reconfigurable antenna system (OPRAS). *IEEE Trans. Antennas Propag.* **60** (2): 1075–1083.
57. Aljonubi, K., AlAmoudi, A.O., Langley, R.J., and Reaney, I. (2013). Reconfigurable antenna using smart material. *Proceedings of 2013 7th European Conference on Antennas and Propagation (EuCAP 2013)*, Gothenburg, Sweden, April 8–12, 2013, pp. 917–918.
58. Apaydin, N., Sertel, K., and Volakis, J.L. (2014). Nonreciprocal and magnetically scanned leaky-wave antenna using coupled CRLH lines. *IEEE Trans. Antenna Propag.* **62** (6): 2954–2961.
59. Aljonubi, K., Langley, R.J., Reaney, I., and AlAmoudi, A.O. (2013). Piezoelectric reconfigurable antenna. *Proceedings of 2013 Loughborough Antennas & Propagation Conference (LAPC 2013)*, Loughborough, UK, November 11–12, 2013, pp. 47–50.
60. Alqurashi, K.Y., Kelly, J.R., Wang, Z. et al. (2019). Liquid metal bandwidth-reconfigurable antenna. *IEEE Antennas Wireless Propag. Lett.* **19** (1): 218–222.
61. Chen, Z., Wong, H., and Kelly, J. (2019). A polarization-reconfigurable glass dielectric resonator antenna using liquid metal. *IEEE Trans. Antennas Propag.* **67** (5): 3427–3432.

8

Beam Pattern Synthesis of Analog Arrays

Given an antenna array and a digital beamformer, one can form individually steerable multi-beams using adaptive beamforming algorithms, but this is achieved at the high cost of both analog and digital hardware, as well as energy consumption. As discussed in Chapter 1, analog arrays can potentially offer significant advantages over digital beamforming arrays such as lower fabrication costs and lower energy consumption. Without direct access to digital array signals, however, analog arrays cannot easily employ adaptive beamforming algorithms to optimize the beam patterns in real time. Consequently, antenna engineers tend to resort to switches for beam-switching or array synthesis methods to form the desired, but often fixed, beam patterns.

It is extremely difficult, if not impossible, to employ one array synthesis algorithm to produce and optimize beam patterns that meet the requirements of all systems. This is partly due to the variety and flexibility of array topologies. For example, there are uniformly distributed and nonuniformly distributed arrays; there are arrays with fixed and reconfigurable elements; and there are arrays for single beam and multi-beam scanning. Each array configuration may have a different set of variables that may change the character of the objective function. Moreover, the effectiveness of different synthesis algorithms is also impacted by various practical constraints. To a large extent, these array issues are similar to the optimization algorithms themselves – the nature of the optimization problem dictates the suitability of certain algorithms. Nevertheless, algorithms for synthesizing analog arrays are expected to play a critical role in current 5G and future 6G and beyond systems.

In this chapter, we shall address a number of array synthesis problems facing the design of 5G, 6G, and beyond arrays. These include the synthesis of thinned arrays to save cost; the synthesis of antenna arrays using element rotations to simplify the feed network; the synthesis of sum and difference patterns for communications with moving platforms; and the synthesis of single input multiple output (SIMO) multi-beam arrays.

8.1 Thinned Antenna Arrays

As noted in Chapter 1, a large number of 5G base stations, especially those at mm-wave frequencies, will employ massive antenna arrays. In theory, a "full" array is needed to maximize their benefits. The advantages of full arrays include high gain, low sidelobe levels (SLLs), and multi-user support. However, the implementation and operation of full massive arrays come with high capital investment and operation costs. In practice, therefore, one might resort to thinned arrays in which a

Advanced Antenna Array Engineering for 6G and Beyond Wireless Communications, First Edition.
Y. Jay Guo and Richard W. Ziolkowski.

significant number of array elements are removed. The thinning can be realized in the design phase in which one removes some array elements while keeping the array dimensions fixed in order to maintain narrow beamwidths and low SLLs. The thinning can also be realized during operations in which some antenna elements are put to "sleep" in order to save power during the off-peak hours. These two scenarios effectively lead to the same optimization problem, though the latter might need a real-time solution.

Given the number of antenna elements and the dimensions of the aperture they occupy, one can optimize the beamwidth and the SLL by optimizing the element positions in a relatively larger aperture. To this end, many element position optimization methods have been developed. These include, for example, the stochastic optimization algorithms [1–4], the sparse array reconstruction algorithms [5–9], the iterative convex optimization algorithms [10–12], and the iterative fast Fourier transform (FFT) algorithms [13–15]. Among them, the iterative FFT algorithm is the most efficient computationally because it utilizes the Fourier transform relationship between the radiation pattern and the aperture excitation distribution and the well-established FFT algorithms. The computational efficiency is further facilitated by the periodicity of the array factors of arrays with equally spaced elements.

The conventional iterative FFT usually starts with the desired radiation pattern and a predefined array thinning or filling factor that defines what percentage of the elements in an array the designer intends to utilize. It then chooses the winning elements in every iteration step by simply discarding those elements with small excitation amplitudes. The iteration stops once the desired thinning factor is achieved and the beam pattern requirements are met. However, the drawback of such algorithms is that the iteration process tends to get trapped into a dead loop [14, 15].

In what follows, we introduce a modified iterative FFT for designing beam-scannable, thinned, and massive antenna arrays [16, 17]. This modified iterative FFT adopts a gradual array thinning strategy which can reduce the likelihood of falling into a dead loop. Furthermore, the periodicity of the radiation pattern of a uniformly spaced array is also explored in order to reduce the computational complexity in synthesizing scannable beam patterns. A 128-element array obtained by thinning a 24×12 full array is synthesized to illustrate the effectiveness of the modified iterative FFT algorithm. The array employs the U-slot microstrip antenna as the array element and operates in the 27.5–28.5 GHz frequency band. The synthesis results show that narrow beams and low sidelobes can indeed be achieved by significantly thinned arrays.

8.1.1 Modified Iterative FFT

Consider a planar array with $M \times N$ elements uniformly arranged in a rectangular grid at distance d_x along the M columns and d_y along the N rows. The array factor (AF) is given by:

$$\mathrm{AF}(u,v) = \sum_{m=0}^{M-1}\sum_{n=0}^{N-1} I_{m,n} e^{j\beta\left(md_x u + nd_y v\right)} \tag{8.1}$$

where

$$u = \sin\theta\cos\varphi - \sin\theta_0\cos\varphi_0 \tag{8.2}$$

$$v = \sin\theta\sin\varphi - \sin\theta_0\sin\varphi_0 \tag{8.3}$$

θ and φ denote the zenith and azimuth angles, respectively; $I_{m,\,n}$ is the excitation coefficient of the $(m,\,n)$-th element; $j = \sqrt{-1}$ is the imaginary unit; $\beta = 2\pi/\lambda$ is the wavenumber in free space with λ as the wavelength; and $(\theta_0,\,\varphi_0)$ is the intended beam-pointing direction.

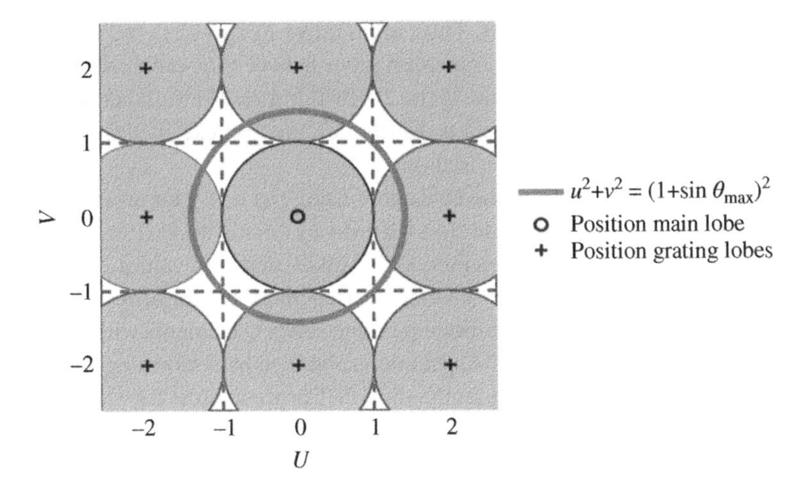

Figure 8.1 The array factor and its periodicity for $d_x = d_y = \lambda/2$, $T_u = T_v = 2$. *Source*: From [17] / with permission of IEEE.

Let us first consider the range of (u, v) for the array factor. When the beam-pointing direction is $\theta_0 = 0$ and $\varphi_0 = 0$, we have $u = \sin\theta\cos\varphi \in [-1, 1]$ and $v = \sin\theta\sin\varphi \in [-1, 1]$. In this situation, the visible region of AF(u, v) can be defined as $\Omega_0 = \{(u, v); \mid u^2 + v^2 \leq 1\}$. For a more general case of beam scanning within the range: $\{(\theta_0, \varphi_0); \mid 0 \leq \theta_0 \leq \theta_{\max} \,\&\, 0 \leq \varphi_0 \leq 2\pi\}$, the variation range of (u, v) is given by

$$\Omega_s = \left\{(u, v); \mid \sqrt{u^2 + v^2} \leq 1 + \sin\theta_{\max}\right\} \tag{8.4}$$

As shown in Figure 8.1, the region Ω_s that is a function of the maximum scanning angle $\theta_{\max}$ may cover some space outside of the square cell $\Omega_{\text{cell}} = \{(u, v) \mid |u| \leq 1 \,\&\, |v| \leq 1\}$.

The array factor for a uniformly spaced planar array has a very interesting property that AF(u, v) is a two-dimensional periodic function with periods of $T_u = \lambda/d_x$ and $T_v = \lambda/d_y$ with respect to u and v, respectively. Hence, it only needs to be evaluated within one period in (u, v)-space. For example, when $d_x = d_y = \lambda/2$, we have $T_u = T_v = 2$. The array factor in this situation only needs to be calculated within the square cell Ω_{cell}. For the part within Ω_s but outside of Ω_{cell}, one can simply use the array factor values obtained from the corresponding part in Ω_{cell}. This property can be used to reduce the computational complexity when we evaluate the array factor.

From Eq. (8.1), it is seen that the array factor defined in (u, v)-space is an inverse two-dimensional (2D) discrete-space Fourier transform (DSFT) of the excitation distribution $\{I_{m,n}\}$. This means that the 2D inverse FFT (2D-IFFT) can be used to speed up the computation of this array factor. As is mentioned above, the scannable array factor for a uniformly spaced planar array can be evaluated from only the information in one period Ω_{cell}. By setting $u = \frac{k\lambda}{Kd_x}\left(k = -\frac{K}{2}, \cdots, \frac{K}{2} - 1\right)$ and $v = \frac{l\lambda}{Ld_y}\left(l = -\frac{L}{2}, \cdots, \frac{L}{2} - 1\right)$, we obtain:

$$\text{AF}(k, l) = \sum_{m=0}^{M-1}\sum_{n=0}^{N-1} I_{m,n} e^{j\frac{2\pi mk}{K}} e^{j\frac{2\pi nl}{L}} \tag{8.5}$$

Clearly, the above summation can be efficiently evaluated using a 2D-IFFT. Conversely, once the array factor in one period Ω_{cell} is known, the excitation distribution $\{I_{m,n}\}$ can be obtained efficiently by performing a 2D-FFT on the array factor. Thus, an iterative FFT (I-FFT) technique can be developed in which the forward and backward transformations between the excitation distribution and array factor are successively performed using the 2D-FFT and 2D-IFFT. In addition, some modifications of the pattern shape and/or excitation distribution are also executed in each iteration to make the synthesis result approach the desired one [13].

From a predefined uniformly spaced array layout, the concept of using the I-FFT for array synthesis can be extended to synthesize a uniform amplitude-thinned array by forcing the excitations of certain elements to be 1 and discarding the remaining elements with small excitation values at each iteration. For example, assume that the goal is to select Q elements from an $M \times N$-element layout. The filling factor is given as $f = Q/MN$. The I-FFT technique needs to select Q elements with uniform excitations in each iteration [14, 15]. In reality, the excitations obtained by performing a 2D-FFT on a given $\{AF(u, v)\}$ are not exactly 1 or 0. Consequently, the I-FFT process forces the Q larger excitations to be 1 and discards the other $(MN - Q)$ elements in each iteration whose excitations are smaller. This synthesis procedure proceeds until a convergence is reached, which is typically defined as the point where the newest distribution of the selected Q "turned-on" elements remains the same as that obtained at the previous iteration.

The algorithm described above has two drawbacks because of its crude operations. First, the iterative synthesis process can easily get trapped in local optima after only a few iterations, typically less than 10. Hence, when using the original I-FFT presented in [14, 15], one usually needs to choose a different initial distribution of element excitations many times and restart the synthesis. It is not unusual that several hundreds or even thousands of reinitiations are needed to obtain a satisfactory pattern with relatively low SLLs, say, less than −13 dB. Second, the original I-FFT has no beamwidth control (BWC) mechanism for the synthesized pattern in the iteration process. Consequently, the beamwidth of the obtained final pattern may be wider than what was desired for a given aperture size.

To overcome these problems, a modified iterative FFT (MI-FFT) was introduced in [17]. The basic concept is to adopt a gradual array-thinning strategy in which more than the targeted number of elements, Q, are chosen at the beginning, and then the number of selected elements are gradually reduced as the iteration proceeds. As a consequence, the distribution of the selected elements can be more easily changed by modifying the radiation pattern in each iteration, thus preventing premature stagnation. Furthermore, beamwidth control is incorporated into the synthesis process by treating the radiation outside of the desired mainlobe region as sidelobes in the pattern modification step in each iteration. The MI-FFT algorithm is detailed as follows:

Algorithm: Modified Iterative FFT for Synthesizing Beam-Scannable Thinned Massive Planar Arrays

Initialization:

1) Set an initial mesh grid with $M \times N$ uniformly spaced potential element positions. Set the maximum beam scanning angle θ_{max} and determine the scanning range Ω_s in the $u - v$ space;
2) Initialize the element excitation distribution of $I_{m,n} (m = 0, 1, \cdots, M-1; n = 0, 1, \cdots, N-1)$ such that each element has the probability p_n to be 1 and the probability $(1 - p_n)$ to be 0. Then set p_n with a value close to 1, e.g., $p_n = 0.95$;
3) Set the number of selected elements with excitation "1" to be Q;
4) Apply a $K \times L$-point IFFT on the distribution of $I_{m,n}$ to obtain the array factor $AF(u, v)$ over the rectangular region Ω_{cell} in the $u - v$ space. By utilizing the periodicity of the array factor, we can obtain all the values of $AF(u, v)$ in the scanning range Ω_s;

Beam pattern shaping:

5) Check whether the pattern levels within Ω_s meet the desired SLL-bound Γ_{SLL}. If not, adjust their magnitudes to an over-suppressed level (OSL) Γ_{OSL} and keep their phases unchanged. Γ_{OSL} is usually set to be lower than Γ_{SLL}, e.g., $\Gamma_{OSL} = \Gamma_{SLL} - 5$ dB;
6) Check whether the angular region covered by the pattern mainlobe is larger than the prescribed region. If so, lower the pattern level out of the prescribed region and keep the phases unchanged. Denote the modified pattern as $AF_{mod}(u, v)$;

Select the "on" elements:

7) Apply a $K \times L$-point FFT on the modified $AF_{mod}(u, v)$ to obtain the new excitation distribution $\widetilde{I}_{m,n}$;
8) Set $Q = Q - \text{integer}[\delta Q]$ where $\delta > 0$ is a small positive value, e.g., $\delta = 0.005$, and set the Q largest excitations of $\left\{\widetilde{I}_{m,n}\right\}$ to 1 and all others to 0;
9) Repeat Steps 4–8 until Q reaches the desired number of elements.

8.1.2 Examples of Thinned Arrays

To illustrate the effectiveness of the MI-FFT algorithm, a thinned massive antenna array that is potentially useful for 5G mm-wave applications has been designed as follows [17]. The thinned antenna array has 128 elements whose positions are optimally selected from a predefined 24 × 12 mesh grid using the algorithm. Each radiating element is a U-slot microstrip antenna with a parasitic-patch. The array works in the frequency band from 25.2 to 31.6 GHz. The performance of the designed thinned array and that of a conventional fully occupied array with the same number of elements are compared below.

For an antenna working at 28.5 GHz for 5G mm-wave communications, broad impedance bandwidth, high gain, and beam scanning ability are generally required. Figure 8.2 shows the

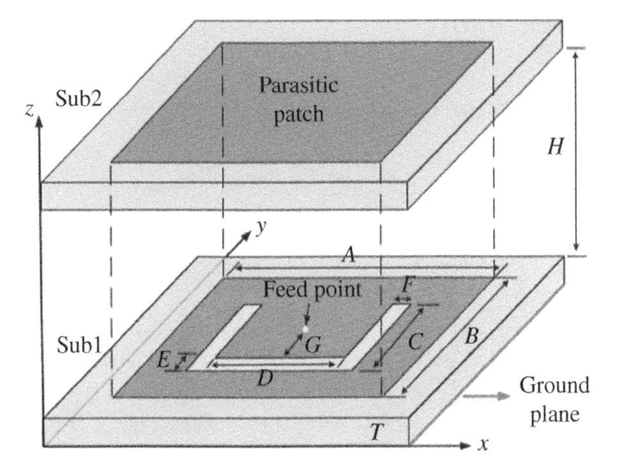

Figure 8.2 Design details of the parasitic-patch-based antenna used in the MI-FFT thinned array. Optimized parameters (in millimeters): $A = 3.10$, $B = 2.37$, $C = 1.67$, $D = 1.49$, $E = 0.14$, $F = 0.19$, $G = 0.58$, $H = 2.25$, and $T = 1.575$. The substrates, Sub1 and Sub2, have the dielectric constant $\varepsilon_r = 2.2$. *Source*: From [17] / with permission of IEEE.

parasitic-patch-based antenna used as the array element and its optimized design. It consists of two substrates: the lower substrate, Sub1, with a U-slot loaded patch working as the driver antenna, and the upper substrate, Sub2, with the same-sized rectangular patch working as the parasitic antenna. Rogers Duroid^{TM} 5880 with a dielectric constant $\varepsilon = 2.2$ and a loss tangent 0.0009 is utilized for both Sub1 and Sub2. It is understood that the performance of the antenna is determined by the size of the patch and the U-slot, and by the distance, H, between Sub1 and Sub2. Specifically, the impedance bandwidth and the gain of the antenna are related to H.

The simulated reflection coefficient, E-plane gain and H-plane gain of the antenna for different values of H are shown in Figures 8.3a–c, respectively. The corresponding results for the U-slot antenna without the parasitic rectangular patch are also included in the figure. As expected, it is observed in Figure 8.3a that the impedance bandwidth is increased after adding the parasitic patch. It is also seen in Figure 8.3b that the E-plane pattern becomes narrower while the maximum gain direction approaches $\theta = 0°$ after the parasitic rectangular patch is added. As can be seen in Figure 8.3c, the gain in the H-plane is increased and the 3 dB beamwidth becomes broader. Broader beamwidth is better for beam scanning. Parameter studies determined that the optimum value of H is 2.25 mm, which results in a fractional impedance bandwidth of 22.5%, covering 25.2–31.6 GHz, and with a gain of 6.53 dBi.

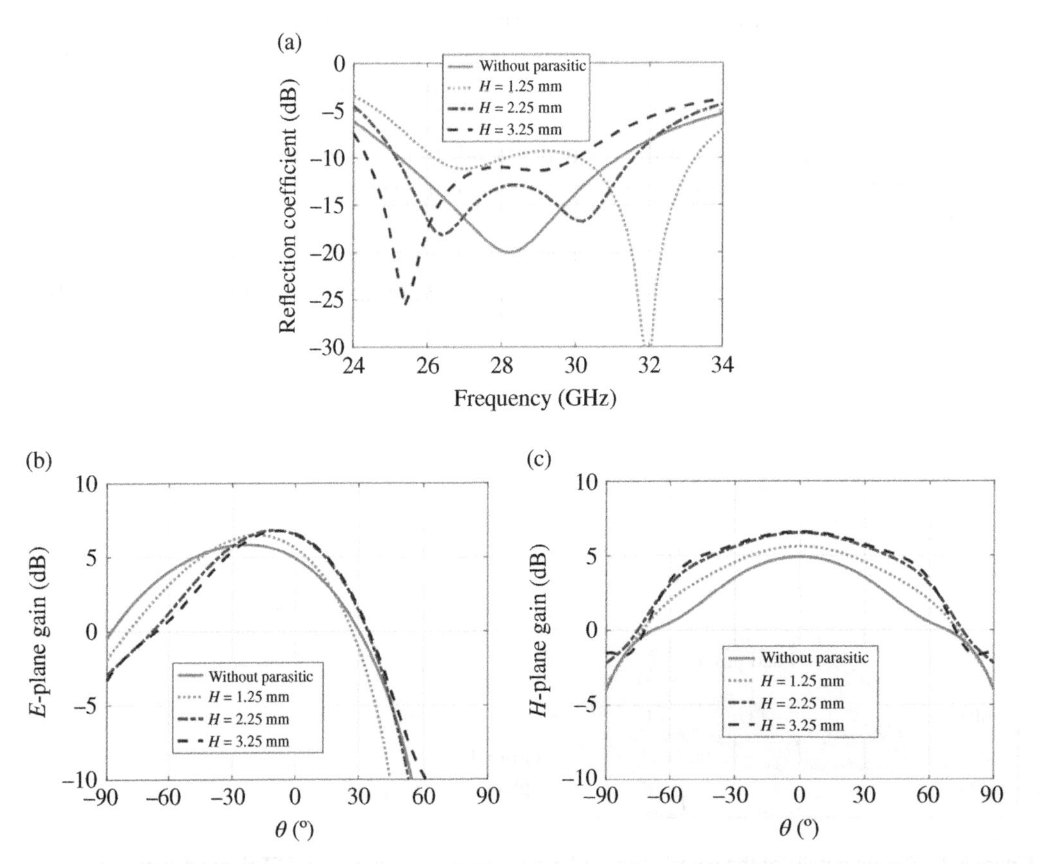

Figure 8.3 Simulated performance characteristics of the parasitic patch-based antenna as functions of H. (a) Reflection coefficient. *Source:* (a) From [17] / with permission of IEEE. (b) E-plane gain pattern. (c) H-plane gain pattern. *Source*: (b and c) Based on [17] / IEEE.

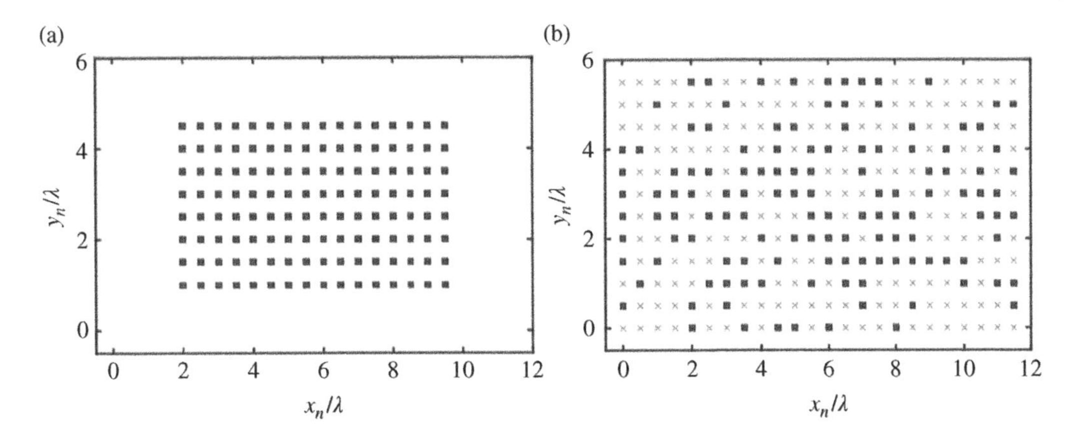

Figure 8.4 Two versions of the 128-element array layout. (a) Conventional fully occupied array. (b) An array thinned from a 24 × 12 grid (squares denote the selected elements). *Source*: From [17] / with permission of IEEE.

A 128-element antenna array was then designed with a beam scanning range of $\{(\theta_0, \varphi_0); \mid |\theta_0| \leq 60° \; \& \; 0 \leq \varphi_0 \leq 360°\}$. As shown in Figure 8.4a, a conventional approach would be to place the 128 antenna elements on a $\lambda/2$-spaced 16×8 mesh grid. As shown in Figures 8.5a and 8.5b, the simulated broadside pattern obtained by multiplying the broadside array factor and element pattern has a maximum SLL of −13.30 dB and the beamwidth is 5.63° in the $\varphi = 0°$-plane and 11.95° in the $\varphi = 90°$-plane. When the pattern is scanned to the direction $(\theta_0 = -60°, \varphi_0 = 0°)$, the maximum SLL is increased to −11.89 dB and the beamwidth in the $\varphi = 0°$-plane is increased to 11.25°. When the pattern is scanned in the $\varphi = 90°$-plane, the SLL and the beamwidth are also increased similarly.

To improve the pattern performance, we choose to optimize the antenna array layout by selecting 128 element positions from a predefined $\lambda/2$-spaced 24×12 mesh grid. It provides a larger aperture with many more degrees of element position freedoms. For comparison, we apply both the original I-FFT and the proposed MI-FFT algorithms to synthesize a thinned array. For both methods, we set $\Gamma_{SLL} = -20$ dB for the desired SLL and $\Gamma_{OSL} = -25$ dB for the over-suppressed level. Since no beamwidth control (BWC) is considered in the original I-FFT, we first do not use the BWC mechanism in the MI-FFT for a fair comparison. It should be noted that in the original I-FFT, only 128 elements are selected and all other potential elements are brutally discarded in each iteration. In contrast, we adopt $p_n = 0.95$ and $\delta_n = 0.005$ for controlling the reduction of the number of selected elements in the MI-FFT approach. In this example, 274 elements were selected at the beginning and then the number of selected elements was gradually reduced to 128. To compare the performance of these two methods comprehensively, they were run for 2000 times by starting from different random initial excitation distributions. Figure 8.6 shows the statistical histograms of the obtained maximum SLLs of the array factors by these two methods over 2000 runs. It is clear that the achievable best and averaged SLLs obtained with the MI-FFT are much lower than the corresponding results obtained using the original I-FFT. This proves the effectiveness of gradually reducing the number of selected elements instead of simply discarding all other elements at the beginning.

We then add the BWC into the MI-FFT process to narrow the obtained beamwidth. The final array layout of the thinned 128-element array is shown in Figure 8.4b. The corresponding patterns obtained by multiplying the array factors with the element pattern in the broadside beam and scanned beam cases are shown in Figures 8.5a–d. The patterns obtained using the MI-FFT without the BWC are also shown therein for comparison. As can be seen, the MI-FFT with/without the

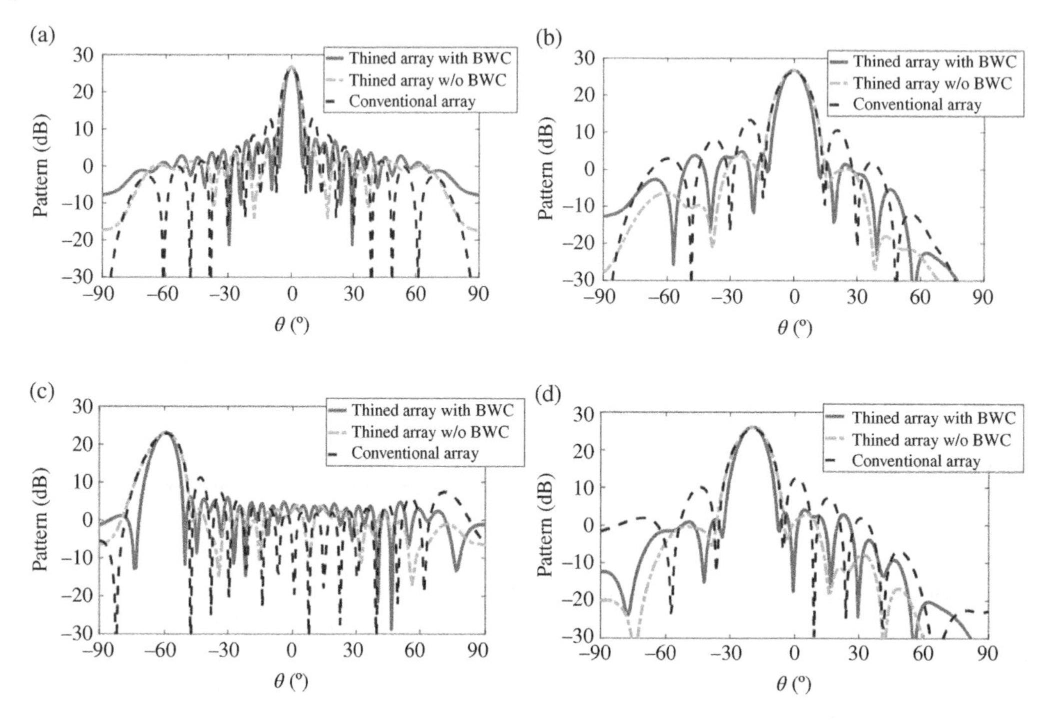

Figure 8.5 The gain patterns of the thinned and full arrays, both with 128 elements. (a) Pattern in the $\varphi = 0°$ plane with the mainbeam maximum at boresight. (b) Pattern in the $\varphi = 90°$ plane with the mainbeam maximum at boresight. (c) Pattern in the $\varphi = 0°$ plane with the mainbeam maximum pointed at ($\theta_0 = -60°$, $\varphi_0 = 0°$). (d) Pattern in the $\varphi = 90°$ plane with mainbeam maximum pointed at ($\theta_0 = -20°$, $\varphi_0 = 90°$). *Source*: From [17] / with permission of IEEE.

BWC can obtain thinned arrays that have much lower SLLs than the conventional fully occupied array. This improvement is due to optimization of the array layout over a larger aperture. Compared with the thinned array obtained without the BWC, the thinned array obtained with it indeed has a significantly reduced beamwidth while the SLL is only increased marginally. The beamwidth of the broadside beam generated with this thinned array is 4.22° in the $\varphi = 0°$-plane and 9.14° in the

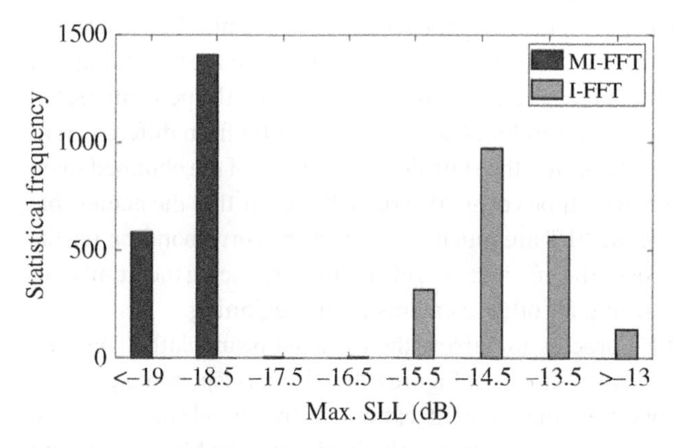

Figure 8.6 A histogram comparison of the maximum SLL obtained for the 128 element thinned array with the I-FFT and MI-FFT algorithms. *Source*: From [17] / with permission of IEEE.

$\varphi = 90°$-plane. These beamwidths are much narrower than those of the conventional array. The relative advantages of the thinned arrays synthesized using the MI-FFT are maintained when the pattern is scanned. We conclude that synthesized thinned arrays have much better overall pattern performance in terms of both narrower beamwidths and lower SLLs than conventional fully occupied arrays. Therefore, they serve as a promising candidate for 5G, 6G, and beyond mm-wave communications systems.

8.2 Arrays with Rotated Elements

The conventional way to synthesize a beam pattern with an analog antenna array is to optimize both the amplitude and phase distributions of the excitation coefficients [18, 19]. However, this may lead to a complicated feeding network in order to realize both the amplitude and phase weightings. Moreover, since multiple unequal power dividers would be required, the design complexity and hardware cost would increase significantly. To avoid the use of unequal power dividers, one could resort to phase-only synthesis methods. However, certain array performance compromises have to be made with the phase-only synthesis approach because it has a limited number of degrees of freedom [20]. Inspired by concepts associated with polarization-reconfigurable antennas, we introduce another degree of freedom for array synthesis, i.e., the rotation of each array element. By optimizing both the distributions of the phase and orientations of the elements in an array, we can obtain synthesized vectorial-shaped power patterns that yield much better performance than simply using the phase-only approach.

It is well understood that rotating an antenna element can change its co-polarized (CoP) and cross-polarized (XP) patterns in a given fixed observation plane. Hence, element rotation can be considered as a way of providing an additional degree of freedom for array pattern synthesis. One of the challenges faced to incorporate element rotation into the synthesis process is the inclusion of mutual coupling. The coupling between any two antenna elements changes with the relative rotation of all of the antenna elements in an array. Since there are no closed form solutions for the mutual coupling between most types of antennas, pattern synthesis is necessarily an iterative process in practice. However, it is computationally too demanding, if not intractable, to conduct a full-wave analysis of a large array every time an antenna element in it is rotated. The developed solution is based on the assumption that only small rotation angles are introduced in each iterative step of the synthesis process.

First, the vectorial active element pattern (VAEP) for each antenna element is adopted in order to include the mutual coupling (MC) effect in the array environment. Second, we assume that the VAEPs of all of the array elements remain the same when some array elements are rotated by small angles. Naturally, the accuracy of such an assumption depends on the rotation angle range for a given element structure and the distribution of the element positions. Third, once a new iteration is completed and a new array configuration is formed, a full array simulation is conducted to determine the new VAEPs of all of the elements. Depending on the rotation increment for each iteration, the VAEP refinement can be done after either every iteration or a few of them.

8.2.1 The Pattern of an Element-Rotated Array

Consider an antenna array with N rotated elements located in the xOy-plane. As an illustration, Figure 8.7a shows a 3×3 element-rotated planar patch array. Suppose it is obtained by separately rotating each element of a conventional array shown in Figure 8.7b in the xOy-plane about the

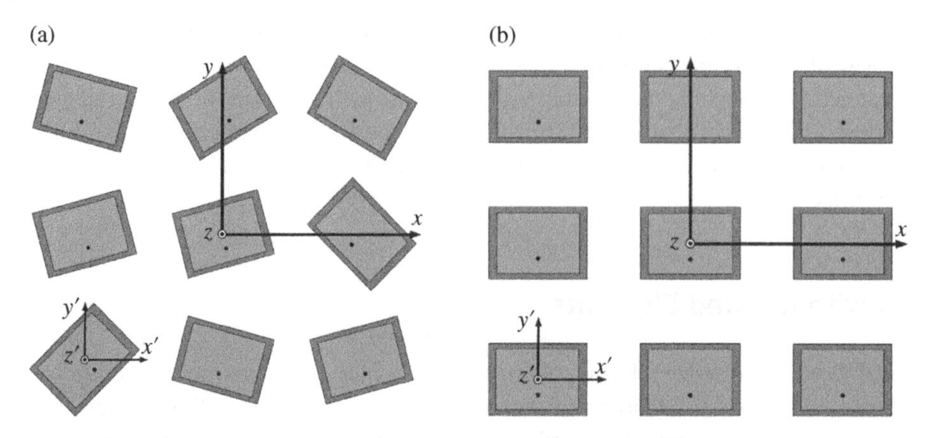

Figure 8.7 Configuration of a 3 × 3 planar array (a) with, and (b) without rotated elements. *Source:* From [21]/with permission of IEEE.

z'-axis of the local coordinate system. Assume the rotation angles are denoted by ξ_n for $n = 1, 2, \ldots, N$. The vectorial array pattern is given as:

$$\vec{F}^{\text{Rot}}(\theta, \varphi) = \sum_{n=1}^{N} \vec{E}_n(\theta, \varphi; \xi_n) e^{j\{\beta \vec{r}_n \cdot \vec{u}(\theta, \varphi) + \alpha_n\}} \tag{8.6}$$

where $\vec{u}(\theta, \varphi) = \sin\theta\cos\varphi\,\hat{e}_x + \sin\theta\sin\varphi\,\hat{e}_y + \cos\theta\,\hat{e}_z$ is the propagation direction vector and α_n is the excitation phase of the n-th element located at $\vec{r}_n$. The phase-adjusted VAEP for the n-th element with a rotation angle of ξ_n (the coordinate origin is located at each element and $\xi_n > 0$ denotes an anticlockwise rotation) is $\vec{E}_n(\theta, \varphi; \xi_n) = \vec{E}_{n,\theta}(\theta, \varphi; \xi_n)\,\hat{e}_\theta + \vec{E}_{n,\varphi}(\theta, \varphi; \xi_n)\,\hat{e}_\varphi$. The AEP is defined as the pattern of the array in which only one element is excited while the remaining elements are connected to matched loads. If the AEP concept were used, the array pattern expression (8.6) would include the mutual coupling even for a complicated array geometry with rotated antenna elements.

The AEPs in an array environment vary across different elements. For an element-rotated array, the AEP of an antenna element depends not only on its rotation angle, but also on the rotation angles of the other elements, especially those nearby. In general, the AEPs can be obtained using full-wave simulations or measurements. However, the element rotations and phases are unknown variables for the synthesis problem to achieve shaped patterns by jointly optimizing the rotations and phases of the elements until all of the elements are dealt with in each iteration step. To overcome this issue, it is assumed that when an element is rotated, the change in the mutual coupling between this element and nearby elements leads to little change in the AEP. Then the pattern of a rotated element can be obtained approximately by mathematically rotating the AEP of this element from its previous rotation state. Mathematically, this means the phase-adjusted AEP for the n-th rotated element is approximated as:

$$\vec{E}_n(\theta, \phi; \xi_n) \approx E_{n,\theta}(\theta, \phi - \xi_n; 0)\,\hat{e}_\theta + E_{n,\varphi}(\theta, \phi - \xi_n; 0)\,\hat{e}_\phi \tag{8.7}$$

where $\vec{E}_{n,\theta}(\theta, \varphi; 0)$ and $\vec{E}_{n,\varphi}(\theta, \varphi; 0)$ represent the $\hat{e}_\theta$ and $\hat{e}_\varphi$ polarization components of the AEP for the n-th element with a rotation angle of "0°." By substituting (8.7) into (8.6) the components of the vectorial array pattern become:

$$F_\theta^{\text{Rot}}(\theta, \phi) \approx \sum_{n=1}^{N} E_{n,\theta}(\theta, \phi - \xi_n; 0) e^{j\{\beta \vec{r}_n \cdot \vec{u}(\theta, \phi) + \alpha_n\}} \tag{8.8}$$

$$F_{\varphi}^{\mathrm{Rot}}(\theta,\phi) \approx \sum_{n=1}^{N} E_{n,\phi}(\theta,\phi-\xi_n;0)\mathrm{e}^{\,j\{\beta \vec{r}_n\cdot\vec{u}(\theta,\phi)+\alpha_n\}} \tag{8.9}$$

Given the element spacings and radiation characteristics, the approximation accuracy of these equations depends on the rotation angles ξ_n. Generally speaking, increasing the rotation angle will reduce the approximation accuracy. A synthesized pattern that is obtained by optimizing the element rotations and phases based on the approximations in (8.8) and (8.9) will deviate from the real array pattern due to changes in the mutual coupling. This pattern discrepancy depends heavily on the permissible range of the rotation angle ξ_n. Naturally, refining steps can be done to reduce the discrepancy. For example, once the optimized element rotations and phases using the approximated expressions are obtained, one can employ a full-wave simulation to acquire all of the real AEPs for the new configuration. Then the elements can be further rotated within a smaller range to reduce the discrepancy between the synthesized and real array patterns.

Clearly, this refined joint rotation/phase optimization can be performed multiple times until the discrepancy between the synthesized and real array patterns becomes negligible or less than a prescribed tolerance. Assume that the rotation angle for the n-th element is $\xi_n^{(0)}$ at the initial rotation step and becomes $\xi_n^{(k)}$ at the k-th refining step ($k=1, 2, \ldots, K$), respectively. The element phase for the n-th element is $\alpha_n^{(0)}$ at the initial rotation step and $\alpha_n^{(k)}$ at the k-th refining step, respectively. Then, the approximated vector components of the array pattern at the k-th refining step can be given as:

$$F_{\theta}^{(k)}(\theta,\varphi) \approx \sum_{n=1}^{N} E_{n,\theta}\left(\theta,\varphi-\xi_n^{(k)};\sum_{l=0}^{k-1}\xi_n^{(l)}\right)\mathrm{e}^{\,j\{\beta \vec{r}_n\cdot\vec{u}(\theta,\varphi)+\alpha_n\}} \tag{8.10}$$

$$F_{\varphi}^{(k)}(\theta,\varphi) \approx \sum_{n=1}^{N} E_{n,\varphi}\left(\theta,\varphi-\xi_n^{(k)};\sum_{l=0}^{k-1}\xi_n^{(l)}\right)\mathrm{e}^{\,j\{\beta \vec{r}_n\cdot\vec{u}(\theta,\varphi)+\alpha_n\}} \tag{8.11}$$

During the successive refining optimization process, the allowed range of $\xi_n^{(k)}$ can be set to be smaller and smaller as k increases. When the allowed range of $\xi_n^{(k)}$ becomes small enough, the synthesized array pattern will agree well with the real one and will have the mutual coupling included.

8.2.2 Vectorial Shaped Pattern Synthesis Using Joint Rotation/Phase Optimization

The goal of the vectorial shaped pattern synthesis using the joint element rotation/phase optimization method is to find the optimal rotations and phases such that the resulting shaped pattern has the CoP component approaching the desired mainlobe shape as close as possible in addition to constraining both the maximum SLL and XP levels (XPL) below the desired level. In practice, it may happen that a user-defined desired polarization, denoted by $\vec{p}_d$, is given as a fixed direction vector, e.g., $\hat{e}_y$. However, the realizable CoP direction radiated by an actual antenna array is always perpendicular to the propagation direction $\vec{u}(\theta,\varphi)$ and it varies with any change of that direction. Thus, if the realizable CoP is viewed in a wide angular range, it is usually different from the fixed user-defined desired polarization.

To facilitate the formulation of the vectorial shaped pattern synthesis problem, a definition of the realizable CoP given in [22] is adopted. The realizable CoP is defined as the projection of $\vec{p}_d$ onto the wavefront plane that is perpendicular to the propagation direction $\vec{u}(\theta,\varphi)$. Referring to Figure 8.8, the CoP is given as:

$$\vec{p}_{\mathrm{co}} = \frac{\vec{p}_d - \left[\vec{p}_d\cdot\vec{u}(\theta,\varphi)\right]\vec{u}(\theta,\varphi)}{\left|\vec{p}_d - \left[\vec{p}_d\cdot\vec{u}(\theta,\varphi)\right]\vec{u}(\theta,\varphi)\right|}. \tag{8.12}$$

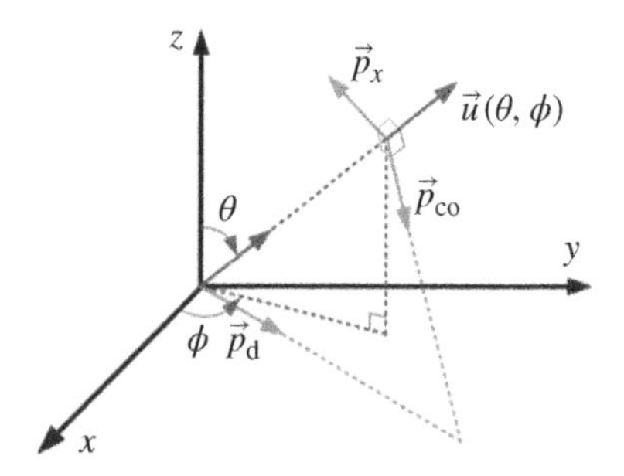

Figure 8.8 Illustration of the definitions of the CoP and XP directions. *Source*: From [21] / with permission of IEEE.

Then the realizable XP direction $\vec{p}_X$ is perpendicular to both $\vec{p}_{co}$ and $\vec{u}(\theta,\varphi)$ and, hence, is given as

$$\vec{p}_X = \vec{p}_{co} \times \vec{u}(\theta,\varphi). \tag{8.13}$$

Note that the so-defined CoP and XP directions can be regarded as an extension of Ludwig's polarization "Definition II" to a more general case of an arbitrarily desired $\vec{p}_d$ [23]. When $\vec{p}_d = \hat{e}_y$, these polarization definitions reduce to the form of Ludwig's Definition II.

By virtue of the above definitions, the approximated CoP and XP patterns can be obtained if the elements of a rotated antenna array are further rotated with angles of $\xi_n^{(k)}$ $(n = 1, 2, \ldots, N)$ at the k-th step. The approximated patterns are given by:

$$F_{co}^{(k)}(\theta,\varphi) \approx \sum\nolimits_{n=1}^{N} E_{n,co}\left(\theta, \varphi - \xi_n^{(k)}; \sum\nolimits_{l=0}^{k-1}\xi_n^{(l)}\right) e^{\,j\{\beta \vec{r}_n \cdot \vec{u}(\theta,\varphi) + \alpha_n\}} \tag{8.14}$$

$$F_{X}^{(k)}(\theta,\varphi) \approx \sum\nolimits_{n=1}^{N} E_{n,X}\left(\theta, \varphi - \xi_n^{(k)}; \sum\nolimits_{l=0}^{k-1}\xi_n^{(l)}\right) e^{\,j\{\beta \vec{r}_n \cdot \vec{u}(\theta,\varphi) + \alpha_n\}} \tag{8.15}$$

where

$$E_{n,co}\left(\theta, \varphi; \sum\nolimits_{l=0}^{k-1}\xi_n^{(l)}\right) = E_{n,\theta}\left(\theta, \varphi; \sum\nolimits_{l=0}^{k-1}\xi_n^{(l)}\right)\hat{e}_\theta \cdot \vec{p}_{co} + E_{n,\varphi}\left(\theta, \varphi; \sum\nolimits_{l=0}^{k-1}\xi_n^{(l)}\right)\hat{e}_\varphi \cdot \vec{p}_{co} \tag{8.16}$$

$$E_{n,X}\left(\theta, \varphi; \sum\nolimits_{l=0}^{k-1}\xi_n^{(l)}\right) = E_{n,\theta}\left(\theta, \varphi; \sum\nolimits_{l=0}^{k-1}\xi_n^{(l)}\right)\hat{e}_\theta \cdot \vec{p}_X + E_{n,\varphi}\left(\theta, \varphi; \sum\nolimits_{l=0}^{k-1}\xi_n^{(l)}\right)\hat{e}_\varphi \cdot \vec{p}_X. \tag{8.17}$$

The terms $E_{n,\theta}\left(\theta, \varphi; \sum_{l=0}^{k-1}\xi_n^{(l)}\right)$ and $E_{n,\varphi}\left(\theta, \varphi; \sum_{l=0}^{k-1}\xi_n^{(l)}\right)$ are obtained from the full-wave simulation of the antenna array after the $(k-1)$-th refining step. To achieve the desired shaped power pattern with constrained SLL and cross-polarization (XPL) for an element-rotated array, the rotation angles $\xi_n^{(k)}$ and excitation phases $\alpha_n^{(k)}$ at each step need to be optimized. A cost function to achieve this function is constructed as follows:

The cost function presented in [24] is extended here to deal with the refined joint rotation/phase optimization problem. Let $P_t(\theta, \varphi)$ denote the desired CoP mainlobe, and let Γ_{SLL} and Γ_{XPL} denote the desired SLL and XPL, respectively. The cost function at the k-th step is then chosen as:

$$f = \frac{W_1}{B}\left\{\left|F_{co}^{(k)}(\theta_b, \varphi_b)\right|^2 - P_t(\theta_b, \varphi_b)\right\}^2 + \frac{W_2}{C}\sum_{c=1}^{C}\frac{1}{2}(X_c + |X_c|)^2 + \frac{W_3}{D}\sum_{d=1}^{D}\frac{1}{2}(Y_d + |Y_d|)^2 \tag{8.18}$$

where

$$\begin{cases} X_c = \left|F_{\text{co}}^{(k)}(\theta_c, \varphi_c)\right|^2 - \Gamma_{\text{SLL}}; \theta_c, \varphi_c \in \text{SLL region} \\ Y_d = \left|F_X^{(k)}(\theta_d, \varphi_d)\right|^2 - \Gamma_{\text{XPL}}; \theta_d, \varphi_d \in \text{XPL region}. \end{cases} \tag{8.19}$$

The angle pairs: (θ_b, φ_b) for $b = 1, 2, ..., B$ and (θ_c, φ_c) for $c = 1, 2, ..., C$ are the sampling angles in the shaped mainlobe region and in the sidelobe region of the CoP pattern, respectively. The angles (θ_d, φ_d) for $d = 1, 2, ..., D$ are the sampling angles in the region in which the XPL needs to be constrained. The terms W_1, W_2, and W_3 are weighting factors. In general, a larger W_1 will lead to a better approximation of the desired mainlobe shape, but large ratios of W_1/W_2 and W_1/W_3 might increase the SLL and XPL. Hence, these parameters need to be chosen carefully in order to achieve a good overall pattern performance.

The minimization of the cost function (8.18) by optimizing the rotation angles and excitation phases is a highly nonlinear problem. Stochastic optimization algorithms that are able to find the global optimum solution would be generally applicable. Here, the Particle Swarm Optimization (PSO) algorithm is adopted to deal with this optimization problem [25]. It is relatively computationally inexpensive in terms of both memory requirements and computing speed [26–29]. In a PSO-based optimization process, a group of particles are randomly generated in the beginning, and each particle represents one solution of the set: $\left\{\left(\xi_n^{(k)}, \alpha_n^{(k)}\right); \mid n = 1, 2, ..., N\right\}$. Then guided by the cost function (8.18), the velocities and positions of these particles will be iteratively updated in the search of better solutions of the rotation angles and phases. Eventually, if the value of the cost function remains unchanged for multiple iterations or the preset maximum iteration number is reached, the optimization procedure will be terminated.

8.2.3 The Algorithm

The overall procedure of the shaped pattern synthesis process using the refined joint rotation/phase optimization method is:

1) Set the initial antenna array configurations including element structure, array dimension, and array configuration. Set the desired mainlobe shape $P_t(\theta_b, \varphi_b)$ and the desired maximum Γ_{SLL} and XPL Γ_{XPL}.
2) Set the parameters for the PSO algorithm and the weighting factors W_1, W_2, and W_3 for the cost function.
3) Find the active element pattern (AEP) for each element by using full-wave simulations or find approximate element patterns by using either analytical solutions or simulations with periodical structures.
4) Set $k = 0$, and initialize the scale factor $s = 1/3$; s determines the angular range of the element rotation. As the iteration process proceeds, finer angular rotation is achieved as shown in step 5 by virtue of s^k.
5) Find the optimized element rotation angles $\xi_n^{(k)} \in s^k[1 + \delta(k)][-\pi/2, \pi/2]$ and phases $\alpha_n^{(k)} \in [0, 2\pi]$ for $n = 1, 2, ..., N$ by minimizing the cost function (8.18). Use the PSO algorithm to maximize the match of the synthesized shaped pattern to the desired one.
6) Update the element rotation angles $\xi_n^{(k)} = \xi_n^{(k)} + \xi_n^{(k-1)}$ for $n = 1, 2, ..., N$.
7) Use full-wave simulations to obtain the real array pattern with the obtained element rotations $\xi_n^{(k)}$ and phases $\alpha_n^{(k)}$. Find all of the AEPs at the current state of rotations.

8) Check if the discrepancy between the synthesized and real array patterns meet the prescribed tolerance in terms of the maximum SLL and mainlobe shape deviation. If so, exit the whole synthesis process. If not, set $k = k + 1$ and loop through Steps 5–8 until they are met.

Note that the rotation range at the k-th step is set as $\xi_n^{(k)} \in s^k[1 + \delta(k)][-\pi/2, \pi/2]$, where the parameter s $(0 < s < 1)$ is a scale factor, and $\delta(k)$ is equal to 1 for $k = 0$ and 0 otherwise. In general, choosing a larger s (e.g., close to 1) requires more refining steps to reduce the discrepancy between the synthesized and real array patterns, and, hence, increases the total time cost. On the other hand, choosing a smaller s leads to faster convergence but may lead to a less than optimal array pattern. The algorithm is further elaborated in the following examples.

8.2.4 Examples of Pattern Synthesis Based on Element Rotation and Phase

8.2.4.1 Flat-Top Pattern Synthesis with a Rotated U-Slot Loaded Microstrip Antenna Array

As the first example, the joint rotation/phase optimization method is employed to synthesize a flat-top-shaped pattern for a 24-element linear array with 0.55λ spacing between its elements. To illustrate the effectiveness of the method for a complicated antenna structure, a U-slot loaded microstrip antenna resonating at 10 GHz as the array element is selected [30]. The antenna model and its detailed parameters are depicted in Figure 8.9.

Assume the user-desired polarization direction $\vec{p}_d = \hat{e}_y$. Then, the CoP and XP for the flat-top-shaped pattern synthesis in the xOz-plane are $\hat{e}_\theta$ and $\hat{e}_\varphi$, respectively. The flat-top mainlobe region is chosen to be $|\theta| \leq 9°$ while the sidelobe region is set at $|\theta| \geq 13°$. Specified values are $\Gamma_{\text{SLL}} = \Gamma_{\text{XPL}} = -16$ dB, $W_1 = 5$, and $W_2 = W_3 = 1$.

The VAEP in the initial step is obtained by simulating the U-slot loaded microstrip antenna with periodic boundaries. All of the elements in the array are individually fed by coaxial ports in the high-frequency structure simulator (HFSS) model. Then the flat-top pattern is synthesized by finding the appropriate rotation angles and excitation phases without considering the variation of the mutual coupling. Clearly, the synthesized array pattern would be different from the real one obtained by full-wave simulation of the rotated array. This point is illustrated in Figure 8.10a. The synthesized SLL, XPL, and mainlobe ripples are −15.85, −15.97, and ±0.45 dB, respectively. They increase to −12.75, −13.11, and ±0.73 dB, respectively, for the real pattern.

To improve the performance of the real pattern, several refining steps are adopted to re-optimize the element rotations and phases. By setting the scale factor $s = 1/3$, the angle range allowed for the element rotation becomes smaller and smaller as the refining steps increase. For example, the

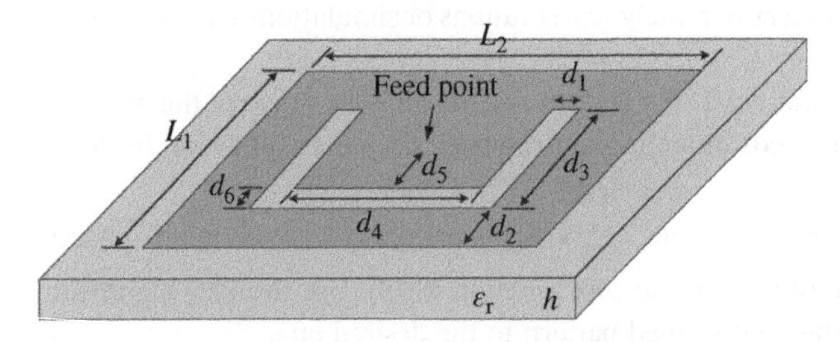

Figure 8.9 Design details of the U-slot-loaded microstrip antenna element utilized in the flat-top pattern synthesis example. Optimized parameters (in millimeters): $d_1 = d_6 = 0.55$ mm, $d_2 = 1.85$ mm, $d_3 = 6.60$ mm, $d_4 = 4.40$ mm, $d_5 = 2.30$ mm, $h = 1.575$ mm, $L_1 = 9.40$ mm, and $L_2 = 9.20$ mm. The substrate has the dielectric constant $\varepsilon_r = 2.2$. *Source*: From [21] / with permission of IEEE.

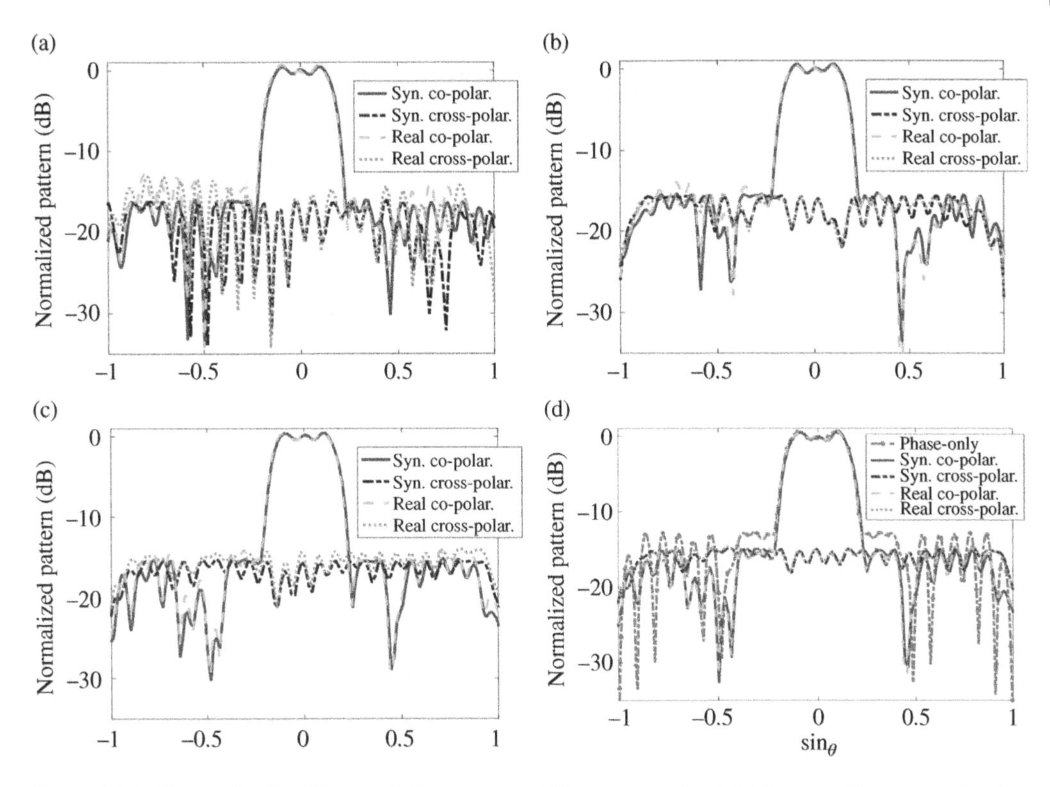

Figure 8.10 The synthesized flat-top CoP patterns and XP patterns at the initial step and the successive three refining steps and corresponding real array patterns obtained using the full-wave simulation for the rotated U-slot loaded antenna array in which each element is individually fed by a coaxial port. (a) Initial step. (b–d) The results at the first, second, and third refining steps, respectively. The synthesized pattern by phase-only optimization is also shown in (d). *Source*: From [21] / with permission of IEEE.

rotation angle range is $\pm\pi/6$, $\pm\pi/18$, and $\pm\pi/54$ for the first, second, and third refining steps. Figures 8.10b–d show the synthesized and real array patterns for those three refining steps, respectively. One observes that the synthesized pattern matches the real pattern better and better as the refining step increases. The obtained SLL and XPL for the real pattern at the third step are −14.58 and −14.57 dB, which are very close to the corresponding synthesized results −14.84 and −14.74 dB.

Figure 8.11a shows the element rotations obtained at the initial step and the three refining steps. Figure 8.11b shows the excitation phases at these steps. The final element rotations and phases obtained after each of the three refining steps are detailed in Table 8.1. Note that some edge elements of the obtained array have much different rotation angles. These rotation angles would be very difficult to find without a systematic approach.

The phase-only optimization method is also used to synthesize the same desired pattern for comparison. The same linear array but with 24 non-rotated U-slot loaded microstrip antenna elements is full-wave simulated to acquire all of the AEPs. Then the excitation phase of every element is optimized using the PSO algorithm to synthesize the same flat-top pattern. The obtained pattern by phase-only optimization is shown in Figure 8.10d. Compared with the final pattern obtained by the refined method, it has almost the same mainlobe shape but with a much higher SLL of −12.77 dB. This further validates the advantage of the joint rotation/phase optimization method.

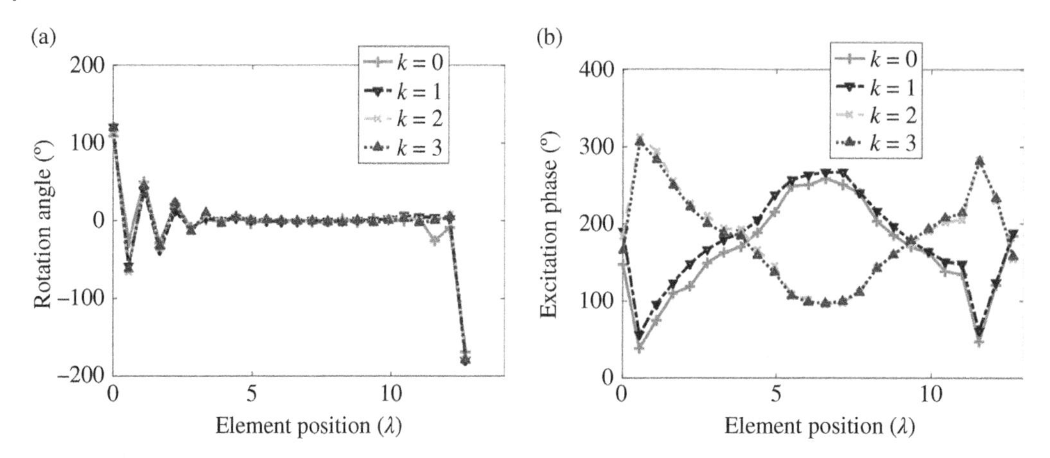

Figure 8.11 The synthesized rotation angles and excitation phases at the initial and three refining steps for the flat-top pattern case. (a) Rotation angles. (b) Excitation phases. *Source*: From [21] / with permission of IEEE.

A prototype of the 24-element U-slot microstrip antenna array with element rotation was fabricated and tested. A photo of the antenna in the anechoic chamber and its bottom and top layers are shown in Figure 8.12. It consists of the rotated U-slot antennas, the RF ground, and the feeding network. The feeding network is designed as a multistage equal power divider followed by phase-shifter lines to provide the required excitation phases. There are 24 metal via holes insulated from the RF ground that connect the feeding lines and the antennas. The upper and lower dielectric layers have the same relative permittivity $\epsilon_r = 2.2$, but with different dimensions. The size of the upper layer dielectric is 429 mm × 131 mm with a thickness of 1.575 mm, and that of the lower layer dielectric is 429 mm × 58 mm with a thickness of 0.508 mm. Since the bottom dielectric layer is thin and long, a hard plastic plate was used to support it in order to keep the whole structure flat.

The antenna array prototype was measured using a far-field measurement system in an antenna anechoic chamber. The measured CoP and XP patterns, along with the HFSS-simulated patterns, are presented together in Figure 8.13. Note that the real patterns depicted in Figure 8.13 were obtained by simulating the element-rotated array fed by the designed feeding network. These results are different from the simulated patterns in Figure 8.10d for the array which was fed by 24 individual coaxial ports in the HFSS model. Thus, there are small deviations between the real patterns in Figure 8.10d and those in Figure 8.13. Figure 8.13 shows the mainlobe ripple of the measured CoP pattern is ±0.84 dB, which is slightly higher than the simulated ripple of ±0.67 dB. The measured SLL and XPL are −13.33 and −12.67 dB, respectively, which are 1.27 and 1.11 dB higher than those of the simulated patterns, respectively. Although there is a small performance degradation, which is presumed due to fabrication errors and a nonideal measurement environment, the measured CoP and XP patterns generally agree well with the full-wave simulation results.

8.2.4.2 Circular Flat-Top Pattern Synthesis for a Planar Array with Rotated Cavity-Backed Patch Antennas

The effectiveness of the refined joint rotation/phase optimization method in synthesizing shaped power patterns for a planar array is illustrated with a second example. A circular flat-top pattern was synthesized in [31] by optimizing the amplitudes and phases of an 11 × 11 $\lambda/2$-spaced array

Table 8.1 The obtained final rotation angles and excitation phases for the flat-top pattern synthesis.

n	Rot. angle (°)	Ext. phase (°)
1	119.73	165.34
2	−61.96	306.40
3	44.40	282.69
4	−32.43	250.32
5	22.65	221.29
6	−13.69	200.44
7	10.22	188.27
8	−2.90	184.39
9	4.45	158.99
10	0.10	136.94
11	1.71	106.78
12	−0.25	98.89
13	−0.31	96.88
14	−0.23	99.48
15	−0.98	111.80
16	−1.91	142.52
17	0.23	160.35
18	−2.17	178.29
19	0.26	193.03
20	2.79	207.40
21	−2.10	214.73
22	0.11	281.43
23	4.91	233.75
24	−179.58	158.09

Source: From [21] / with permission of IEEE.

without considering the mutual coupling between its elements. This pattern had a circular flat-top mainlobe in the region: $\{\theta \leq 9^\circ$ and $\varphi \in (0^\circ, 360^\circ)\}$, and the maximum SLL is less than −10 dB in the region: $\{\theta \geq 20^\circ$ and $\varphi \in (0^\circ, 360^\circ)\}$ as shown in Figure 4a of [31]. The planar array considered here has the same overall size but adopts the cavity-backed patch antenna presented in [32] as the array elements.

The user-desired polarization direction is still assumed to be $\vec{p}_d = \hat{e}_y$. However, different from the linear array cases, $\hat{e}_\theta$ and $\hat{e}_\varphi$ are no longer the CoP and XP directions for this planar array. The realizable CoP and XP directions change with the propagation direction (θ, φ) according to (8.12) and (8.13). The same circular flat-top function is selected as the desired mainlobe shape for this array. The terms $\Gamma_{\text{SLL}} = \Gamma_{\text{XPL}} = -11$ dB and the other parameters, including W_1, W_2, and W_3, to be the same as those in the first example.

Figure 8.12 The fabricated 24-element U-slot microstrip antenna array with rotated elements along with is supportive plastic plate are shown as the antenna under test in the anechoic chamber. Photos of the ground plane on the bottom of its bottom dielectric layer and the array along with its feeding network on its upper surface of the upper substrate are shared in the subfigures. *Source*: From [21] / with permission of IEEE.

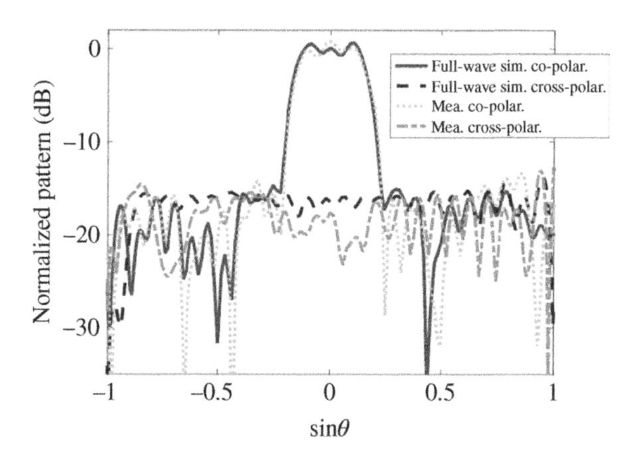

Figure 8.13 The measured CoP and XP patterns as well as the full-wave simulation results for the 24-element, antenna array with rotated U-slot microstrip antenna elements integrated with its feeding network. *Source*: From [21] / with permission of IEEE.

Table 8.2 lists the simulated maximum SLL, XPL, and mainlobe ripple for both the synthesized and real array patterns obtained at the initial and next three refining steps. The SLL and XPL for the real pattern decrease as the refining steps are performed. The real SLL, XPL, and mainlobe ripple values at the 3rd refining step are −10.32, −10.18, and ±1.23 dB, respectively. Figures 8.14a–d show the top views of the CoP and XP components of the synthesized and full-wave simulated real array patterns at that third step. The final synthesized patterns clearly agree well with the corresponding EM-simulated patterns for both the CoP and XP components. Compared with the result in Figure 4a of [29], which did not include mutual coupling effects, the obtained patterns have better

Table 8.2 The maximum SLL, XPL, and mainlobe ripple of the synthesized and real array patterns at the initial and three refining steps for the planar array with rotated cavity-backed patch antenna elements.

	Synthesized results (dB)			Simulated results (dB)		
k – th iteration	SLL	XPL	Ripple	SLL	XPL	Ripple
0	−10.62	−10.83	±1.14	−7.57	−8.25	±1.60
1	−10.77	−10.63	±1.09	−9.25	−9.88	±1.19
2	−10.30	−10.20	±1.14	−9.83	−10.39	±1.17
3	−10.32	−10.32	±1.17	−10.32	−10.18	±1.23

Source: From [21] / with permission of IEEE.

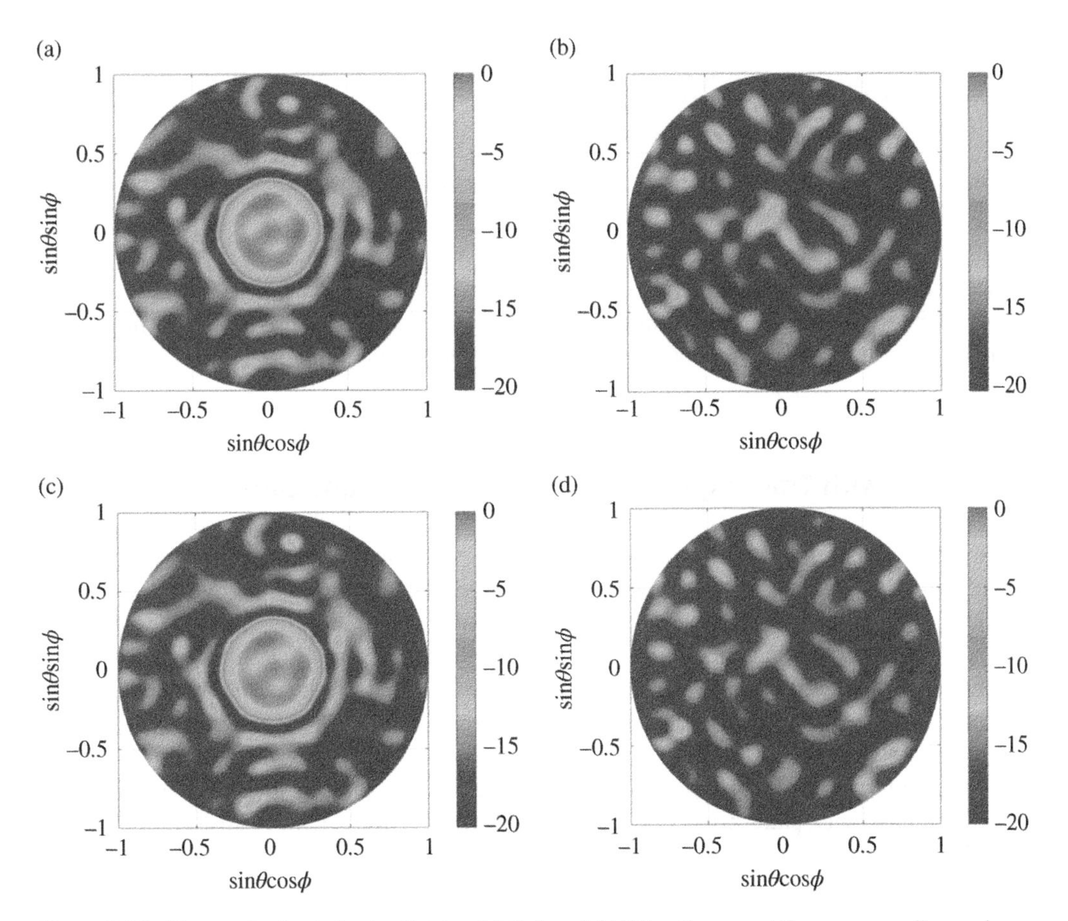

Figure 8.14 The synthesized circular flat-top (a) CoP and (b) XP patterns and the corresponding real array patterns (c) and (d) obtained by full-wave simulation. *Source*: From [21] / with permission of IEEE.

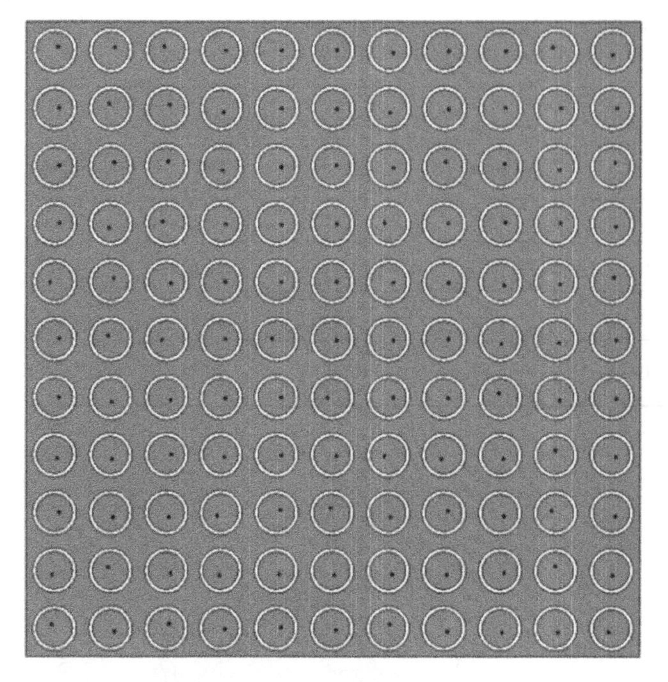

Figure 8.15 The synthesized element-rotated planar array that radiates the specified circular flat-top pattern. *Source*: From [21] / with permission of IEEE.

sidelobe performance even though the mutual coupling effect has now been included. Figure 8.15 shows the array arrangement with the optimized element rotations. Since this array does not utilize amplitude weighting, the number of unequal power dividers is minimal.

8.3 Arrays with Tracking Abilities Employing Sum and Difference Patterns

For communications with a moving platform, the beam radiated by an antenna array needs to follow its movements so that the quality of the wireless link can be maintained. If the position of the moving platform is known, one can easily calculate the linear phase distribution required for the array to produce the beam to sustain a link in real-time. If the position of the moving platform is unknown, however, one would need to have a difference antenna pattern to track the incoming signal. Therefore, it is vitally important for practical analog antenna arrays to have the ability to form both a sum and a difference pattern.

The sum pattern is typically a focused beam having its peak gain in the intended direction, whereas the difference pattern has a null in that direction. Together they can form a solution for mobile point-to-point wireless communications. Applications include satellite on the move systems, as well as monopulse radar systems [33]. In order to achieve high accuracy in the measurement of the incoming signal direction, the difference pattern is generally required to have high directivity, low SLLs, and steep slopes around its nulls. These requirements generally result in an array with a nonuniform amplitude distribution which, in practice, would necessitate multiple unequal power dividers. The resulting system leads to relatively complicated beamforming networks (BFNs). To avoid having to use nonuniform amplitudes, one can resort to rotating the antenna orientations as demonstrated in the previous section.

Detailed synthesis algorithms are presented in the following subsections to illustrate the advantages of an array with rotated elements to meet such wireless communication system requirements. Two examples of synthesizing sum and difference patterns using rotated dipole array elements are examined to demonstrate the efficacy of the approach. Their synthesis results demonstrate that satisfactory sum and difference patterns with the desired SLLs, XPLs, and slopes are obtained without using nonuniform amplitude weighting.

8.3.1 Nonuniformly Spaced Dipole-Rotated Linear Array

Consider a linear array comprised of $2N$ nonuniformly spaced elements. This assumption that the element number is even is made for convenience in deriving the following formulas. Nevertheless, they are also applicable with some care in rewriting them, when the element number is odd. Figure 8.16 shows how the sum and difference patterns are produced by a linear array with a specialized two-section BFN. Different from a conventional array radiating a single beam pattern, the sum and difference array is generally divided into the indicated two halves. The two halves of the array are fed in-phase to produce an in-phase composition of all the element patterns and, hence, to obtain the sum pattern. In contrast, they are fed in an antiphase manner to generate the difference beam. The excitation amplitudes are generally optimized to obtain sum and difference patterns with as low SLLs as possible. Therefore, unequal power dividers are required in the BFN in addition to the indispensable π-phase shifter.

The joint optimization of element rotations and positions instead of using nonuniform excitation amplitudes can improve the performance of the sum and difference patterns. This will not only simplify the BFN design, but it will also decrease the cost and weight of the whole system. Suppose each element in Figure 8.16 is rotated with an arbitrary angle denoted by $\xi_n \in (-\pi, \pi)$ for $n = -N$, $-N+1, \ldots, -1, 1, \ldots, N$. The vectorial sum and difference power patterns of this array in the principal plane, i.e., the $x0y$-plane, can be expressed as:

$$\begin{cases} \vec{F}_{\Sigma,\theta}(\varphi) = \vec{E}_{1,\theta}(\varphi) + \vec{E}_{2,\theta}(\varphi) \\ \vec{F}_{\Sigma,\varphi}(\varphi) = \vec{E}_{1,\varphi}(\varphi) + \vec{E}_{2,\varphi}(\varphi) \end{cases} \tag{8.20}$$

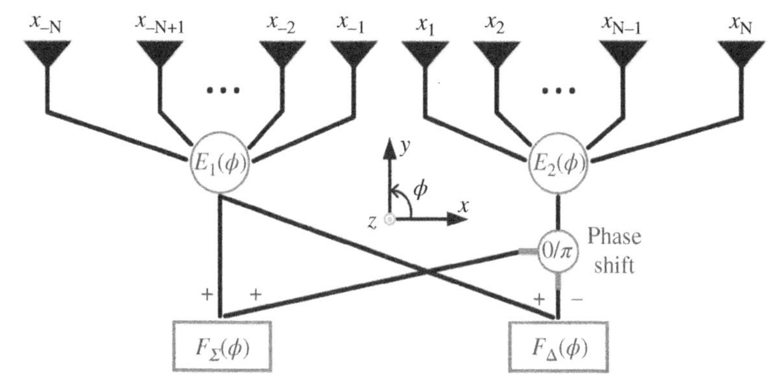

Figure 8.16 Schematic diagram of a nonuniformly spaced linear array and the BFN that facilitates it radiating a sum and a difference pattern. *Source*: From [34] / with permission of IEEE.

$$\begin{cases} \vec{F}_{\Delta,\theta}(\varphi) = \vec{E}_{1,\theta}(\varphi) - \vec{E}_{2,\theta}(\varphi) \\ \vec{F}_{\Delta,\varphi}(\varphi) = \vec{E}_{1,\varphi}(\varphi) - \vec{E}_{2,\varphi}(\varphi) \end{cases} \tag{8.21}$$

where

$$\begin{cases} \vec{E}_{1,\theta}(\varphi) = \sum_{n=-N}^{-1} a_{n,\theta}(\varphi;\xi_n)\, e^{j\beta x_n \cos\varphi} \hat{e}_\theta \\ \vec{E}_{2,\theta}(\varphi) = \sum_{n=1}^{N} a_{n,\theta}(\varphi;\xi_n)\, e^{j\beta x_n \cos\varphi} \hat{e}_\theta \end{cases} \tag{8.22}$$

$$\begin{cases} \vec{E}_{1,\varphi}(\varphi) = \sum_{n=-N}^{-1} a_{n,\varphi}(\varphi;\xi_n)\, e^{j\beta x_n \cos\varphi} \hat{e}_\varphi \\ \vec{E}_{2,\varphi}(\varphi) = \sum_{n=1}^{N} a_{n,\varphi}(\varphi;\xi_n)\, e^{j\beta x_n \cos\varphi} \hat{e}_\varphi \end{cases} \tag{8.23}$$

Simply note that the observation angle in the azimuth direction ($x0y$ plane) is restricted to $\varphi \in (0, \pi)$. The vector $\vec{a}_n(\varphi;\xi_n) = a_{n,\theta}(\varphi;\xi_n)\,\hat{e}_\theta + a_{n,\varphi}(\varphi;\xi_n)\,\hat{e}_\varphi$ is the rotated vectorial pattern of the n-th element located at x_n. $\vec{F}_{\Sigma,\theta}(\varphi)$ and $\vec{F}_{\Sigma,\varphi}(\varphi)$ are the θ and φ components of the sum pattern, respectively, whereas $\vec{F}_{\Delta,\theta}(\varphi)$ and $\vec{F}_{\Delta,\varphi}(\varphi)$ are the θ and φ components of the difference pattern, respectively.

Linear dipole arrays are considered here to illustrate the feasibility of synthesizing sum and difference patterns by optimizing the rotation angles and positions of their elements. Initially, suppose that all of the dipoles are positioned along the x-axis in the $x0z$ plane and are oriented perpendicular to that axis. Then, as Figure 8.17 indicates, assume every dipole is rotated with an arbitrary angle ξ_n about its center in the $x0z$ plane, i.e., the rotation occurs about the y-axis. A local coordinate system x'-y'-z' is then introduced to facilitate the formulation of the element pattern with the z'-axis coinciding with the orientation of the rotated dipole and the y'-axis being parallel to the fixed y-axis. The vectorial element patterns of the rotated dipole in the local x'-y'-z' coordinates are obtained in a straightforward manner. The θ- and φ-polarized patterns of the rotated dipole in the global coordinate xyz can then be obtained by using a coordinate transformation. The θ- and φ-polarized patterns in the principal $x0y$-plane obtained in this manner in [24] are:

$$a_{n,\theta}(\varphi;\xi_n) = \frac{\cos\xi_n \cos\left(\frac{\pi}{2}\sin\xi_n\cos\varphi\right)}{1 - sin^2\xi_n\ \cos^2\varphi} \tag{8.24}$$

$$a_{n,\varphi}(\varphi;\xi_n) = \frac{\sin\xi_n \sin\varphi\ \cos\left(\frac{\pi}{2}\sin\xi_n\cos\varphi\right)}{sin^2\xi_n\ \cos^2\varphi - 1} \tag{8.25}$$

Substituting (8.24) and (8.25) into (8.20)–(8.23), one obtains the vectorial sum and difference patterns of the nonuniformly spaced element-rotated dipole array.

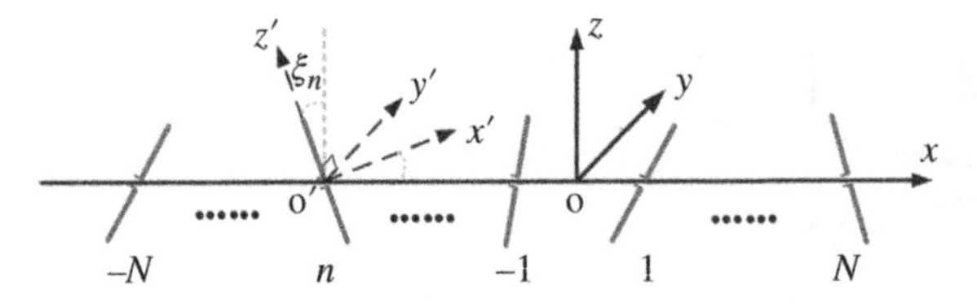

Figure 8.17 The element-rotated dipole array with both the global and local coordinate systems depicted. *Source*: From [34] / with permission of IEEE.

8.3.2 PSO-Based Element Rotation and Position Optimization

The problem is to determine an optimal solution of the common element rotations ξ_n and positions x_n ($n = -N, -N+1, \ldots, -1, 1, \ldots, N$) so that both the sum and difference patterns have low SLLs and XPLs. Moreover, the slope of the difference pattern in the target direction must be as steep as possible. Note that in order to obtain a symmetrical BFN, the element positions x_n are assumed to be symmetrical about the center of the array. Thus, only N element positions for one half of the array are optimized. However, the rotation angles of all of the elements are optimized to obtain as large a number of degrees of freedom in the synthesis process as possible. Therefore, a total number of $3N$ variables must be optimized.

The following cost function, which comprises several terms to minimize the SLLs and XPLs of both the sum and difference patterns, as well as the slope of the difference pattern, was constructed:

$$f_c = \frac{W_1}{B}\sum_{b=1}^{B}\frac{1}{2}(X_b + |X_b|)^2 + \frac{W_2}{C}\sum_{c=1}^{C}\frac{1}{2}(Y_c + |Y_c|)^2 + \frac{W_3}{D}\sum_{d=1}^{D}\frac{1}{2}(Z_{\Sigma,d} + |Z_{\Sigma,d}|)^2 + \frac{W_4}{D}\sum_{d=1}^{D}\frac{1}{2}(Z_{\Delta,d} + |Z_{\Delta,d}|)^2 + \frac{W_5}{2}(|S| - S)^2 \tag{8.26}$$

where

$$\begin{cases} X_b = |F_{\Sigma,\theta}(\varphi_b)|^2 - \Gamma_{\text{SLL1}} \\ Y_c = |F_{\Delta,\theta}(\varphi_c)|^2 - \Gamma_{\text{SLL2}} \\ Z_{\Sigma,d} = \left|F_{\Sigma,\varphi}(\varphi_d)\right|^2 - \Gamma_{\text{XPL1}} \\ Z_{\Delta,d} = \left|F_{\Delta,\varphi}(\varphi_d)\right|^2 - \Gamma_{\text{XPL2}} \\ S = \left|\left.\frac{\partial F_{\Delta,\theta}(\varphi)}{\partial\varphi}\right|_{\varphi=\varphi_0}\right|^2 - \eta \end{cases} \tag{8.27}$$

The terms W_1, W_2, W_3, W_4, and W_5 are the weighting factors. The terms Γ_{SLL1} and Γ_{SLL2} are the desired SLLs for the CoP sum and difference patterns $F_{\Sigma,\theta}(\varphi)$ and $F_{\Delta,\theta}(\varphi)$, respectively. The terms Γ_{XPL1} and Γ_{XPL2} are the desired XPLs for the XP patterns $F_{\Sigma,\varphi}(\varphi)$ and $F_{\Delta,\varphi}(\varphi)$, respectively. The term η denotes the desired slope of the difference pattern in the target direction. Note that the θ-polarized component of the electrical field pattern is considered as the CoP component here since it is the dominant component when the dipole is not rotated. The terms φ_b ($b = 1, 2, \ldots, B$) and φ_c ($c = 1, 2, \ldots, C$) are the sampling angles of the SLL regions for $F_{\Sigma,\theta}(\varphi)$ and $F_{\Delta,\theta}(\varphi)$, respectively. The term φ_d ($d = 1, 2, \ldots, D$) is the sampling angle in the full space of $\varphi \in [0, \pi]$. The term φ_0 is the angle of the target direction, which is chosen as $\varphi_0 = \pi/2$ here. Consequently, the slope of the difference pattern at $\varphi_0 = \pi/2$ is obtained as:

$$\left.\frac{\partial F_{\Delta,\theta}(\varphi)}{\partial\varphi}\right|_{\varphi=\pi/2} = \sum_{n=-N}^{N} \text{sgn}\,(n) j\beta x_n \cos\xi_n \tag{8.28}$$

where sgn(n) is the sign function of n. It should be noted that the multi-region SLLs can also be obtained in some applications by slightly modifying the cost function (8.28). The PSO algorithm presented in Section 8.2 is used here to synthesize the sum and difference patterns by optimizing the element rotations and positions.

8.3.3 Examples

8.3.3.1 Synthesis of a 56-Element Sparse Linear Dipole Array

The rotation-/position-based synthesis process will be used as a first example to generate the sum and difference patterns presented in Figures 6 and 7 of [35], which were obtained by optimizing the excitation amplitudes and positions of a 56-element linear array. The obtained SLLs of the sum and difference patterns in [35] are −20 and −16.5 dB, respectively. The required dynamic range ratio (DRR) of the excitation amplitudes is about 3.65, whereas the optimized positions are symmetrical with the element spacings varying from 0.43λ to 1.28λ. The same sum and difference patterns with the same SLL requirements are synthesized here by optimizing the element rotations and positions of a 56-element sparse dipole array.

Suppose the element rotation angle range is $\xi_n \in [-\pi, \pi]$, and the element spacing is restricted in the range (0.5λ, 1.28λ). Note that the minimum element spacing is chosen as 0.5λ, which is slightly larger than the 0.43λ value used in [35], to avoid the strong mutual coupling among neighboring elements. The SLLs and XPLs for the sum and difference patterns in this example are set to the same values as those in [35], i.e., $\Gamma_{SLL1} = \Gamma_{XPL1} = -20$ dB for the sum pattern and $\Gamma_{SLL2} = \Gamma_{XPL2} = -16.5$ dB for the difference pattern. The desired slope of the difference pattern remains set at $\eta = 45$ dB. The weighting factors in the cost function are chosen as: $W_1 = W_2 = W_3 = W_4 = 1$ and $W_5 = 5$ for the synthesis process.

The obtained sum and difference patterns are compared with those in [35] in Figures 8.18a–b. It is seen that the SLL and XPL values of the sum pattern obtained with the rotation-position synthesis method are −20.01 and −20.03 dB, respectively. On the other hand, the obtained SLL and XPL values of the difference pattern are −16.50 and −16.51 dB, respectively. While the characteristics of these patterns are comparable to those in [35], it is noted that the synthesized array obtained by using the joint rotation-position optimization method does not use nonuniform amplitudes. Consequently, no unequal power dividers would be required in its BFN. The synthesized element rotations and positions are listed in the left two columns of Table 8.3. In this example, the minimum, maximum, and average element spacings of the synthesized array are 0.5λ, 1.07λ, and 0.88λ, respectively. The element saving is 42.27% when compared with a λ/2-spaced array occupying the same aperture.

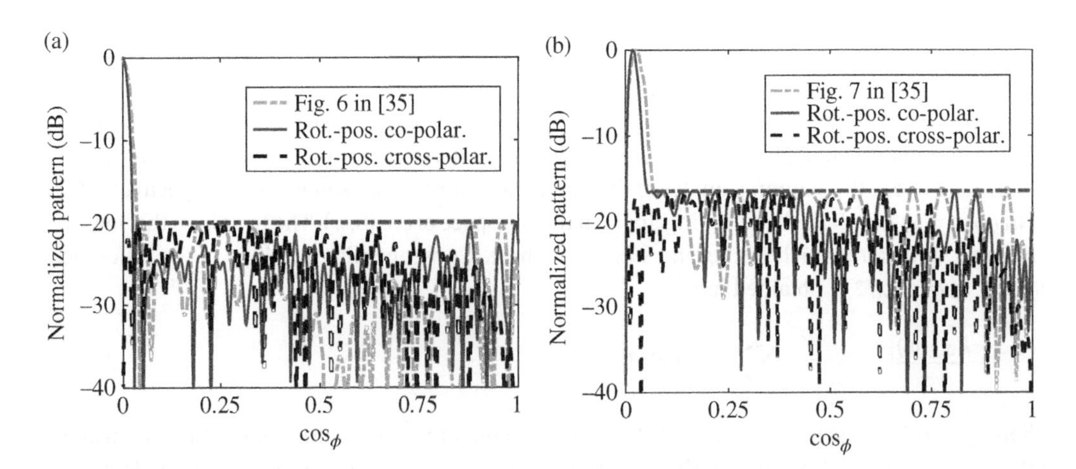

Figure 8.18 The sum and difference patterns obtained by the joint rotation and position optimization method are compared to the corresponding patterns obtained in [35] for a 56-element linear array. (a) Sum patterns. (b) Difference patterns. *Source*: From [34] / with permission of IEEE.

Table 8.3 The rotation angles and element positions ($x_{-n} = -x_n$) obtained by the joint rotation/phase optimization method for the two linear dipole array examples.

	Example 1		Example 2	
n	Rot. angle (°)	Pos. (λ)	Rot. angle (°)	Pos. (λ)
1	24.72	−24.10	−58.70	−24.08
2	−83.14	−23.23	91.32	−22.90
3	152.22	−22.17	−29.12	−21.91
4	10.77	−21.10	71.44	−20.93
5	−66.37	−20.03	94.69	−19.87
6	−23.75	−19.04	60.82	−19.02
7	71.53	−18.09	48.28	−18.16
8	−61.46	−17.33	23.68	−17.18
9	−35.25	−16.38	59.38	−16.29
10	−42.79	−15.32	−33.20	−15.32
11	20.44	−14.44	−10.64	−14.41
12	2.38	−13.53	−5.25	−13.50
13	−17.80	−12.68	−4.94	−12.58
14	−3.38	−11.71	4.90	−11.74
15	6.52	−10.85	4.07	−10.84
16	15.16	−9.96	20.32	−10.06
17	−4.56	−9.18	−4.38	−9.28
18	−3.21	−8.46	4.48	−8.45
19	−10.39	−7.60	3.40	−7.68
20	−9.93	−6.80	−7.84	−6.80
21	0.60	−5.89	1.86	−5.93
22	5.71	−5.05	−10.03	−5.05
23	3.80	−3.99	9.82	−4.27
24	−5.30	−3.15	−15.60	−3.39
25	35.09	−2.21	5.90	−2.36
26	−4.83	−1.40	−8.25	−1.54
27	−0.13	−0.75	0.24	−0.80
28	2.80	−0.25	−0.86	−0.25
29	−1.41	0.25	−0.61	0.25
30	3.54	0.75	−1.17	0.80
31	4.53	1.40	−7.78	1.54
32	−2.61	2.21	6.03	2.36
33	−2.56	3.15	−1.43	3.39
34	2.01	3.99	18.74	4.27
35	−1.23	5.05	−9.34	5.05

(Continued)

Table 8.3 (Continued)

n	Example 1		Example 2	
	Rot. angle (°)	Pos. (λ)	Rot. angle (°)	Pos. (λ)
36	−0.29	5.89	3.56	5.93
37	11.08	6.80	−12.19	6.80
38	−3.14	7.60	−0.89	7.68
39	−15.88	8.46	10.76	8.45
40	10.68	9.18	7.04	9.28
41	6.05	9.96	4.71	10.06
42	−3.86	10.85	1.86	10.84
43	−4.59	11.71	9.70	11.74
44	−16.40	12.68	−2.38	12.58
45	9.71	13.53	−1.84	13.50
46	18.59	14.44	−31.45	14.41
47	−1.01	15.32	−37.48	15.32
48	−41.83	16.38	41.17	16.29
49	−46.77	17.33	63.99	17.18
50	−55.55	18.09	−0.54	18.16
51	83.30	19.04	−29.74	19.02
52	55.06	20.03	−109.26	19.87
53	84.40	21.10	26.57	20.93
54	39.19	22.17	119.04	21.91
55	−80.79	23.23	−55.72	22.90
56	77.53	24.10	60.95	24.08

Source: From [34] / with permission of IEEE.

8.3.3.2 Synthesizing Sum and Difference Patterns with Multi-Region SLL and XPL Constraints

As a second example, the effectiveness of the rotation-position synthesis method for realizing sum and difference patterns with multi-region SLL and XPL constraints is demonstrated. This application is associated with known issues of wanting lower SLLs to cope with jammings from specific directions in various complicated electromagnetic environments. For simplicity, we consider synthesizing the same 56-element dipole array with the same sum and difference patterns, but with additional SLL and XPL requirements of −30 dB in the regions $\varphi \in [40°, 45°] \cup [135°, 140°]$ for both the sum and difference patterns.

The parameter settings: the available rotation angles ξ_n, the element spacing restriction, the desired slope of the difference pattern η, and the weighting factors for the cost function, are all the same as those in the first example. Figures 8.19a and 8.19b show the obtained sum and difference patterns. Note that even with the complicated additional multi-region SLL and XPL requirements, the obtained patterns are still satisfactory except for the fact that the first sidelobe of the sum pattern is a little bit above the desired level (about 0.55 dB higher). Furthermore, the SLLs and XPLs for both the sum and difference patterns in $\varphi \in [40°, 45°] \cup [135°, 140°]$ are below −30 dB. These

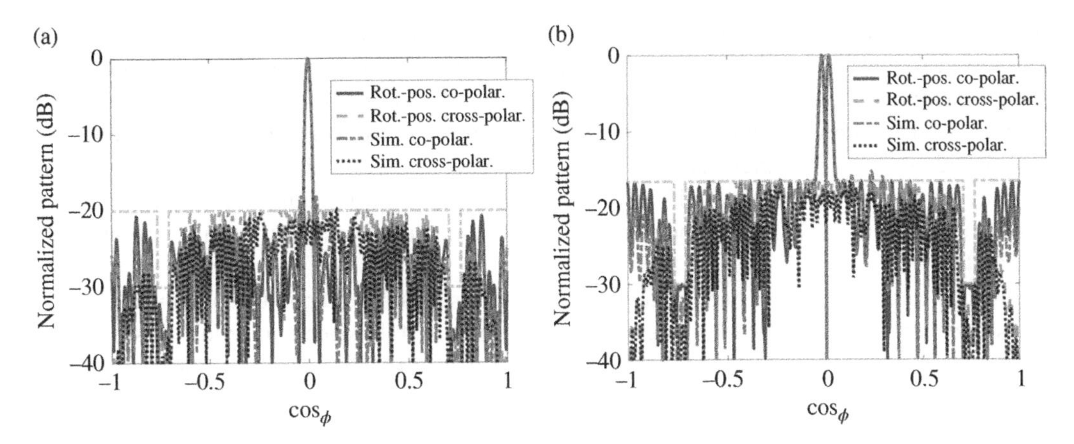

Figure 8.19 The synthesized and full-wave simulated sum and difference patterns obtained by the joint rotation-position optimization method for the 56-element linear array when the multi-region SLL and XPL constraints are imposed. (a) Sum patterns. (b) Difference patterns. *Source*: From [34] / with permission of IEEE.

results further demonstrate the effectiveness of the joint rotation-position optimization method. The obtained rotation angles and element positions are listed in the right two columns of Table 8.3. The minimum, maximum, and average element spacings of this array are 0.5λ, 1.18λ, and 0.88λ, respectively. In this example, an element number savings of about 42.27% was realized when compared with a uniformly spaced array with $\lambda/2$ spacing occupying the same aperture.

The performance of the obtained sum and difference patterns when mutual coupling is included is also illustrated. The synthesized sparse dipole-rotated array is modeled and simulated directly with HFSS. The half-wavelength dipole element utilized in the simulation is designed to operate at 3 GHz with a diameter of 1.0 mm and a total length of 48.0 mm. The full-wave simulated sum and difference patterns are also included in Figure 8.19. It is seen that the simulated patterns agree well with the synthesized ones except for some SLL degradation at a few particular angles. The maximum SLLs of the simulated sum and difference patterns are increased by 1.57 and 1.28 dB, respectively, while the obtained XPLs of the two patterns still remain below the desired threshold. The good agreement between the synthesized and simulated results is mainly attributed to the actual weak mutual coupling among elements due to the relatively large element spacings.

It should be noted that the element rotations obtained by the synthesis technique are optimal for the desired sum and difference patterns with a prefixed beam direction. When the beam direction is scanned far from the prefixed direction by using additional progressive linear phases, the performance of the obtained sum and difference patterns may degrade to a certain degree. The rotation-position synthesis method can then be further extended to handle this situation. One can determine the common optimal element positions/rotations for the scanned sum and difference patterns by incorporating the requirements for multiple different beam directions into the objective function defined by Eqs. (8.26) and (8.27). This is one of our current research directions.

8.4 Synthesis of SIMO Arrays

Future wireless communications systems including 5G, 6G, and beyond will service various applications including intelligent transport, drone networking, and Internet of Things (IoT)s [36, 37]. They all require multi-beam antennas with each beam individually scannable to connect all users.

Whist multi-beams can be easily realized by using digital beamforming arrays, analog phase arrays can serve as a low-cost and low-energy-consumption alternative for many applications. One application example is a point-to-multipoint information distribution system in which the same information, such as entertainment services, is required to be broadcast to a number of geographically distributed users [38]. Another example is a drone or robot networking in which each entity needs to share the same information with its neighboring ones [39]. A third example is more general wireless networking in which different beams need to carry different information for different intended users [40].

Analog phased arrays have been widely used in wireless sensing and point-to-point wireless systems associated, for example, with satellite communications [41]. When employing phased arrays for multi-user communications, different design strategies are required. A systematic study on the design of analog SIMO multi-beam antenna arrays is presented in this section. It is first demonstrated how multi-beams can be generated using both phase- and magnitude-controlled analog linear arrays. It will then be shown how multi-beam arrays can be produced using phase-only analog arrays. This step is important because their implementations would be easier. Two strategies for real-time analog multi-beamforming are presented to realize the desired scanning functionality.

8.4.1 Analog Dual-Beam Antenna Arrays with Linear Phase Distribution

The discussion is focused initially on a dual-beam linear antenna model. Assume there are N antenna elements with an inter-element spacing d. To produce two beams in the θ_A and θ_B directions, as was shown in Figure 8.14, one simple set of antenna weights can be the following:

$$a_n = e^{-jknd\cos\theta_A} + e^{-jknd\cos\theta_B}\ (n = 0, 1,\ 2, ...N-1) \tag{8.29}$$

where a_n represents the weight on the n-th antenna element. Let

$$\theta_C = (\theta_A + \theta_B)/2 \tag{8.30}$$

$$\Delta\theta = (\theta_A - \theta_B)/2 \tag{8.31}$$

One then has:

$$\theta_1 = \theta_c + \Delta\theta$$

$$\theta_2 = \theta_c - \Delta\theta$$

With some mathematical manipulations, Eq. (8.29) becomes:

$$a_n = 2\cos\left[knd\ \sin\theta_c\ \sin(\Delta\theta)\right]e^{-j\,knd\ \cos\theta_c\cos(\Delta\theta)} \tag{8.32}$$

The corresponding array factor is given by:

$$F(\theta) = F_A(\theta) + F_B(\theta) \tag{8.33}$$

where:

$$F_{A/B}(\theta) = \sum_{n=1}^{N} e^{-jknd\cos\theta_{A/B}}\,e^{j\beta nd\cos\theta} \tag{8.34}$$

Eq. (8.33) is the simple addition of two classic beam patterns. However, Eq. (8.32) tells us that if one simply adds two sets of linearly phased weights with a uniform magnitude distribution, the combined weights will have a new linear phase *and* nonuniform magnitude distribution.

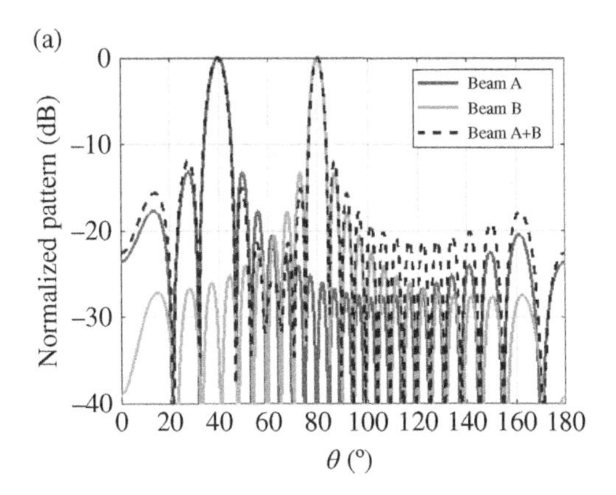

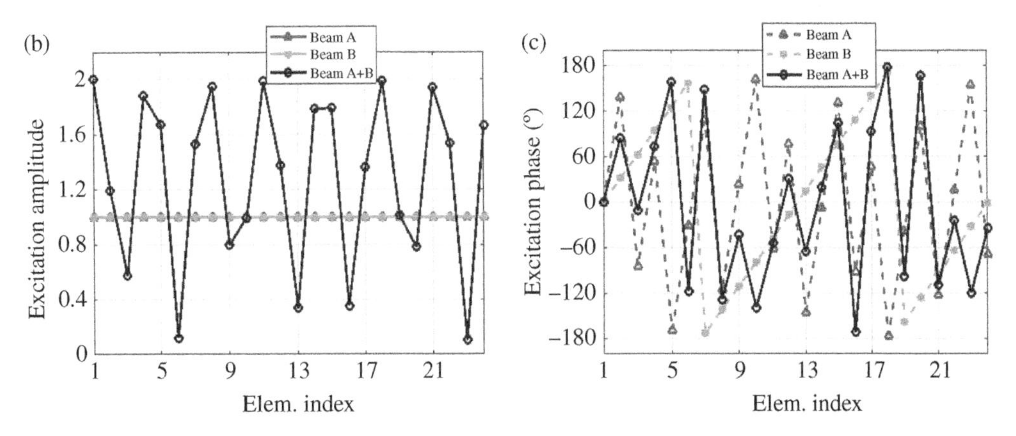

Figure 8.20 An example of the antenna radiation patterns for a linear array with 24 elements when one beam is pointed at 40° and the other pointed 80°. (a) Sum patterns. (b) Excitation amplitudes. (c) Excitation phases.

Figure 8.20 shows an example of the antenna radiation pattern for an array with 24 elements when one beam is pointed at 40° and the other pointed at 80°. It is seen that the magnitudes of the weights fluctuate violently from one element to another. It would be quite difficult in practice to realize such a fluctuating amplitude distribution. Instead, it is more desirable to achieve scannable multi-beams by only employing a uniform amplitude distribution with optimized nonlinear phases.

8.4.2 Phase-Only Optimization of Multi-Beam Arrays

There are a number of practical disadvantages in trying to adjust the magnitudes of the antenna weights on the fly. A better option would be to only change the phase distribution across the array adaptively. In contrast to conventional arrays in which only a linear phase change is required to steer their beams, a multi-beam antenna array requires a complicated phase distribution across its elements.

In general, the array's pattern can be written as:

$$F(\theta;\boldsymbol{\varphi}) = \sum_{i=1}^{N} a_n e^{j\varphi_n} e^{j\beta z_n \cos\theta} \tag{8.35}$$

where $z_n = nd$ ($n = 1, 2, ..., N$), and a_n and φ_n are the amplitude and phase of the n-th element, respectively. Let us introduce the excitation amplitude vector $\boldsymbol{a} = [a_1, a_2, ..., a_N]$, the excitation phase vector $\boldsymbol{\varphi} = [\varphi_1, \varphi_2, ..., \varphi_N]$, and the element position vector $\mathbf{z} = [z_1, z_2, ..., z_N]$. A phase-only iterative perturbation method based on convex optimization starts with the basic idea that a small phase perturbation vector $\boldsymbol{\delta}$ can be added to the current phase excitation vector $\boldsymbol{\varphi}$ at each iteration step. The optimal $\boldsymbol{\delta}$ is then determined iteratively until the desired multi-beam pattern performance is reached or the pattern remains the same over multiple iterations.

Consider the k-th iteration and its phase distribution $\boldsymbol{\varphi}^{(k)} = \boldsymbol{\varphi}^{(k-1)} + \boldsymbol{\delta}^{(k)}$. The pattern function $F(\theta, \boldsymbol{\varphi}^{(k)})$ is given by:

$$F\left(\theta; \boldsymbol{\varphi}^{(k)}\right) = \sum_{i=1}^{N} a_n e^{j\varphi_n^{(k)}} e^{j\beta z_n \cos\theta} = \sum_{i=1}^{N} a_n e^{j\left(\varphi_n^{(k-1)} + \delta_n^{(k)}\right)} e^{j\beta z_n \cos\theta} \tag{8.36}$$

The phase perturbation $\boldsymbol{\delta}^{(k)}$ in this expression can be linearized by using the Taylor expansion method. It leads to the approximate relation:

$$F\left(\theta; \boldsymbol{\varphi}^{(k)}\right) \approx \sum_{i=1}^{N} a_n e^{j\varphi_n^{(k-1)}} e^{j\beta z_n \cos\theta}\left(1 + j\delta_n^{(k)}\right) = F\left(\theta; \boldsymbol{\varphi}^{(k-1)}\right) + \sum_{n=1}^{N} j\delta_n^{(k)} a_n e^{j\varphi_n^{(k-1)}} e^{j\beta z_n \cos\theta} \tag{8.37}$$

While only the dual-beam synthesis case is considered in the following, the described methodology is also suitable for the synthesis of three or even more beams. Assume that the desired dual-beam pattern has the following characteristics: (i) the power pattern reaches its maximal points exactly at the specified directions θ_{ML_1} and θ_{ML_2}; (ii) the maximum pattern powers for θ_{ML_1} and θ_{ML_2} should be equal to each other; and (iii) the pattern level and null in the sidelobe region Ω_{SL} should be reduced as much as possible. Based on these considerations, the optimization objective is set as:

$$\min_{\delta^{(k)}} \left\{ W_1 \sum_{m=1}^{2} \left| \frac{\Delta\theta\left(\left|F\left(\theta_{\mathrm{ML}_1}; \boldsymbol{\varphi}^{(k)}\right)\right|^2\right)'}{\left|F\left(\theta_{\mathrm{ML}_m}; \boldsymbol{\varphi}^{(k)}\right)\right|^2} \right| + W_2 \left| \frac{\Delta F^{(k)}}{\overline{F}^{(k-1)}} \right| + W_3 \max_{\theta \in \Omega_{\mathrm{SL}}} \left| \frac{F\left(\theta_{\mathrm{SL}}; \boldsymbol{\varphi}^{(k)}\right)}{\overline{F}^{(k-1)}} \right| + W_4 \left| \frac{F\left(\theta_{\mathrm{Null}}; \boldsymbol{\varphi}^{(k)}\right)}{\overline{F}^{(k-1)}} \right| \right\} \tag{8.38}$$

where W_1, W_2, W_3,and W_4 are the non-negative weights, and

$$\left(\left|F\left(\theta_{\mathrm{ML}_m}; \boldsymbol{\varphi}^{(k)}\right)\right|^2\right)' = \left[\partial\left(\left|F\left(\theta; \boldsymbol{\varphi}^{(k)}\right)\right|^2\right)/\partial\theta\right]_{\theta = \theta_{\mathrm{ML}m}} \quad (m = 1, 2) \tag{8.39}$$

$$\Delta F^{(k)} = F\left(\theta_{\mathrm{ML}_1}; \boldsymbol{\varphi}^{(k)}\right) e^{-j\angle F\left(\theta_{\mathrm{ML}_1}; \boldsymbol{\varphi}^{(k-1)}\right)} - F\left(\theta_{\mathrm{ML}_2}; \boldsymbol{\varphi}^{(k)}\right) e^{-j\angle F\left(\theta_{\mathrm{ML}_2}; \boldsymbol{\varphi}^{(k-1)}\right)} \tag{8.40}$$

$$\overline{F}^{(k-1)} = \frac{1}{2}\left(\left|F\left(\theta_{\mathrm{ML}_1}; \boldsymbol{\varphi}^{(k-1)}\right)\right| + \left|F\left(\theta_{\mathrm{ML}_2}; \boldsymbol{\varphi}^{(k-1)}\right)\right|\right) \tag{8.41}$$

The first item of the objective function (8.38) guarantees the accuracy of the synthesized dual-beam directions by forcing the two partial derivatives of the power pattern $\left[\partial\left(\left|F(\theta; \boldsymbol{\varphi}^{(k)})\right|^2\right)/\partial\theta\right]_{\theta = \theta_{\mathrm{ML}_1}}$ and $\left[\partial\left(\left|F(\theta; \boldsymbol{\varphi}^{(k)})\right|^2\right)/\partial\theta\right]_{\theta = \theta_{\mathrm{ML}_2}}$ to be close to zero. The second item is used to make the powers at these two directions equal to each other. The remaining two items are used to suppress the sidelobe and null levels.

In general, the partial derivative of the power pattern with respect to θ after the introduction of the perturbed excitation phases can be written as:

$$\frac{\partial|F(\theta;\boldsymbol{\varphi}+\boldsymbol{\delta})|^2}{\partial\theta} = \frac{\partial\,\mathrm{Re}^2(F(\theta;\boldsymbol{\varphi}+\boldsymbol{\delta}))}{\partial\theta} + \frac{\partial\mathrm{Im}^2(F(\theta;\boldsymbol{\varphi}+\boldsymbol{\delta}))}{\partial\theta} = 2\,\mathrm{Re}\,(F(\theta;\boldsymbol{\varphi}+\boldsymbol{\delta}))\frac{\partial\,\mathrm{Re}\,(F(\theta;\boldsymbol{\varphi}+\boldsymbol{\delta}))}{\partial\theta}$$
$$+\,2\mathrm{Im}(F(\theta;\boldsymbol{\varphi}+\boldsymbol{\delta}))\frac{\partial\mathrm{Im}(F(\theta;\boldsymbol{\varphi}+\boldsymbol{\delta}))}{\partial\theta} \tag{8.42}$$

where

$$\mathrm{Re}\,(F(\theta;\boldsymbol{\varphi}+\boldsymbol{\delta})) = \sum_{n=1}^{N} a_n\cos(\beta z_n\cos\theta-\varphi_n) + \sum_{n=1}^{N} a_n\sin(\beta z_n\cos\theta-\varphi_n) \tag{8.43}$$

$$\mathrm{Im}(F(\theta;\boldsymbol{\varphi}+\boldsymbol{\delta})) = -\sum_{n=1}^{N} a_n\sin(\beta z_n\cos\theta-\varphi_n) + \sum_{n=1}^{N} a_n\cos(\beta z_n\cos\theta-\varphi_n) \tag{8.44}$$

$$\frac{\partial\,\mathrm{Re}\,(F(\theta;\boldsymbol{\varphi}+\boldsymbol{\delta}))}{\partial\theta} = \sum_{n=1}^{N} a_n\beta z_n\sin\theta\sin(\beta z_n\cos\theta-\varphi_n) - \sum_{n=1}^{N}\delta_n a_n\beta z_n\sin\theta\,\cos(\beta z_n\cos\theta-\varphi_n) \tag{8.45}$$

$$\frac{\partial\mathrm{Im}(F(\theta;\boldsymbol{\varphi}+\boldsymbol{\delta}))}{\partial\theta} = \sum_{n=1}^{N} a_n\beta z_n\sin\theta\,\cos(\beta z_n\cos\theta-\varphi_n) - \sum_{n=1}^{N}\delta_n a_n\beta z_n\sin\theta\,\sin(\beta z_n\cos\theta-\varphi_n) \tag{8.46}$$

By substituting (8.43)–(8.46) into (8.42) and then ignoring all of the quadratic terms of the perturbation $\boldsymbol{\delta}$, the following approximate expression is obtained:

$$\frac{\partial|F(\theta;\boldsymbol{\varphi}+\boldsymbol{\delta})|^2}{\partial\theta} \approx 2\boldsymbol{\delta}^T(\mathrm{C}_1\mathbf{Y}_1+\mathrm{C}_2\mathbf{Y}_2+\mathrm{C}_3\mathbf{Y}_3+\mathrm{C}_4\mathbf{Y}_4) + 2(\mathrm{C}_1\mathrm{C}_2+\mathrm{C}_3\mathrm{C}_4) \tag{8.47}$$

where

$$C_1 = \sum_{n=1}^{N} a_n\cos(\beta z_n\cos\theta-\varphi_n) \tag{8.48a}$$

$$C_2 = \sum_{n=1}^{N} a_n\beta z_n\,\sin\theta\,\sin(\beta z_n\cos\theta-\varphi_n) \tag{8.48b}$$

$$C_3 = \sum_{n=1}^{N} a_n\sin(\beta z_n\cos\theta-\varphi_n) \tag{8.48c}$$

$$C_4 = \sum_{n=1}^{N} a_n\beta z_n\sin\theta\,\cos(\beta z_n\cos\theta-\varphi_n) \tag{8.48d}$$

$$\mathbf{Y}_1 = \boldsymbol{a}\circ\mathbf{z}\circ\cos(\beta\mathbf{z}\,\cos\theta-\boldsymbol{\varphi})\beta\sin\theta \tag{8.49a}$$

$$\mathbf{Y}_2 = \boldsymbol{a}\circ\sin(\beta\mathbf{z}\,\cos\theta-\boldsymbol{\varphi}) \tag{8.49b}$$

$$\mathbf{Y}_3 = \boldsymbol{a}\circ\mathbf{z}\circ\sin(\beta\mathbf{z}\,\cos\theta-\boldsymbol{\varphi})\beta\sin\theta \tag{8.49c}$$

$$\mathbf{Y}_4 = \boldsymbol{a}\circ\cos(\beta\mathbf{z}\,\cos\theta-\boldsymbol{\varphi}) \tag{8.49d}$$

where the superscript "T" denotes the transpose of a matrix, and "∘" represents the Hadamard product of two vectors.

8.4.3 The Algorithm

The perturbation method needs an initial phase distribution to be prescribed first. Recall from Section 8.4.1 that the dual-beam pattern which points at θ_{ML_1} and θ_{ML_2} can be generated by using the summation of two individual excitation vectors: $e^{j\beta z\cos\theta_{ML_1}}$ and $e^{j\beta z\cos\theta_{ML_2}}$. Note that adding a constant phase to one of the two excitation vectors would still lead to a dual-beam pattern. This means that a more general initial phase distribution for producing a dual-beam pattern can be given as:

$$\boldsymbol{\varphi}^{(0)} = \text{angle}\{e^{j\beta \mathbf{z}\cos\theta_{ML_1}} + e^{j\xi}e^{j\beta \mathbf{z}\cos\theta_{ML_2}}\} \tag{8.50}$$

where $\xi \in [0, 2\pi]$ is a real number. If ξ can take a number of different values within $[0, 2\pi]$, e.g., $\xi = \xi_1$, ξ_2, ..., ξ_P, one gets P sets of different initial excitation phase distributions, i.e., $\boldsymbol{\varphi}(\xi_1)$,$\boldsymbol{\varphi}(\xi_2)$, ...,$\boldsymbol{\varphi}(\xi_P)$. As an iterative optimization algorithm, a different initial phase distribution may give rise to a different dual-beam pattern performance. The strategy is to pick the best one (i.e., with the lowest SLL) among all of the different results.

Based on these issues, an iterative convex optimization procedure for synthesizing dual-beam patterns is described as follows:

1) Set the array configurations including the element count, inter-element spacing, and working frequency. Specify the desired dual-beam pattern characteristics including the dual-beam-pointing directions θ_{ML_1} and θ_{ML_2}, the sidelobe region Ω_{SL}, and the null position θ_{Null}.
2) Set the weight coefficients w_1, w_2, w_3, and w_4 and set the maximum iteration number K.
3) Set a fixed amplitude distribution $\mathbf{a}$ as desired and construct P initial phase distribution vectors $\boldsymbol{\varphi}(\xi_1)$, $\boldsymbol{\varphi}(\xi_2)$, ..., $\boldsymbol{\varphi}(\xi_P)$.
4) Set $p = 1$ and $k = 1$.
5) Choose $\boldsymbol{\varphi}^0 = \boldsymbol{\varphi}(\xi_p)$.
6) Apply a convex optimization algorithm to minimize the objective function in (8.4.8), and find the optimal solution of $\boldsymbol{\delta}^{(k)}$.
7) Update $\boldsymbol{\varphi}^{(k)} = \boldsymbol{\varphi}^{(k-1)} + \boldsymbol{\delta}^{(k)}$ and the count $k = k + 1$.
8) Repeat Steps 6 and 7 until k reaches its maximum number K. Then record the obtained phase distribution $\boldsymbol{\varphi}^{(K)}(\xi_p)$.
9) Update the count $p = p + 1$ and repeat Steps 5–8 until $p = P$. Record all of the optimized phase vectors: $\boldsymbol{\varphi}^{(K)}(\xi_1)$, $\boldsymbol{\varphi}^{(K)}(\xi_2)$, ..., $\boldsymbol{\varphi}^{(K)}(\xi_P)$ that correspond to the different initial distributions: $\boldsymbol{\varphi}(\xi_1)$, $\boldsymbol{\varphi}(\xi_2)$, ..., $\boldsymbol{\varphi}(\xi_P)$.
10) Choose the excitation phase distribution with the best dual-beam pattern performance, for instance, the one with the lowest SLL and nulls.

8.4.4 Simulation Examples

A dual-beam pattern is synthesized as the first example in which one beam is pointed at 70° and the other is pointed at 130°, and with a null located at 30° by optimizing the excitation phases of a 24-element linear array. The excitation amplitudes are set to 1. Based on some experiments, the weighting factors for the objective function of the phase-only iterative perturbation method are chosen to be: $W_1 = 1$, $W_2 = 1$, $W_3 = 15$, and $W_4 = 50$. Figure 8.21a shows the obtained dual-beam pattern. As is expected, the dual beams are pointed exactly at their desired directions and the maximum SLL is about −14 dB. The null level is −46 dB. The corresponding phase distribution is shown in Figure 8.21b.

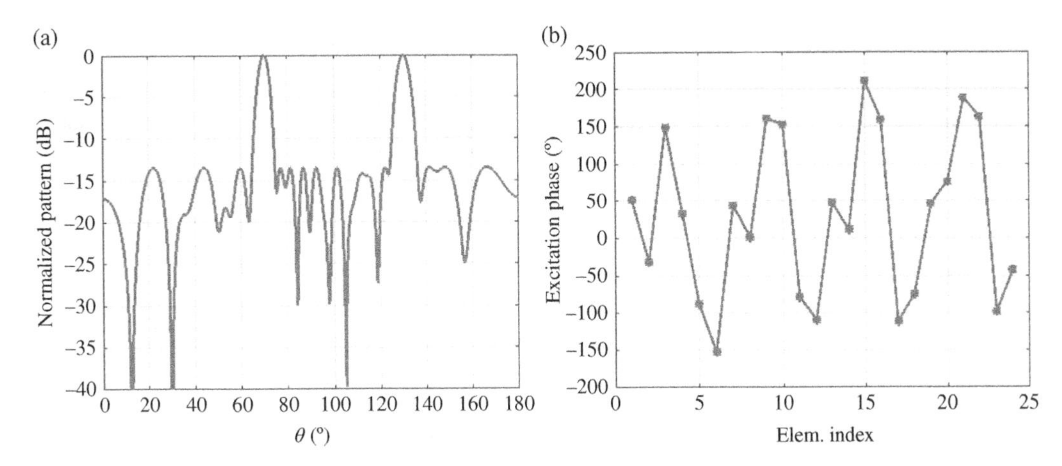

Figure 8.21 The dual-beam for the 24-element linear array synthesized with the phase-only iterative perturbation method using convex optimization. (a) Normalized pattern. (b) Excitation phase distribution.

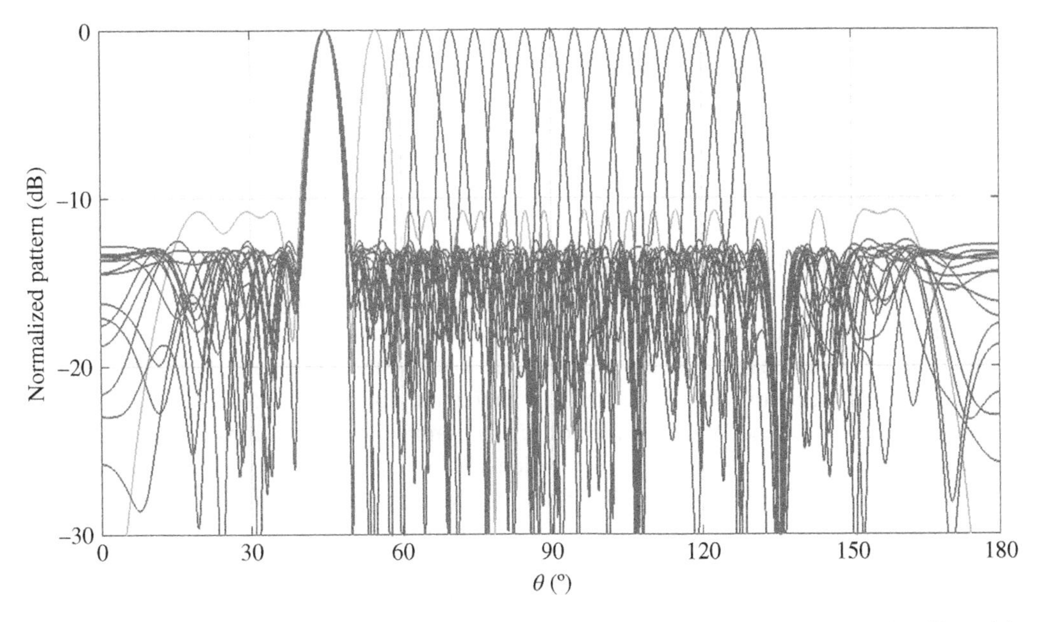

Figure 8.22 The dual-beam pattern of the 32-element linear array includes one beam fixed at 45° and the other beam scanned from 55 to 130° at an interval of 5°. The null is fixed at 135°.

A second example demonstrates that one can control the scanning of a beam in the dual-beam pattern with the phase-only optimization method for a 32-element linear array. Figure 8.22 shows the dual-beam pattern with one beam fixed at 45° and the other beam scanned from 55° to 130°. A null is chosen to be fixed at 135° in the beam scanning synthesis process. The corresponding SLLs for different beam scanning directions are listed in Table 8.4. Except for the case of the dual beams pointing at 45° and 55°, the obtained SLLs for the other cases reach about −13 dB. This is usually acceptable for many applications. For the case of dual-beam directions of 45° and 55°, the obtained

Table 8.4 The maximum SLLs for the dual-beam pattern with one beam fixed at 45° and the other beam that scanned from 55 to 130°.

Dual-beam direction	Max. SLL (dB)	Dual-beam direction	Max. SLL (dB)
(45°, 55°)	-10.79	(45°, 100°)	-13.05
(45°, 60°)	-12.54	(45°, 105°)	-12.92
(45°, 65°)	-13.02	(45°, 110°)	-13.68
(45°, 70°)	-12.81	(45°, 115°)	-13.56
(45°, 75°)	-12.58	(45°, 120°)	-13.20
(45°, 80°)	-13.00	(45°, 125°)	-13.19
(45°, 85°)	-12.97	(45°, 130°)	-13.56
(45°, 90°)	-13.02	(45°, 135°)	-13.22
(45°, 95°)	-12.89		

The null is fixed at 135°.

SLL is only −10.79 dB, which is worse than in the other cases. This difference indicates that the performance in the SLL reduction by the phase-only optimization approach may degrade when the dual-beam directions get close to each other. This is a typical problem for dual-beam pattern synthesis.

8.5 Conclusions

Traditional analog antenna arrays for wireless communications are either phased arrays for beam scanning or fixed multi-beam arrays employing an analog beamforming network as outlined in Chapter 2. With the advancement of mobile wireless communications and sensing networks, new requirements are being constantly imposed on antenna array system designers. These requirements can range from thinned arrays for cost-saving to scanning multiple beams. Moreover, they demand new array synthesis solutions. While research on digital beamforming algorithms has been going on for several decades, algorithms for synthesizing various analog beams are still being developed within the antenna research community. Combining new algorithms with new antenna array configurations, one faces new challenges which potentially can lead to novel system solutions. In this chapter, we have introduced a number of exciting topics in analog array synthesis. Although these topics are of great research interest, the solutions presented can be expected to find applications in both fixed and moving platforms for 5G systems in the near term and for future 6G and beyond systems.

References

1. Akdagli, A. and Guney, K. (2003). Shaped-beam pattern synthesis of equally and unequally spaced linear antenna arrays using a modified tabu search algorithm. *Microw. Opt. Technol. Lett.* **36** (1): 16–20.
2. Kurup, D.G., Himdi, M., and Rydberg, A. (2003). Synthesis of uniform amplitude unequally spaced antenna arrays using the differential evolution algorithm. *IEEE Trans. Antennas Propag.* **51** (9): 2210–2217.

3. Bhattacharya, R., Bhattacharya, T.K., and Garg, R. (2012). Position mutated hierarchical particle swarm optimization and its application in synthesis of unequally spaced antenna arrays. *IEEE Trans. Antennas Propag.* **60** (7): 3174–3181.
4. Darvish, A. and Ebrahimzadeh, A. (2018). Improved fruit-fly optimization algorithm and its applications in antenna arrays synthesis. *IEEE Trans. Antennas Propag.* **66** (4): 1756–1766.
5. Liu, Y., Nie, Z., and Liu, Q.H. (2008). Reducing the number of elements in a linear antenna array by the matrix pencil method. *IEEE Trans. Antennas Propag.* **56** (9): 2955–2962.
6. Liu, Y., Liu, Q.H., and Nie, Z. (2010). Reducing the number of elements in the synthesis of shaped-beam patterns by the forward-backward matrix pencil method. *IEEE Trans. Antennas Propag.* **58** (2): 604–608.
7. Liu, Y., Liu, Q.H., and Nie, Z. (2014). Reducing the number of elements in multiple-pattern linear arrays by the extended matrix pencil methods. *IEEE Trans. Antennas Propag.* **62** (2): 652–660.
8. Shen, H. and Wang, B. (2018). Two-dimensional unitary matrix pencil method for synthesizing sparse planar arrays. *Digit. Signal Process.* **73**: 40–46.
9. Viani, F., Oliveri, G., and Massa, A. (2013). Compressive sensing pattern matching techniques for synthesizing planar sparse arrays. *IEEE Trans. Antennas Propag.* **61** (9): 4577–4587.
10. Prisco, G. and D'Urso, M. (2012). Maximally sparse arrays via sequential convex optimizations. *IEEE Antennas Wireless Propag. Lett.* **11**: 192–195.
11. Fuchs, B. (2012). Synthesis of sparse arrays with focused or shaped beampattern via sequential convex optimizations. *IEEE Trans. Antennas Propag.* **60** (7): 3499–3503.
12. You, P., Liu, Y., Chen, S.L. et al. (2017). Synthesis of unequally spaced linear antenna arrays with minimum element spacing constraint by alternating convex optimization. *IEEE Antennas Wireless Propag. Lett.* **16**: 3126–3130.
13. Keizer, W.P.M.N. (2007). Fast low-sidelobe synthesis for large planar array antennas utilizing successive fast fourier transforms of the array factor. *IEEE Trans. Antennas Propagat.* **55** (3): 715–722.
14. Keizer, W.P.M.N. (2008). Linear array thinning using iterative FFT techniques. *IEEE Trans. Antennas Propag.* **56** (8): 2757–2760.
15. Keizer, W.P.M.N. (2009). Large planar array thinning using iterative FFT techniques. *IEEE Trans. Antennas Propag.* **57** (10): 3359–3362.
16. Liu, Y., Luo, Q., Li, M., and Guo, Y.J. (2019). Thinned massive antenna array for 5G millimeter-wave communications. *2019 13th European Conference on Antennas and Propagation (EuCAP)*, Krakow, Poland, 2019, pp. 1–4.
17. Liu, Y., Zheng, J., Li, M. et al. Synthesizing beam-scannable thinned massive antenna array utilizing modified iterative FFT for millimeter-wave communication. *IEEE Antennas Wireless Propag. Lett.* **19** (11): 1983–1987.
18. Fuchs, B. (2014). Application of convex relaxation to array synthesis problems. *IEEE Trans. Antennas Propag.* **62** (2): 634–640.
19. Li, J.-Y., Qi, Y.-X., and Zhou, S.-G. (2017). Shaped beam synthesis based on superposition principle and Taylor method. *IEEE Trans. Antennas Propag.* **65** (11): 6157–6160.
20. Liang, J., Fan, X., Fan, W. et al. (2017). Phase-only pattern synthesis for linear antenna arrays. *IEEE Antennas Wireless Propag. Lett.* **16**: 3232–3235.
21. Liu, Y., Li, M., Haupt, R.L., and Guo, Y.J. (2020). Synthesizing shaped power patterns for linear and planar antenna arrays including mutual coupling by refined joint rotation/phase optimization. *IEEE Trans. Antennas Propag.* **68** (6): 4648–4657.
22. Liu, Y., Bai, J., Xu, K.D. et al. (2018). Linearly polarized shaped power pattern synthesis with sidelobe and cross-polarization control by using semidefinite relaxation. *IEEE Trans. Antennas Propag.* **66** (6): 3207–3212.

23. Ludwig, A. (1973). The definition of cross polarization. *IEEE Trans. Antennas Propag.* **21** (1): 116–119.
24. Li, M., Liu, Y., and Guo, Y.J. (2018). Shaped power pattern synthesis of a linear dipole array by element rotation and phase optimization using dynamic differential evolution. *IEEE Antennas Wireless Propag. Lett.* **17** (4): 697–701.
25. Eberhart, R.C., and Shi, Y. (2001). Particle swarm optimization: developments, applications and resources. *Proceedings of the 2001 Congress on Evolutionary Computation*, Seoul, South Korea, 2001, Vol. 1, pp. 81–86.
26. Kennedy, J., and Eberhart, R. (1995). Particle swarm optimization. *Proceedings of International Conference on Neural Networks*, Perth, WA, Australia, 1995, pp. 1942–1948.
27. Robinson, J. and Rahmat-Samii, Y. (2004). Particle swarm optimization in electromagnetics. *IEEE Trans. Antennas Propag.* **52** (2): 397–407.
28. Poli, L., Rocca, P., Manica, L., and Massa, A. (2010). Handling sideband radiations in time-modulated arrays through particle swarm optimization. *IEEE Trans. Antennas Propag.* **58** (4): 1408–1411.
29. Yang, S.-H. and Kiang, J.-F. (2014). Adjustment of beamwidth and side-lobe level of large phased-arrays using particle swarm optimization technique. *IEEE Trans. Antennas Propag.* **62** (1): 138–144.
30. Weigand, S., Huff, G.H., Pan, K.H., and Bernhard, J.T. (2003). Analysis and design of broad-band single-layer rectangular U-slot microstrip patch antennas. *IEEE Trans. Antennas Propag.* **51** (3): 457–468.
31. Fuchs, B., Skrivervik, A., and Mosig, J.R. (2013). Shaped beam synthesis of arrays via sequential convex optimizations. *IEEE Antennas Wireless Propag. Lett.* **12**: 1049–1052.
32. Echeveste, J.I., de Aza, M.A.G., Rubio, J., and Zapata, J. (2016). Near-optimal shaped-beam synthesis of real and coupled antenna arrays via 3-D-FEM and phase retrieval. *IEEE Trans. Antennas Propag.* **64** (6): 2189–2196.
33. Skolnik, M.I. (2008). *Radar Handbook*. New York, NY: McGraw-Hill.
34. Li, M., Liu, Y., and Guo, Y.J. (2021). Design of sum and difference patterns by optimizing element rotations and positions for linear dipole array. *IEEE Trans. Antennas Propag.* **69** (5): 3027–3032.
35. Morabito, A.F. and Rocca, P. (2015). Reducing the number of elements in phase-only reconfigurable arrays generating sum and difference patterns. *IEEE Antennas Wireless Propag. Lett.* **14**: 1338–1341.
36. Andrews, J.G., Buzzi, S., Choi, W. et al. (2014). What will 5G be. *IEEE J. Sel. Area. Comm.* **32** (6): 1065–1082.
37. Bhushan, N., Li, J., Malladi, D. et al. (2014). Network densification: the dominant theme for wireless evolution into 5G. *IEEE Commun Mag* **52** (2): 82–89.
38. Araniti, G., Condoluci, M., Scopelliti, P. et al. (2017). Multicasting over emerging 5G networks: challenges and perspectives. *IEEE Network* **31** (2): 80–89.
39. Roh, W., Seol, J.-Y., Park, J. et al. (2014). Millimeter-wave beamforming as an enabling technology for 5G cellular communications: theoretical feasibility and prototype results. *IEEE Commun. Mag.* **52** (2): 106–113.
40. Ho, W.W.L. and Liang, Y. (2009). Optimal resource allocation for multiuser MIMO-OFDM systems with user rate constraints. *IEEE Trans. Veh. Technol.* **58** (3): 1190–1203.
41. Sarkar, T.K., Mailloux, R., Oliner, A.A. et al. (2006). *History of Wireless*, vol. **177**. Wiley.

Index

NOTE: Page numbers followed by f refer to figures.

Advanced Antenna Array Engineering for 6G and Beyond Wireless Communications, First Edition.
Y. Jay Guo and Richard W. Ziolkowski.